Optimal Control of Partial Differential Equations

International Conference in Chemnitz, Germany, April 20–25, 1998

Edited by

K.-H. Hoffmann
G. Leugering
F. Tröltzsch

Birkhäuser Verlag
Basel · Boston · Berlin

Editors:

Karl-Heinz Hoffmann
Zentrum Mathematik
Technische Universität München
Arcisstraße 21
80333 München
Germany

and

Stiftung Caesar
Friedensplatz 16
53111 Bonn
Germany

Günter Leugering
Mathematisches Institut
Universität Bayreuth
95440 Bayreuth
Germany

Fredi Tröltzsch
Fakultät für Mathematik
TU Chemnitz
09107 Chemnitz
Germany

1991 Mathematics Subject Classification 49-06, 93-06

A CIP catalogue record for this book is available from the Library of Congress, Washington D.C., USA

Deutsche Bibliothek Cataloging-in-Publication Data
Optimal control of partial differential equations : international conference in Chemnitz, Germany, April 20 - 25, 1998 / ed. by K.-H. Hoffmann ... - Basel ; Boston ; Berlin : Birkhäuser, 1999
(International series of numerical mathematics ; Vol. 133)
ISBN 3-7643-6151-4 (Basel ...)
ISBN 0-8176-6151-4 (Boston)

Printed on acid-free paper produced of chlorine-free pulp. TCF ∞
Cover design: Heinz Hiltbrunner, Basel
Printed in Germany
ISBN 3-7643-6151-4
ISBN 0-8176-6151-4

9 8 7 6 5 4 3 2 1

Preface

The international *Conference on Optimal Control of Partial Differential Equations* was held at the *Wasserschloss Klaffenbach*, a recently renovated moated castle near Chemnitz (Germany), from April 20 to 25, 1998. This conference has been an activity of the *IFIP working group WG7.2.* The organizers greatfully acknowledge the generous financial support by the *Deutsche Forschungsgemeinschaft* (DFG), the *Sächsisches Staatsministerium für Wissenschaft und Kunst*, the *European Community*, the *LUK GmbH& Co.* (Bühl, Germany), and the *Technische Universität Chemnitz.*

The conference has been attended by more than 60 scientists from 15 countries. The scientific program included 40 invited talks in the area of optimization, optimal control and controllability, shape optimization and related mechanical modelling. The underlying idea was to enhance the flow of information in this vast area of important industrial applications between theorists and numerical analysts, and, hence, the topics of the talks have been selected towards a balanced presentation of the overall field. This volume contains 27 refereed original papers which can be classified as follows: 9 papers in controllablity, stabilizability and identifiability, 7 papers in optimal control, 6 papers in shape optimization and 5 papers on general modelling and qualitative issues related to partial differential equations. About one half of the papers can be classified as theoretical work, while the other part is devoted to algorithmic procedures and their numerical implementation.

It has become apparent that the theory of optimal control and optimal shapes as well as a concise numerical analysis in this area is extremely important for industrial applications. Moreover, complex processes such as fluid-structure interactions, noise-reduction, structural mechanics, smart materials etc., in many cases imply a modelling based on partial differential equations. Typically, as large rather than infinitesimal motion becomes more and more relevant, those equations have to be taken nonlinear. It is fair to say that a concise unifying mathematical theory in this area is still lacking. Nervertheless, in recent years there has been a tremendous effort to go beyond the more classical 'linearized theories'. This is particularily true for the problem of optimal control of fluids and the control of fluid-structure interactions, where one tries e.g. to reduce the drag or the tendency of transitions from laminar to turbulent flow regimes, or for noise-reduction problems and problems of optimal control of vibrations in large flexible structures in general, where one wants to actively extinguish unwanted oscillations, or for problems related to opimal melting and cristalization procedures. In many of these applications the optimization starts right on the design level, giving rise to problem of finding shapes of structures that are optimal in some sense. Moreover, in order to handle complicated dynamical behaviour numerically it is sometimes necessary to reduce the complexity of the underlying structures e.g. by applying the technique of homogenisation or to reduce the dimension of its representation e.g. by using Karhunen-Loeve approximations.

It is apparant that problems in this area constitute a grand challenge for mathematicians, both for theorists and numerical analysts. The papers collected in this volume provide a competent 'state of the art' presentation. The organizers hope that this book will contribute to the exciting ongoing discussion and express their gratitude to the authors, to the publisher and last but not least all who have worked hard to make the conference happen and to turn it into a success, notably A. Rösch, A. Unger, and G. Weise (Chemnitz). The organizers also greatfully acknowledge the work of W. Rathmann (Bayreuth) who was involved in the technical preparation of this volume.

München, Bayreuth and Chemnitz, 20.02.1999

K.-H. Hoffmann
G. Leugering
F. Tröltzsch

Contents

International Series of Numerical Mathematics
Vol. 133, © 1999 Birkhäuser Verlag Basel/Switzerland

Well-posedness of Semilinear Heat Equations with Iterated Logarithms

Paolo Albano, Piermarco Cannarsa and Vilmos Komornik

Abstract. A global existence and uniqueness result is obtained for a semilinear parabolic equation of the form

$$u_t = \Delta u + f(u) + h.$$

The nonlinear term f is assumed to satisfy a suitable growth condition at ∞, that allows superlinear growth of $|f|$. An example is given to show the optimality of our growth assumption on f.

1. Introduction

Consider the problem

$$\begin{cases} u_t = \Delta u + f(u) + h \quad \text{in} \quad \Omega \times (0,T), \\ u = 0 \quad \text{on} \quad \Gamma \times (0,T), \\ u(0) = u_0 \quad \text{in} \quad \Omega, \end{cases} \tag{1}$$

where

- Ω is a nonempty bounded open domain in $\mathbb{R}^N$ having a boundary Γ of class C^2;
- $f : \mathbb{R} \to \mathbb{R}$ is a function of class C^1;
- T is a positive number.

We study the existence of a unique global solution

$$u \in C([0,T]; L^2(\Omega)) \tag{2}$$

for every given

$$u_0 \in L^2(\Omega), \quad \text{and} \quad h \in L^2(0,T; L^2(\Omega)). \tag{3}$$

It is well known that, if f has sublinear growth at ∞, then problem (1) has a unique global solution u. If, on the other hand, f has a faster growth at ∞, then the solution of (1) may blow up in finite time, unless suitable sign conditions are assumed on f. Such conditions are usually of the form

$$sf(s) \leq C(1 + s^2), \tag{4}$$

for all $s \in \mathbb{R}$ and some $C > 0$. This paper aims to obtain a global existence and uniqueness result for problem (1), allowing the right-hand side of condition (4) to grow faster than quadratically at ∞. For this purpose, we will use a method introduced in [2] for hyperbolic problems.

To be more precise, let us introduce the *iterated logarithm* functions $\log_j$ defined by the formulae

$$\log_0 s := s \quad \text{and} \quad \log_j s := \log(\log_{j-1} s), \quad j = 1, 2, \ldots .$$

Let us also define the iterated exponential functions $\exp_j : \mathbb{R} \to \mathbb{R}$ as

$$\exp_0 s = s \quad \text{and} \quad \exp_j s = \exp(\exp_{j-1} s), \ j = 1, 2, \ldots .$$

Given an integer $k \geq 0$, we set $e_k = \exp_k 1$. Moreover, for any $k \geq 0$, we define

$$L_k(s) = \prod_{j=0}^{k} \log_j(e_j + |s|), \quad (s \in \mathbb{R}). \tag{5}$$

Then, we will show that problem (1) has a unique global solution u, provided that an integer $k \geq 0$ exists such that

$$sf(s) \leq C(1 + |s|)L_k(s), \quad \forall s \in \mathbb{R}, \tag{6}$$

for some $C \geq 0$. Moreover, if $N > 2$ we will also assume the following growth condition on f':

$$|f'(s)| \leq C(1 + |s|^{\frac{2}{N}}), \quad \forall s \in \mathbb{R}. \tag{7}$$

We observe that no growth condition is required on f' in the case of $N = 1$. If $N = 2$, then the exponent $\frac{2}{N}$ in (7) can be replaced by any positive number.

We will also prove that the growth conditions (6)–(7) are optimal in the sense that, if an integer $k \geq 1$ and real numbers $p > 1$ and $\gamma, c > 0$ exist such that

$$f(s) \geq \gamma L_{k-1}(s) \log_k^p s \quad \text{for all} \quad s > c, \tag{8}$$

then one can find an initial condition $u_0 \in L^2(\Omega)$ for which (1) has no global solution.

Finally, we would like to add that, using the well-posedness result proved in this paper, one can try to obtain null and approximate controllability for equation (1), even in the case of a locally distributed control, i.e. when h has support in a subdomain $\omega \subset \Omega$. These controllability issues have already been addressed in [5] if f grows slower than $s \log |s|$ at ∞. We leave such a topic to a future work.

2. Abstract Formulation and Preliminaries

In this section we recast problem (1) as a semilinear Cauchy problem in the Hilbert space $X = L^2(\Omega)$, denoting by $\|\cdot\|$ and $\langle \cdot\,, \cdot \rangle$ the usual norm and the scalar product in $L^2(\Omega)$, respectively. Let us consider the evolution equation

$$\begin{cases} u'(t) = Au(t) + F(u(t)) + h(t), & t \in (0, T) \\ u(0) = u_0 \in X \,, \end{cases} \tag{9}$$

where $A : D(A) \subset X \to X$ is defined by

$$D(A) = H^2(\Omega) \cap H_0^1(\Omega)$$
$$Au(x) = \Delta u(x) \quad (x \in \Omega, u \in D(A)).$$

It is well known that A is the infinitesimal generator of an analytic semigroup of bounded linear operators on X, e^{tA}. Moreover, the *fractional powers* of $-A$, denoted by $(-A)^\theta$, are well defined for any $\theta \in [0,1]$ and satisfy

$$\|(-A)^\theta e^{tA} u\| \leq \frac{M_\theta}{t^\theta} \|u\| \quad (u \in X, t > 0) \tag{10}$$

for a suitable constant $M_\theta > 0$. As is well known, in our case, these spaces can be characterized as follows

$$D((-A)^\theta) = \begin{cases} H^{2\theta}(\Omega) & \text{if} \quad 0 \leq \theta < \frac{1}{4} \\ \{u \in H^{2\theta}(\Omega) \, : \, u = 0 \quad \text{on} \quad \Gamma\} & \text{if} \quad \frac{1}{4} < \theta \leq 1. \end{cases}$$

Next, let $f : \mathbb{R} \to \mathbb{R}$ be a function of class C^1, satisfying the growth condition (7). Then, one can show that the composition operator

$$F(u)(x) = f(u(x)) \quad (x \in \Omega, u \in X) \tag{11}$$

is well defined from $D((-A)^{\frac{1}{2}})$ into X. In fact, the following result holds.

Lemma 2.1. *Let f be a function of class C^1, satisfying (7). Then the composition operator F is continuous from $D((-A)^{\frac{1}{2}})$ into X, and*

$$\|F(u) - F(v)\| \leq C_N \left(1 + \|u\| + \|v\|\right)^{\frac{2}{N}} \|(-A)^{\frac{1}{2}}(u - v)\| \tag{12}$$

for any $u, v \in D((-A)^{\frac{1}{2}})$.

Proof. As far as (12) is concerned, we note that such an estimate is well known in the literature on parabolic equations, see e.g. [6]. We give a proof for completeness, in the case of $N \geq 3$. Recalling the Sobolev Embedding Theorem, we have that

$$D((-A)^{\frac{1}{2}}) = H_0^1(\Omega) \hookrightarrow L^{\frac{2N}{N-2}}(\Omega).$$

So, the fact that F maps $D((-A)^{\frac{1}{2}})$ into X is a direct consequence of the growth assumption (7). Moreover,

$$\begin{aligned} \|F(u) - F(v)\|^2 &\leq \int_\Omega |u - v|^2 \left[\int_0^1 f'(\lambda u + (1-\lambda)v)\, d\lambda\right]^2 dx \\ &\leq C^2 \int_\Omega |u - v|^2 \left[\int_0^1 (1 + |\lambda u + (1-\lambda)v|^{\frac{2}{N}})\, d\lambda\right]^2 dx \end{aligned} \tag{13}$$

in light of assumption (7). Hence, from the Hölder inequality we deduce that

$$(13) \leq C \left(\int_\Omega |u - v|^{\frac{2N}{N-2}}\, dx\right)^{\frac{N-2}{N}} \left(\int_\Omega [1 + |u|^2 + |v|^2]\, dx\right)^{\frac{2}{N}}. \tag{14}$$

Then, using once again the Sobolev Embedding Theorem, (13) and (14) yield (12). □

The next estimate is given in terms of the function L_k, defined in (5).

Lemma 2.2. *Let f be a function of class C^1 satisfying (6). Then, for any $\varepsilon > 0$ a constant $C_\varepsilon > 0$ exists such that*

$$\langle F(u), u\rangle \leq \varepsilon \|(-A)^{\frac{1}{2}} u\|^2 + C_\varepsilon L_k(\|u\|^2) \tag{15}$$

for any $u \in D((-A)^{\frac{1}{2}})$.

Proof. We note that, since $1 + |s| \leq L_k(s)$, condition (6) yields

$$\langle F(u), u\rangle \leq C \int_\Omega (1 + |u|) L_k(u) dx \leq C \int_\Omega L_k^2(u) dx\,.$$

On the other hand, as is shown in [2], for any $\varepsilon > 0$ a constant $C_\varepsilon > 0$ exists such that

$$\int_\Omega L_k^2(u) dx \leq \varepsilon \|(-A)^{\frac{1}{2}} u\|^2 + C_\varepsilon L_k(\|u\|^2).$$

From the above estimates the conclusion follows. □

Finally, we define solutions of problem (9).

Definition 2.3. *A function $u \in C([0,T]; X) \cap L^2(0,T; D((-A)^{1/2}))$ is a* mild solution *of (9) if*

$$u(t) = e^{tA} u_0 + \int_0^t e^{(t-s)A} [F(u(s)) + h(s)]\, ds\,, \quad \forall t \in [0,T]\,. \tag{16}$$

Notice that the convolution term in (16) is well defined in light of Lemma 2.1. We conclude this section recalling a well-known result about existence and uniqueness of solution to problem (9) with regular data (see e.g. [6], [8] and [9]). We say that u is a *strict solution* of problem (9) in $[0,T]$ if

$$\begin{cases} u \in H^1(0,T; X) \cap C([0,T]; D((-A)^{\frac{1}{2}})) \cap L^2(0,T; D(A)), \\ u'(t) = Au(t) + F(u(t)) + h(t) \quad \text{a. e. in } [0,T] \\ u(0) = u_0 \end{cases} \tag{17}$$

Proposition 2.4. *Suppose that f satisfies assumptions (6), (7) and let $h \in L^2(0,T; X)$. Then, for any $u_0 \in D((-A)^{\frac{1}{2}})$ a positive number $\tau = \tau(u_0) \leq T$ exists so that problem (9) admits a unique strict solution in the interval $[0,\tau]$.*

The above result is certainly known and widely used in the control theoretical literature, see e.g. [7]. On the other hand, the results contained in most references on evolution equations, such as [6], [8] and [9], yield the existence of strict solutions to (9) under some continuity assumption on h. The same kind of arguments, however, can be used to show the existence of a solution satisfying (17). Indeed, such a result is well known in the linear case, i.e. for $F \equiv 0$, see e.g. [1]. In the general case it suffices to apply a fixed point argument in the space $H^1(0,T; X) \cap C([0,T]; D((-A)^{\frac{1}{2}})) \cap L^2(0,T; D(A))$.

3. Well-posedness

In this section, we will prove the following

Theorem 3.1. *Let f be a function of class C^1, satisfying (6) and (7) for some $k > 0$. Then, for any $u_0 \in X$ and $h \in L^2(0,T;X)$, problem (9) possesses a unique mild solution.*

In order to prove Theorem 3.1 we need the following Gronwall type inequalities.

Lemma 3.2. *Let $\varphi : [0,T] \to \mathbb{R}$ be a nonnegative continuous function, satisfying for some numbers $A, B > 0$ and k the inequalities*

$$\varphi(t) \leq A + B \int_0^t L_k(\varphi(s))\, ds \quad \textit{for all} \quad t \in [0,T],$$

for L_k defined as in (5). Then φ is bounded on $[0,T]$ by a constant depending only on L_k, A, B and on T.

Lemma 3.3. *Let $\varphi : [0,T] \to \mathbb{R}$ be a nonnegative function satisfying, for some numbers $A, B \geq 0$ and $0 \leq \beta < 1$, the inequality*

$$\varphi(t) \leq A + B \int_0^t \frac{\varphi(s)}{(t-s)^\beta}\, ds \quad \textit{for all} \quad t \in [0,T].$$

Then, a constant $C(\beta, B, T)$ exists such that

$$\varphi(t) \leq AC(\beta, B, T) \quad (t \in [0,T]).$$

Remark 3.4. (i) In the case of $k = 1$, Lemma 3.2 is proved in [3]. Moreover, as is shown in [2], a stronger form of this lemma holds true replacing L_k with a non-decreasing continuous function $g : [0,\infty) \to \mathbb{R}$, if one assumes that a positive number $c > 0$ exists such that

$$g > 0 \ \text{ in } (c,\infty) \quad \text{and} \quad \int_c^\infty \frac{ds}{g(s)} = \infty\,.$$

(ii) Lemma 3.3 is a special case of a well-known result, see e.g. [6].

A crucial step in the proof of Theorem 3.1 is the following a priori estimate.

Lemma 3.5. *In addition to the assumptions of Theorem 3.1 suppose that $u_0 \in D((-A)^{\frac{1}{2}})$, and let u be the solution of (9) given by Proposition 2.4 on the interval $0 \leq t < \tau \leq T$. Then,*

$$\sup_{0 \leq t \leq \tau} \|(-A)^{\frac{1}{2}} u(t)\|^2 \leq C_T\,, \tag{18}$$

where $C_T = C(T, \|u_0\|, \int_0^T \|h\|^2\, ds)$.

Remark 3.6. We note that, in particular, estimate (19) implies that the solution of (9) with initial conditions in $D((-A)^{\frac{1}{2}})$ is global, i.e. $\tau = T$ for any $u_0 \in D((-A)^{\frac{1}{2}})$.

Proof. First we prove that

$$\sup_{0\le t\le \tau} \|u(t)\|^2 + \int_0^\tau \|(-A)^{\frac{1}{2}} u(s)\|^2\, ds \le c_1 \tag{19}$$

for a suitable constant $c_1 = c_1(T, \|u_0\|, \int_0^T \|h\|^2 \; ds)$. Indeed, taking the scalar product of the equation in (9) with u, we obtain

$$\frac{1}{2}\frac{d}{dt}\|u(t)\|^2 + \|(-A)^{\frac{1}{2}} u\|^2 = \langle F(u) + h\,,\, u\rangle \quad t \in [0, \tau[\,.$$

Hence, applying Lemma 2.2 we have that for any $\varepsilon > 0$ a constant $c_\varepsilon > 0$ exists such that

$$\begin{aligned}
&\|u(t)\|^2 + 2\int_0^t \|(-A)^{\frac{1}{2}} u\|^2 \; ds \\
&\le \; \|u_0\|^2 + 2\int_0^t \langle F(u), u\rangle \; ds + 2\int_0^t \|u\| \; \|h\| \; ds \\
&\le \; \|u_0\|^2 + \varepsilon c_2 \int_0^t \|(-A)^{\frac{1}{2}} u\|^2 \; ds + c_\varepsilon \int_0^t L_k(\|u\|^2) \; ds + \frac{1}{\varepsilon}\int_0^t \|h\|^2 \; ds,
\end{aligned}$$

where c_2 is a positive constant independent of ε. Choosing $\varepsilon = 1/(c_2+2)$ we have that

$$\begin{aligned}
&\|u(t)\|^2 + \int_0^t \|(-A)^{\frac{1}{2}} u\|^2 \; ds \\
&\le \|u_0\|^2 + (c_2+2)\int_0^T \|h\|^2 \; ds + c_3 \int_0^t L_k\left(\|u\|^2 + \int_0^s \|(-A)^{\frac{1}{2}} u\|^2 \; d\sigma\right) \; ds,
\end{aligned}$$

as L_k is increasing. Then, applying Lemma 3.2 we get

$$\|u(t)\|^2 + \int_0^t \|(-A)^{\frac{1}{2}} u(s)\|^2\, ds \le c_1 \quad (t \in [0, \tau[)\,, \tag{20}$$

for a suitable constant $c_1 = c_1(T, \|u_0\|, \int_0^T \|h\|^2 \; ds)$. This proves our claim (19). Now, applying $(-A)^{\frac{1}{2}}$ to (16) and recalling that

$$t \mapsto (-A)^{\frac{1}{2}} \int_0^t e^{(t-s)A} h(s)\, ds$$

in bounded in $[0, T]$ (see [1]), we have that

$$\begin{aligned}
\|(-A)^{\frac{1}{2}} u(t)\| \;&\le\; \|(-A)^{\frac{1}{2}} e^{tA} u_0\| + \|(-A)^{\frac{1}{2}} \int_0^t e^{(t-s)A} h(s)\, ds\| \\
&\quad + \|(-A)^{\frac{1}{2}} \int_0^t e^{(t-s)A} F(u(s))\, ds\| \\
&\le\; M_0 \|(-A)^{\frac{1}{2}} u_0\| + c_4 + \int_0^t \frac{c_5}{\sqrt{t-s}} \|(-A)^{\frac{1}{2}} u(s)\|\, ds
\end{aligned}$$

where $c_5 = M_{\frac{1}{2}}[\|F(0)\| + C_N \sup_{0\le t\le T}(1 + \|u(t)\|)^{\frac{2}{N}}]$. Then, (18) follows from Lemma 3.3. □

We now turn to the proof of Theorem 3.1.

Proof. **Existence:** we take a sequence $u_n \in D((-A)^{\frac{1}{2}})$ such that u_n strongly converges to u_0 in X and $\|u_n\| \le \|u_0\|$. For any $n \ge 1$, let $u_n(\cdot)$ be the classical solution of (9) with initial condition u_n, given by Proposition 2.4. Recalling Remark 3.6 we have that

$$u_n(t) = e^{tA}u_n + \int_0^t e^{(t-s)A}[F(u_n(s)) + h(s)]\, ds\,, \tag{21}$$

for any $t \in [0, T]$. Lemma 3.5 implies that

$$\sup_{0\le t\le T} \|u_n(t)\|^2 + \int_0^T \|(-A)^{\frac{1}{2}}u_n(s)\|^2\, ds \le c_1\,. \tag{22}$$

First, we will show that

$$\{u_n(\cdot)\}_{n\in\mathbb{N}} \quad \text{is a Cauchy sequence in} \quad C([0,T];X)\,. \tag{23}$$

Fix $m, n \in \mathbb{N}$. Then, the difference of the two equations satisfied by $u_n(\cdot)$ and $u_m(\cdot)$ reads as follows

$$\begin{cases} u_n'(t) - u_m'(t) = A(u_n(t) - u_m(t)) + F(u_n(t)) - F(u_m(t)), & t \in (0,T) \\ u_n(0) - u_m(0) = u_n - u_m\,. \end{cases}$$

Taking the scalar product of the above equation with $u_n(\cdot) - u_m(\cdot)$, and then applying (12) and (22), we get

$$\begin{aligned} &\frac{1}{2}\ \frac{d}{dt}\|u_n(t) - u_m(t)\|^2 \\ \le\ & -\|(-A)^{\frac{1}{2}}(u_n(t) - u_m(t))\|^2 + \frac{1}{2\varepsilon}\|u_n(t) - u_m(t)\|^2 \\ & + \frac{\varepsilon}{2}\|F(u_n(t)) - F(u_m(t))\|^2 \\ \le\ & -\|(-A)^{\frac{1}{2}}(u_n(t) - u_m(t))\|^2 + \frac{1}{2\varepsilon}\|u_n(t) - u_m(t)\|^2 \\ & + \varepsilon\frac{(1 + 2\sqrt{c_1})^{\frac{4}{N}}}{2}\|(-A)^{\frac{1}{2}}(u_n(t) - u_m(t))\|^2\,. \end{aligned} \tag{24}$$

Choosing $\varepsilon > 0$ such that $1 - \varepsilon\frac{(1+2\sqrt{c_1})^{\frac{4}{N}}}{2} < 0$, the above inequality and the Gronwall Lemma yield

$$\|u_n(t) - u_m(t)\|^2 \le e^{\frac{t}{\varepsilon}}\|u_n - u_m\|^2\,, \quad (0 \le t \le T)\,. \tag{25}$$

We have thus proved (23). In order to obtain a solution of (9), it suffices to show that

$$\{u_n(\cdot)\}_{n\in\mathbb{N}} \quad \text{is a Cauchy sequence in} \quad L^2(0,T;D((-A)^{\frac{1}{2}}))\,. \tag{26}$$

In fact, recalling (12) once again , it is easy to see that the above convergence is sufficient to pass to the limit in (21). Now, in order to prove (26), we integrate (24). Then, using (25), we conclude that

$$\int_0^T \|(-A)^{\frac{1}{2}}(u_n(t)-u_m(t))\|^2\,dt \le c_6\|u_n-u_m\|^2$$

for some $c_6 = c_6(T) > 0$, and (26) follows.
Uniqueness: assume that there exist two solutions, u and v, of equation (9). Taking the difference we get

$$u(t)-v(t) = \int_0^t e^{(t-s)A}[F(u(s))-F(v(s))]\,ds \quad (t\in[0,T]).$$

Then, applying $(-A)^{\frac{1}{2}}$ to $u(t)-v(t)$, ($t\in[0,T]$ a.e.) and recalling (12), we obtain

$$\begin{aligned}&\|(-A)^{\frac{1}{2}}(u(t)-v(t))\|\\ &\le \int_0^t \frac{M_{\frac{1}{2}}}{(t-s)^{\frac{1}{2}}}C_N(1+\|u(s)\|+\|v(s)\|)^{\frac{2}{N}}\|(-A)^{\frac{1}{2}}(u(s)-v(s))\|\,ds\\ &\le \int_0^t \frac{c_7}{(t-s)^{\frac{1}{2}}}\|(-A)^{\frac{1}{2}}(u(s)-v(s))\|\,ds \quad t\in[0,T]\text{ a.e.}\end{aligned}$$

where

$$c_7 = M_{\frac{1}{2}}C_N \sup_{0\le t\le T}(1+\|u(t)\|+\|v(t)\|)^{\frac{2}{N}}.$$

Hence, by Lemma 3.3, $\|(-A)^{\frac{1}{2}}(u(t)-v(t))\| \equiv 0$. □

4. Blow-up

In this section we will show that the growth conditions (6) and (7) are, in some sense, optimal. For $k\ge 1$, $p>1$ and $s>0$, define

$$g_k(s) := \begin{cases} s\log^p(e_1+s) & \text{if } k=1,\\ s\prod_{j=1}^{k-1}\log_j(e_k+s)\log_k^p(e_k+s) & \text{if } k\ge 2.\end{cases}$$

We will prove the following result.

Theorem 4.1. *Let f be a function of class C^1, such that $f(0)=0$, satisfying, for some integer $k\ge 1$ and positive numbers $c,\gamma\in\mathbb{R}$,*

$$f(s)\ge \gamma g_k(s), \quad \forall s>c. \tag{27}$$

Then, for any $T>0$, a function $u_0\in C_0^\infty(\Omega)$ exists such that the solution of equation (1) with $h=0$ blows up before time T, i.e.

$$\lim_{t\to T}\|u(t)\| = \infty.$$

Remark 4.2. Since $g_k(s) = O(L_{k-1}(s)\log_k^p s)$ as $s\to+\infty$, the above theorem shows that assumption (6) is optimal.

Lemma 4.3. *The above defined function $g_k(s)$ is convex and increasing.*

Proof. It sufficies to show that g_k' is increasing. To this end, let us note that

$$\frac{d}{ds}\log_j(e_k+s) = \frac{1}{e_k+s}\quad\frac{1}{\prod_{i=1}^{j-1}\log_i(e_k+s)}.$$

Then,

$$\begin{aligned} g_k'(s) &= \prod_{j=1}^{k-1}\log_j(e_k+s)\log_k^p(e_k+s)+ \\ &+ p\frac{s}{e_k+s}\log_k^{p-1}(e_k+s)+\frac{s}{e_k+s}\log_k^p(e_k+s)\sum_{i=1}^{k-1}\prod_{j=i+1}^{k-1}\log_j(e_k+s) \end{aligned}$$

is increasing, being the sum of increasing functions. □

Notice that the above lemma applies also to

$$g(s) := \frac{\gamma}{2}g_k\left(\frac{s}{2}\right). \tag{28}$$

For the proof of Theorem 4.1 we adapt a known technique (see [4]).

Proof. Let $\varphi \in H^2(\Omega)\cap H_0^1(\Omega)$ be the first eigenfunction of $-\Delta$ (with eigenvalue $\lambda > 0$), normalized so that

$$\begin{cases} -\Delta\varphi = \lambda\varphi & \text{in}\,\Omega \\ \varphi = 0 & \text{on}\,\Gamma \\ \varphi > 0 & \text{in}\,\Omega \\ \|\varphi\|_{L^\infty} = 1\,. \end{cases}$$

Let $\alpha \geq 0$ be such that

$$f(s) \geq g(s) - \alpha s\,, \qquad \forall s \geq 0, \tag{29}$$

where g is the function defined in (28). Let $T > 0$ be fixed and choose $c > 0$ so that

$$g(s) \geq 2(\lambda+\alpha)s \qquad \forall s \geq c\,, \tag{30}$$

and

$$\int_c^{+\infty}\frac{ds}{g(s)} < \frac{T}{2}\,. \tag{31}$$

Let $u_0 \in C_0^\infty(\Omega)$ be a fixed non-negative function such that

$$\int_\Omega \varphi(x)u_0(x)\,dx = c\,. \tag{32}$$

Let u be the mild solution of

$$\begin{cases} u' = \Delta u + f(u) & \text{in }\ \Omega\times(0,T) \\ u = 0 & \text{in }\ \Gamma\times(0,T) \\ u(0) = u_0\,. \end{cases} \tag{33}$$

A well-known regularity result implies that u is continuous in $\Omega \times [0, T]$, see e.g. [8] . Observe that, by the maximum principle,

$$u(t, x) > 0 \qquad \text{in } \Omega \times (0, T) .$$

Define

$$\phi(t) = \frac{1}{|\Omega|} \int_\Omega u(t, x)\, \varphi(x)\, dx$$

where $|\Omega|$ is the Lebesgue measure of the set Ω. Then, $\phi(t) > 0$ and we deduce, from (29), (33), that

$$\phi'(t) \geq \frac{1}{|\Omega|} \int_\Omega (u\Delta\varphi + g(u)\varphi - \alpha u\varphi)\, dx = -(\lambda + \alpha)\phi(t) + \frac{1}{|\Omega|} \int_\Omega g(u)\varphi\, dx\, . \quad (34)$$

Recall Lemma 4.3, Jensen's inequality, and the fact that $\|\varphi\|_{L^\infty} = 1$ to derive

$$g(\phi(t)) \leq \frac{1}{|\Omega|} \int_\Omega g(u(t, x)\varphi(x))\, dx \leq \frac{1}{|\Omega|} \int_\Omega g(u(t, x))\varphi(x)\, dx\, . \qquad (35)$$

Inequalities (34) and (35) yield

$$\phi'(t) \geq -(\lambda + \alpha)\phi(t) + g(\phi(t))\, . \qquad (36)$$

So, by a standard argument,

$$\phi(t) \geq c\, , \qquad \forall t \in [0, T]\, . \qquad (37)$$

Now, using (36), (30) and (37) we conclude that

$$\phi'(t) \geq \frac{1}{2} g(\phi(t)) - (\lambda + \alpha)\phi(t) + \frac{1}{2} g(\phi(t)) \geq \frac{1}{2} g(\phi(t)) \qquad (38)$$

for all $t \in [0, T]$. Define

$$G(r) = \int_{\phi(0)}^r \frac{ds}{g(s)} = \int_c^r \frac{ds}{g(s)}, \qquad \forall r \geq c\, .$$

Then,

$$\frac{d}{dt} G(\phi(t)) = \frac{\phi'(t)}{g(\phi(t))} \geq \frac{1}{2}$$

so that

$$G(\phi(T)) \geq \frac{T}{2} + G(\phi(0)) = \frac{T}{2}\, .$$

We have thus obtained

$$\int_c^\infty \frac{ds}{g(s)} \geq \int_c^{\phi(T)} \frac{ds}{g(s)} \geq \frac{T}{2}$$

which is in contradiction with (31). □

References

[1] A. Bensoussan, G. Da Prato, M.C. Delfour and S.K. Mitter, *Represantation and control of infinite dimensional systems*, Volume I, 1992, Birkhäuser, Boston.

[2] P. Cannarsa, V. Komornik and P. Loreti, *Well-posedness and control of semilinear wave equations with iterated logarithms*, to appear in ESAIM: Control, Optimization and Calculus of Variations.

[3] T. Cazenave and A. Haraux, *Équations d'évolution avec non linéarité logarithmique*, Ann. Fac. Sci. Toulouse, **2** (1980), 21–51.

[4] T. Cazenave and A. Haraux, *Introduction aux problèmes d'évolution semi-linéaires*, Mathématiques et applications, **1** 1990, Ellipses et SMAI, Paris.

[5] E. Fernández-Cara, *Null controllability of the semilinear heat equation*, ESAIM: Control, Optimization and Calculus of Variations, **2** (1997), 87–103.

[6] D. Henry, *Geometric theory of semilinear parabolic equations*, Lecture Notes in Mathematics, **840** 1981, Springer-Verlag, Berlin.

[7] X. Li and J. Yong, *Optimal control theory for infinite dimensional systems*, 1995, Birkhäuser, Boston.

[8] A. Lunardi, *Analytic semigroups and optimal regularity in parabolic problems*, PNLDE, **16** 1995, Birkhäuser, Basel.

[9] A. Pazy, *Semigroups of linear operators and applications to partial differential equations*, Applied Mathematical Sciences, **44** 1983, Springer-Verlag, New York.

Paolo Albano
Dipartimento di Matematica,
Università di Roma Tor Vergata,
Via della Ricerca Scientifica
I-00133 Roma, Italy
E-mail address: `albano@axp.mat.uniroma2.it`

Piermarco Cannarsa
Dipartimento di Matematica,
Università di Roma Tor Vergata,
Via della Ricerca Scientifica
I-00133 Roma, Italy
E-mail address: `cannarsa@axp.mat.uniroma2.it`

Vilmos Komornik
Institut de Recherche Mathématique Avancée,
Université Louis Pasteur et CNRS,
7, rue René Descartes
F-67084 Strasbourg Cédex, France
E-mail address: `komornik@math.u-strasbg.fr`

International Series of Numerical Mathematics
Vol. 133,

Uniform Stability of Nonlinear Thermoelastic Plates with Free Boundary Conditions

George Avalos[1], Irena Lasiecka[2] and Roberto Triggiani[2]

Abstract. In this work, we derive stability properties for a nonlinear thermoelastic plate system in which the higher order "free" boundary conditions are enforced on the displacement of the plate. The class of nonlinearities under consideration here include those seen in classical models in mechanics; such as von Kármán systems, the quasilinear Berger's equation which models extensible plates, and Euler–Bernoulli semilinear plates. This paper focuses on the case that rotational inertia is unaccounted for in the model, which corresponds to the absence of the parameter γ (i.e., $\gamma = 0$). In this case, we show that for initial data in the basic space of well-posedness, solutions of the PDE system decay to zero uniformly as time gets large. In our proof of uniform decay for $\gamma = 0$, absolutely critical use is made of the recently discovered fact that the linearization of the thermoelastic model generates an analytic semigroup.

1. Introduction

1.1. The model

Let Ω be a bounded open domain in $\mathbb{R}^2$ with sufficiently smooth boundary Γ. On this geometry we consider the following thermoelastic plate system (see [14]):

$$\begin{cases} \omega_{tt} - \gamma\Delta\omega_{tt} + \Delta^2\omega + \alpha\Delta\theta + F(\omega) = 0 & \text{in } \Omega \times (0,\infty); \\ \theta_t - \Delta\theta - \alpha\Delta\omega_t = 0 & \text{in } \Omega \times (0,\infty), \end{cases} \tag{1}$$

with the following "free" boundary conditions for ω, and Robin boundary condition for θ prescribed on the boundary $\Gamma \times (0,\infty)$ (here, τ and ν are respectively, the unit (counter clockwise) tangent vector and the outward normal vector):

$$\begin{cases} \Delta\omega + (1-\mu)B_1\omega + \alpha\theta = 0; \\ \dfrac{\partial\Delta\omega}{\partial\nu} + (1-\mu)\dfrac{\partial B_2\omega}{\partial\tau} - \eta\omega - \gamma\dfrac{\partial\omega_{tt}}{\partial\nu} + \alpha\dfrac{\partial\theta}{\partial\nu} = 0; \\ \dfrac{\partial\theta}{\partial\nu} + \lambda\theta = 0, \end{cases} \tag{2}$$

[1]Research supported in part by the National Science Foundation Grant No. DMS–9710981.
[2]Research supported in part by the National Science Foundation Grant No. DMS–9504822, and by the Army Research Office Grant DAAH04–96–1–0059.

where the boundary operators B_1 and B_2 are given by (see [14])

$$B_1\omega = 2\nu_1\nu_2\frac{\partial^2\omega}{\partial x\partial y} - \nu_1^2\frac{\partial^2\omega}{\partial y^2} - \nu_2^2\frac{\partial^2\omega}{\partial x^2};$$

$$B_2\omega = \left(\nu_1^2 - \nu_2^2\right)\frac{\partial^2\omega}{\partial x\partial y} + \nu_1\nu_2\left(\frac{\partial^2\omega}{\partial y^2} - \frac{\partial^2\omega}{\partial x^2}\right). \tag{3}$$

Above, $\mu \in (0,1)$ is Poisson's modulus, $\gamma \geq 0$, $\alpha > 0$, $\eta \geq 0$ and $\lambda \geq 0$ are constants. With this model we associate the initial conditions

$$\omega(\cdot,0) = \omega_0 \in H^2(\Omega);\ \omega_t(\cdot,0) = \omega_1 \in H^1_\gamma(\Omega);\ \theta(\cdot,0) = \theta_0 \in L^2(\Omega), \tag{4}$$

where

$$H^1_\gamma(\Omega) = \begin{cases} H^1(\Omega), \text{ if } \gamma > 0 \\ L^2(\Omega), \text{ if } \gamma = 0. \end{cases} \tag{5}$$

Of all the physical constants in the model (1)–(2), the most critical one is γ: For $\gamma > 0$, the model accounts for rotational inertia, and for $\gamma = 0$ it does not. Correspondingly, the two cases $\gamma > 0$ and $\gamma = 0$ produce drastically different dynamical properties, as described below. The main contribution of the present paper–the stabilization result of Theorem 1.3 below–refers to the case $\gamma = 0$. In the case $\gamma > 0$, Theorem 1.3 was already proved in [5] in the nonlinear case, after the key contribution in [3] in the linear case, and is included here so as to complete our treatment and to contrast the two radically different cases. Indeed, to appreciate this contrast, it suffices to specialize to the linear dynamics (1)–(2) when $F = 0$: Then, as recently discovered, the s.c. semigroup generated by (1)–(2) for $F = 0$ is *analytic* (holomorphic) if $\gamma = 0$ [23, 25], and is *group-dominated* if $\gamma > 0$ [24, 9]. More precisely, the following result from [23], [25] (Chapter 3, Appendix I) forms the critical basis for our analysis below in the case $\gamma = 0$.

Theorem 1.1. *Let $\gamma = 0$ and $F = 0$ in (1)–(2). Then, the linear problem (1)–(2) determines a s.c. contraction semigroup $e^{\mathcal{A}t}$, defined by*

$$y_0 \equiv \{w_0, w_1, \theta_0\} \in H_0 \equiv H^2(\Omega) \times L_2(\Omega) \times L_2(\Omega)$$
$$\rightarrow e^{\mathcal{A}t}y_0 \equiv \{w(t), w_t(t), \theta(t)\} \in C([0,\infty]; H_0)$$

which moreover is analytic and uniformly stable in H_0.

For the group-dominated character of the semigroup generated by the linear model (1)–(2) (i.e. $F = 0$), when instead $\gamma > 0$, we refer to [24], as results concerning that situation will not be needed here explicitly.

Goal: The main goal of this paper is to show that the system described above, with either $\gamma > 0$ or $\gamma = 0$, is uniformly stable for a large class of "dissipative" nonlinear operators $F(\cdot)$. This class, defined explicitly below, will include classical models in mechanics such as the von Kármán system, the quasilinear Berger's equation which models extensible plates, and semilinear Euler–Bernoulli plate models.

We impose throughout that the nonlinearity $F(\cdot)$ satisfy the following conditions:

Assumption 1: The mapping $F : H^2(\Omega) \to [H^1(\Omega)]'$ is locally Lipschitz continuous, with $F(0) = 0$, and is moreover Fréchet differentiable at the point 0.

Assumption 2: The mapping F further satisfies the following relations for all $\omega \in H^2(\Omega)$ and $\varpi \in W^{1,2}(0,T; H^2(\Omega))$:

$$\int_\Omega F(\omega)\omega d\Omega \geq 0; \quad \text{and} \quad \int_\Omega F(\varpi(t))\varpi_t(t)d\Omega = \frac{1}{2}\frac{d}{dt}E_F(\varpi(t)) \quad \text{for all } t > 0, \tag{6}$$

for some functional E_F on $H^2(\Omega)$ satisfying

$$0 \leq E_F(\omega)) \leq C\left(\|\omega\|^2_{H^2(\Omega)}\right)\|\omega\|^2_{H^2(\Omega)}. \tag{7}$$

The above two Assumptions 1 and 2 on F are all that are needed to obtain the main new stabilization result for this paper (Theorem 1.3 below), which refers to the case $\gamma = 0$. However, for the sake of completeness we include also the result for the case $\gamma > 0$, where, as recalled above, Theorem 1.3 was recently proved in [3] in the critical linear case, which in turn served as a basis for the nonlinear case considered in [5]. For $\gamma > 0$, we need an additional assumption:

Assumption 3: In the case $\gamma > 0$, F is weakly continuous from $H^2(\Omega)$ into $\mathcal{D}'(\Omega)$.

Throughout, $C(\cdot)$ always denotes a generic function which is bounded for bounded values of its argument. As usual, C will denote a generic constant, taking different values in its various occurrences.

It can be shown directly that examples of nonlinearities which meet the abstract Assumptions 1–3 above include the following models:

1. (**The von Kármán model**, see [12], [11] and [26])

$$F(\omega) = -[\mathcal{F}(\omega), \omega];$$

where $[\omega, \varpi] \equiv \omega_{xx}\varpi_{yy} + \omega_{yy}\varpi_{xx} - 2\omega_{xy}\varpi_{xy}$, and the (Airy stress) function $\mathcal{F}$ satisfies

$$\Delta^2\mathcal{F}(\omega) = -[\omega, \omega] \text{ in } \Omega; \quad \mathcal{F}(\omega) = \frac{\partial\mathcal{F}(\omega)}{\partial\nu} = 0 \text{ on } \Gamma.$$

2. (**Berger's model**, see [7], [32] and [34]; often referred to in the literature as the extensible plate equation)

$$F(\omega) = -\Delta\omega \cdot f\left(\int_\Omega |\nabla\omega|^2 d\Omega\right), \quad \text{where } f \in C^1(\mathbb{R}) \text{ with } f' \geq 0.$$

3. (**Semilinear Euler–Bernoulli plates)**

$$F(\omega) = |\omega|^p \omega, \quad \text{where } p \geq 0.$$

1.2. Main Results

1.2.1. Well-posedness The model introduced above is well posed in the state space

$$\mathbf{H}_\gamma \equiv H^2(\Omega) \times H^1_\gamma(\Omega) \times L^2(\Omega), \tag{8}$$

where $H^1_\gamma(\Omega)$ is as defined in (5). More precisely, we have the following:

Proposition 1.2. *Assume $\gamma \geq 0$ and Assumptions 1–3.*

(i) For arbitrary initial data $[\omega_0, \omega_1, \theta_0] \in \mathbf{H}_\gamma$, there exists a unique solution $[\omega, \omega_t, \theta] \in C([0,\infty); \mathbf{H}_\gamma)$ for problem (1)–(2). In the case that $\gamma = 0$, we have the following additional regularity: For any $T > 0$

$$\omega \in L^2(0,T; H^3(\Omega));\ \omega_t \in L^2(0,T; H^1(\Omega));\ \theta \in L^2(0,T; H^1(\Omega)). \tag{9}$$

(ii) If, in the case $\gamma = 0$, we additionally assume that $F : H^2(\Omega) \to L^2(\Omega)$ is a bounded mapping (which is the case for the three examples cited above), and that $[\omega_0, \omega_1, \theta_0] \in H^3(\Omega) \times H^1(\Omega) \times H^1(\Omega)$, then

$$\omega \in L^2(0,T; H^4(\Omega));\ \omega_t \in L^2(0,T; H^2(\Omega));\ \theta \in L^2(0,T; H^2(\Omega)). \tag{10}$$

These results, except for the extra regularity in (9), follow through the classical procedure of first establishing local solutions, and then extending them globally, by virtue of an *a priori* bound (see [2]). In the case $\gamma = 0$, critical use is made of the analyticity of the underlying linear semigroup, shown recently in [23], [25] (Chapter 3). The proof of the additional regularity stated in (9) will be carried out below (see Lemmas 2.1 and 3.1).

1.2.2. Stabilization As stated earlier, our principal intent here is to show that the solution to (1)–(2), provided for by Proposition 1.2, decays uniformly to zero as time $t \uparrow \infty$.

In order to formulate our results concisely, we define the energy functional for the model (1). It is the sum of two parts: For the solution $[\omega, \omega_t, \theta]$ of (1)–(2), we set

$$E(t) \equiv E_\gamma\left(\begin{bmatrix} \omega(t) \\ \omega_t(t) \\ \theta(t) \end{bmatrix}\right) + E_F(\omega(t)), \tag{11}$$

where E_F is the functional assumed in Assumption 2, and

$$E_\gamma\left(\begin{bmatrix} \omega(t) \\ \omega_t(t) \\ \theta(t) \end{bmatrix}\right) \equiv a(\omega(t),\omega(t)) + \|\omega_t(t)\|^2_{L^2(\Omega)} + \gamma\,\|\nabla\omega_t(t)\|^2_{L^2(\Omega)} + \|\theta(t)\|^2_{L^2(\Omega)}, \tag{12}$$

with the bilinear form $a(\cdot,\cdot)$ being defined by [14]

$$\begin{aligned} a(\omega,\widehat{\omega}) \equiv & \int_\Omega [\omega_{xx}\widehat{\omega}_{xx} + \omega_{yy}\widehat{\omega}_{yy} + \mu(\omega_{xx}\widehat{\omega}_{yy} + \omega_{yy}\widehat{\omega}_{xx}) + 2(1-\mu)\omega_{xy}\widehat{\omega}_{xy}]\, d\Omega \\ & + \eta\int_\Gamma \omega\widehat{\omega}\, d\Gamma. \end{aligned}$$

With this bilinear form, we have the following "Green's formula" (see [14]) for functions ω and $\widehat{\omega}$ smooth enough:

$$\int_\Omega \Delta^2 \omega \widehat{\omega} d\Omega + \eta \int_\Gamma \omega \widehat{\omega} d\Gamma \tag{13}$$
$$= a(\omega, \widehat{\omega}) + \int_\Gamma \left(\frac{\partial \Delta \omega}{\partial \nu} + (1-\mu) \frac{\partial B_2 \omega}{\partial \tau} \right) \widehat{\omega} d\Gamma - \int_\Gamma (\Delta \omega + (1-\mu) B_1 \omega) \widehat{\omega} d\Gamma.$$

(note that $a(\cdot, \cdot)$ is $H^2(\Omega)$–elliptic when $\eta > 0$).

Throughout, we will further assume that η and λ are both strictly positive, in which case one can readily show that the energy as defined in (11) is topologically equivalent to the $\mathbf{H}_\gamma$–norm. We are now in a position to state our main result.

Theorem 1.3. *Besides Assumptions 1–3, take the parameters η and λ to be positive. Then for arbitrary initial data $[\omega_0, \omega_1, \theta_0] \in \mathbf{H}_\gamma$, there exist positive constants C and ρ, possibly depending upon γ and $E(0)$, such that*

$$E(t) \leq Ce^{-\rho t} E(0) \text{ for } t \geq 0.$$

Remark 1.4. *The assumption that η and λ are positive is not all essential. It is simply a convenience here, so as to avoid steady states for the problem. Almost identical results would be obtained by imposing homogeneous clamped boundary conditions on some nonvoid segment Γ_0 of Γ, in addition to the boundary conditions (2) on $\Gamma \backslash \Gamma_0$. This new specification of boundary values would not introduce any additional difficulties.*

1.2.3. Comments on Theorem 1.3 *The Case* $\gamma = 0$. Theorem 1.3 is new in the case $\gamma = 0$ (whereby Assumption 3 is then empty), where it thus generalizes the stability statement of Theorem 1.1 from the linear to the nonlinear model. In the linear case, the proof of analyticity of the s.c. semigroup of Theorem 1.1 readily implies its uniform stability [23], [25] (Chapter 3, Appendix I) for $\eta > 0$. Our proof below in the nonlinear case makes critical use of the analyticity of the linear case.

The Case $\gamma > 0$. When the parameter γ is positive, Theorem 1.3 was first proved in the linear case in [3], and the technique used there served as the basis to prove the extension to the nonlinear case in [5].

Relation to other boundary conditions. In the case of lower order boundary conditions (B.C.'s), such as *clamped* or *hinged*, the result of Theorem 1.3 is already known in full generality (see [5]); the specific case of the von Kármán model with *hinged* and *clamped* B.C.'s is also given in [16] and [8]. Moreover, for the said case of *clamped* and *hinged* B.C.'s, references [16], [4] (the linear case) showed in addition that the decay rates stated in Theorem 1.3 are independent of the parameter γ. However, for the case of free B.C.'s, which is being considered here, the analysis in [4] falters at a certain point, owing to the coupling on the boundary between the thermal variable θ and the mechanical ω. In fact, it turns out that in order to obtain the appropriate estimates in the free case, "sharp" regularity results for the traces of the mechanical variable had to be established. With the parameter $\gamma > 0$ (the "hyperbolic" case), this resolution of the mechanical traces was done in [3] (which handles the linear model) and in [5], [2] (which treat the nonlinear version) by utilizing the deep trace results available for solutions of second order hyperbolic equations (see [35] and related results in [17, 18, 19]). An analogous trace result is not available for $\gamma = 0$, and so the analysis of [2], [3] and [5] cannot be invoked for this case.

The strategy in this paper is to devise a *time-domain* technique for $\gamma = 0$ (as opposed to the *resolvent [frequency] domain* technique used in [23, 25] to obtain the analyticity in the linear case $F = 0$) which, while using the analyticity of the underlying linearization (see below), still provides another avenue for showing stability estimates in the time domain for $F = 0$, and is sufficiently flexible so as to be extendable to the nonlinear case $F \neq 0$. In the present proof, even at the linear level, the analyticity of the semigroup of the linear part is used to obtain higher regularity properties of the solutions, hence giving a well-posedness to certain boundary traces (which in the case $\gamma > 0$ came about as a result of hyperbolicity, as mentioned before).

1.3. Literature

Linear thermoelastic plates have been studied extensively in recent years, with most of the papers dealing with the case of clamped or hinged boundary conditions, and with the case that the parameter $\gamma = 0$. In fact, for these particular models, the exponential stability of the corresponding solutions was shown by several authors (see [13], [29], [30], [4] and [33]). Later, and in fact very recently, [27] proved the stronger and more desirable result that when the lower order clamped (or hinged) mechanical/Dirichlet thermal boundary conditions are present, and $\gamma = 0$, the semigroup associated with the linear thermoelastic plate is analytic. Generalizations including the more challenging case of coupled hinged/Neumann boundary conditions were given in [20], [21] and [22]; see also [30] which treats a case of uncoupled boundary conditions. As a consequence of analyticity, the exponential decay of the energy is readily deduced. The situation is even more challenging in the case of *free* boundary conditions. These B.C.'s involve a strong coupling on the boundary between the thermal and mechanical variable, and it is this coupling and the fact that the Lopatinski conditions are not satisfied, which have been major complications in the attempt to prove analyticity. In fact, the question as to whether the semigroup corresponding to the linear thermoelastic plate under free B.C.'s is analytic remained an open one, until it was recently answered in the affirmative (see [23] and [20]). Thus, in the free case, we know additionally that the corresponding linear semigroup is uniformly stable. With the uniform decay of the linear model having been established, an important follow up problem is that of generating explicit estimates which can be carried over to nonlinear variations of the model. These stability estimates were obtained for all cases but that of free B.C.'s (see [4]), by means of an appropriate operator multiplier which was constructed in [4]. However, the situation is drastically different in the case of free B.C's. Indeed, the same multiplier "almost" gives the sought after inequalities, these, however, being "polluted" by the appearance of certain boundary traces of the solutions which are not bounded by the energy functional, or determined by some "hidden" regularity (as is the case for clamped B.C.'s; see [4]). A similar problem has already occurred in those linear thermoelastic plate models with free B.C.'s, and which account for rotational inertia (i.e., $\gamma > 0$). In this (hyperbolic) case, deep and sharp regularity results for wave traces (see [35, 17]) are ultimately responsible for the procurement of the estimates in [3]. In the case when rotational inertia terms are not present (i.e., $\gamma = 0$), there is no hyperbolicity of the corresponding mechanical part of the system, and so the aforementioned trace theory does not apply. In order to derive estimates analogous to those in [3], one would have to prove trace results for the Euler–Bernoulli model in the case when the Lopatinski conditions are not satisfied (an analog of theory developed in [17], [18], [19] and [35] for the wave equation). However, this analysis has not yet been done, and its undertaking would constitute a very difficult and technical project. On the face of it, this missing trace ingredient seemed indispensible for any successful attainment of explicit decay rates for linear thermoelastic systems with $\gamma = 0$ and with free boundary conditions.

This impasse had remained unchanged until recently, when new results on the analyticity of the underlying semigroup (for $\gamma = 0$) were proved in [23] and [20] for the linear model. The purpose of this paper is to show how the extra regularity implied by this newfound analyticity can be exploited so as to successfully obtain the explicit estimates for the decay rates of the energy. A positive benefit of this method is that it can be used to treat nonlinear models as well, as is so done in this paper. However, even in the linear case the posted methodology is new.

Exponential decay rates of the energy for the same class of nonlinear problems as above but with *clamped/hinged* B.C.'s and/or with $\gamma > 0$ are given in [5], [2] and [16] (see also [24] and [9] for the linear case). Therefore, the present paper completes the picture by asserting that all thermoelastic nonlinear models (subject to Assumptions 1–3) are exponentially stable, with or without the presence of rotational inertia, and regardless of the boundary conditions being considered.

The remainder of this paper is devoted to the proof of Theorem 1.3 for the case $\gamma = 0$. In [2], it is shown that there are smooth, approximating solutions to (1). Accordingly, we can assume that the solution $[\omega, \omega_t, \theta]$ has the regularity needed to justify the computations performed below.

2. Case $\gamma = 0$: Multiplier Estimates

We begin with a fundamental energy relation which shows the dissipative nature of the problem.

Lemma 2.1. *Let $[\omega, \omega_t, \theta]$ be a solution of (1)–(2). Then, for all $t, s \geq 0$, $E(t)$ in (11) satisfies*

$$E(t) + 2\int_s^t \|\nabla\theta(\tau)\|^2_{L^2(\Omega)}\,d\tau + 2\lambda\int_s^t \|\theta(\tau)\|^2_{L_2(\Gamma)}\,d\tau = E(s);$$
$$E(t) \leq E(s) \text{ for } t \geq s. \tag{14}$$

The proof of Lemma 2.1 is routine. We multiply the first equation of (1) by ω_t, the second by θ, integrate the resulting quantities in time and space, integrate by parts and evoke (14) and (6).

The following preliminary estimate is the crucial result here.

Lemma 2.2. *Let $[\omega, \omega_t, \theta]$ be a solution of (1)–(2). Then for any $T > 0$ we have*

$$\int_0^T E(t)dt \leq C(E(0)) \left[E(T) + E(0) + \int_0^T \|\theta\|^2_{H^1(\Omega)}\, dt + \int_0^T \|\omega_t\|^2_{H^{-\frac{1}{2}}(\Gamma)}\, dt \right]. \tag{15}$$

Proof. We begin by defining the following operators [25]:

$$\begin{aligned} A_D &: L^2(\Omega) \to L^2(\Omega), \;\; A_D \equiv -\Delta; \;\; D(A_D) \equiv H^2(\Omega) \cap H^1_0(\Omega); \\ D &: D \in H^s(\Gamma) \to H^{s+\frac{1}{2}}(\Omega); \\ & \quad h = Dg \Leftrightarrow \Delta h = 0 \text{ on } \Omega \text{ and } h = g \text{ on } \Gamma, \end{aligned} \tag{16}$$

for $s \in \mathbb{R}$. By elliptic theory $A_D^{-1} \in \mathcal{L}\left(L^2(\Omega), D(A_D)\right)$ and $D \in \mathcal{L}(H^s(\Gamma),$-$H^{s+\frac{1}{2}}(\Omega))$. Moreover we have the relation

$$\Delta u = \Delta(u - Du|_\Gamma) = -A_D(u - Du|_\Gamma) \Rightarrow A_D^{-1}\Delta u = -u + Du|_\Gamma \;\; \forall u \in H^2(\Omega). \tag{17}$$

STEP 1. The key point here is to apply the operator theoretic multiplier $A_D^{-1}\theta$ to the first equation of (1). We note that this same multiplier was used in [3], [4] and [5]. We shall provide partial computations: First off, we have

$$\begin{aligned} & \int_0^T \left(\omega_{tt}, A_D^{-1}\theta\right)_{L^2(\Omega)} dt \qquad\qquad (18) \\ = \; & \left[\left(\omega_t, A_D^{-1}\theta\right)_{L^2(\Omega)} \right]_0^T + \int_0^T \left[\alpha \|\omega_t\|^2_{L^2(\Omega)} + \left(\omega_t, \theta - D\left[\theta|_\Gamma + \alpha\, \omega_t|_\Gamma\right]\right)_{L^2(\Omega)} \right] dt, \end{aligned}$$

where here we have used the heat equation in (1) and the relation (17) applied to θ and ω_t. Moreover, we note by the regularity of the map D in (16) for $s = -\frac{1}{2}$ that

$$\begin{aligned} & \left| \int_0^T \left[\left(\omega_t, \theta - D\,\theta|_\Gamma\right)_{L^2(\Omega)} - \alpha \left(\omega_t, D\,\omega_t|_\Gamma\right)_{L^2(\Omega)} \right] dt \right| \\ \leq \; & \epsilon \int_0^T \|\omega_t\|^2_{L^2(\Omega)}\, dt + C_\epsilon \int_0^T \|\theta\|^2_{H^1(\Omega)}\, dt + C_\epsilon \int_0^T \|\omega_t\|^2_{H^{-\frac{1}{2}}(\Gamma)}\, dt. \end{aligned} \tag{19}$$

Secondly, by using the Green's formula (14) (with $\widehat{\omega} = A_D^{-1}\theta$, where $A_D^{-1}\theta\big|_\Gamma = 0$ since $A_D^{-1}\theta \in D(A_D)$), and the first B.C. in (2), we obtain

$$\begin{aligned} \int_0^T \left(\Delta^2\omega, A_D^{-1}\theta\right)_{L^2(\Omega)} dt \;&=\; \int_0^T \left[a\left(\omega, A_D^{-1}\theta\right) + \alpha\left(\theta, \frac{\partial A_D^{-1}\theta}{\partial\nu}\right)_{L^2(\Gamma)} \right] dt \\ &\leq\; \epsilon \int_0^T \|\omega\|^2_{H^2(\Omega)}\, dt + C_\epsilon \int_0^T \|\theta\|^2_{H^1(\Omega)}\, dt, \end{aligned} \tag{20}$$

where in the last step we have also used the continuity of $a(\cdot,\cdot)$ in the $H^2(\Omega)\times H^2(\Omega)$ topology, as well as trace theory. In addition, applying (17) to θ, we also find that

$$\int_0^T \alpha\left(\Delta\theta, A_D^{-1}\theta\right)_{L^2(\Omega)} dt = \int_0^T \alpha\left(\theta, -\theta + D\,\theta|_\Gamma\right)_{L^2(\Omega)} dt \le C\int_0^T \|\theta\|^2_{H^1(\Omega)}\,dt. \tag{21}$$

Moreover, by the continuity of F in Assumption 1, and as $E(t)\le E(0)$ for t positive (by Lemma 2.1), we have for $t\ge 0$

$$\|F(\omega(t))\|_{[H^1(\Omega)]'} \le C\left(\|\omega(t)\|^2_{H^2(\Omega)}\right)\|\omega(t)\|_{H^2(\Omega)} \le C\left(E(0)\right)\|\omega(t)\|_{H^2(\Omega)}. \tag{22}$$

With this inequality and Lemma 2.1 we thereby obtain

$$\begin{aligned}\int_0^T \left(F(\omega), A_D^{-1}\theta\right)_{L^2(\Omega)} dt &\le C\int_0^T \|F(\omega)\|_{[H^1(\Omega)]'}\|\theta\|_{H^1(\Omega)}\,dt\\ &\le \int_0^T C\left(E(0)\right)\|\omega\|_{H^2(\Omega)}\|\theta\|_{H^1(\Omega)}\,dt\\ &\le \epsilon\int_0^T \|\omega\|^2_{H^2(\Omega)}\,dt + C_\epsilon(E(0))\int_0^T \|\theta\|^2_{H^1(\Omega)}\,dt.\end{aligned} \tag{23}$$

Collecting (19)–(24) and performing rather straightforward majorizations (which involve the use of trace theory and the regularity of the mapping D) yield

$$\begin{aligned}(\alpha-\epsilon_0)\int_0^T \|\omega_t\|^2_{L^2(\Omega)}\,dt &\le C\left[E(0)+E(T)\right] + \epsilon\int_0^T \|\omega\|^2_{H^2(\Omega)}\,dt\\ &+C_{\epsilon,\epsilon_0}(E(0))\int_0^T \|\theta\|^2_{H^1(\Omega)}\,dt + C_{\epsilon_0}\int_0^T \|\omega_t\|^2_{H^{-\frac{1}{2}}(\Gamma)}\,dt.\end{aligned} \tag{24}$$

STEP 2. We now apply the "classic" multiplier ω. Multiplication of the first equation in (1) by ω, followed by an integration in time and space, an integration by parts and the use of the Green's formula (14) (in which we account for the B.C. in (2)) gives

$$\int_0^T \left[a(\omega,\omega) + (F(\omega),\omega)_{L^2(\Omega)}\right] dt = -\left[(\omega_t,\omega)_{L^2(\Omega)}\right]_0^T \tag{25}$$

$$\begin{aligned}&+\int_0^T \|\omega_t\|^2_{L^2(\Omega)}\,dt - \alpha\int_0^T \left(\theta, \frac{\partial\omega}{\partial\nu}\right)_{L^2(\Gamma)} dt + \alpha\int_0^T (\nabla\theta,\nabla\omega)_{L^2(\Omega)}\,dt\\ \le\;& C\left[E(0)+E(T)\right]\\ &+\epsilon\int_0^T \|\omega\|^2_{H^2(\Omega)}\,dt + C_\epsilon\int_0^T \|\theta\|^2_{H^1(\Omega)}\,dt + \int_0^T \|\omega_t\|^2_{L^2(\Omega)}\,dt.\end{aligned} \tag{26}$$

STEP 3. We now add the term $\int_0^T \|\omega_t\|^2_{L^2(\Omega)}\,dt$ to both sides of (26); and thereafter invoke the estimate (24) to the right-hand side of the resulting inequality to get (since $(F(\omega),\omega)_{L^2(\Omega)} \geq 0$, by Assumption 1)

$$\begin{aligned}
&\int_0^T \left[a(\omega,\omega) + \|\omega_t\|^2_{L^2(\Omega)}\right] dt \\
\leq\ & \int_0^T \left[a(\omega,\omega) + \|\omega_t\|^2_{L^2(\Omega)} + (F(\omega),\omega)_{L^2(\Omega)}\right] dt \\
\leq\ & C\left[E(0)+E(T)\right] + C(E(0)) \int_0^T \|\theta\|^2_{H^1(\Omega)}\, dt + \\
& \epsilon \int_0^T \|\omega\|^2_{H^2(\Omega)}\, dt + C_\epsilon \int_0^T \|\omega_t\|^2_{H^{-\frac{1}{2}}(\Gamma)}\, dt. \qquad (27)
\end{aligned}$$

Recalling the $H^2(\Omega)$–ellipticity of $a(\cdot,\cdot)$ in (14) when the parameter η of (2) is strictly positive, and taking ϵ sufficiently small in (27) yields

$$\begin{aligned}
\int_0^T \left[a(\omega,\omega) + \|\omega_t\|^2_{L^2(\Omega)}\right] dt \ \leq\ & C\left[E(0)+E(T)\right] + C(E(0)) \int_0^T \|\theta\|^2_{H^1(\Omega)}\, dt \\
& + C \int_0^T \|\omega_t\|^2_{H^{-\frac{1}{2}}(\Gamma)}\, dt. \qquad (28)
\end{aligned}$$

Noticing that (7) and the fact that $E(t) \leq E(0)$ (by Lemma 2.1) imply that

$$\int_0^T E_F(\omega) dt \leq C(E(0)) \int_0^T \|\omega\|^2_{H^2(\Omega)}\, dt \leq C(E(0)) \int_0^T a(\omega,\omega) dt, \qquad (29)$$

we can then couple this with (28) to arrive at the inequality

$$\begin{aligned}
\int_0^T E(t) dt \ =\ & \int_0^T E_0(t) dt + \int_0^T E_F(t) dt \qquad (30)\\
\leq\ & C(E(0)) \left[E(0) + E(T) + \int_0^T \|\theta\|^2_{H^1(\Omega)}\, dt + \int_0^T \|\omega_t\|^2_{H^{-\frac{1}{2}}(\Gamma)}\, dt\right].
\end{aligned}$$

This completes the proof of Lemma 2.2. □

At this point we note that the "bad" term in the inequality (15) is $\int_0^T \|\omega_t\|^2_{H^{-\frac{1}{2}}(\Gamma)}\, dt$, inasmuch as it can not be majorized by the energy functional. It is at the level of coping with this trace term that the analyticity of the semigroup associated with the linearization of (1)–(2) must be invoked. We note explicitly that this term has arisen as a result of our use of the expression (17) (with $u = \omega_t$ therein) in the derivation of the estimate (19).

3. Case $\gamma = 0$: Analytic Estimates (Proof of (9))

The main goal in this section is to prove the additional regularity of the solution component $[\omega, \omega_t]$ to (1)–(2), as stated in (9) of Proposition 1.2 (the $L^2(0,T;H^1(\Omega))$–regularity of θ has already been shown via Lemma 2.1).

Lemma 3.1. *Let* $[\omega, \omega_t, \theta]$ *be the solution to (1)–(2), corresponding to initial data* $[\omega_0, \omega_1, \theta_0] \in \mathbf{H}_0$. *Then* $[\omega, \omega_t, \theta] \in L^2(0,T;H^3(\Omega)) \times L^2(0,T;H^1(\Omega)) \times L^2(0,T;H^1(\Omega))$, *with the following valid estimate:*

$$\int_0^T \left[\|\omega\|^2_{H^3(\Omega)} + \|\omega_t\|^2_{H^1(\Omega)} + \|\theta\|^2_{H^1(\Omega)}\right] dt \le CE(0) + C(E(0)) \int_0^T \|\omega\|^2_{L^2(\Omega)}\, dt, \tag{31}$$

where the constants are independent of T.

Proof. We write the solution $[\omega, \omega_t, \theta]$ to the equation (1)–(2) via the variation of parameters formula. To this end, let $\mathcal{A}$ be the generator corresponding to the linear part of (1)–(2). In other words,

$$\begin{aligned} &\mathcal{A}\left[\omega_0, \omega_1, \theta_0\right] = \left[\omega_1, -\Delta^2\omega_0 - \alpha\Delta\theta_0, \Delta\theta_0 + \alpha\Delta\omega_1\right], \text{ with} \\ &D(\mathcal{A}) = \left\{[\omega_0, \omega_1, \theta_0] \in H^4(\Omega) \times H^2(\Omega) \times H^2(\Omega)\right. \\ &\qquad\qquad \left.\text{such that the B.C.'s in (2) are satisfied}\right\}. \end{aligned}$$

It is known from [23] that $\mathcal{A}$ generates an s.c. analytic semigroup $\left\{e^{\mathcal{A}t}\right\}_{t\ge 0}$ of contractions on $\mathbf{H}_0$, and is furthermore boundedly invertible for $\eta > 0$. Given these dynamics we can then write the solution (1)–(2) explicitly as

$$\begin{bmatrix} \omega(t) \\ \omega_t(t) \\ \theta(t) \end{bmatrix} = e^{\mathcal{A}t} \begin{bmatrix} \omega_0 \\ \omega_1 \\ \theta_0 \end{bmatrix} + \int_0^t e^{\mathcal{A}(t-s)} \begin{bmatrix} 0 \\ F(\omega) \\ 0 \end{bmatrix} ds. \tag{32}$$

Using the equality

$$\mathcal{A}^{\frac{1}{2}} \begin{bmatrix} \omega(t) \\ \omega_t(t) \\ \theta(t) \end{bmatrix} = \mathcal{A}^{\frac{1}{2}} e^{\mathcal{A}t} \begin{bmatrix} \omega_0 \\ \omega_1 \\ \theta_0 \end{bmatrix} + \int_0^t \mathcal{A} e^{\mathcal{A}(t-s)} \mathcal{A}^{-\frac{1}{2}} \begin{bmatrix} 0 \\ F(\omega) \\ 0 \end{bmatrix} ds, \tag{33}$$

we then obtain the estimate

$$\left\| \mathcal{A}^{\frac{1}{2}} \begin{bmatrix} \omega(t) \\ \omega_t(t) \\ \theta(t) \end{bmatrix} \right\|_{L^2(0,T;\mathbf{H}_0)} \le C \left\| \begin{bmatrix} \omega_0 \\ \omega_1 \\ \theta_0 \end{bmatrix} \right\|_{\mathbf{H}_0} + C \left\| \mathcal{A}^{-\frac{1}{2}} \begin{bmatrix} 0 \\ F(\omega) \\ 0 \end{bmatrix} \right\|_{L^2(0,T;\mathbf{H}_0)}. \tag{34}$$

The justification for the estimate (34) rests ultimately on the following three properties: (i) The quantity $\left\{e^{\mathcal{A}t}\right\}_{t\ge 0}$ is a s.c. contraction semigroup on $\mathbf{H}_0$ (so that the operator $(-\mathcal{A})$ is accretive here); (ii) it is moreover analytic on $\mathbf{H}_0$; and (iii) it is exponentially stable on $\mathbf{H}_0$ (so that, in particular, $\mathcal{A}^{-1}$ is bounded on $\mathbf{H}_0$). In fact, the integral term in (33) yields its corresponding estimate in (34) by virtue of a standard property of stable, analytic semigroups [15]. What is, however, a more delicate issue is the assertion that the term based on the initial data is in $L^2(0,\infty;\mathbf{H}_0)$, consequently yielding up its corresponding estimate in (34). To show this, we first note that since $(-\mathcal{A})$ is accretive and $\mathcal{A}^{-1}$ is bounded, then $D((-\mathcal{A})^{\frac{1}{2}}) = [D(\mathcal{A}), \mathbf{H}_0]_{\frac{1}{2}} = (D(\mathcal{A}), \mathbf{H}_0)_{\theta=\frac{1}{2},p=2}$ (see e.g., [6], Vol. I, Proposition

6.1; or [31], p. 164). With this characterization, one can then appeal to the estimate in [31], Lemma 3.5 (or see [6], Vol. II) to obtain (34).

On the other hand, we know by the definition of $D(\mathcal{A})$ $(= D(\mathcal{A}^*))$ and interpolation that

$$D((\mathcal{A}^*)^{\frac{1}{2}}),\ D(\mathcal{A}^{\frac{1}{2}}) \subset H^3(\Omega) \times H^1(\Omega) \times H^1(\Omega); \tag{35}$$

and so by duality with respect to the $\mathbf{H}_0$–topology, we have

$$\left[H^3(\Omega) \times H^1(\Omega) \times H^1(\Omega)\right]' \subset \left[D((\mathcal{A}^*)^{\frac{1}{2}})\right]'. \tag{36}$$

In particular, the second component space of $D((\mathcal{A}^*)^{\frac{1}{2}}),\ D(\mathcal{A}^{\frac{1}{2}})$ is equal to $H^1(\Omega)$, so that the second component space of $\left[D((\mathcal{A}^*)^{\frac{1}{2}})\right]'$ is equal to $\left[H^1(\Omega)\right]'$. (This is so, since the B.C. in (2) couple the first and third coordinates only, and not the second.) It thus follows from (36) and the fact that $F(\omega(t)) \in \left[H^1(\Omega)\right]'$ (from Assumption 1) that

$$\begin{aligned}
&\left\| \mathcal{A}^{-\frac{1}{2}} \begin{bmatrix} 0 \\ F(\omega(t)) \\ 0 \end{bmatrix} \right\|_{\mathbf{H}_0} = \left\| \begin{bmatrix} 0 \\ F(\omega(t)) \\ 0 \end{bmatrix} \right\|_{\left[D((\mathcal{A}^*)^{\frac{1}{2}})\right]'} \\
&\leq C \left\| F(\omega(t)) \right\|_{[H^1(\Omega)]'} \leq C(E(0)) \left\| \omega(t) \right\|_{H^2(\Omega)},
\end{aligned} \tag{37}$$

where for the last inequality we have used (22). Splicing together (35) with the estimates (34) and (37), we have

$$\begin{aligned}
&\int_0^T \left[\|\omega\|^2_{H^3(\Omega)} + \|\omega_t\|^2_{H^1(\Omega)} + \|\theta\|^2_{H^1(\Omega)} \right] dt \leq \int_0^T \left\| [\omega, \omega_t, \theta] \right\|^2_{D(\mathcal{A}^{\frac{1}{2}})} dt \\
&\leq CE(0) + C(E(0)) \int_0^T \|\omega\|^2_{H^2(\Omega)} dt \\
&\quad \leq CE(0) + \epsilon \int_0^T \|\omega\|^2_{H^3(\Omega)} dt + C_\epsilon(E(0)) \int_0^T \|\omega\|^2_{L^2(\Omega)} dt,
\end{aligned} \tag{38}$$

where in obtaining the third inequality above, we have used a classical moment inequality. Now taking $\epsilon > 0$ small enough gives

$$\int_0^T \left[\|\omega\|^2_{H^3(\Omega)} + \|\omega_t\|^2_{H^1(\Omega)} + \|\theta\|^2_{H^1(\Omega)} \right] dt \leq CE(0) + C(E(0)) \int_0^T \|\omega\|^2_{L^2(\Omega)} dt, \tag{39}$$

which completes the proof of Lemma 3.1. □

By Lemma 3.1 and trace theory, we have that $\omega_t|_\Gamma \in L^2(0, T; H^{\frac{1}{2}}(\Gamma))$, this term being bounded above by the right-hand side of (31). Using this fact, the dissipativity inherent in energy relation (14) (viz., for $t \geq s$, $E(t) \leq E(s)$), and the estimate (15) of Lemma 2.2, we obtain

$$TE(T) \le \int_0^T E(t)dt \le C(E(0)) \qquad [E(0) + E(T) + \int_0^T \|\theta\|^2_{H^1(\Omega)}\, dt + \int_0^T \|\omega\|^2_{L^2(\Omega)}\, dt\bigg].$$

From here we can eliminate the term $E(0)$ by applying the energy relation (14) in Lemma 2.1 once more (with $t = T$ and $s = 0$) to the right-hand side (40). Doing this and taking $T > 2C(E(0))$ leads to a preliminary observability inequality:

Lemma 3.2. *Let $[\omega, \omega_t, \theta]$ be the solution of (1)–(2), corresponding to initial data $[\omega_0, \omega_1, \theta_0]$. Let $T = T(E(0))$ be large enough. Then we have the estimate*

$$E(T) \le C(E(0)) \left[\int_0^T \|\theta\|^2_{H^1(\Omega)}\, dt + \int_0^T \|\omega\|^2_{L^2(\Omega)}\, dt\right]. \tag{40}$$

Remark 3.3. *The requirement that T be "large enough" so as to secure the observability estimate (40) seems an unnatural one for problems of analytic type. As a matter of fact, one can obtain the same estimate for arbitrary T. This however requires a more subtle analysis. Since this time dependence of $C(E(0))$ has no implications on the stability result Theorem 1.2, the proof of which is the ultimate goal of this paper, we find the extra effort unwarranted in the present context.*

Our final step is to eliminate the lower order term $\int_0^T \|\omega\|^2_{L^2(\Omega)}\, dt$ from the inequality (40). This is undertaken in the next section.

4. Absorption of the Lower Order Term and Completion of the Proof of Theorem 1.3

Compactness/uniqueness arguments are by now rather standard in linear stability problems. However, the general nature of the nonlinearity F present in the thermoelastic PDE under consideration here necessitates a nonstandard approach, in which the analyticity of the underlying linearization is a key ingredient. The main result of this section is

Lemma 4.1. *Let $[\omega, \omega_t, \theta]$ be the solution of (1)–(2). Then the existence of the inequality (40) implies that there exists a constant $C(E(0))$ such that the following estimate holds:*

$$\int_0^T \|\omega\|^2_{L^2(\Omega)}\, dt \le C(E(0)) \int_0^T \|\theta\|^2_{H^1(\Omega)}\, dt. \tag{41}$$

Proof. The argument is, as usual, by contradiction. The key ingredients are the following: The uniqueness of the corresponding static problem (due to the structural Assumption 1), and the compactness of the embeddings which are inherent in the analytic estimate (31). The explicit details of this argument are given here. If the lemma is false, then there exist a sequence of initial data $\left\{\omega_0^{(n)}, \omega_1^{(n)}, \theta_0^{(n)}\right\}_{n=1}^{\infty} \in \mathbf{H}_0$, and a corresponding solution sequence $\left\{\omega^{(n)}, \omega_t^{(n)}, \theta^{(n)}\right\}_{n=1}^{\infty}$ to (1)–(2) which satisfy collectively

$$\lim_{n\to\infty} \frac{\int_0^T \left\|\omega^{(n)}\right\|_{L^2(\Omega)}^2 dt}{\int_0^T \left\|\theta^{(n)}\right\|_{H^1(\Omega)}^2 dt} = \infty; \;\; E^{(n)}(0) \le \text{constant}, \forall\, n. \tag{42}$$

As $E^{(m)}(t) \le E^{(m)}(0)$ by (14) of Lemma 2.1, then the sequence $\left\{E^{(n)}(t)\right\}_{n=1}^{\infty}$ is then also bounded uniformly in n for $0 \le t \le T$. This boundedness and the estimate (31) in turn gives the uniform bound

$$\int_0^T \left[\left\|\omega^{(n)}\right\|_{H^3(\Omega)}^2 + \left\|\omega_t^{(n)}\right\|_{H^1(\Omega)}^2 + \left\|\theta^{(n)}\right\|_{H^1(\Omega)}^2\right] dt \le C, \tag{43}$$

where the dependence of the constant C on T is not noted here. Consequently there exists a subsequence, still denoted here as $\left\{\left[\omega^{(n)}, \omega_t^{(n)}, \theta^{(n)}\right]\right\}_{n=1}^{\infty}$, and $[\omega, \omega_t, \theta] \in L^2(0,T; H^3(\Omega)) \times L^2(0,T; H^1(\Omega)) \times L^2(0,T; H^1(\Omega))$ such that

$$\begin{aligned} \omega^{(n)} &\to \omega \text{ weakly in } L^2(0,T; H^3(\Omega)); \\ \omega_t^{(n)} &\to \omega_t \text{ weakly in } L^2(0,T; H^1(\Omega)); \\ \theta^{(n)} &\to \theta \text{ weakly in } L^2(0,T; H^1(\Omega)). \end{aligned} \tag{44}$$

In addition, we can use the heat equation in (1), the fact that $\Delta \in \mathcal{L}\left(H^1(\Omega), H^{-1}(\Omega)\right)$ and the estimate (31) on $\theta^{(n)}$ to deduce that $\left\{\theta_t^{(n)}\right\}_{n=1}^{\infty}$ is a bounded sequence in $L^2(0,T; H^{-1}(\Omega))$. By using the plate equation in (1) and the estimate (31), we likewise have that $\left\{\omega_{tt}^{(n)}\right\}_{n=1}^{\infty}$ is a bounded sequence in $L^2(0,T; H^{-1}(\Omega))$. Consequently, by a compactness result of Aubin in [1], we have that

$$\left[\omega^{(n)}, \omega_t^{(n)}, \theta^{(n)}\right] \to [\omega, \omega_t, \theta] \text{ strongly in } L^2(0,T; \mathbf{H}_0) \tag{45}$$

(note that it is critical here that we have justified the *strong*, not just weak, convergence of $\omega^{(n)}$ in $L^2(0,T;\, H^2(\Omega))$).

We now consider two possibilities.

Case I. $\lim_{n\to\infty} \omega^{(n)} = \omega \ne 0$ (in $L^2(0,T; H^2(\Omega))$). In this event, the limit (42) implies that $\left\|\theta^{(n)}\right\|_{L^2(0,T;H^1(\Omega))} \to 0$. With this convergence, and those posted in (44) and (45), and further recalling Assumption 1, we can pass to the limit in the coupled system (1)–(2) to find the limit ω satisfies:

$$\begin{cases} \omega_{tt} + \Delta^2\omega + F(\omega) = 0 \\ \Delta\omega_t = 0. \end{cases} \quad \text{on } (0,T) \times \Omega \tag{46}$$

This implies then that ω_t satisfies:

$$\Delta^2\omega_t = 0 \quad \text{on } (0,T)\times\Omega;$$
$$\begin{cases} \Delta\omega_t + (1-\mu)B_1\omega_t = 0 \\ \dfrac{\partial\Delta\omega_t}{\partial\nu} + (1-\mu)\dfrac{\partial B_2\omega_t}{\partial\tau} - \eta\omega_t = 0. \end{cases} \quad \text{on } (0,T)\times\Gamma \tag{47}$$

By elliptic theory we then have that $\omega_t = 0$ (recall the ellipticity of the bilinear form $a(\cdot,\cdot)$ defined in (13), with $\eta > 0$). In turn, going back to the first equation of (46), we have $\Delta^2\omega + F(\omega) = 0$. Now recalling Assumption 2 (viz., $(F(\omega),\omega)_{L^2(\Omega)} \geq 0$) and the ellipticity of $a(\cdot,\cdot)$, we deduce that $\omega = 0$, which is a contradiction.

Case II. $\lim_{n\to\infty}\omega^{(n)} = \omega = 0$ (in $L^2(0,T;H^2(\Omega))$). In this case, denoting

$$C_n \equiv \left\|\omega^{(n)}\right\|_{L^2(0,T;L^2(\Omega))}; \quad \left[\widetilde{\omega}^{(n)},\widetilde{\theta}^{(n)}\right] \equiv \frac{1}{C_n}\left[\omega^{(n)},\theta^{(n)}\right], \tag{48}$$

then

$$\left\|\widetilde{\omega}^{(n)}\right\|_{L^2(0,T;L^2(\Omega))} = 1 \quad \text{for every } n, \tag{49}$$
$$C_n \to 0 \quad \text{(from (45) and the present assumption of Case II)}; \tag{50}$$
$$\int_0^T \left\|\widetilde{\theta}^{(n)}\right\|^2_{H^1(\Omega)} dt \to 0 \quad \text{(from (48) and (42))}. \tag{51}$$

Furthermore, $\left[\widetilde{\omega}^{(n)},\widetilde{\theta}^{(n)}\right]$ satisfies (after dividing (1),(2) by C_n)

$$\begin{cases} \widetilde{\omega}^{(n)}_{tt} + \Delta^2\widetilde{\omega}^{(n)} + \alpha\Delta\widetilde{\theta}^{(n)} + \frac{1}{C_n}F(\omega^{(n)}) = 0 \\ \widetilde{\theta}^{(n)}_t - \Delta\widetilde{\theta}^{(n)} - \alpha\Delta\widetilde{\omega}^{(n)}_t = 0 \end{cases} \quad \text{in } \Omega\times(0,T) \tag{52}$$

$$\begin{cases} \Delta\widetilde{\omega}^{(n)} + (1-\mu)B_1\widetilde{\omega}^{(n)} + \alpha\widetilde{\theta}^{(n)} = 0; \\ \dfrac{\partial\Delta\widetilde{\omega}^{(n)}}{\partial\nu} + (1-\mu)\dfrac{\partial B_2\widetilde{\omega}^{(n)}}{\partial\tau} - \eta\omega^{(n)} + \alpha\dfrac{\partial\widetilde{\theta}^{(n)}}{\partial\nu} = 0 \\ \dfrac{\partial\widetilde{\theta}^{(n)}}{\partial\nu} + \lambda\widetilde{\theta}^{(n)} = 0. \end{cases} \quad \text{in } \Gamma\times(0,T) \tag{53}$$

Moreover, using the analytic estimate (31) and the energy relation (14), one has the inequality:

$$\int_0^T \left[\left\|\omega^{(n)}\right\|^2_{H^3(\Omega)} + \left\|\omega^{(n)}_t\right\|^2_{H^1(\Omega)}\right] dt$$
$$\leq C(E^{(n)}(0))\left(E^{(n)}(T) + \int_0^T \left\|\theta^{(n)}\right\|^2_{H^1(\Omega)} dt + \int_0^T \left\|\omega^{(n)}\right\|^2_{L^2(\Omega)} dt\right) \tag{54}$$

Dividing both sides of this inequality by C_n^2, recalling the estimate (40) for $\left\{E^{(n)}(T)\right\}_{n=1}^\infty$, the limit (51), and the fact that $\left\{E^{(n)}(0)\right\}_{n=1}^\infty$ is uniformly bounded (see (42)) gives the bound

$$\int_0^T \left[\left\|\widetilde{\omega}^{(n)}\right\|^2_{H^3(\Omega)} + \left\|\widetilde{\omega}^{(n)}_t\right\|^2_{H^1(\Omega)}\right] dt \leq C \quad \text{for all } n. \tag{55}$$

Consequently, there exists a converging subsequence $\left\{\left[\widetilde{\omega}^{(n)}, \widetilde{\omega}_t^{(n)}\right]\right\}_{n=1}^{\infty}$ and $[\widetilde{\omega}, \widetilde{\omega}_t]$-$\in L^2(0,T;H^3(\Omega)) \times L^2(0,T;H^1(\Omega))$ such that

$$\left[\widetilde{\omega}^{(n)}, \widetilde{\omega}_t^{(n)}\right] \to [\widetilde{\omega}, \widetilde{\omega}_t] \text{ weakly in } L^2(0,T;H^3(\Omega)) \times L^2(0,T;H^1(\Omega)). \tag{56}$$

By Aubin's compactness result and (49), we then have, similar to Case I above, that

$$\widetilde{\omega}^{(n)} \to \widetilde{\omega} \text{ strongly in } L^2(0,T;H^{3-\epsilon}(\Omega)); \tag{57}$$

$$\|\widetilde{\omega}\|_{L^2(0,T;L^2(\Omega))} = 1. \tag{58}$$

In addition, by the energy relation (14) and the estimate (40),

$$\begin{aligned} E^{(n)}(t) &\leq E^{(n)}(0) \\ &= E^{(n)}(T) + 2\int_0^T \left\|\nabla\theta^{(n)}\right\|_{L^2(\Omega)}^2 dt + 2\lambda \int_0^T \left\|\theta^{(n)}\right\|_{L^2(\Gamma)}^2 dt \\ &\leq (C(E^{(n)}(0)) \left[\int_0^T \left\|\theta^{(n)}\right\|_{H^1(\Omega)}^2 dt + \int_0^T \left\|\omega^{(n)}\right\|_{L^2(\Omega)}^2 dt\right]. \end{aligned} \tag{59}$$

This and an argument identical to that used to derive (55) then gives

$$\left\|\widetilde{\omega}^{(n)}\right\|_{L^\infty(0,T;H^2(\Omega))}^2 \leq C \text{ for all } n. \tag{60}$$

It is now our intent to pass to the limit in equation (52)–(53), and the only troublesome term in this connection is $\frac{1}{C_n}F(\omega^{(n)})$. To deal with it, we invoke Assumption 1. By the definition of Fréchet differentiability and the fact that $F(0)=0$, we have for (almost) all $t \in (0,T)$

$$\frac{1}{C_n}F(\omega^{(n)}(t)) = \frac{1}{C_n}\left[F(\omega^{(n)}(t)) - F(0)\right] = F'(0)\widetilde{\omega}^{(n)}(t) + \frac{1}{C_n}R(\omega^{(n)}(t)), \tag{61}$$

where $F'(0) \in \mathcal{L}(H^2(\Omega), [H^1(\Omega)]')$ denotes the Fréchet derivative of F at the point 0, and the mapping $R(\cdot): H^2(\Omega) \to [H^1(\Omega)]'$ has the property that

$$\lim_{\|\phi\|_{H^2(\Omega)}\to 0} \frac{\|R(\varphi)\|_{[H^1(\Omega)]'}}{\|\phi\|_{H^2(\Omega)}} = 0. \tag{62}$$

By the assumption of Case II, $\omega^{(n)} \to 0$ in $L^2(0,T;H^2(\Omega))$; hence, by (62) and (60), we have the almost everywhere convergence in time

$$\lim_{n\to\infty} \frac{1}{C_n}\left\|R(\omega^{(n)}(t))\right\|_{[H^1(\Omega)]'} = \lim_{n\to\infty} \frac{\left\|R(\omega^{(n)}(t))\right\|_{[H^1(\Omega)]'}}{\left\|\omega^{(n)}(t)\right\|_{H^2(\Omega)}} \cdot \left\|\widetilde{\omega}^{(n)}(t)\right\|_{H^2(\Omega)} = 0. \tag{63}$$

Moreover, by (61) and the Lipschitz continuity of F in Assumption 1, we have almost everywhere in time

$$\frac{1}{C_n}\left\|R(\omega^{(n)}(t))\right\|_{[H^1(\Omega)]'} = \frac{1}{C_n}F(\omega^{(n)}(t)) - F'(0)\widetilde{\omega}^{(n)}(t)$$
$$\leq C(E^{(n)}(0))\frac{\left\|\omega^{(n)}(t)\right\|_{H^2(\Omega)}}{C_n} + \|F'(0)\|\left\|\widetilde{\omega}^{(n)}(t)\right\|_{H^2(\Omega)} \leq C_T\,,\ \forall n,$$

where in the last step we have recalled (60). We can thus apply the Lebesgue Dominated Convergence Theorem and the convergence (57) to the right-hand side of (61) to thereby obtain

$$\lim_{n\to\infty}\frac{1}{C_n}F(\omega^{(n)}) = F'(0)\widetilde{\omega}\ \text{ in } L^2(0,T;[H^1(\Omega)]'). \tag{64}$$

Using the convergences posted in (51), (56) and (64), we can now freely pass to the limit in (52)–(53) to have that the limit $\widetilde{\omega}$ satisfies:

$$\begin{cases} \widetilde{\omega}_{tt} + \Delta^2\widetilde{\omega} + F'(0)\widetilde{\omega} = 0 \\ \Delta\widetilde{\omega}_t = 0; \end{cases} \quad \text{in } (0,T)\times\Omega$$
$$\begin{cases} \Delta\widetilde{\omega} + (1-\mu)B_1\widetilde{\omega} = 0 \\ \dfrac{\partial\Delta\widetilde{\omega}}{\partial\nu} + (1-\mu)\dfrac{\partial B_2\widetilde{\omega}}{\partial\tau} - \eta\widetilde{\omega} = 0. \end{cases} \quad \text{on } (0,T)\times\Gamma \tag{65}$$

Similar to Case I, we obtain from this system that $\widetilde{\omega}_t = 0$, which in turn implies that $\widetilde{\omega}$ satisfies

$$\Delta^2\widetilde{\omega} + F'(0)\widetilde{\omega} = 0$$
$$\begin{cases} \Delta\widetilde{\omega} + (1-\mu)B_1\widetilde{\omega} = 0 \\ \dfrac{\partial\Delta\widetilde{\omega}}{\partial\nu} + (1-\mu)\dfrac{\partial B_2\widetilde{\omega}}{\partial\tau} - \eta\widetilde{\omega} = 0. \end{cases} \quad \text{on } (0,T)\times\Gamma \tag{66}$$

To now finish the proof, the assumption that $\int_\Omega F(\omega^{(n)})\omega^{(n)}d\Omega \geq 0$, and the convergences in (57) and (64) imply that $\int_\Omega F'(0)\widetilde{\omega}\widetilde{\omega}d\Omega \geq 0$, which combined with the ellipticity of the bilinear form $a(\cdot,\cdot)$ yield that $\widetilde{\omega} = 0$. But this contradicts the equality (58), thereby completing the lemma. □

Completion of the Proof of Theorem 1.3. Combining the inequalities in Lemmas 3.2 and 4.1 yield

$$E(T) \leq C(E(0))\int_0^T \|\theta\|^2_{H^1(\Omega)}\,dt. \tag{67}$$

Using the energy relation (14) once more,

$$E(T) \leq C(E(0))\left[E(0) - E(T)\right],$$

or

$$E(T) \leq \frac{C(E(0))}{1+C(E(0))}E(0) < 1\cdot E(0). \tag{68}$$

This and the usual semigroup argument completes the proof of the Theorem. □

References

[1] J. P. Aubin, *Un théorème de compacité*, C. R. Acad. Sci. 256 (1963), pp. 5042–5044.

[2] G. Avalos, *Well-posedness and decay of nonlinear thermoelastic systems*, preprint, 1999.

[3] G. Avalos and I. Lasiecka, *Exponential stability of a thermoelastic system with free boundary conditions without mechanical dissipation*, SIAM Journal of Mathematical Analysis, Vol. 29, No. 1 (January 1998), pp. 155–182.

[4] G. Avalos and I. Lasiecka, *Exponential stability of a thermoelastic system without mechanical dissipation*, Rendiconti Di Istituto Di Matematica Dell'Università di Trieste, Suppl. Vol. XXVIII (1997), pp. 1–28.

[5] G. Avalos and I. Lasiecka, *Uniform decays in nonlinear thermoelastic systems*, Optimal Control: Theory, Algorithms and Applications, W. W. Hagar and P. M. Pardalos (Editors), Kluwer Academic Publishers, Boston (1998), pp. 1–23.

[6] A. Bensoussan, G. Da Prato, M. C. Delfour and S. K. Mitter, *Representation and control of infinite dimensional systems Volumes I and II*, Birkhäuser, Boston, 1992.

[7] H. M. Berger, *A new approach to the analysis of large deflection of plates*, J. Appl. Mech. Trans ASME **22** (1955), pp. 465–472.

[8] E. Bisognin, V. Bisognin, P. Menzala and E. Zuazua, *On the exponential stability for von Kármán equation in the presence of thermal effects*, to appear in Mathematical Models and Methods in the Applied Sciences.

[9] S.K. Chang and R. Triggiani, *Spectral Analysis of thermo–elastic plates with rotational forces*, Optimal Control: Theory, Algorithms and Applications, W. W. Hager and P. Pardalos (Editors), Kluwer Academic Publishers, Boston (1998), pp. 84–113.

[10] S. Chen and R. Triggiani, *Characterization of domains of fractional powers of certain operators arising in elastic systems, and applications*, Journal of Differential Equations, Vol 64 (1990), pp. 26–42.

[11] G. Duvaut and J. L. Lions, *Les inéquations en mécanique et en physique*, Dunod, Paris (1972).

[12] T. von Kármán, *Festigkeitsprobleme im Maschinenbau*, Encyklopedie der Mathematischen Wissenschaften, Vol. 4 (1910), pp. 314–385.

[13] J. Kim, *On the energy decay of a linear thermoelastic bar and plate,* SIAM J. Math. Anal. 23 (1992), pp. 889–899.

[14] J. Lagnese, *Boundary stabilization of thin plates, SIAM Stud. Appl. Math.,* **10** (1989).

[15] I. Lasiecka, *A unified theory for abstract parabolic boundary problems–a semigroup approach*, Appl. Math. Optim., Vol. 6 (1980), pp. 287–333.

[16] I. Lasiecka, *Control and stabilization of interactive structures*, Systems and Control in the Twenty- First Century, Birkhäuser, 1997 pp. 245-263.

[17] I. Lasiecka and R. Triggiani, *Sharp regularity results for mixed second order hyperbolic equations of Neumann type. Part I: The L^2 boundary case*, Annali di Matematica Pura ed Applicata, Vol. 157 (1990), pp. 285–367.

[18] I. Lasiecka and R. Triggiani, *Sharp regularity results for mixed second order hyperbolic equations of Neumann type. Part II: General boundary data*, J. Diff. Equations, Vol. 94 (1991), pp. 112–164.

[19] I. Lasiecka and R. Triggiani, *Recent advances in regularity of second–order hyperbolic mixed problems and applications*, Dynamics Reported–Expositions in Dynamical Systems, Vol. 3 (1994), pp. 25–104.

[20] I. Lasiecka and R. Triggiani, *Analyticity and lack thereof, of thermoelastic semigroups*, to appear in ESAIM (1998).

[21] I. Lasiecka and R. Triggiani, *Two direct proofs on the analyticity of the S.C. semigroup arising in abstract thermo-elastic equations*, to appear in Advances in Differential Equations (1998).

[22] I. Lasiecka and R. Triggiani, *Analyticity of thermo-elastic semigroups with coupled/hinged Neumann B.C.*, to appear in Abstract and Applied Analysis, Vol.3 No. 2 (1998), pp. 153–169.

[23] I. Lasiecka and R. Triggiani, *Analyticity of thermo-elastic semigroups with free boundary conditions*, Annali di Scuola Normale Superiore di Pisa, Cl. Sci. (14), Vol. XXVII (1998).

[24] I. Lasiecka and R. Triggiani, *Structural decomposition of thermoelastic semigroups with rotational forces.* To appear in Semigroup Forum.

[25] I. Lasiecka and R. Triggiani, *Control Theory for Partial Differential Equations.* Cambridge University Press, Encyclopedia of Mathematics and its Applications, to appear in 1999.

[26] J. L. Lions, *Quelques méthodes de résolution des problèmes aux limites non linéaires*, Dunod, Paris (1969).

[27] Z. Liu and M. Renardy, *A note on the equations of thermoelastic plate*, Appl. Math. Lett., vol. 8 no. 3 (1995), pp. 1–6.

[28] Z. Liu and S. Zheng, *Exponential stability of semigroups associated with thermoelastic systems*, Quarterly of Applied Mathematics, Vol. 52 (1993), pp. 535–545.

[29] Z. Liu and S. Zheng, *Exponential stability of the Kirchoff plate with thermal or viscoelastic damping*, Quarterly of Applied Mathematics, Vol. 55 (1997), pp. 551–564.

[30] K. Liu and Z. Liu, *Exponential stability and analyticity of abstract linear thermoelastic systems*, ZAMP, 48, (1997) pp. 885–904.

[31] A. Lunardi, *On the Ornstein–Uhlenbeck operator in L^2 spaces with respect to invariant measures*, Transactions of the American Mathematical Society, Vol. 349, No. 1, January (1997), pp. 155–169.

[32] J. R. Modeer and W. A. Nash, *Certain approximate analysis of the nonlinear behaviour of plates and shallow shells*, in the Proceedings of the Symposium on the Theory of Thin Elastic Shells, at the Technological University of Delft (1959), edited by W. T. Koiter, North Holland, New York (1960).

[33] J. E. M. Rivera, *Energy decay rates in linear thermoelasticity*, Funkcial. Ekvac., Vol. 35 (1992), pp. 19–30.

[34] T. Wah, *Large amplitude flexural vibrations of rectangular plates*, Int. J. Mech. Sci., Vol. 5 (1963), pp. 425–438.

[35] D. Tataru, *On the regularity of boundary traces for the wave equation*, Annali di Scuola Normale di Pisa, to appear.

George Avalos
Department of Mathematics and Statistics
Texas Tech University
Lubbock
Texas 79409 (USA).

Irena Lasiecka and Roberto Triggiani
Department of Mathematics
University of Virginia
Kerchof Hall
Charlottesville
Virginia 22903 (USA).

International Series of Numerical Mathematics
Vol. 133, © 1999 Birkhäuser Verlag Basel/Switzerland

Exponential Bases in Sobolev Spaces in Control and Observation Problems

Sergei A. Avdonin, Sergei A. Ivanov and David L. Russell [1]

Abstract. The Fourier method in control systems reduces the study of controllability/observability to the study of related exponential families. In this paper we present examples of such systems, specifically those for which we can prove that the related exponential families form a Riesz basis in corresponding appropriately defined Sobolev spaces. This makes it possible to choose 'natural' pairs of spaces: the state space / observability space and the control space / state space, depending on whether an observation or a control problem is studied, respectively, so that the observation and control operators are isomorphisms.

1. Introduction

We study applications of exponential families $\{e^{i\lambda_k t}\}$ to problems of control and observation for certain systems governed by partial differential equations of hyperbolic type.

Families of exponentials appear in various fields of mathematics such as the theory of nonselfadjoint operators, the Regge problem for resonance scattering, the theory of linear initial boundary value problems for partial differential equations, the related control theory, and signal processing. One of the central problems arising in all of these applications is the question of the Riesz basis property of an exponential family. The modern approach and applications to control theory are presented in [2]. Properties of Riesz bases of exponentials have been extensively exploited in the control theory of distributed parameter systems. The reason is the following. If the exponential family arising by the moment approach forms a Riesz basis, the control system possesses:

(i) exact controllability in the corresponding state space,
(ii) the control transferring the system to a given state is unique,
(iii) the norm of the control is equivalent to the norm of the state.

Almost all papers deal with exponential families in L^2. Therefore, if we can check the Riesz basis properties in spaces different from L^2, we can prove such

[1]This work was partially supported by the US National Science Foundation (grant #DMS-95-01036) and by the Russian Fundamental Research Foundation (grant # 97-01-01115).

controllability in more general situations. The natural extensions of L^2 space are the Sobolev spaces. The first result in this direction has been obtained by Russell [11] for integer m. See also papers [5], [9].

The present paper is based on the following recent result [3].

Theorem 1.1. *Let F (the generating function) be an entire function of exponential type, with indicator diagram of width a, whose zero set $\{\lambda_k\}$ lies in a strip $|Im\, z| <$ const and satisfies the separation condition*

$$\inf_{k \neq j} |\lambda_k - \lambda_j| > 0.$$

If for some real h and s

$$|F(x+ih)| \asymp (1+|x|)^s, \quad x \in \mathbb{R},$$

then the family $\{e^{i\lambda_k t}/(1+|\lambda_k|)^s\}$ forms a Riesz basis in $H^s(0,a)$ for $s \notin \{-\mathbb{N}+\frac{1}{2}\}$ and in $[H_{00}^{n-1/2}(0,a)]'$ for $s = -n+\frac{1}{2}$, $n \in \mathbb{N}$.

Here $[H_{00}^{n-1/2}(0,a)]'$ is the dual space to $H_{00}^{n-1/2}(0,a)$ relative to $L^2(0,a)$; it is a space of distributions, i.e., linear functionals, on $H^{n-1/2}(\mathbb{R})$ supported on $[0,a]$.

The case $\alpha = 0$ is the well-known theorem of Levin and Golovin [7], [4], which was the first result on Riesz bases of exponentials given in terms of the generating function F. We need also in the following result [1], [2].

Theorem 1.2. *Let $|F(x+ih)| \asymp (|x|+1)^\alpha$. Then the exponential family $\{e^{i\lambda_k t}\}$ forms a Riesz basis in $L^2(0,a)$ if and only if $-1/2 < \alpha < 1/2$.*

In this paper we study control and observation problems for a circular membrane (Section 3) and for a vector wave equation (Section 4). For the last case we find subspaces of the state space, which may be recovered by the boundary observations for time not sufficient to recover the whole state. The purpose of Section 2 is to show how exponential bases in Sobolev spaces (in this case of integer order) arise naturally in formulation of sharp (in the sense of control/observation time) controllability/observability results.

2. Controllability and Observability Problems for a Regular String Equation

We start with a study of observability for the Neumann problem. Let $\rho(x)$ be a smooth positive function on $[0,l]$, $T = 2\int_0^l \sqrt{\rho(x)}\, dx$, and $u(x,t)$ be a solution of the following initial–boundary value problem:

$$\begin{aligned}
&\rho(x) u_{tt}(x,t) = u_{xx}(x,t), \quad 0 < x < l, \quad 0 < t < T, \\
&u_x(0,t) = u_x(l,t) = 0, \quad 0 < t < T, \\
&u(x,0) = u_0(x), \quad u_t(x,0) = u_1(x), \quad 0 < x < l.
\end{aligned}$$

We assume a boundary observation

$$y(t) := u(0,t), \quad 0 < t < T,$$

and show that we can use it to recover the initial data (u_0, u_1), supposed such that $(u_0, u_1) \in E := H^1(0,l) \times L^2(0,l)$:

$$\|y\|^2_{H^1(0,T)} \asymp \|u_0\|^2_{H^1(0,l)} + \|u_1\|^2_{L^2(0,l)}. \tag{1}$$

Thus, *we have observability of the system in time T with the state space E and observation space $H^1(0,T)$.*

Proof. We denote by λ_n and φ_n the eigenvalues and eigenfunctions for the corresponding boundary value problem and represent the initial data in the form

$$u_0(x) = \sum_{n=0}^{\infty} a_n^0 \varphi_n(x), u_1(x) = \sum_{n=0}^{\infty} a_n^1 \varphi_n(x).$$

Then the solution can be written as

$$u(x,t) = (a_0^0 + a_0^1 t)\varphi_0(x) + \sum_{n=1}^{\infty} \left[a_n^0 \cos \omega_n t + \frac{a_n^1}{\omega_n} \sin \omega_n t \right] \varphi_n(x)$$

and the observation has the form

$$y(t) = \varphi_0(0)(a_0^0 + a_0^1 t) + \sum_{n=1}^{\infty} \varphi_n(0) \left[a_n^0 \cos \omega_n t + \frac{a_n^1}{\omega_n} \sin \omega_n t \right]. \tag{2}$$

Using the Euler formulas, the last expression may be rewritten in the more convenient form

$$y(t) = b_0 + b_0^1 t + \sum_{k \in \mathbb{K}} b_k e^{i\omega_k t}, \quad \mathbb{K} = \{-\mathbb{N}\} \cup \{+\mathbb{N}\}. \tag{3}$$

where

$$\begin{aligned} b_0 &= a_0^0 \varphi_0(0), \quad b_0^1 = a_0^1 \varphi_0(0), & \omega_{-|k|} &:= -\omega_{|k|} \\ b_k &= \tfrac{\varphi_{|k|}(0)}{2} \left(a_{|k|}^0 + \tfrac{a_{|k|}^1}{i\omega_{|k|}} \right), & k &\in \mathbb{K}. \end{aligned}$$

The family $\mathcal{E} = \{1\} \cup \{t\} \cup \{e^{i\omega_k t}/\omega_k\}$, $k \in \mathbb{K}$, forms a Riesz basis in $H^1(0,T)$ [11] and we have

$$\begin{aligned} \|y\|_Y^2 &\asymp |b_0|^2 + |b_0^1|^2 + \sum_{k \in \mathbb{K}} |b_k|^2 \omega_k^2 \\ &\asymp |a_0|^2 + |a_0^1|^2 + \sum_{n=1}^{\infty} \left(|a_n^0|^2 \omega_n^2 + |a_n^1|^2 \right) \end{aligned}$$

which gives (1). □

The dual result expresses exact controllability and can be formulated as follows. Let us consider the problem

$$\begin{aligned}&\rho(x)v_{tt}(x,t) = v_{xx}(x,t), \quad 0 < x < l, \quad 0 < t < T,\\ &v_x(0,t) = f(t), \ v_x(l,t) = 0; \quad 0 < t < T,\\ &v(x,0) = v_0(x), \quad v_t(x,0) = v_1(x), \quad 0 < x < l.\end{aligned}$$

For any $(v_0, v_1) \in W := L^2(0,l) \times [H^1(0,l)]'$ *there exists a unique* $f \in F := [H^1(0,T)]'$ *such that the solution satisfies equalities* $v(x,T) = v_t(x,T) = 0$ *and estimates* $\|f\|_F^2 \asymp \|(v_0,v_1)\|_W^2$.

Let us compare this result with the case of controls from $L^2(0,T)$. In this case we have not even approximate controllability in the critical time T, and we obtain lack of observability for this time or, more precisely, observability "up to a constant".

The Dirichlet problem can be treated similarly.

3. Observability and Controllability Problems for a Circular Membrane

In this section we study a boundary observability and controllability problem for a circular membrane in the case of rotational symmetry. The problem is reduced to a Riesz basis property for a set of exponentials in a Sobolev space of half–integer order.

Let $\Omega = \{(x,y) | x^2 + y^2 < 1\}$, $\Gamma = \partial\Omega$, $T = 2$, $Q = \Omega \times (0,T)$, $\Sigma = \Gamma \times (0,T)$, and ν is the unit exterior normal vector to Γ.

We consider the initial boundary value problem

$$\begin{aligned}&u_{tt} = \Delta u \ \text{ in } Q, \quad \tfrac{\partial u}{\partial \nu}\Big|_\Sigma = 0,\\ &u|_{t=0} = u_0, \quad u_t|_{t=0} = u_1 \ \text{ in } \Omega\end{aligned}$$

with observation

$$y = u|_\Sigma.$$

We suppose that u_0 and u_1 do not depend on polar angle θ, $u_0(x,y) = w_0(r)$, $u_1(x,y) = w_1(r)$.

Theorem 3.1. *The system is observable in time* $T = 2$ *with respect to the state space consisting of the subspace of rotationally symmetric functions in* $H^{1/2}(\Omega) \times H^{-1/2}(\Omega)$ *and the space of observations* $H^{1/2}(0,2)$. *Further,*

$$\|y\|^2_{H^{1/2}(0,2)} \asymp \|u_0\|^2_{H^{1/2}(\Omega)} + \|u_1\|^2_{H^{-1/2}(\Omega)}.$$

Proof. Setting $w(r,t) = u(x,y,t)$ we have the equation with one spatial variable

$$w_{tt} = w_{rr} + \frac{1}{r} w_r, \quad 0 < r < 1, \quad 0 < t < 2.$$

The "elliptical" part of the equation is the Bessel equation. The eigenfrequencies of this operator, $\omega_n = \sqrt{\lambda_n}$, are nonnegative zeros of the derivative of the zero-order Bessel function $J_0'(z)$ and the normalized eigenfunctions are

$$\varphi_n(r) = \sqrt{2} J_0(\omega_n r)/|J_0(\omega_n)|, \quad n = 1, 2, \dots ; \quad \omega_0 = 0.$$

We present the initial data in the form of the series

$$w_0(r) = \sum_{n=0}^{\infty} a_n^0 \varphi_n(r), \quad w_1(r) = \sum_{n=0}^{\infty} a_n^1 \varphi_n(r),$$

and suppose that

$$\sum_{n=1}^{\infty} (|a_n^0|^2 \omega_n + |a_n^1|^2 \omega_n^{-1}) < \infty.$$

In terms of the original problem, this means $u_0 \in H^{1/2}(\Omega)$, $u_1 \in H^{-1/2}(\Omega)$. The observation may be written similarly to (2), (3):

$$\begin{aligned} y(t) &= \varphi_0(1)(a_0^0 + a_0^1 t) + \textstyle\sum_{n=1}^{\infty} \varphi_n(1) \left[a_n^0 \cos \omega_n t + \frac{a_n^1}{\omega_n} \sin \omega_n t\right] \\ &= b_0 + b_0^1 t + \textstyle\sum_{k \in \mathbb{K}} b_k e^{i\omega_k t}. \end{aligned}$$

The exponential family

$$\mathcal{E} = \left\{ e^{i\omega_k t}/\omega_k^{1/2} \right\}_{k \in \mathbb{K}} \cup \{1\} \cup \{t\}$$

is generated by the function $F(z) = zJ_0'(z)$. It is an entire function of exponential type with indicator diagram of width 2. On lines $z = x + ih$ it satisfies the relation

$$|F(x + ih)| \asymp (1 + |x|)^{1/2}, \quad x \in \mathbb{R}.$$

From Theorem 1.1 it follows that the family $\mathcal{E}$ forms a Riesz basis in $H^{1/2}(0, 2)$. This fact completes the proof taking into account the representation of $y(t)$ in a series similar to (3). □

Thus, the spaces of control and observation are intrinsic to the problem. If we take as the observation space $L^2(0, 2)$ then

$$\|y\|^2_{L^2(0,2)} \prec \|u_0\|^2_{L^2(\Omega)} + \|u_1\|^2_{[H^1(\Omega)]'}.$$

On the other hand, no estimate

$$\|y\|^2_{L^2(0,2)} \succ \|u_0\|^2_{L^2(\Omega)} + \|u_1\|^2_{[H^1(\Omega)]'}.$$

in the other direction can valid. Indeed, both estimates together would imply that $\mathcal{E}$ is a Riesz basis in $L^2(0, 2)$ which is, however, impossible because of Theorem 1.2.

Let us now formulate Theorem 3.1 in a dual form and consider the initial–boundary value problem

$$v_{tt} = \Delta v \ \text{ in } Q, \quad \left.\frac{\partial v}{\partial \nu}\right|_{\Sigma} = f,$$

$$v|_{t=0} = v_0, \quad v_t|_{t=0} = v_1.$$

Theorem 3.2. *For any rotationally symmetric initial data* $(v_0, v_1) \in H^{1/2}(\Omega) \times H^{-1/2}(\Omega)$, *there exists a unique control* f *depending on* t *alone,* $f \in H^{-1/2}(0,2)$, *such that we have*

$$v(x,2) = v_t(x,2) = 0$$

and

$$\|f\|^2_{H^{-1/2}(0,2)} \asymp \|v_0\|^2_{H^{1/2}(\Omega)} + \|v_1\|^2_{H^{-1/2}(\Omega)}.$$

For $f \in L^2(0,2)$ the system is not exactly controllable with respect to the (natural for such controls) space of rotationally symmetric functions in $H^1(\Omega) \times L^2(\Omega)$, because the exponential family is not even minimal in $L^2(0,2)$, which in turn follows (see [10]) from the fact that

$$\int_{\mathbb{R}} |F(x+ih)|^2/(1+x^2)dx = \infty.$$

4. A Vector Wave Equation

Let $U(x,t)$ be a vector function with two components:

$$U(x,t) = \begin{pmatrix} u_1(x,t) \\ u_2(x,t) \end{pmatrix}, \quad 0 < x < \pi, \quad t > 0.$$

Let us consider the following initial boundary value problem

$$\begin{gathered} \frac{\partial^2 U}{\partial t^2} = \frac{\partial^2 U}{\partial x^2}, \\ U_x|_{x=0} = 0, \quad U_t|_{x=\pi} = AU_x|_{x=\pi}, \\ U|_{t=0} = U_0, \quad U_t|_{t=0} = U_1. \end{gathered} \tag{4}$$

Here $A = \begin{pmatrix} 0 & \alpha \\ -\alpha & 0 \end{pmatrix}$, $\alpha > 0$. It is convenient to replace (4) by the corresponding first order system. Setting $W = U_x$, $V = U_t$, we consider the problem

$$\begin{gathered} \frac{\partial}{\partial t}\begin{pmatrix} W \\ V \end{pmatrix} = \begin{pmatrix} 0 & 1 \\ 1 & 0 \end{pmatrix} \frac{\partial}{\partial x}\begin{pmatrix} W \\ V \end{pmatrix}, \\ W(0,t) = 0, \quad V(\pi,t) = AW(\pi,t), \\ W(x,0) = W_0(x), \quad V(x,0) = V_0(x). \end{gathered} \tag{5}$$

Spaces for the initial data W_0, V_0 will be specified below. The vector functions $W(x,t)$ and $V(x,t)$ have two components $\begin{pmatrix} w_1(x,t) \\ w_2(x,t) \end{pmatrix}$ and $\begin{pmatrix} v_1(x,t) \\ v_2(x,t) \end{pmatrix}$. We associate with the problem (5) the observation operator $\mathcal{O}$:

$$\left[\mathcal{O}\begin{pmatrix} W_0 \\ V_0 \end{pmatrix}\right](t) = v_1(0,t), \quad 0 \le t \le 2\pi. \tag{6}$$

It is (almost) evident that this observation is insufficient for recovery of the initial data; for estimation of the complete initial data either this observation (6) on the interval $(0, 4\pi)$ or observation of the vector function $V(0,t)$, on $(0, 2\pi)$ is sufficient. We will prove that with use of $v_1(0,t)$, $t \in [0, 2\pi]$ we can estimate initial data in a (naturally described) subspace.

To construct the solution of (5) we introduce the operator $\mathcal{B}$ whose action on $L^2(0,\pi,\mathbb{C}^4)$ is described by

$$\mathcal{B}\begin{pmatrix}\varphi\\ \psi\end{pmatrix}=\begin{pmatrix}0&1\\1&0\end{pmatrix}\frac{d}{dx}\begin{pmatrix}\varphi\\ \psi\end{pmatrix}$$

together with the boundary conditions

$$\varphi(0)=0,\quad \psi(\pi)=A\varphi(\pi). \tag{7}$$

It is easy to check that $i\mathcal{B}$ is a selfadjoint operator with a basis family of eigenfunctions. The spectrum of $\mathcal{B}$ consists of

$$\lambda=i(k+1/2\pm\gamma),\quad k\in\mathbb{Z},\ \gamma:=\frac{1}{2\pi i}\log\frac{1+i\alpha}{1-i\alpha}\in(0,1/2)$$

and the eigenfunctions of $\mathcal{B}$ are

$$\begin{pmatrix}\varphi\\ \psi\end{pmatrix}(x)=\begin{pmatrix}\begin{pmatrix}1\\ \pm i\end{pmatrix} i\sin(k+1/2\pm\gamma)x\\ \begin{pmatrix}1\\ \pm i\end{pmatrix}\cos(k+1/2\pm\gamma)x\end{pmatrix}. \tag{8}$$

In the work to follow it is convenient to write (8) as the union, $\Phi\cup\Psi$, of sets of 4-dimensional vector function families. Here

$$\Phi=\{\Phi_n^+\}\cup\{\Phi_n^-\},\quad \Psi=\{\Psi_n^+\}\cup\{\Psi_n^-\},\quad n=1,2,\dots;$$

$$\Phi_n^+=\begin{pmatrix}\begin{pmatrix}1\\ -i\end{pmatrix} i\sin(n-1/2-\gamma)x\\ \begin{pmatrix}1\\ -i\end{pmatrix}\cos(n-1/2-\gamma)x\end{pmatrix},\quad \Phi_n^-=\begin{pmatrix}-\begin{pmatrix}1\\ i\end{pmatrix} i\sin(n-1/2-\gamma)x\\ \begin{pmatrix}1\\ i\end{pmatrix}\cos(n-1/2-\gamma)x\end{pmatrix}; \tag{9}$$

$$\Psi_n^+=\begin{pmatrix}\begin{pmatrix}1\\ i\end{pmatrix} i\sin(n-1/2+\gamma)x\\ \begin{pmatrix}1\\ i\end{pmatrix}\cos(n-1/2+\gamma)x\end{pmatrix},\quad \Psi_n^-=\begin{pmatrix}-\begin{pmatrix}1\\ -i\end{pmatrix} i\sin(n-1/2+\gamma)x\\ \begin{pmatrix}1\\ -i\end{pmatrix}\cos(n-1/2+\gamma)x\end{pmatrix}.$$

This representation is natural because each of the two families Φ and Ψ consists of complex conjugate pairs of functions.

Let us represent the initial data $\begin{pmatrix}W_0\\ V_0\end{pmatrix}$ (see (5)) in the form of a series

$$\begin{pmatrix}W_0\\ V_0\end{pmatrix}=\sum_{n=1}^{\infty}(a_n^+\Phi_n^++a_n^-\Phi_n^-)+\sum_{n=1}^{\infty}(b_n^+\Psi_n^++b_n^-\Psi_n^-).$$

Then the solution of (5) has the form

$$\begin{aligned}\begin{pmatrix}W\\ V\end{pmatrix}(x,t)&=\textstyle\sum_{n=1}^{\infty}(a_n^+e^{i(n-1/2-\gamma)t}\Phi_n^+(x)+a_n^-e^{-i(n-1/2-\gamma)t}\Phi_n^-(x))\\ &\quad+\textstyle\sum_{n=1}^{\infty}(b_n^+e^{i(n-1/2+\gamma)t}\Psi_n^+(x)+b_n^-e^{-i(n-1/2+\gamma)t}\Psi_n^-(x)).\end{aligned} \tag{10}$$

Problem 4.1. *Let $b_n^\pm = 0$, $n = 1, 2, \ldots$. Given the observation (6), to recover the initial data (i.e. the coefficients $a_n^\pm$).*

Problem 4.2. *Let $a_n^\pm = 0$, $n = 1, 2, \ldots$. The problem is to recover the coefficients $b_n^\pm$ using the same observation (6).*

Problem 4.3. *Given the observation $v_1(0,t)$ for $0 \le t \le 4\pi$, to recover $a_n^\pm$ and $b_n^\pm$.*

Problem 4.4. *Given the observation $V(0,t) = \begin{pmatrix} v_1(0,t) \\ v_2(0,t) \end{pmatrix}$, $0 \le t \le 2\pi$, to recover the initial data $\begin{pmatrix} W_0 \\ V_0 \end{pmatrix}$, i.e. $a_n^\pm$ and $b_n^\pm$.*

Spaces appropriate for the observation functions and the initial data are specified in the process of solving these problems.

Let us define the spaces

$$\mathcal{D}(\mathcal{B}^s) =: H_s = \{f = \sum_{k \in \mathbb{Z}} c_k \Theta_k \;\Big|\; \|f\|_s^2 := \sum_{k \in \mathbb{Z}} |c_k|^2 |\lambda_k|^{2s} < \infty\},$$

where $\{\Theta_k\}$ is the family of eigenfunctions of the operator $\mathcal{B}$ and λ_k are its eigenvalues. It is clear that $\mathcal{D}(\mathcal{B}^{-s}) = [\mathcal{D}(\mathcal{B}^s)]'$; they are dual spaces with respect to $\mathcal{D}(\mathcal{B}^0) = L^2(0, \pi; \mathbb{C}^4)$.

The following lemma describes connections between the spaces H_s and the standard Sobolev spaces for $s \neq \frac{1}{2}$. The space $H_{1/2}$ has a more complicated description (cf. [8]).

Lemma 4.5.
(i) $H_s = H^s(0, \pi; \mathbb{C}^4)$ *for* $0 \le s < 1/2$,
(ii) $H_s = \{f \in H^s(0, \pi; \mathbb{C}^4) \,|\, \text{with the conditions (7)}\}$ *for* $1/2 < s \le 1$.

Solution of Problem 1.
We will use the notation $\mathrm{Cl}_X \mathrm{Lin}\Phi$ to denote the closure in the space X of the linear span of the elements in the set Φ.

Theorem 4.6.
(i) For any $\gamma \in (0, 1/2)$ the system (5), (6) is observable on the subspace of initial data $\mathrm{Cl}_{H_s} \mathrm{Lin}\Phi$, $s = 2\gamma$ and

$$\|y\|^2_{H^s(0,2\pi)} \asymp \sum_{n=1}^{\infty} (|a_n^+|^2 |\lambda_n^+|^{2s} + |a_n^-|^2 |\lambda_n^-|^{2s})$$
$$\asymp \|(W_0, V_0)\|_s^2, \quad \lambda_n^\pm := \pm(n - 1/2 - \gamma).$$

(ii) For $\gamma \in (0, 1/4)$ the system is also observable on $\mathrm{Cl}_{L^2} \mathrm{Lin}\Phi$:

$$\|y\|^2_{L^2(0,2\pi)} \asymp \|(W_0, V_0)\|^2_{L^2(0,\pi;\mathbb{C}^4)}.$$

(iii) For $\gamma \in [1/4, 1/2)$ the system is not L^2–observable.

Proof. (i) From (9) and (10) we have

$$\mathcal{O}\begin{pmatrix} W_0 \\ V_0 \end{pmatrix} = v_1(0,t) = \sum_{n=1}^{\infty} \left[a_n^+ e^{i(n-1/2-\gamma)t} + a_n^- e^{-i(n-1/2-\gamma)t} \right].$$

Let us set $\mathcal{E} = \{e_n^\pm\}_{n=1,2,\ldots}$, $e_n^\pm(t) = e^{\pm i(n-1/2-\gamma)t}$. After normalization the family $\mathcal{E}$ forms a Riesz basis in $H^s(0,2\pi)$. Indeed, its generating function has the form

$$F(z) = \prod_{n=1}^{\infty} \left[1 - \frac{z^2}{(n-1/2-\gamma)^2} \right].$$

According to [1]

$$|F(x+ih)| \asymp (1+|x|)^{2\gamma}, \quad x \in \mathbb{R}, \tag{11}$$

and we can use Theorem 1.1.

(ii) Since $|F(x+ih)|^2$ satisfies the Muckenhoupt condition (see (11)) for $\gamma \in (0,1/4)$ the family $\mathcal{E}$ is a Riesz basis in $L^2(0,2\pi)$ (this also follows from Kadec's "1/4" theorem [6] – the family $\mathcal{E}$ in this case is a "small" perturbation of a standard basis $\left\{e^{\pm(n-\frac{1}{2})t}\right\}_{n\in\mathbb{N}}$).

(iii) For $\gamma \in [1/4, 1/2)$

$$\int_{\mathbb{R}} \frac{|F(x+ih)|^2}{1+x^2} dx = \infty$$

and the family $\mathcal{E}$ is not minimal in $L^2(0,2\pi)$ [10]. Therefore the estimate

$$\|y\|^2_{L^2(0,2\pi)} \succ \|(W_0,V_0)\|^2_{L^2(0,\pi;\mathbb{C}^4)}$$

is not valid. □

Actually, a stronger negative result is true: for $\gamma \in (1/4, 1/2)$ the system is not L^2–weakly observable; there exists a nonzero initial state $(W_0, V_0) \in L^2(0,\pi;\mathbb{C}^4)$ such that $y = 0$ in $L^2(0,2\pi)$.

Problems 2–4 can be studied similarly.

References

[1] S.A. Avdonin, *On Riesz bases of exponentials in L^2*, Vestnik Leningrad Univ., Ser. Mat., Mekh., Astron., **13** (1974), 5–12 (Russian); English transl. in Vestnik Leningrad Univ. Math., **7** (1979), 203–211.

[2] S.A. Avdonin and S.A. Ivanov, *Families of Exponentials. The Method of Moments in Controllability Problems for Distributed Parameter Systems* Cambridge University Press, New York, 1995.

[3] S.A. Avdonin and S.A. Ivanov, *Levin-Golovin theorem for the Sobolev spaces*, submitted to Matemat. Zametki (Russian).

[4] V.D. Golovin, *Biorthogonal decompositions of linear combinations of exponential functions in L^2 space*, Zap. Mekh. Fakul't., Khar'kov. Gosud. Univers. i Khar'k. Mat. Obshch., **30** (1964), 18–24 (Russian).

[5] L.P. Ho, *Uniform basis properties of exponential solutions of functional differential equations of retarded type*, Proceedings of the Royal Society of Edinburgh, **96A**, (1984), 79–94.

[6] M. I. Kadeč, *The exact value of the Paley-Wiener constant*, Soviet Math., **5** (1964), 559–561.

[7] B.Ya. Levin, *On Riesz bases of exponentials in L^2*, Zapiski math. otd. phys.-math. facul. Khark. univ., **27**, Ser. 4 (1961), 39–48 (Russian).

[8] J.-L. Lions and E. Magenes, *Problèmes aux limites nonhomogènes et applications*, T. 1, 2 (1968), Dunod, Paris.

[9] K. Narukawa and T. Suzuki, *Nonharmonic Fourier series and its applications.* Appl. Math. Optim. **14** (1986), 249–264.

[10] R.E.A.C. Paley and N. Wiener, *Fourier Transform in the Complex Domain*, AMS Coll. Publ. (1934), **XIX**, New York.

[11] D.L. Russell, *On exponential bases for the Sobolev spaces over an interval,* J. Math. Anal. Appl., **87** (1982), 528–550.

S. A. Avdonin
Department of Applied Mathematics and Control,
St. Petersburg State University, Bibliotechnaya sq. 2,
198904 St. Petersburg, Russia
and
Department Mathematics and Statistics,
The Flinders University of South Australia,
GPO Box 2100, Adelaide SA 5001, Australia
E-mail address: avdonin@ist.flinders.edu.au

S. A. Ivanov
Russian Center of Laser Physics,
St. Petersburg State University, Ul'yanovskaya 1,
198904 St. Petersburg, Russia
E-mail address: Sergei.Ivanov@pobox.spbu.ru

D. L. Russell
Department of Mathematics,
Virginia Polytechnic Institute and State University,
Blacksburg, VA 24061, USA
E-mail address: russell@calvin.math.vt.edu

International Series of Numerical Mathematics
Vol. 133, © 1999 Birkhäuser Verlag Basel/Switzerland

Sampling and Interpolation of Functions with Multi-Band Spectra and Controllability Problems

Sergei Avdonin and William Moran [1]

Abstract. We reduce the problem of constructing an exponential Riesz basis in L^2 on the union of several intervals and the equivalent problem of constructing a sampling and interpolating set for the space of functions with limited multi-band spectra to a controllability problem for a model dynamical system. As a model we consider the wave equation with piecewise constant density and boundary control supported on the same union of intervals. For the case of two intervals we construct a controllable system and, as a consequence, a Riesz basis of exponentials produced by the spectrum of the system.

1. Introduction

Let E be a union of a finite number of disjoint intervals:

$$E = \bigcup_{j=1}^{N} I_j, \quad I_j = [a_j,\, b_j], \quad 0 = a_1 < b_1 < a_2 < b_2 < \ldots < a_N < b_N. \qquad (1)$$

Several papers ([4, 7, 12, 16]) have recently appeared which discuss Riesz bases of exponentials in $L^2(E)$. All of them emphasize the importance of this problem in communication theory – if $\{e^{i\lambda_k t}\}$ forms a Riesz basis in $L^2(E)$ then Λ is a sampling and interpolating set for corresponding multi-band signals. In other words, the interpolation problem

$$f(\lambda_k) = \alpha_k, \quad \lambda_k \in \Lambda, \quad f \in L^2_E,$$

has a unique solution for each $\{\alpha_k\} \in l^2$. Here L^2_E is the space of entire functions of the form

$$f(\lambda) = \int_E e^{i\lambda t} \phi(t)\, dt, \quad \phi \in L^2(E),$$

endowed with the $L^2(\mathbb{R})$ norm. The equivalence of these two problems is well known; it follows from standard duality arguments (see, for example [6, 11]). We

[1] This work was partially supported by the Australian Research Council and by the Russian Fundamental Research Foundation (grant # 97-01-01115).

mention also [5] where various problems of sampling theory are considered from the viewpoint of the Riesz basis setting.

In the case of one interval ($N = 1$) necessary and sufficient conditions for $\{e^{i\lambda_k t}\}$ to form a Riesz basis in $L^2(E)$ were obtained by Pavlov [14]. Note that [2] and [6] give the complete characterization of such Λ's. There are only a few results concerning more general sets E. In [9] the case of two intervals of the same length was considered. The method of that paper, as pointed out in [12] allows the construction of a sampling and interpolating sequence for any set E which is a union of a finite number of intervals having commensurable lengths. This result can also be proved by means of another approach suggested in [4]. The results of [16] are free of arithmetic restrictions on E; in particular, they give a construction of a real sampling and interpolating sequence for an arbitrary E consisting of two intervals. In the case $N > 2$ rather severe quantitative restrictions are required to relative lengths of the intervals in E and the gaps between them. In the recent paper [12] a sampling and interpolating set located in a strip parallel to the real axis is constructed for an arbitrary E consisting of n intervals. For the case of two intervals sampling and interpolating sequences consisting of two arithmetics progressions are studied. More details about the history of the problem of constructing a Riesz basis from exponentials in $L^2(E)$ and methods of its solution may be found in [7, 11].

In this paper we propose another approach to the problem. It based on connections between controllability of a dynamical system described by a linear PDE and the Riesz basis property of a corresponding exponential family. These connections are well known and widely exploited in control theory; see, for example, an excellent survey paper [15] and the book [2]. The problem of constructing an exponential basis on several intervals gives rise to a new type of control problems with the boundary control supported on these intervals of time. More exactly, we reduce our original problem to the investigation of controllability of the wave equation with piecewise constant density and boundary control with support on prescribed intervals of time. We construct a density in such a way that each state of the system is reachable from zero and the corresponding control is unique. This fact is equivalent to the Riesz basis property in $L^2(E)$ of exponentials constructed by eigen-frequencies of the system.

2. The Control Problem in $L^2([0,\ T])$

Let

$$0 = x_0 < x_1 < x_2 < \cdots < x_{N-1} < x_N = l$$

and $\rho(x)$ be a piecewise C^2 (PC^2) positive function on $[0,\ l]$ with jump discontinuities at points x_j, $j = 1,\ 2,\ \ldots,\ N-1$.

Consider the following boundary value problem

$$\begin{gathered} \phi''(x) + \lambda^2\rho^2(x)\phi(x) = 0, \quad x_{j-1} < x < x_j, \quad j = 1,\ 2,\ \ldots,\ N, \\ \phi(0) = \phi'(l) = 0, \quad [\phi(x_j)] = [\phi'(x_j)] = 0, \quad j = 1,\ 2,\ \ldots,\ N-1, \end{gathered} \tag{2}$$

where $[\phi(x_j)] := \phi(x_j+0)-\phi(x_j-0)$. Problems of such types have been investigated by many authors; see, for example, [8, 13] and [2], Chap. VII. It is known that the eigen-functions $\phi_n(x)$ $(n \in \mathbb{N})$ form an orthonormal basis in $L^2([0,\ l];\rho^2\ dx)$ and

$$0 < \inf_{n\in\mathbb{N}} \left|\frac{\phi_n'(0)}{\lambda_n}\right| \le \sup_{n\in\mathbb{N}} \left|\frac{\phi_n'(0)}{\lambda_n}\right| < \infty. \tag{3}$$

Let us introduce a function $\Phi(x,\lambda)$ as a solution to the Cauchy problem

$$\begin{gathered}\Phi''(x,\lambda)+\lambda^2\rho^2(x)\Phi(x,\lambda)=0,\quad x_{j-1}<x<x_j,\quad j=1,\ 2,\ \dots,\ N,\\ \Phi(0,\lambda)=0,\ \Phi'(0,\lambda)=\lambda,\ [\Phi(x_j,\lambda)]=[\Phi'(x_j,\lambda)]=0,\ j=1,\ 2,\ \dots,\ N-1,\end{gathered}$$

(Φ' means $\partial_x\Phi$). Using the approach of [2], Sec. VII.2, VII.3, it can be proved that the function $F(\lambda) := \Phi'(l,\lambda)/\lambda$ is a sine type function (see the definition, for example, in [6]) with width of indicator diagram equal to $2L$, $L := \int_0^l \rho(x)\ dx$. Eigenvalues λ_n^2 of problem (2) are obtained from zeros $\pm\lambda_n$ of the function $F(\lambda)$. It follows from the Levin-Golovin theorem (see, for example [6] and [2], Sec. II.4) that the family of exponentials $\{e^{\pm i\lambda_n t}\}$ forms a Riesz basis in $L^2([0,\ 2L])$. We will obtain this result below in other way as a consequence of controllability of the wave equation corresponding to problem (2) . Then we turn to the Riesz basis property of the exponential family $\{e^{\pm i\lambda_n t}\}$ in $L^2(E)$.

Consider the following initial boundary value problem

$$\begin{gathered}\rho^2(x)u_{tt}(x,t)=u_{xx}(x,t),\quad x_{j-1}<x<x_j,\quad j=1,\ 2,\ \dots,\ N,\quad 0<t<T,\\ u(0,t)=f(t),\quad u_x(l,t)=0,\quad f\in L^2([0,\ T]),\\ [u(x_j,t)]=[u_x(x_j,t)]=0,\ j=1,\ 2,\ \dots,\ N-1,\quad 0<t<T,\\ u(x,0)=u_t(x,0)=0,\quad 0<x<l.\end{gathered} \tag{4}$$

It can be proved in a standard way using the Fourier method (see [2], Sec. III.2, V.2) that the solution of problem (4) satisfies the relations

$$u\in C([0,\ T];L^2([0,\ l])),\qquad u_t\in C([0,\ T];H_{-1}([0,\ l])), \tag{5}$$

where $H_{-1}([0,\ l])$ is the dual of $\{v : v\in H^1([0,\ l]), v(0)=0\}$.

The Fourier representation of $u(x,T)$ has the form

$$u(x,T)=\sum_{n=1}^{\infty} a_n\phi_n(x),\qquad u_t(x,T)=\sum_{n=1}^{\infty} b_n\phi_n(x), \tag{6}$$

where

$$a_n=\phi_n'(0)\int_0^T f(t)\frac{\sin\lambda_n(T-t)}{\lambda_n}\ dt, \tag{7}$$

$$b_n=\phi_n'(0)\int_0^T f(t)\cos\lambda_n(T-t)\ dt. \tag{8}$$

Equalities (7), (8) can be written in the form

$$c_k=\int_0^T f(t)e^{i\lambda_k(T-t)}\ dt,\quad k\in\mathbb{K}, \tag{9}$$

where

$$\mathbb{K} = \mathbb{Z}\backslash\{0\}, \quad c_{\pm n} := \frac{1}{\phi_n'(0)}\big[\pm i\lambda_n a_n + b_n\big], \quad \lambda_{-n} = \lambda_n.$$

It follows from (3), (6) that the map of

$$\{u(\cdot, T),\ u_t(\cdot, T)\} \mapsto \{c_k\}_{k\in\mathbb{K}}$$

is an isomorphism of spaces $L^2([0,\ l]) \times H_{-1}([0,\ l])$ and l^2 (see also similar proposition in [2], Sec. III.2, for more general equations).

System (4) is said to be controllable in time T, if for any $u_0 \in L^2([0,\ l])$, $u_1 \in H_{-1}([0,\ l])$, there exists $f \in L^2([0,\ T])$ such that

$$u(\cdot, T) = u_0, \quad u_t(\cdot, T) = u_1.$$

If f is unique we say that the system is uniquely controllable. It follows from (9) (see [15] and [2], Sec. III.3 for a detailed proof) that controllability of (4) in time T is equivalent to the fact that the family $\{e^{i\lambda_k t}\}$ forms an $\mathcal{L}$-basis (that is, a Riesz basis in the closure of its linear span) in $L^2([0,\ T])$. Completeness of $\{e^{i\lambda_k t}\}$ is evidently equivalent to uniqueness of control. Therefore controllability and uniqueness of control are equivalent to the fact that $\{e^{i\lambda_k t}\}$ forms a Riesz basis in $L^2([0,\ T])$.

We have already mentioned that the Riesz basis property of the family $\{e^{i\lambda_k t}\}$ in $L^2([0,\ 2L])$ can be proved using properties of the entire function $F(\lambda)$. Let us give another proof of this fact based on controllability of system (4). For this, consider separately the following two control problems for system (4): for given $u_0 \in L^2([0,\ l])$ and $u_1 \in H_{-1}([0,\ l])$

(i) find f such that $u(x, L) = u_0(x),\ \ 0 \le x \le l$;

(ii) find f such that $u_t(x, L) = u_1(x),\ \ 0 \le x \le l$.

In the domain $(x,\ t) \in [0,\ l] \times [0,\ L]$ we can use the fact that the variables x and t are changeable. Namely, to solve (i) we consider the equation

$$\rho^2(x) v_{xx}(t, x) = v_{tt}(t, x), \quad x_{j-1} < x < x_j, \quad j = 1,\ 2,\ \ldots,\ N, \quad 0 < t < L,$$

with compatibility conditions

$$[v(t, x_j)] = [v_x(t, x_j)] = 0, \quad j = 1,\ 2,\ \ldots,\ N-1, \quad 0 < t < L,$$

initial conditions

$$v(t, l) = v_x(t, l) = 0, \quad 0 < t < L,$$

and boundary condition

$$v(L, x) = u_0(x), \quad 0 < x < l.$$

It can be easily proved that this initial–boundary value problem has a unique solution $v(t, x)$ such that $v \in C([0,\ l];\ L^2([0,\ L]))$. The function $f(t) := v(t, 0)$ gives us the unique solution of problem (i).

To solve problem (ii) we change the boundary condition $v(L, x) = u_0(x)$ to $v_t(L, x) = u_1(x)$.

Unique solvability of problem (i) together with (7) imply the fact that the family $\{\sin \lambda_n t\}$ forms a Riesz basis in $L^2([0,\ L])$. Similarly (ii) and (8) give that $\{\cos \lambda_n t\}$ is a Riesz basis in $L^2([0,\ L])$. Using Euler's formula and evenness (oddness) properties of cos (sin) we get that $\{e^{i\lambda_k t}\}_{k\in\mathbb{K}}$ forms a Riesz basis in $L^2([-L,\ L])$ and so in $L^2([0,\ 2L])$ (see [1] for details). The last fact implies controllability of (4) in time $2L$.

3. The Control Problem in $L^2(E)$

Let E be a union of intervals as described in (1), $l_j := b_j - a_j$, $\text{meas}\, E = \sum_{j=1}^N l_j$.

Consider the system (4) with $T = b_N$, control $f \in L^2(0,T)$ such that $\text{supp} f = E$; denote this system by S_E. From (9) it follows that the controllability of S_E is equivalent to solvability of the moment problem

$$c_k = \int_E f(t) e^{i\lambda_k (T-t)}\, dt, \quad k \in \mathbb{K}.$$

Therefore the Riesz basis property of $\{e^{i\lambda_k t}\}_{k\in\mathbb{K}}$ in $L^2(E)$ is equivalent to the fact that for any given terminal state

$$u(\cdot,T) = u_0 \in L^2([0,\ l]),\quad u_t(\cdot,T) = u_1 \in H_{-1}([0,\ l]) \tag{10}$$

the control problem (4), (10) has a unique solution $f \in L^2(E)$.

The problem is to construct the function $\rho(x)$ in such a way that system S_E is uniquely controllable. It is easy to obtain a necessary condition for this in the form of the equality

$$L := \int_0^l \rho(x)\, dx = \frac{1}{2}\text{meas}\, E. \tag{11}$$

Indeed, it is well known that, for the spectral problem (2),

$$\lim_{k\to\infty} \frac{\lambda_k}{k} = \frac{\pi}{L}.$$

On the other hand, a necessary condition for a sequence λ_k to be a sampling and interpolating sequence for E and, equivalently, for $\{e^{i\lambda_k t}\}_{k\in\mathbb{K}}$ to form a Riesz basis in $L^2(E)$ is the equality (see the paper of Landau [10])

$$\lim_{k\to\infty} \frac{\lambda_k}{k} = \frac{2\pi}{\text{meas}\, E}$$

which gives us (11).

Therefore, in what follows we consider system S_E with optical length L equal to $\frac{1}{2}\text{meas}\, E$. Certainly, this equality is far from sufficient for the Riesz basis property of $\{e^{i\lambda_k t}\}_{k\in\mathbb{K}}$ in $L^2(E)$. To prove this, we consider the case of two intervals, $N = 2$, and show that the standard orthogonal basis $\{e^{i\frac{\pi}{L}(k+\frac{1}{2})}\}_{k\in\mathbb{Z}}$ in $L^2([0,\ 2L])$ forms a Riesz basis in $L^2([a_1, b_1] \cup [a_2, b_2])$ if and only if $b_2 = 2Lm$ for some $m \in \mathbb{N}$. The same statement is also true for a family $\{e^{i\frac{\pi}{L}kt}\}_{k\in\mathbb{Z}}$.

Let us take in problems (2) and (4) $\rho(x) = 1$. Then $l = L$ and

$$\lambda_n^2 = \left[\frac{\pi}{L}\left(n - \frac{1}{2}\right)\right]^2, \quad n \in \mathbb{N},$$

so

$$\lambda_k = \operatorname{sgn}(k)\frac{\pi}{L}\left(k - \frac{1}{2}\right), \quad k \in \mathbb{K}.$$

We show that corresponding system S_E is controllable if and only if $T = 2Lm$.

For $\rho(x) = 1$ it is easy to see that the solution of (4) has the form of

$$u(x,t) = \sum_{n=0}^{\infty}(-1)^n f(t - 2Ln - x) + \sum_{n=1}^{\infty}(-1)^{n-1} f(t - 2Ln + x). \tag{12}$$

We consider here that $f \in L^2(\mathbb{R})$ and $\operatorname{supp} f = E$. Unique controllability of the system means bijectivity of the map

$$f(t)|_{t\in E} \mapsto (u(x,T),\ u_t(x,T))\,|_{x\in[0,\ L]} \tag{13}$$

considered as a map from $L^2(E)$ to $L^2([0,\ l]) \times H_{-1}([0,\ l])$ (see (5)). Because of time–reversibility of the system S_E (change of variables $t \mapsto T - t$) we can assume that $l_1 \le l_2$. Using the method of characteristics (see e.g. [15, 17]), it can be proved that bijectivity of the map (13) is equivalent to bijectivity of the map

$$f(t)|_{t\in[0,\ l_1]} \mapsto u(L,t)|_{t\in[T-L,\ T-L+l_1]} \tag{14}$$

from $L^2([0,l_1])$ to $L^2([T-L, T-L+l_1])$. From (12) it follows that this property is equivalent to bijectivity of the map

$$f(t)|_{t\in[0,\ l_1]} \mapsto f(t - 2Lm + L)|_{t\in[T-L,\ T-L+l_1]}$$

for some $m \in \mathbb{N}$. The last property is true if and only if $T = 2Lm$.

To make the system S_E controllable for E described in (1) we construct $\rho(x)$ as a piecewise constant function such that

$$\rho(x) = \rho_j, \ \text{ for } \ x_{j-1} < x < x_j; \ \ 0 < \rho_j < \infty, \ \ j = 1,\ 2,\ \ldots,\ N; \tag{15}$$

$$\rho_j \ne \rho_{j+1}, \ \ j = 1,\ 2,\ \ldots,\ N-1; \tag{16}$$

$$\rho_j(x_j - x_{j-1}) = l_j/2, \ \ j = 1,\ 2,\ \ldots,\ N. \tag{17}$$

Note that (17) implies (11).

If we set

$$\xi(x) = \int_0^x \rho(s)\,ds, \quad \xi_j = \xi(x_j)$$

the initial boundary problem (4) takes the form

$$\begin{gathered} u_{tt}(\xi,t) = u_{\xi,\xi}(\xi,t), \ \ \xi_{j-1} < \xi < \xi_j, \ \ j = 1,\ 2,\ \ldots,\ N, \ \ 0 < t < T, \\ u(0,t) = f(t), \ \ u_\xi(L,t) = 0, \ \ f \in L^2(0,\ T), \\ [u(\xi_j,t)] = 0, \ \ u_\xi(\xi_j + 0,t) = \nu_j u_\xi(\xi_j - 0,t), \ j = 1,\ 2,\ \ldots,\ N-1, \\ u(\xi,0) = u_t(\xi,0) = 0, \ \ 0 < \xi < L. \end{gathered} \tag{18}$$

Here $\nu_j := \rho_j/\rho_{j-1}$. From (15), (16) it follows that

$$\nu_j \in (0,\ 1) \cup (1,\ \infty), \quad j = 1,\ 2,\ \dots,\ N-1. \tag{19}$$

These inclusions tell us that all reflection and transmission coefficients corresponding to the lines $\xi = \xi_j$, $j = 1,\ 2,\ \dots,\ N-1$, are not equal to zero. Equalities (17) imply

$$\xi_j - \xi_{j-1} = l_j/2, \quad j = 1,\ 2,\ \dots,\ N, \tag{20}$$

and so $\xi_N = L = \frac{1}{2}\text{meas } E$.

It is easy to see that for all ν_j satisfying (19) conditions (20) ensure that the system S_E satisfies the necessary geometric controllability conditions (cf. [3]): for any $\zeta \in [0,\ L]$ both rays started from the point (ζ, T) along the characteristics $\xi + t = \zeta + T$ and $\xi - t = \zeta - T$ meet the set E. To ensure the sufficient controllability conditions for the system S_E we can vary the coefficients ν_j.

Main Principle *Let E be a multiband set of the form (1) and conditions (20) are satisfied. Then, for all values of coefficients ν_j, $j = 1,\ 2,\ \dots,\ N$, satisfying (19), except for a thin set, the system S_E is uniquely controllable.*

We have been inprecise about the nature of this "thin set", because at the moment we are unclear as to its size. Examples with commensurable endpoints a_j, b_j lead us to suspect that it may even be empty, but at the very least it should be a set of first category.

We have confirmed this principle in many particular cases but at the present moment we do not have a complete proof. Similar statement in other terms has been formulated by Katsnelson [7].

From this principle it follows that exponential family $\{e^{\pm i\lambda_n t}\}_{n\in\mathbb{N}}$ forms a Riesz basis in $L^2(E)$ where λ_n^2 are eigenvalues of the boundary problem (2) and $\rho(x)$ satisfies conditions (15)–(17). It is important for applications that the sampling and interpolating set $\{\pm\lambda_n\}$ is real.

As an example we prove the principle for the case of two–band set with a short gap. Let E be a union of two intervals, $E = [a_1,\ b_1] \cup [a_2,\ b_2]$, and ν $(:= \nu_1)$ be a positive number, $\nu \neq 1$. One can check that the solution of (18) in this case takes the form

$$u(\xi, t) = \begin{cases} u_1(\xi, t), & 0 \le \xi \le l_1/2; \\ u_2(\xi, t), & l_1/2 < \xi \le L; \end{cases} \tag{21}$$

where

$$u_1(\xi, t) = \textstyle\sum_{r=0}^{\infty} \sum_{k=0}^{\infty} \sum_{q=0}^{k} A(r, k, q)[F_1(t, \xi, r, k, q) + F_2(t, \xi, r, k, q) \\ + \lambda F_3(t, \xi, r, k, q) + \lambda F_4(t, \xi, r, k, q)];$$

$$u_2(\xi, t) = \frac{2\nu}{\nu + 1} \sum_{r=0}^{\infty} \sum_{k=0}^{\infty} \sum_{q=0}^{k} A(r, k, q)[F_1(t, \xi, r, k, q) + F_2(t, \xi, r, k, q)];$$

$$A(r, k, q) = (-1)^{r+k} \lambda^k \frac{(r+k)!}{r!q!(k-q)!}, \quad \lambda = \frac{\nu - 1}{\nu + 1};$$

$$F_i(t,\xi,r,k,q) = f(t - w_i(\xi,r,k,q)), \quad i = 1,2,3,4;$$
$$w_1(\xi,r,k,q) = w(r,k,q) + \xi; \quad w_2(\xi,r,k,q) = w(r,k,q) + 2L - \xi;$$
$$w_3(\xi,r,k,q) = w(r,k,q) + l_2 + \xi; \quad w_4(\xi,r,k,q) = w(r,k,q) + l_1 - \xi;$$
$$w(r,k,q) = 2Lr + l_1(k-q) + l_2 q.$$

For the same reason as above we can assume that $l_1 \le l_2$. Following in a similar way to the case $\rho(x) = 1$ and using controllability results proved in Section 2, one can prove that bijectivity of the map (13) is equivalent to the bijectivity of the map (14). From (21) it follows that

$$u(L,t) = \frac{4\nu}{\nu+1} \sum_{r=0}^{\infty} \sum_{k=0}^{\infty} \sum_{q=0}^{k} A(r,k,q) f(t - w(r,k,q) - L). \tag{22}$$

Let us suppose now that the gap between the intervals I_1 and I_2 is not large, more exactly, $G := a_2 - b_1 \le l_1$. In this case the restriction of the right-hand side of (22) to the interval $[T-L,\ T-L+l_1]$ contains only three nonzero terms. Taring into account that $T - L = L + G$, we reduce the problem to the question of invertibility of the map

$$f(t)|_{t\in[0,\ l_1]} \mapsto f(t-L) - \lambda f(t-L-l_1)) - \lambda f(t-L-l_2)|_{t\in[L+G,\ L+G+l_1]},$$

which is equivalent to invertibility of the map

$$f(t)|_{t\in[0,\ l_1]} \mapsto f(t) - \lambda f(t-l_1)) - \lambda f(t-l_2)|_{t\in[G,\ G+l_1]}.$$

The last map is evidently invertible as a "triangular" map.

It follows that the family $\{e^{\pm i\lambda_n t}\}_{n\in\mathbb{N}}$ forms a Riesz basis in $L^2([a_1,\ b_1] \cup [a_2,\ b_2])$ where λ_n^2 are eigenvalues of the boundary problem (2) and $\rho(x)$ satisfies conditions (15)–(17).

References

[1] S.A. Avdonin, M.I. Belishev and S.A. Ivanov, *Boundary control and matrix inverse problem for the equation* $u_{tt} - u_{xx} + V(x)u = 0$, Math. USSR Sbornik, **72** (1992), 287–310.

[2] S.A. Avdonin and S.A. Ivanov, *Families of Exponentials. The Method of Moments in Controllability Problems for Distributed Parameter Systems*, Cambridge University Press, New York, 1995.

[3] C. Bardos, G. Lebeau and J. Rauch, *Sharp sufficient condition for the observation, control and observation of waves from the boundary*, SIAM J. Control Optim., **30** (1992), 1024–1065.

[4] L. Bezuglaya and V. Katsnelson, *The sampling theorem for functions with limited multi-band spectrum, I*, Z. Anal. Anwendungen, **12** (1993), 511–534.

[5] J.R. Higgins, *Sampling theory for Pale–Wiener spaces in the Riesz basis setting*, Proc. Royal Irish Acad., Sect. A, **94** (1994), 219–236.

[6] S.V. Hruščev, N.K. Nikol'skii and B.S. Pavlov, *Unconditional bases of exponentials and reproducing kernels*, Complex Analysis and Spectral Theory, Lecture Notes Math., **864**, (1981), 214–335, Springer–Verlag, Berlin Heidelberg.

[7] V.E. Katsnelson, *Sampling and interpolation for functions with multi-band spectrum: the mean-periodic continuation method*, Wiener–Symposium (Grossbothen, 1994), 91–132, Synerg. Syntropie Nichtlineare Syst., 4, Verlag Wiss. Leipzig, Leipzig, 1996.

[8] I.S. Katz and M.G. Krein, *On spectral functions of a string*, Appendix in: F.A. Atkinson, *Discrete and Continuous Boundary Problems* (Russian edition), Mir, Moscow, 1968.

[9] A. Kohlenberg, *Exact interpolation of band-limited functions*, J. Appl. Phys., **24** (1953), 1432–1436.

[10] H.J. Landau, *Necessary density conditions for sampling and interpolation of certain intire functions*, Acta Math., **117**, 37–52.

[11] Yu. Lyubarskii and K. Seip, *Sampling and interpolating sequences for multiband-limited functions and exponential bases on disconnected sets*, J. Fourier Analysis Appl., **3** (1997), 597–615.

[12] Yu. Lyubarskii and I. Spitkovsky, *Sampling and interpolation for a lacunary spectrum*, Proc. Royal. Soc. Edinburgh, **126 A** (1996), 77–87.

[13] M.A. Naimark, *Linear Differential Operators*, v. 1, Ungar, New York, 1967.

[14] B.S. Pavlov, *Basicity of an exponential system and Muckenhoupt's condition*, Soviet Math. Dokl., **20** (1979), 655–659.

[15] D.L. Russell, *Controllability and stabilizability theory for linear partial differential equations*, SIAM Review, **20** (1978), 639–739.

[16] K. Seip, *A simple construction of exponential bases in L^2 of the union of several intervals*, Proc. Edinburgh Math. Soc., **38** (1995), 171–177.

[17] A.N. Tikhonov and A.A. Samarskii, *Equations of Mathematical Physics*, McMillan, New York, 1963.

S. Avdonin
Department of Applied Mathematics and Control,
St. Petersburg State University,
Bibliotechnaya sq. 2, 198904 St. Petersburg, Russia
and
Department Mathematics and Statistics,
The Flinders University of South Australia,
GPO Box 2100, Adelaide SA 5001, Australia
E-mail address: avdonin@ist.flinders.edu.au

W. Moran
Department Mathematics and Statistics,
The Flinders University of South Australia,
GPO Box 2100, Adelaide SA 5001, Australia
E-mail address: bill@ist.flinders.edu.au

International Series of Numerical Mathematics
Vol. 133,

Discretization of the Controllability Grammian in View of Exact Boundary Control: the Case of Thin Plates

Frédéric Bourquin, José Urquiza and Rabah Namar

Abstract. We expose, in the context of a plate-like equation, an approximation method for the exact boundary controllability problem. It is based on a Fourier analysis on the boundary and does not require high-order discretizations. The method is very general and can be applied to a wide set of partial differential equations.

1. Introduction

The control theory of infinite-dimensional systems has given rise to many works in the last decades and several approaches have proved successful in extending the finite-dimensional control theory to the continuous case. Among them, the PDE approach [27, 28], [37] remains "close" to the continuous mechanical description of the problem.

In the design of efficient closed-loop feedback laws, the question of controllability is of crucial importance. As a matter of fact, exact controllability is a necessary condition for the existence of a feedback law yielding uniform exponential decay of the response. On the other hand, this property is also sufficient, when supplemented with other continuity properties that are easier to establish. In particular, null controllability implies the existence of at least one control such that the infinite horizon cost functional is finite, thus leading to a well-posed minimization problem, the solution of which is given pointwise in time as a function of the state through the positive definite solution of an operator algebraic Riccati equation [24].

Based on, following, or independently of the initial work of J-L. Lions [30], many mathematical developments have addressed the controllability of a wide set of structures. See e.g. [12], [20], [21], [23], [25], [26], [29], [36], for plates, [13, 14], [32], [38], for shells, [22] for trusses and frames ... Moreover, necessary and sufficient conditions for the exact controllability of the wave equation have been derived in [4]. See [9] for plates.

On the other hand, the status of the numerical approximation of these problems seems less satisfactory since, for example, good approximation schemes for the dynamics may fail for the boundary control: negative mathematical results [8], [19], [34] as well as numerous negative numerical experiments do exist in this direction. This partly explains why many efforts in control theory are devoted to the design of numerical schemes that keep the essential properties of the controlled continuous system [2], [11], [24]. Such approximation schemes are expected to preserve the uniform stability with respect to the discretization parameter. In the absence of damping, this property does not necessarily hold with low-order elements which, however, are suitable for assessing the dynamics. So far, spectral methods, high-order splines or mixed methods have proved successful but the theory is more difficult for the boundary control of the undamped wave equation [24]. For the boundary exact controllability, special smoothing tricks must be introduced to smear the unwanted oscillations when standard discretization techniques are used [15, 16, 17, 18].

However, in some engineering applications, control design should build upon a given discrete model, *i.e.* a given finite element approximation of the ideal continuous model. Therefore, a natural question arises: assume the available numerical tools can correctly predict the statics and the slow dynamics of the structure, can these tools be used for boundary control also? The purpose of our contribution is to propose such a method of approximation that may resort to any kind of discretization, including low-order finite elements, and that can be easily implemented in any general purpose finite element software. It is developed here for thin plates. It can be put to work in closed-loop as well but we shall not focus on this point. The next section is devoted to background information on exact controllability. The numerical method is explained in Section 3 and a test is presented in Section 4.

2. Background information on exact controllability

Let $\omega \subset \mathbb{R}^2$ denote a bounded open set with a piecewise smooth boundary $\partial\omega = \bar{\gamma}_0 \bigcup \bar{\gamma}_1$, where $\gamma_0 \cap \gamma_1 = \emptyset$, and $meas\,(\gamma_0) \neq 0$. Let T stand for the horizon of controllability. We set $\Sigma = \partial\omega \times [0,T]$, $\Sigma_0 = \gamma_0 \times [0,T]$, $\Sigma_1 = \gamma_1 \times [0,T]$. If we assume that T is greater than the diameter of ω and that γ_0 coincides with the shadow cone of ω with respect to some point x_0, then the genuine Hilbert Uniquenes Method of Lions [28, 29, 30] resorting to multiplier techniques yields the existence of a control $v \in L^2(\Sigma_0)$ such that for every pair $\{y^0, y^1\} \in L^2(\omega) \times H^{-2}(\omega)$ of initial conditions, the very weak solution $y(x,t;v) \in L^\infty(0,T;L^2(\omega)) \cap W^{1,\infty}(0,T;H^{-2}(\omega))$ of the problem

$$\left\{\begin{array}{rcll} \dfrac{\partial^2 y}{\partial t^2} + \Delta^2 y & = & 0 & \text{in } \omega \times [0,T], \\ y = 0, \quad \dfrac{\partial y}{\partial \nu} & = & v & \text{on } \Sigma_0, \\ y = 0, \quad \dfrac{\partial y}{\partial \nu} & = & 0 & \text{on } \Sigma_1, \\ y(x,0) = y^0(x), \quad \dfrac{\partial y}{\partial t}(x,0) & = & y^1(x) & \text{in } \omega, \end{array}\right. \tag{1}$$

satisfies $y(T, v) = y'(T, v) = 0$. Also the control v continuously depends on the initial conditions. This is what exact controllability means, and the fonction v is uniquely defined if it minimizes the functional $\int_{\Sigma_0} v^2$. As shown by Zuazua in [29], T can be chosen arbitrarily small. For the sake of simplicity, we concentrate on Sophie Germain's plate model controlled through the rotation only, but of course, other plate models and other boundary controls like any combination of displacement, rotation, moment or shear force can be chosen and the controllability results may be established under different geometric assumptions regarding ω and Σ_0 [22], [25], [26], [29].

The controllability result we use here relies on an inverse estimate of the controllability grammian defined by $a\left(\{\varphi^0, \varphi^1\}, \{\hat{\varphi}^0, \hat{\varphi}^1\}\right) = \int_{\Sigma_0} \Delta\varphi\Delta\hat{\varphi}$, where φ solves

$$\left\{\begin{array}{rcll} \dfrac{\partial^2\varphi}{\partial t^2} + \Delta^2\varphi & = & 0 & \text{in } \omega \times [0, T], \\ \varphi = 0, \quad \dfrac{\partial\varphi}{\partial\nu} & = & 0 & \text{on } \Sigma, \\ \varphi(x, 0) = \varphi^0(x), \quad \dfrac{\partial\varphi}{\partial t}(x, 0) & = & \varphi^1(x) & \text{in } \omega, \end{array}\right. \tag{2}$$

and $\hat{\varphi}$ solves the same boundary value problem starting from $\{\hat{\varphi}^0, \hat{\varphi}^1\}$. The energy associated with φ is conserved. It writes

$$E\left(\varphi^0, \varphi^1\right) = \frac{1}{2}\left\{\int_\omega \varphi^1\varphi^1 + \int_\omega \Delta\varphi^0\Delta\varphi^0\right\}.$$

The regularity result $a\left(\{\varphi^0, \varphi^1\}, \{\varphi^0, \varphi^1\}\right) \leq CE\left(\varphi^0, \varphi^1\right)$, which is valid for any T and any γ_0, cannot be viewed as a consequence of trace theorems. But the hard property is the reverse one which is a strong observability property. It does hold when geometric conditions are satisfied.

Notice that in contrast with the energy $E\left(\varphi^0, \varphi^1\right)$, the controllability grammian lacks a clear mechanical meaning since this is a product of stresses, more precisely a product of bending moments, along the region of control. Therefore, the grammian is neither homogeneous to a virtual work, nor to an energy. Moreover, above strong observability holds for solutions φ which satisfy the conservation of linear momentum (2) exactly. Since standard Galerkin approximations aim at computing virtual works or energies, but violate this conservation, such an observability inequality may not hold at the discrete or semi-discrete level uniformly with respect to the discretization parameter. To our mind, this fact underlies the numerical difficulties encountered in this field [16], [19], [34]: strong observability may not be robust with respect to Galerkin aproximations.
Nevertheless, by using the Hilbert Uniqueness Method (HUM), observability leads to the controllability of all initial data $\{y^0, y^1\} \in L^2(\omega) \times H^{-2}(\omega)$. Moreover HUM suggests the following algorithm:

Step 1: find $\left\{\varphi^0, \varphi^1\right\} \in H_0^2(\omega) \times L^2(\omega)$ such that

$$a\left(\left\{\varphi^0, \varphi^1\right\}, \left\{\hat{\varphi}^0, \hat{\varphi}^1\right\}\right) = \left\langle\left\{y^1, -y^0\right\}, \left\{\hat{\varphi}^0, \hat{\varphi}^1\right\}\right\rangle \ \forall \left\{\hat{\varphi}^0, \hat{\varphi}^1\right\} \in H_0^2(\omega) \times L^2(\omega) \tag{3}$$

where $\langle ., . \rangle$ denotes the corresponding duality product.

Step 2: solve equation (2) starting from $\left\{\varphi^0, \varphi^1\right\}$ computed in step 1.

Step 3: set and compute the exact control $v = \Delta\varphi \ on \ \Sigma_0$.

Steps 2 and 3 are postprocessing steps and we will focus on step 1. Of course, (3) is well posed whenever strong observability holds.

3. Numerical approximation

The pioneering method of [15, 16, 17, 18] consists in solving (3) by a preconditioned conjugate gradient algorithm. At each step, the computation of the gradient amounts to solve two subsequent plate equations. To this end, finite differences or finite elements and time integration are used. The resulting oscillations of the control can be filtered out by means of Tychonov regularization, grid coarsening or mixed finite elements. See [1], [31], [33] for extensions or applications, [19], [35] for some aspects of the convergence analysis.

Our method is rather a Galerkin method as in [10, 14] for 1D problems, but with a general and careful numerical integration of the corrresponding stiffness matrix which works for many controllable problems in any dimension. See e.g. [3, 5, 7] for the wave equation, [6] for beams. Actually, to our knowledge, our general method is the only one which can be easily developped for plate-like structures and which can accomodate for any kind of discretization.

3.1. Internal Fourier-Galerkin approximation

Let $(\lambda_i, \psi_i)_{i=1}^{+\infty}$ denote the family of eigenpairs of the *clamped* plate, *i.e.* of the operator $\left(\Delta^2, \omega, Dir\right)$, and $V_N = Span\left\{\psi_1, \ldots, \psi_N\right\}$.

Theorem 3.1. *The problem: find* $\{\varphi_N^0, \varphi_N^1\} \in V_N \times V_N$ *such that*

$$a(\{\varphi_N^0, \varphi_N^1\}, \{\hat{\varphi}_N^0, \hat{\varphi}_N^1\}) = \langle\{y^1, -y^0\}, \{\hat{\varphi}_N^0, \hat{\varphi}_N^1\}\rangle \ \forall \{\hat{\varphi}_N^0, \hat{\varphi}_N^1\} \in V_N \times V_N, \tag{4}$$

admits a unique solution and

$$\lim_{N \to +\infty} \{\varphi_N^0, \varphi_N^1\} = \{\varphi^0, \varphi^1\} \ in \ H_0^2(\omega) \times L^2(\omega).$$

This is a direct consequence of standard spectral analysis of compact self-adjoint operators: the family of eigenfunctions is dense in both spaces involved. Moreover, error estimates can be proved as in [5] if $\left\{y^1, -y^0\right\}$ is sufficiently smooth.

The problem (4) can be put in an equivalent matrix form

$$Kx = b, \tag{5}$$

with

$$K = \begin{pmatrix} K^{00} & K^{01} \\ K^{10} & K^{11} \end{pmatrix}, \; K_{ij}^{\ell m} = f^{\ell m}(\lambda_i, \lambda_j, T) \int_{\gamma_0} \mathcal{M}\psi_i \mathcal{M}\psi_j, \; \mathcal{M}\psi_i = \Delta\psi_{i|\gamma_0} \quad (6)$$

where $f^{\ell m}$ stands for a smooth explicit function.

In the sequel, h stands for a discretization parameter associated with a Galerkin method. In the case of a finite element method, h would represent the mesh size.

In view of the discretization of the stiffness matrix K, deriving an accurate approximation of the end moments $\mathcal{M}\psi_i$ is essential. The problem is to approximate $\Delta\psi_{j|\gamma_0}$ in $L^2(\gamma_0)$ when we at most expect $H^2(\omega)$ convergence of the discrete modes ψ_j^h. It turns out that $\Delta\psi_{j|\gamma_0}^h$ may not be the right choice. As matter of fact, the corresponding expression for the Laplacian yields an extremely poor approximation, even for conforming finite elements.

3.2. Numerical integration of the stiffness matrix by means of a Fourier analysis on the boundary

The idea is to filter out the high-frequency content of the moments $\mathcal{M}\psi_i$ at the continuous level, since these functions are reasonnably smooth, and to compute the low-frequency content by a standard Galerkin method. To this end, we need an orthonormal basis $(z_\ell)_{\ell=1}^{+\infty}$ of $L^2(\gamma_0)$ such that

H1: $z_\ell \in H_{00}^{1/2}(\gamma_0)$.

Then there exists $\tilde{z}_\ell = Rz_\ell \in H^2(\omega)$, where $R : H_{00}^{1/2}(\gamma_0) \longrightarrow H^2(\omega)$, $\tilde{v} = Rv$ is defined by

$$\left\{ \begin{array}{rcll} \Delta^2\tilde{v} + \tilde{v} & = & 0 & \text{in } \omega, \\ \tilde{v} = 0, \quad \dfrac{\partial\tilde{v}}{\partial\nu} & = & v & \text{on } \gamma_0, \\ \tilde{v} = 0, \quad \dfrac{\partial\tilde{v}}{\partial\nu} & = & 0 & \text{on } \gamma_1. \end{array} \right. \quad (7)$$

H2: there exists a computable approximation $\tilde{z}_\ell^h$ of $\tilde{z}_\ell$ such that $\tilde{z}_\ell^h \xrightarrow{H^2(\omega)} \tilde{z}_\ell$.

In this setting, we have $\mathcal{M}\psi_j = \sum_{\ell=1}^{+\infty}(\int_{\gamma_0} \Delta\psi_j z_\ell) z_\ell$. Then, for any $M \in \mathbb{N}$, we are led to set

$$\mathcal{M}_M\psi_j = \sum_{\ell=1}^{M}(\int_{\gamma_0} \Delta\psi_j z_\ell) z_\ell \quad (8)$$

and by just integrating by parts, we have

$$\mathcal{M}_M\psi_j = -(1+\lambda_j)\sum_{\ell=1}^{M}(\int_{\omega} \psi_j \tilde{z}_\ell) z_\ell. \quad (9)$$

Therefore we are led to introduce the approximate bending moment

$$\mathcal{M}_M^h \psi_j = -(1+\lambda_j^h)\sum_{\ell=1}^{M}\Big(\int_\omega \psi_j^h z_\ell^h\Big) z_\ell^h \tag{10}$$

and the approximate stiffness matrix $K^{M,h} = \begin{pmatrix} K^{M,h,00} & K^{M,h,01} \\ K^{M,h,10} & K^{M,h,11} \end{pmatrix}$, where

$$K_{ij}^{M,h,\ell m} = f^{\ell m}(\lambda_i^h, \lambda_j^h, T)(1+\lambda_i^h)(1+\lambda_j^h)\sum_{\ell=1}^{M}\int_\omega \psi_i^h z_\ell^h \int_\omega \psi_j^h z_\ell^h. \tag{11}$$

Theorem 3.2. *When* $h \longrightarrow 0$ *first and then* $M \longrightarrow +\infty$, $K^{M,h} \longrightarrow K$.

The proof consists of rewriting above lines bottom up. Moreover, it turns out that the convergence should be fast w.r.t. h since the eigenvalues can be accurately determined by finite elements as well as the mode shapes in $L^2(\omega)$. The convergence w.r.t. M will depend on the smoothness of the continuous mode shapes, but it is very fast in practice. Therefore, it is possible to reconstruct a good approximation of the initial data $\{\varphi^0, \varphi^1\}$, and close-form solutions for the boundary control can be easily derived.

3.3. How to construct z_ℓ ?

Let us simply use the spectral basis of a compact symmetric operator acting on the region of control. For example, the capacitance operator $\mathcal{T} : H_{00}^{1/2}(\gamma_0) \longrightarrow (H_{00}^{1/2}(\gamma_0))'$, $v \longrightarrow (\Delta\tilde{v})_{|\gamma_0}$ enjoys the right properties. If $(\lambda_{\gamma_0\ell}, u_{\gamma_0\ell})_{\ell=1}^{+\infty}$ denotes the family of eigenpairs of $\mathcal{T}$, then it remains to chose $z_\ell = Ru_{\gamma_0\ell}$. As for H2, the approximation of the eigenmodes of the capacitance operator associated to the Laplacian has been addressed and can be extended to our case. Note that, again, only $L^2(\omega)$ convergence is needed, therefore a very good accuracy is expected. On the other hand, a to fine mesh will neither lead to spurious oscillations nor to any other problem.

More importantly, the computation of these modes may rely on a Lanczos procedure, at each step n of which the simple "Neumann" problem

$$\left\{\begin{array}{rcll} \Delta^2\tilde{v}+\tilde{v} &=& 0 & \text{in } \omega, \\ \tilde{v}=0, \quad \Delta\tilde{v} &=& w_n & \text{on } \gamma_0, \\ \tilde{v}=0, \quad \dfrac{\partial\tilde{v}}{\partial\nu} &=& 0 & \text{on } \gamma_1, \end{array}\right. \qquad w_{n+1} = \left.\frac{\partial w_n}{\partial\nu}\right|_{\gamma_0}, \tag{12}$$

has to be solved. Therefore the whole procedure appeals to very standard finite element routines and can be implemented in any software.

4. Numerical tests

Most of the procedure has been implemented in the finite element code CESAR-LCPC where the quantities appearing in (3.7) are computed. The rest is implemented in Matlab. We consider a plate of arbitrary shape, controlled along one edge. We use general shell elements. We chose $N = 20$, $M = 30$. The controlled response is expanded on the first 30 modes. The history of the controlled vibration is shown on fig 1. The energy is reduced by several orders of magnitude as shown on fig 2.

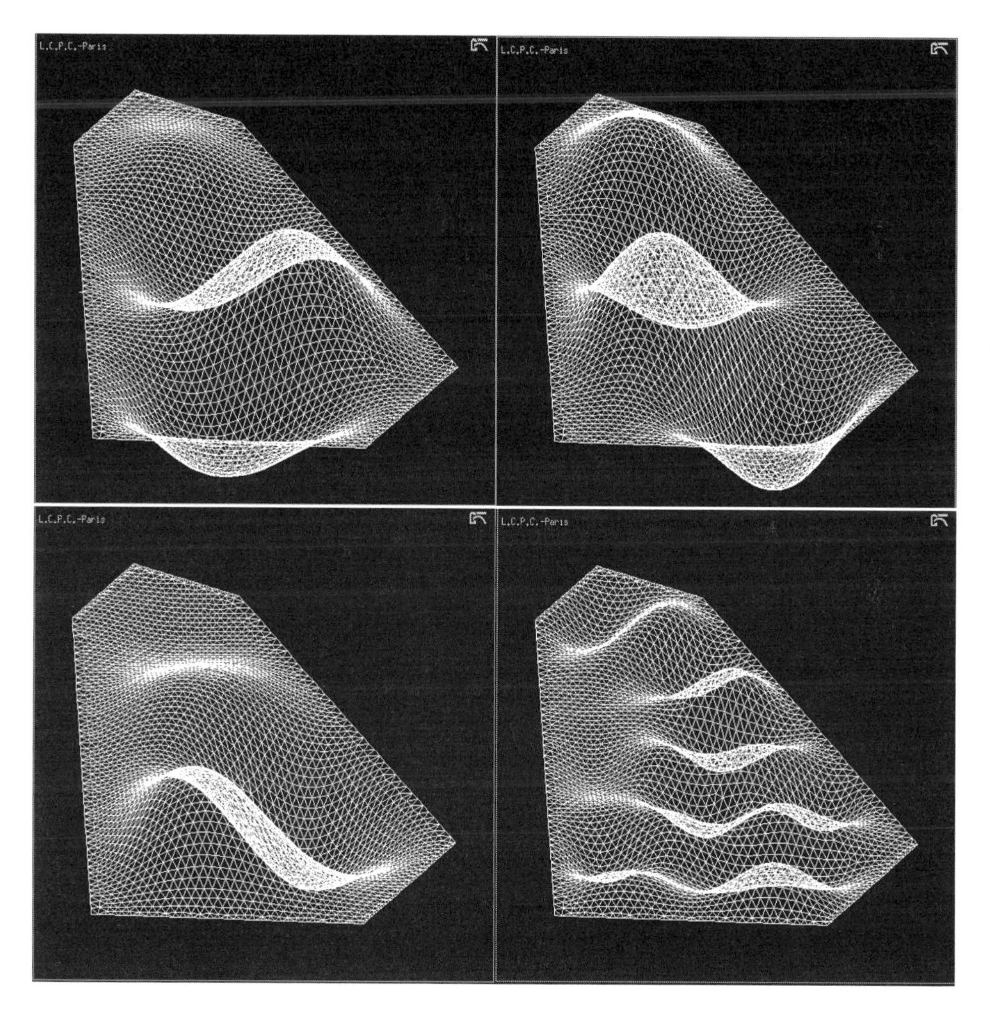

FIGURE 1. Displacement of the plate at times 0, 0.4, 0.9 and 1 with scales (resp.) 1/1, 1/25, 1/50 and 1/2500000

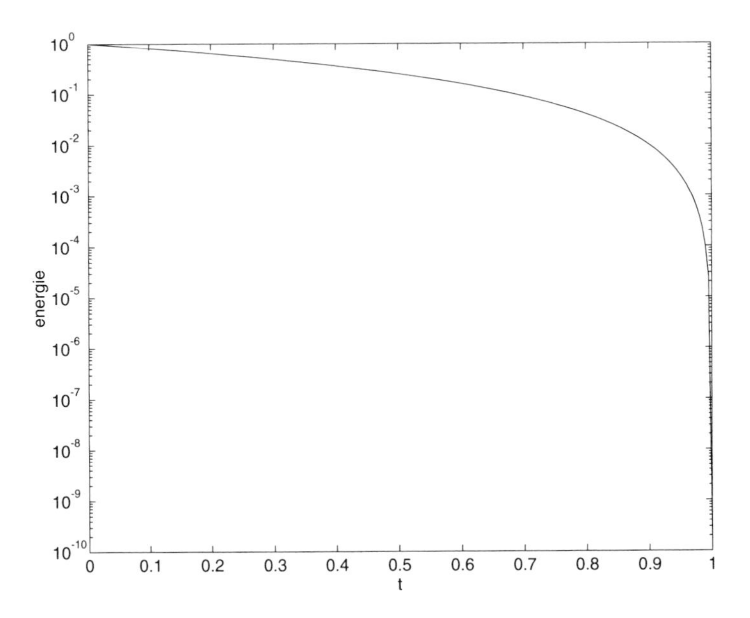

FIGURE 2. Decay of the energy of the system

References

[1] M. Asch and G. Lebeau, *Geometrical aspects of exact controllability for the wave equation: a numerical study,* ESAIM-COCV, **3** (1998), 163–212.

[2] H.T. Banks and K. Ito and Y. Wang, *Well-posedness and approximation for damped second-order systems with unbounded input operators,* Diff. Integral Equations, **3** (1995), 587–606.

[3] C. Bardos and F. Bourquin and G. Lebeau, *Calcul de dérivées normales et méthode de Galerkin appliquée au problème de la contrôlabilité exacte,* C.R. Acad. Sci., série 1, **313** (1991), 757–760.

[4] C. Bardos and G. Lebeau and J. Rauch, *Sharp sufficient conditions for the observability, control and stabilization of waves from the boundary,* SIAM J. Cont. Optim., **30** (1992), 1024–1065.

[5] F. Bourquin, *Approximation Theory for the problem of exact controllability of the wave equation,* Proc. Second SIAM Conference on Mathematical and Numerical Aspects of Wave Propagation, Newark (Delaware), 1993, 103–112.

[6] F. Bourquin, *A numerical approach to the exact controllability of Euler-Navier-Bernoulli beams,* Proceedings of the First World Conference on Structural Control, Pasadena (California), 1994, 120–129.

[7] F. Bourquin and J-S. Briffaut and J. Urquiza, *Contrôle et stabilisation des structures: aspects numériques,* Actes de l'Ecole CEA EDF INRIA sur les matériaux intelligents, 1997.

[8] J.A. Burns and K. Ito and G. Propst, *On nonconvergence of adjoint semigroups for control systems with delays,* SIAM J. Cont. Optim., **26** (1988), 1442–1454.

[9] N. Burq, Thèse de l'Université Paris Sud, Orsay, (1994).

[10] A. Eljendy, *Numerical approach to the exact controllability of hyperbolic systems,* Lecture Notes in Control and Information Science, vol. 178, J.-P. Zolesio ed., (1991).

[11] R. Fabiano, *Preserving exponential stability under approximation for distributed parameter systems,* Computation and Control IV, Proc. fourth Bozeman Conf., Bozeman, MT, USA, 1994, K.L. Bowers et al. ed., (1995), 143–154.

[12] I. Figueiredo and E. Zuazua, *Exact controllability and asymptotic limit for thin plates,* Asympt. Analysis, **12** (1996), 213–252.

[13] G. Geymonat and P. Loreti and V. Valente, *Contrôlabilité exacte d'un modèle de coque mince,* C.R. Acad. Sci., série 1, **313** (1991), 81–86.

[14] G. Geymonat and P. Loreti and V. Valente, *Exact controllability of a shallow shell model,* Int. Series of Num. Math., **107**, Basel, Birkhäuser, (1992).

[15] R. Glowinski and C.H. Li, *On the numerical implementation of the Hilbert Uniqueness Method for the exact boundary controllability of the wave equation,* C.R. Acad. Sci., série 1, **311** (1990), 135–142.

[16] R. Glowinski and C.H. Li and J.-L. Lions, *A numerical approach to the exact boundary controllability of the wave equation (I) Dirichlet control: description of the numerical methods,* Japan J. of Applied Math., **7** (1990), 1–76.

[17] R. Glowinski and J.-L. Lions, *Exact and approximate controllability for distributed-parameter systems I,* Acta Numerica, **3** (1994), 269–378.

[18] R. Glowinski and J.-L. Lions, *Exact and approximate controllability for distributed-parameter systems II,* Acta Numerica, **5** (1996), 159–333.

[19] J.-A. Infante and E. Zuazua, *Boundary observability for the space-discretizations of the 1-d wave equation,* C. R. Acad. Sci., série 1, **326** (1998), 713–718.

[20] S. Jaffard, *Contrôle interne exact des vibrations d'une plaque rectangulaire,* Port. Math., **47** (1990), 423–429.

[21] J. Lagnese, *Stabilization of Thin Plates,* SIAM Studies in Appl. Math., Philadelphia, 1989.

[22] J. Lagnese and G. Leugering and E.J.P.G. Schmidt, *Modeling, Analysis and Control of Dynamic Elastic Multi-Link Structures,* 1994, Birkhäuser, Boston.

[23] J. Lagnese and J.-L. Lions, *Modelling, Analysis and Control of Thin Plates,* 1992, Masson.

[24] I. Lasiecka and R. Triggiani, *Differential and Algebraic Riccati Equations with Applications to Boundary/Pint Control Problems: Continuous Theory and Approximation Theory,* Lecture Notes in Control and Information Sciences, **164**.1991, Springer Verlag.

[25] I. Lasiecka and R. Triggiani, *Sharp trace estimates for solutions to Kirchhoff and Euler-Bernoulli equations,* ppl. Math. Optim., **28** (1993), 277–306.

[26] G. Lebeau, *Contrôle de l'équation de Schrödinger,* J. Math. Pures Appl., **71** (1992), 267–291.

[27] J.-L. Lions, *Contrôle Optimal des Systèmes Gouvernés par des Equations aux Dérivées Partielles,* 1968, Dunod, Paris.

[28] J.-L. Lions, *Exact controllability, stabilization and perturbations for distributed systems,* The John Von Neumann Lecture, SIAM national meeting (1986).

[29] J.-L. Lions, *Contrôlabilité Exacte, Perturbations et Stabilisation de Systèmes Distribués, vol1* 1988, RMA, Masson, Paris.

[30] J.-L. Lions, *Exact controllability, stabilization and perturbations for distributed systems,* SIAM Review, **30** (1988), 1–68.

[31] R. Luce, Thèse de l'Université de Compiegne, France, (1994).

[32] B. Miara, *Exact controllability of shells,* in preparation.

[33] A. Osses and J.-P. Puel, *Some extensions of approximate controllability results to inverse problems,* ESAIM Proc., **2** (1997), 133–143.

[34] G.H. Peichl and C. Wang, *Asymptotic analysis of stabilizability of a control system for a discretized boundary damped wave equation,* Numer. Funct. Anal. and Optim., **19** (1998), 91–113.

[35] M.M. Potapov, *Strong convergence of difference approximations for problems of boundary control and observation for the wave equation,* Comp. Mathematics and Math. Physics, **38** (1998), 373–383.

[36] J.-P. Puel and M. Tucsnak, *Boundary stabilization for the Von Karman equations,* SIAM J. Cont. Optim., **33** (1995), 255–273.

[37] D. Russell, *Controllability and stabilizability theory for linear partial differential equations: recent progress and open questions,* SIAM Review, **20** (1978), 639–740.

[38] V. Valente, *Relaxed exact spectral controllability of membrane shells,* J. Math. Pures et Appl., **76** (1997), 551–562.

Laboratoire des Matériaux et Structures du Génie Civil
UMR113 LCPC/CNRS
2 allée Kepler, 77420 Champs sur Marne
France
E-mail address: bourquin@lcpc.fr, urquiza@lcpc.fr, namar@lcpc.fr

International Series of Numerical Mathematics
Vol. 133,

Stability of Holomorphic Semigroup Systems under Nonlinear Boundary Perturbations

Francesca Bucci [1]

Abstract. In this paper we re-study, by a different approach, the absolute stability of a class of holomorphic semigroup systems with nonlinear boundary feedback. A frequency criterion for equi-asymptotic stability in the large of the equilibrium of the corresponding closed loop has been recently derived in [3], by means of the Lyapunov function method. That stability result requires that the 'infinite-sector' nonlinearities satisfy a suitable growth condition. In this paper we show that the previous restriction is unnecessary and that furthermore the result still holds true under a weaker frequency domain condition.

1. Introduction

In this paper we return to the stability problem, described below, which was the object of the recent study [3]. In that work an extension to the boundary control setting of a frequency criterion for absolute stability of the closed loop of a semigroup system with nonlinear feedback law has been provided. The problem is described as follows: let X be a Hilbert space and consider the abstract system

$$x' = Ax + bu, \tag{1}$$

coupled with control

$$u = \varphi(\sigma), \quad \sigma' = \langle c, x\rangle - \rho u, \tag{2}$$

(the precise assumptions are specified below). This model was examined in [10, 11], when $b \in X$, which refers to the case of distributed systems with *distributed* control. With the view of modelling *boundary* control systems, here b is in the dual space of the domain of A^*, $(D(A^*))'$. We shall use the notation $b = -Ad$, so that the the state equation reads as

$$x' = A(x - du). \tag{3}$$

[1]This research was partially supported by the Italian Ministero dell'Università e della Ricerca Scientifica e Tecnologica within the program of G.N.A.F.A.–C.N.R. Part of this research was performed while the author was visiting the Centre Émile Borel of the Institut Henri Poincaré in Paris (IHP: UMS 839 CNRS/UPMC).

The dynamics is subject to the following assumptions, to be maintained throughout the paper:

$$(H1)\begin{cases} (i) & A : D(A) \subset X \to X \textit{ is the generator of a strongly continuous} \\ & \textit{semigroup } e^{tA} \textit{ on } X,\ t \geq 0, \textit{ which is analytic on } X \textit{ for } t > 0, \\ & \textit{and which, moreover, is exponentially stable;} \\ (ii) & d \in D((-A)^{\alpha}) \textit{ for some } \alpha \in (0,1). \end{cases}$$

Under condition $(H1)$ the abstract model (3) covers parabolic (parabolic-like) partial differential equations on a bounded domain, with scalar control acted through the boundary (see, e.g., [1], [6]).
With reference to (2), it is assumed that

$(H2)\qquad c \in X, \quad \rho \geq 0.$

In view of the applications, φ is a sector-bound nonlinearity: more precisely, is subject to the condition

$$r \cdot \varphi(r) > 0 \qquad \text{for } r \neq 0, \tag{4}$$

which is usually referred to as φ *belongs to the sector* $]0, +\infty[$ (see, e.g., [9]).

As it is well known, the absolute stability problem can be stated as follows: *to find conditions under which the trivial solution of the closed loop of (3) with (2) is globally asymptotically stable, for* **every** φ *which belongs to the sector* $]0, +\infty[$.

The absolute stability problem of feedback systems is an old problem, with an extensive literature. We only recall the papers [8], [12] and [5], as crucial contributions in a finite dimensional context. Extensions to distributed systems are given in [10, 11]. For a description of different methods to deal with feedback systems in Hilbert spaces and a larger list of references, see [2].

The frequency criterion provided in [11] for the closed loop of (1) and (2) has been extended to holomorphic semigroup systems with unbounded input operator in [3]. Both papers are based on the Lyapunov direct method, and the Lyapunov function for the closed loop is constructed by using a solution to a suitable operator Riccati Equation/Dissipation Inequality arising in quadratic regulator problems with indefinite cost functional. Those stability results require, besides an appropriate frequency condition (see (38) in Section 4), that the nonlinearities satisfy

$$\lim_{|\sigma| \to +\infty} \int_0^{\sigma} \varphi(s)\, \mathrm{d}s = +\infty. \tag{5}$$

In contrast, the present paper is based on former Popov's integral equation approach, combined with regularity results for the solutions to the closed loop. This way, we are able to show that the restriction to the subclass of nonlinearities satisfying (5) is unnecessary, and that in fact absolute stability still holds true under weaker conditions, i.e. assumptions (H3) in Section 3. In conclusion, a new frequency criterion for stability of holomorphic semigroup systems under nonlinear boundary perturbations is provided, which improves the criterion formulated in

[3, Theorem 4.1], and that generalizes [10, Theorem 1] to the boundary control case, see Remark 3.8.

The paper is organized as follows. In Section 2 we study the regularity of the solutions to the closed loop of (3) with (2), and provide some estimates to be used in the proof of the stability results in Section 3. In Section 3, preliminary results, detailing the steps of global attractivity and of global stability of the equilibrium of the closed loop, eventually lead to the main result of the paper, Theorem 3.6, with its Corollary 3.7. A comparison between the assumptions of the present paper and those of [3] is made in Section 4.

2. Regularity of the solutions

Consider the closed loop of the state equation (3) with the feedback (2), namely

$$\begin{cases} x' = A(x - d\varphi(\sigma)) \\ \sigma' = \langle c, x \rangle - \rho\varphi(\sigma). \end{cases} \tag{6}$$

In this section we investigate existence, uniqueness and regularity properties of the solutions $(x(t), \sigma(t))$ to the system (6), to be used in the proof of the stability results in Section 3.

Remark 2.1. We recall from [3] the following notation: the class of locally Lipschitz continuous functions $\varphi : \mathbb{R} \to \mathbb{R}$ which satisfy the sector condition (4) will be denoted by $\{\mathbf{SNL}\}$ (sectorial nonlinearities).

2.1. Preliminaries

The following properties will be used throughout the paper:

- since $\varphi : \mathbb{R} \to \mathbb{R}$ satisfies a local Lipschitz condition, there exists a strictly increasing, continuous function $l : \mathbb{R}^+ \to \mathbb{R}^+$, such that
$$r > 0,\ |\sigma_1|, |\sigma_2| \le r \Longrightarrow |\varphi(\sigma_1) - \varphi(\sigma_2)| \le l(r)|\sigma_1 - \sigma_2|; \tag{7}$$
- from analiticity and exponential stability of the semigroup e^{tA} in $(H1)$ it follows that
$$\exists M \ge 1,\ \omega > 0 : ||(-A)^\beta\, e^{At}|| \le M\frac{e^{-\omega t}}{t^\beta}, \quad t > 0,\ \beta \in [0, 1]. \tag{8}$$
As a consequence, with reference to the operator L defined by
$$L : u \to (Lu)(t) := -A\int_0^t e^{(t-s)A} d\, u(s) \mathrm{d}s, \tag{9}$$
it is readily verified that
$$L : \text{ continuous } L^\infty(0, \tau; \mathbb{R}) \to L^\infty(0, \tau; X) \quad \forall \tau > 0,$$
$$||L||_{\mathcal{L}(L^\infty(0,\tau;U), L^\infty(0,\tau;X))} \le C := M \cdot I_\alpha \cdot c_\alpha, \text{ with} \tag{10}$$
$$c_\alpha := |(-A)^\alpha d|, \quad I_\alpha := \int_0^\infty \frac{e^{-\omega s}}{s^{1-\alpha}} \mathrm{d}s. \tag{11}$$

Associated with the differential system (6) is the initial value problem

$$\begin{cases} x' = A(x - d\varphi(\sigma)) & x(0) = x_0 \\ \sigma' = \langle c, x\rangle - \rho\varphi(\sigma) & \sigma(0) = \sigma_0, \end{cases} \tag{12}$$

which leads to the following system of integral equations:

$$x(t) = e^{tA}x_0 - A\int_0^t e^{(t-s)A}d\varphi(\sigma(s))\,\mathrm{d}s, \tag{13}$$

$$\sigma(t) = \sigma_0 + \langle c, e^{tA}A^{-1}x_0 - A^{-1}x_0\rangle + \int_0^t [\langle c, (I - e^{(t-s)A})d\rangle - \rho]\varphi(\sigma(s))\,\mathrm{d}s. \tag{14}$$

It is natural to seek for at least continuous solutions of the system (13)–(14):

Definition 2.2. *A* continuous *solution* (x, σ) *of the system (13)–(14) will be called a* mild *solution to the initial value problem (12).*

2.2. Basic regularity of solutions

We begin with the analysis of the latter equation between (13) and (14). Existence of maximal solutions to the nonlinear integral equation (14) was already stated in [3], making reference to general results based on fixed point arguments. Here we state local existence and uniqueness for the solutions $(x(t), \sigma(t))$ to the system (6), complemented with explicit uniform bounds on $|x|$ and $|\sigma|$ for all solutions (x, σ) with initial data in a ball, which will play a key role in the proof of the stability results in Section 3.

Lemma 2.3. *Assume (H1)–(H2). Let* $\varphi : \mathbb{R} \to \mathbb{R}$ *be a locally Lipschitz continuous function. Then for every* $r > 0$ *there is a* $\tau(r) > 0$ *such that for any initial data* $(x_0, \sigma_0) \in X \times \mathbb{R}$, *with* $|(x_0, \sigma_0)| \le r$, *the integral equation (14) has a unique continuous solution* σ *on* $[0, \tau(r)]$. *More precisely, if we set*

$$\nu_1(r) := 2r(1 + |c|(1 + M)||A||^{-1}), \tag{15}$$

then

$$\tau(r) = \frac{1}{2l(\nu_1(r))(|c|(1 + M)|d| + \rho)}, \tag{16}$$

and $|\sigma(t)| \le \nu_1(r)$ *for all* t *in* $[0, \tau(r)]$.

Proof. It is sufficient to proceed as in [7, Ch. 6, Theorem 1.4]). □

Remark 2.4. Note that $\tau : \mathbb{R}^+ \to \mathbb{R}^+$ is a strictly decreasing, continuous function, with

$$\lim_{r\downarrow 0} \tau(r) = L \in]0, +\infty].$$

It is useful to recall the well-known definition:

Definition 2.5. *A function* $\psi : \mathbb{R}_+ \to \mathbb{R}_+$ *is of class* K *if it is continuous, strictly increasing, and* $\psi(0) = 0$.

Theorem 2.6. *Assume (H1)–(H2). Let $\varphi : \mathbb{R} \to \mathbb{R}$ be a locally Lipschitz continuous function. Then there exist two functions $\mu_1, \nu_1 : \mathbb{R}_+ \to \mathbb{R}_+$ of class K such that, for every $r > 0$, and any $(x_0, \sigma_0) \in X \times \mathbb{R}$, with $|(x_0, \sigma_0)| \le r$, system (12) has a unique mild solution $(x(t), \sigma(t))$ defined on $[0, \tau(r)]$, with*

$$|x(t)| \le \mu_1(r), \quad |\sigma(t)| \le \nu_1(r), \quad t \in [0, \tau(r)], \tag{17}$$

where $\tau(r)$ is given by formula (16).

Proof. For $r > 0$, and an $(x_0, \sigma_0) \in X \times \mathbb{R}$, with $|(x_0, \sigma_0)| \le r$, let $\nu_1(r)$, $\tau(r)$ and $\sigma \in C([0, \tau(r)]; X)$ as from Lemma 2.3: it is obvious that ν_1 in (15) is a function of class K. Moreover, $|\varphi(\sigma(t))| \le l(\nu_1(r))\, \nu_1(r)$ for all $t \in [0, \tau(r)]$, due to the Lipschitz condition (7).

Insert $\varphi \circ \sigma$ in (13) to derive the corresponding $x(t)$, namely

$$x(t) = e^{tA} x_0 + (L(\varphi \circ \sigma))(t), \quad t \in [0, \tau(r)]. \tag{18}$$

It can be easily shown that since $\varphi \circ \sigma$ is continuous on $[0, \tau(r)]$, then the same property holds for x (see [3, Lemma 2.2]). Moreover, by using the estimate (10) in (18), it is readily verified that

$$|x(t)| \le Mr + C\, l(\nu_1(r))\, \nu_1(r) =: \mu_1(r), \quad t \in [0, \tau(r)],$$

and μ_1 is a function of class K. □

Further regularity properties of the solutions $(x(t), \sigma(t))$ to the system (12) were already shown in [3], in the form of Theorem 2.7 and Theorem 2.8 below, exploiting both smoothing properties of the operator L defined in (9), and differentiability almost everywhere of φ.

Theorem 2.7. (cf. [3]) *Assume (H1)–(H2) and let $\varphi : \mathbb{R} \to \mathbb{R}$ be a locally Lipschitz continuous function. Let $(x_0, \sigma_0) \in X \times \mathbb{R}$ and let $\sigma \in C([0, \tau]; \mathbb{R})$ be the solution to (14) as from Lemma 2.3. Then $\sigma \in C^1([0, \tau]; \mathbb{R})$ and*

$$\sigma'(t) = \langle c, x(t) \rangle - \rho\, \varphi(\sigma(t)), \quad 0 \le t \le \tau. \tag{19}$$

Theorem 2.8. (cf. [3]) *Assume hypotheses (H1)–(H2) and let $\varphi : \mathbb{R} \to \mathbb{R}$ be a locally Lipschitz continuous function. Let $(x_0, \sigma_0) \in X \times \mathbb{R}$ and let (x, σ) be the solution to the system (12) defined on $[0, \tau]$. Then*

$$x(t) = \xi(t) + d\varphi(\sigma(t)), \qquad t \in [0, \tau] \tag{20}$$

where ξ is a classical solution in $L^\infty_{\mathrm{loc}}(0, \tau; X)$ of the problem

$$\begin{cases} \xi'(t) = A\xi(t) - d\varphi'(\sigma(t))\sigma'(t) & a.e.\ in\ ]0, \tau[\\ \xi(0) = x_0 - d\varphi(\sigma_0). \end{cases} \tag{21}$$

Moreover x is differentiable almost everywhere on $]0, \tau[$, with $x'(t) = A\xi(t)$ a.e. on $]0, \tau[$.

2.3. Additional properties of the solutions

We return to the integral equation (14), with a view to tackling the problem of absolute stability. Let us note that if we introduce the functions

$$h(t) := \sigma_0 + \langle c, e^{tA} A^{-1} x_0 \rangle - \langle c, A^{-1} x_0 \rangle, \tag{22}$$

$$k(t) := \langle c, d \rangle - \langle c, e^{tA} d \rangle - \rho, \tag{23}$$

then equation (14) can be rewritten more concisely as follows:

$$\sigma(t) = h(t) + \int_0^t k(t-s)\varphi(\sigma(s))\, ds. \tag{24}$$

Extensions of Popov's former frequency criterion, dealing with solutions to integral equations of the form (24), are well known. We recall the following formulation.

Theorem 2.9. (cf. [4, Theorem 3.1) *Consider the integral equation (24) under the following assumptions:*

1. $h', h'' \in L^1(\mathbb{R}^+)$;
2. $k(t) = k_0(t) - \rho_0$, *with* $\rho_0 > 0$ *and* k_0, $k_0' \in L^1(\mathbb{R}^+)$;
3. $\varphi : \mathbb{R} \to \mathbb{R}$ *is a continuous function such that*
$$\sigma\varphi(\sigma) > 0 \text{ for } \sigma \neq 0;$$
4. *there exists* $q \geq 0$ *such that*
$$Re\big((1 - isq)G(s)\big) \leq 0, \quad s \neq 0,$$
where
$$G(s) = \int_0^\infty e^{ist} k_0(t) dt + \rho_0 (is)^{-1}.$$

Then every solution $\sigma(t)$ *of equation (24) is defined for all* $t \geq 0$ *and tends to zero as* $t \to +\infty$.

It is easily seen that Theorem 2.9 is **not directly applicable** to the equation under consideration (24), when h and k are given by (22)–(23). In fact, trying to check the conditions required, it is readily verified that $h(t)$ is twice differentiable for $t > 0$, for every $x_0 \in X$, with $h'(t) = -\langle c, e^{tA} x_0 \rangle \in L^1(\mathbb{R}^+)$, whereas

$$h''(t) = -\langle c, A e^{tA} x_0 \rangle = O(1/t), \quad t \to 0^+.$$

Therefore the first hypothesis of Theorem 2.9 is **not** satisfied.

Remark 2.10. This fact was pointed out in a similar situation in [10], where a class of *differentiable* semigroup systems with *distributed* control (i.e. with $b \in X$) is examined. In that paper this problem was overcome by using the following idea: it was first observed that if $x_0 \in D(A)$, then assumption 1. is satisfied; next, the solution $x(t)$ to the equation (1) belongs to $D(A)$, when $t \in]0, \tau]$ (local existence interval) due to differentiability of the semigroup, so that $x(\tau) \in D(A)$. Thus, Theorem 2.9 is applied to an integral equation of the same form (24), where initial data x_0 and σ_0 in h are replaced by $\hat{x}_0 := x(\tau)$ and $\hat{\sigma}_0 := \sigma(\tau)$, respectively.

The same argument does not apply to the present situation, since here $x(t)$ is not in $D(A)$ any more, even if $t > 0$, due to lower regularity properties of the solutions to equation (1), when $b \in D(A^*)'$. However, an intermediate regularity result can be shown, which will play a crucial role in the sequel.

Theorem 2.11. *Assume hypotheses (H1)–(H2) and let $\varphi : \mathbb{R} \to \mathbb{R}$ be a locally Lipschitz continuous function. There exists a function $\mu_2 : \mathbb{R}_+ \to \mathbb{R}_+$ of class K such that, for every $r > 0$, and any $(x_0, \sigma_0) \in X \times \mathbb{R}$, with $|(x_0, \sigma_0)| \le r$, the first component $x(t)$ of the solution $(x(t), \sigma(t))$ to the system (13)–(14), defined on $[0, \tau(r)]$ (as from Theorem 2.6), satisfies*

$$x(\tau(r)) \in D((-A)^\alpha), \qquad |(-A)^\alpha x(\tau(r))| \le \mu_2(r). \tag{25}$$

Proof. For $r > 0$, and an $(x_0, \sigma_0) \in X \times \mathbb{R}$, with $|(x_0, \sigma_0)| \le r$, let $\tau = \tau(r)$, $(x, \sigma) : [0, \tau] \to X \times \mathbb{R}$ as from Theorem 2.6. That $x(\tau(r)) \in D((-A)^\alpha)$ follows immediately from formula (20) of Theorem 2.8, which shows that x can be viewed as $x(t) = \xi(t) + d\varphi(\sigma(t))$, with $\xi \in C^1(]0, \tau]; X) \cap C(]0, \tau]; D(A))$.

Let us write down the explicit expression of ξ, namely

$$\xi(t) = e^{tA}(x_0 - d\varphi(\sigma_0)) - \int_0^t e^{(t-s)A} d\varphi'(\sigma(s))\sigma'(s)\, \mathrm{d}s, \tag{26}$$

and let us note preliminarly that since σ satisfies the equation (19), then by using the estimates (17) we obtain

$$|\sigma'(t)| \le |c|\, \mu_1(r) + \rho\, l(\nu_1(r))\, \nu_1(r), \quad t \in [0, \tau(r)]. \tag{27}$$

Thus, from (26) it follows

$$\begin{aligned} |(-A)^\alpha \xi(t)| \;\; &\le \;\; ||(-A)^\alpha e^{tA}|||x_0| + ||e^{tA}|||(-A)^\alpha d||\varphi(\sigma_0)| \\ &\quad + \int_0^t ||e^{(t-s)A}||\, |(-A)^\alpha\, d|\, |\varphi'(\sigma(s))||\sigma'(s)|\mathrm{d}s, \quad t \in [0, \tau(r)], \end{aligned}$$

so that by using once more local Lipschitz continuity of φ, the estimate (8) and the bound (27) for $|\sigma'(\cdot)|$, one easily gets

$$|(-A)^\alpha \xi(\tau(r))| \le \psi(r),$$

where we have set

$$\psi(r) := M\frac{e^{-\omega\tau(r)}}{[\tau(r)]^\alpha} r + M\, c_\alpha l(r) r + \frac{M}{\omega} c_\alpha\, l(\nu_1(r))\big(|c|\mu_1(r) + \rho\, l(\nu_1(r))\nu_1(r)\big),$$

with c_α defined by (11). Note that the first summand in the right-hand side is a continuous, strictly increasing function of r, which tends to zero as $r \to 0^+$, see Remark 2.4. Since the same holds true for the other summands, ψ is a function of class K.

This finally leads to

$$\begin{aligned} |(-A)^\alpha x(\tau(r))| &\le |(-A)^\alpha \xi(\tau(r))| + |(-A)^\alpha d||\varphi(\sigma(\tau(r)))| \\ &\le \psi(r) + c_\alpha\, l(\nu_1(r))\nu_1(r) =: \mu_2(r), \end{aligned}$$

which completes the proof. □

3. The frequency domain criterion

In this section we come to the core of the present paper, that is to show that global asymptotic stability of the zero solution to the system (6) still holds true, under weaker conditions than the ones assumed in [3]. Our method owes a great deal to the adaptation of Popov's integral equation approach derived in [10].

We consider separately the issues of global attractivity and of global stability. Let us begin with the analysis of the asymptotic behaviour of the solutions to (14), when $x_0 \in D((-A)^\alpha)$.

Lemma 3.1. *Assume hypotheses* $(H1)$–$(H2)$. *If, in addition,*

$$(H3)\ \begin{cases} (i) & \rho > \langle c, d\rangle, \\ (ii) & \exists q \geq 0: \ Re[(1-isq)\langle c, (is-A)^{-1}d\rangle] + q(\rho - \langle c, d\rangle) \geq 0, \ s \neq 0, \end{cases}$$

then for any given $\varphi \in \{SNL\}$ *and any* $(x_0, \sigma_0) \in D((-A)^\alpha) \times \mathbb{R}$, *the solution* $\sigma(t)$ *to the integral equation (14) is defined for all* $t \geq 0$ *and tends to zero, as* $t \to +\infty$.

Proof. Let $\varphi \in \{SNL\}$ and $(x_0, \sigma_0) \in D((-A)^\alpha) \times \mathbb{R}$ be given. Local existence and uniqueness of the solution to equation (14) is guaranteed by Theorem 2.3. With the view of applying Theorem 2.9 to (14), let us check all the assumptions required therein. It was already observed in Section 2 that h is twice differentiable for $t > 0$, with $h''(t) = -\langle c, A\, e^{tA} x_0\rangle$. Since now $x_0 \in D((-A)^\alpha)$, then

$$h''(t) = -\langle c, (-A)^{1-\alpha} e^{tA} (-A)^\alpha x_0\rangle = O(1/t^{1-\alpha}), \quad t \to 0^+.$$

This, combined with analiticity and exponential stability of the semigroup via (8), forces $h'' \in L^1(\mathbb{R}^+)$, hence condition 1. of Theorem 2.9 is satisfied.

Next, the kernel k is of the form $k(t) = k_0(t) - \rho_0$, with

$$k_0(t) = -\langle c, e^{tA} d\rangle, \quad \rho_0 = \langle c, d\rangle - \rho.$$

It is readily verified that $k_0 \in L^1(\mathbb{R}^+)$, and that there exists $k_0'(t) = -\langle c, Ae^{tA}d\rangle$ for $t > 0$. Moreover, since $d \in D((-A)^\alpha)$, with the same arguments as above we get $k_0' \in L^1(\mathbb{R}^+)$, so that hypothesis 2. is satisfied, provided that

$$\rho_0 = \rho - \langle c, d\rangle > 0.$$

This shows how assumption $(H3)(i)$ arises.

Therefore the local solution $\sigma(t)$ to the equation (24) is in fact defined for all $t \geq 0$ and moreover tends to zero, as $t \to +\infty$, if the frequency domain condition 4. of Theorem 2.9 holds. In order to rewrite that condition in terms of the actual coefficients of the problem, let us compute the Fourier transform

$$\int_0^\infty e^{ist} k_0(t)dt = -\int_0^\infty e^{ist}\langle c, e^{tA}d\rangle dt = -\langle c, (is-A)^{-1}d\rangle,$$

which yields

$$Re[(1-isq)G(s)] = -Re[(1-isq)\langle c,(is-A)^{-1}d\rangle] + Re\big((1-isq)\frac{\rho-\langle c,d\rangle}{is}\big)$$
$$= -Re[(1-isq)\langle c,(is-A)^{-1}d\rangle] - q(\rho-\langle c,d\rangle).$$

Consequently, condition 4. of Theorem 2.9 is satisfied if and only if $(H3)(ii)$ holds. The proof is complete. □

Theorem 3.2. (Global attractivity) *Under the assumptions* $(H1)$, $(H2)$ *and* $(H3)$, *any solution* $(x(t),\sigma(t))$ *to the system (6) is defined for all* $t \geq 0$ *and tends to zero, as* $t \to +\infty$, *for every* $\varphi \in \{SNL\}$.

Proof. In order to extend the validity of Lemma 3.1 to all solutions to the equation (14), when $x_0 \in X$, we use a similar argument as in [10]. Fix $\varphi \in \{SNL\}$ and $(x_0,\sigma_0) \in X \times \mathbb{R}$, and let $(x(t),\sigma(t))$ be the local solution to the system (13)–(14), defined on a suitable interval $[0,\tau]$. From Theorem 2.11 it follows that $x(\tau) \in D((-A)^\alpha)$. Set $\hat{x}_0 := x(\tau)$, $\hat{\sigma}_0 := \sigma(\tau)$, and let us consider then the solution $(\hat{x}(t),\hat{\sigma}(t))$ to the system (13)–(14), with initial data $(\hat{x}_0,\hat{\sigma}_0)$. Thus $\hat{\sigma}$ satisfies an integral equation of the form (24), with h, k given by (22), (23), respectively, but with (x_0,σ_0) replaced by $(\hat{x}_0,\hat{\sigma}_0)$. Since by construction $\hat{x}_0$ is in $D((-A)^\alpha)$, then Lemma 3.1 applies, yielding global existence of $\hat{\sigma}$, and $\hat{\sigma}(t) \to 0$, as $t \to +\infty$.

It is now sufficient to observe that since the system is autonomous $\sigma(t) = \hat{\sigma}(t-\tau)$ for $t \geq \tau$, which implies that $\sigma(t)$ is defined for all $t \geq 0$ and that $\sigma(t) \to 0$, as well, when $t \to +\infty$.

To conclude the proof, we would only need to show that $x(t)$ is defined for all $t \geq 0$ and tends to zero, as $t \to +\infty$. This readily follows by applying the dominated convergence theorem in the integral contained in formula (13). □

Let us turn now to the issue of global stability. To this end, we recall the following definition of stability which involves *comparison functions*.

Definition 3.3. *The zero solution to (6) is* globally stable *if*

- *for every* $(x_0,\sigma_0) \in X \times \mathbb{R}$ *there exists a unique solution* $(x(t),\sigma(t))$ *to (12) defined for all* $t \geq 0$;
- *there exists a function* $\pi : \mathbb{R}_+ \to \mathbb{R}_+$ *of class* K *such that, with reference to the solution* (x,σ) *to the system (12), we have:*

$$\text{if } r > 0,\ |(x_0,\sigma_0)| \leq r, \qquad \text{then} \quad |(x(t),\sigma(t))| \leq \pi(r), \quad \forall t \geq 0.$$

We return to the integral equation (24), with h, k defined by (22)–(23). Let us introduce the integral function

$$F(t) := \int_0^t \varphi(\sigma(s))\,\mathrm{d}s, \qquad t \geq 0. \tag{28}$$

and use integration by parts to rewrite the equation (24) in the following way:

$$\begin{aligned}\sigma(t) &= h(t) - \rho F(t) - \langle c, A\int_0^t e^{(t-s)A} d\,F(s)\mathrm{d}s\rangle \\ &= h(t) - \rho F(t) + \langle c, (L\,F)(t)\rangle. \end{aligned} \tag{29}$$

It is clear then in order to get a comparison function for σ it will be sufficient to produce bounds for $|h(t)|$ and $|F(t)|$ by means of suitable functions of class K.

Theorem 3.4. (Global stability) *Assume (H1), (H2) and (H3). Then the zero solution to the system (6) is globally stable, for every* $\varphi \in \{SNL\}$.

Proof. Let an arbitrary $\varphi \in \{SNL\}$ be given: our goal is to produce a comparison function π as in Definition 3.3.

For $r > 0$, let $(x_0, \sigma_0) \in X \times \mathbb{R}$ such that $|(x_0, \sigma_0)| \le r$, and let $(x(t), \sigma(t))$ be the solution to the system (13)–(14), whose existence on $[0, \tau(r)]$ was provided in Theorem 2.6, and which is in fact defined for all $t \ge 0$, due to Theorem 3.2. Set $\hat{x}_0 := x(\tau(r))$, $\hat{\sigma}_0 := \sigma(\tau(r))$, and consider then the solution $(\hat{x}(t), \hat{\sigma}(t))$ to the system (13)–(14), corresponding to initial data $(\hat{x}_0, \hat{\sigma}_0)$. As a consequence of Theorems 2.6 and 2.11, there exist three functions of class K, say μ_1, ν_1, and μ_2, such that

$$|\hat{x}_0| \le \mu_1(r),\ |\hat{\sigma}_0| \le \nu_1(r); \qquad \hat{x}_0 \in D((-A)^\alpha)),\ |(-A)^\alpha \hat{x}_0| \le \mu_2(r). \tag{30}$$

Then, regarding

$$\hat{h}(t) = \hat{\sigma}_0 + \langle c, e^{tA}A^{-1}\hat{x}_0\rangle - \langle c, e^{tA}\hat{x}_0\rangle,$$

we have

$$\begin{aligned}&|\hat{h}(0)| = |\hat{\sigma}_0| \le \nu_1(r), \\ &|\hat{h}(t)| \le \nu_1(r) + |c|\,||A^{-1}||(M+1)\mu_1(r) =: \nu_2(r), \quad t > 0. \end{aligned} \tag{31}$$

As for $F(t)$ defined in (28), it turns out from the proof of Theorem 3.1 in [4] (with an entire change of notation) that

$$|F(t)| \le g(\Phi(\hat{h}(0)), \quad \forall t \ge 0, \quad \text{with} \tag{32}$$

$$\Phi(\sigma) := \int_0^\sigma \varphi(s)\mathrm{d}s, \quad \text{and} \tag{33}$$

$$g(u) := \frac{\lambda}{\rho_0} + [(\lambda/\rho_0)^2 + 2qu/\rho_0]^{1/2}, \quad u \ge 0, \quad \text{where} \tag{34}$$

$$\lambda := ||\hat{h}||_{L^\infty} + q||\hat{h}'||_{L^\infty} + ||\hat{h}'||_{L^1} + q||\hat{h}''||_{L^1}. \tag{35}$$

Note that g is a monotone increasing function for $u \ge 0$.

First of all, using again the fact that φ is sectorial, we obtain

$$\begin{aligned} 0 \le \Phi(\hat{h}(0)) = \int_0^{\hat{\sigma}_0} \varphi(s)\,\mathrm{d}s \quad &\le \quad \nu_1(r)\sup\{|\varphi(s)| : \ |s| \le |\hat{\sigma}_0|\} \\ &\le \quad [\nu_1(r)]^2 l(\nu_1(r)) := m(r), \end{aligned}$$

with m of class K. Hence by using monotonicity of g in (32) we obtain $|F(t| \leq g(m(r))$ for $t \geq 0$, which explicitly reads as

$$|F(t)| \leq \frac{\lambda}{\rho_0} + [(\lambda/\rho_0)^2 + 2qm(r)/\rho_0]^{1/2}, \qquad t \geq 0. \tag{36}$$

Let as assume fot he moment that

Claim 3.5. *With reference to λ defined in (35), there exists a function ν_3 of class K such that $\lambda \leq \nu_3(r)$.*

We then have via (36):

$$|F(t)| \leq \frac{\nu_3(r)}{\rho_0} + [(\nu_3(r)/\rho_0)^2 + 2qm(r)/\rho_0]^{1/2} =: \nu_4(r). \tag{37}$$

Thus, let us go back to formula (29) and combine (31), (37) and (10) to get

$$|\hat{\sigma}(t)| \leq \nu_2(r) + \rho\nu_4(r) + |c|\, C\, \nu_4(r) =: \nu_5(r), \quad t \geq 0,$$

which in turn implies

$$|\hat{x}(t)| \leq M\big(r + C\, l(\nu_5(r))\nu_5(r)\big) =: \mu_3(r), \quad t \geq 0.$$

Notice that the second hand sides are both functions of class K.

We return to the original solution $(x(t), \sigma(t))$ of the system (13)–(14) and observe – as in the proof of Theorem 3.2 – that by construction

$$(x(t), \sigma(t)) = (\hat{x}(t - \tau(r)), \hat{\sigma}(t - \tau(r))) \quad \text{ for } t \geq \tau(r).$$

Therefore we obtain $|x(t)| \leq \mu_4(r) =: \max\{\mu_1(r), \mu_3(r)\}$ and $|\sigma(t)| \leq \nu_5(r)$ for all $t \geq 0$. Consequently, we have

$$|(x(t), \sigma(t)| \leq \sqrt{(\mu_4(r))^2 + (\nu_5(r))^2} =: \pi(r),$$

as desired, and the zero solution to the closed loop (6) is stable in the large. Since φ is arbitrary, this extends to all of $\varphi \in \{SNL\}$.

To conclude the proof, it remains to be shown that Claim 3.5 is correct. We need to provide bounds for all terms in the formula (35). We have seen in (31) that $|\hat{h}(t)| \leq \nu_2(r)$ for $t \geq 0$, which means $||\hat{h}||_{L^\infty} \leq \nu_2(r)$. As for $h'(t) = -\langle c, e^{tA}\hat{x}_0\rangle$, it is readily verified that

$$||\hat{h}'||_{L^\infty} \leq M|c|\mu_1(r), \quad ||\hat{h}'||_{L^1} \leq |c|\, M \int_0^\infty e^{-\omega t}\mathrm{d}t\, |\hat{x}_0| \leq |c| M\omega^{-1}\mu_1(r).$$

As for $\hat{h}''(t)$, after using the key estimate (30), we obtain

$$||\hat{h}''||_{L^1} \leq |c|\, M\, I_\alpha\, \mu_2(r).$$

Conclusion follows by setting

$$\nu_3(r) := \nu_2(r) + M\,|c|\big((q + \omega^{-1})\mu_1(r) + qI_\alpha\, \mu_2(r)\big).$$

□

We can finally state the following absolute stability criterion:

Theorem 3.6. *Assume* $(H1)$, $(H2)$ *and* $(H3)$. *Then system (6) is globally asymptotically stable for every* $\varphi \in \{SNL\}$.

As a result, we obtain the following Corollary (cf. [3, Theorem 4.1, Step 2]).

Corollary 3.7. *Assume* $(H1)$, $(H2)$ *and* $(H3)$. *Then the zero solution to system (6) is equi-asymptotically stable in the large for every* $\varphi \in \{SNL\}$.

Remark 3.8. Notice that Corollary 3.7 generalizes Theorem 1 in [10] to the boundary control setting, since its statement – in the special case $\rho = 0$ – readily reduces to that one when referred to $b \in X$ in (1), (i.e. to the case of holomorphic semigroup systems with *distributed* control).

4. A comparison between frequency criteria

Let us recall from [3] that the subclass of $\varphi \in \{SNL\}$ which satisfy the growth condition (5) was denoted by {**SSNL**} (special sectorial nonlinearities). The frequency criterion below was derived in [3].

Theorem 4.1. **([3, Theorem 4.1])** *Assume* $(H1)$–$(H2)$. *If, in addition,*

$$\exists \delta > 0: \quad \rho + Re\langle c, A\,(is - A)^{-1}\, d\rangle \geq \delta \quad \forall s \in \mathbb{R} \tag{38}$$

then the zero solution to system (6) is equi-asymptotically stable in the large for every $\varphi \in \{SSNL\}$.

With Theorem 3.6 we have provided a new criterion for absolute stability of the closed loop (6), when φ varies merely in $\{SNL\}$. This shows, in the first place, that the restriction to the subclass $\{SSNL\}$ has been removed. Secondly, notice that now $\rho \geq 0$ is allowed, unlike under (38), which implies the technical condition $\rho > 0$. To see this, let $|s| \to +\infty$ in (38), and use $(H1)$ to get $\rho \geq \delta$.

In the following proposition it is shown that, in addition, the assumptions $(H3)$ are implied by (38).

Proposition 4.2. *Assume* $(H1)$–$(H2)$. *If the frequency condition (38) holds, then* $\rho > 0$ *and conditions* $(H3)(i)$-(ii) *are satisfied.*

Proof. Assume that (38) holds. When $s = 0$, this condition reads as $\rho - \langle c, d\rangle > 0$, which is nothing but $(H3)(i)$. Let now $s \neq 0$. Multiplyig by an arbitrary $q \geq 0$ in (38), we readily get

$$q\,\rho + q\,\mathrm{Re}\langle c, A\,(is - A)^{-1}\, d\rangle \geq q\,\delta, \quad s \neq 0,$$

that is, by using the well-known property of the resolvent operator $A\,(is - A)^{-1} = is\,(is - A)^{-1} - I$,

$$q\,\rho + q\,\mathrm{Re}\langle c, is(is - A)^{-1}\, d\rangle - q\langle c, d\rangle \geq q\,\delta, \quad s \neq 0.$$

This in turn is rewritten as

$$q\,(\rho - \langle c, d\rangle) - q\,\mathrm{Re}\, is\langle c, (is - A)^{-1}\, d\rangle \geq q\,\delta, \quad s \neq 0.$$

Thus, adding and subtracting the term $\mathrm{Re}\,\langle c,(is-A)^{-1}\,d\rangle$, we obtain

$$q\,(\rho-\langle c,d\rangle)+\mathrm{Re}(1-isq)\langle c,(is-A)^{-1}\,d\rangle-\mathrm{Re}\langle c,(is-A)^{-1}\,d\rangle\geq q\,\delta,\quad s\neq 0. \tag{39}$$

Observe now that if

$$\exists q\geq 0:\quad q\delta+\mathrm{Re}\langle c,(is-A)^{-1}\,d\rangle\geq 0,\quad s\neq 0, \tag{40}$$

then from (39) it follows

$$\exists q\geq 0:\quad q\,(\rho-\langle c,d\rangle)+\mathrm{Re}(1-isq)\langle c,(is-A)^{-1}\,d\rangle\geq 0,\quad s\neq 0,$$

i.e. $(H3)(ii)$ holds.

It remains to be shown that (40) holds true. This is readily verified, since $|(is-A)^{-1}\,d|$ is uniformly bounded due to (H1), which implies that there exists a constant C such that

$$\mathrm{Re}\langle c,(is-A)^{-1}\,d\rangle\geq C\quad\forall s\neq 0.$$

Finally, $q\geq 0$ can be chosen in such a way that $-q\delta\leq C$, and (40) holds true, as desired. □

References

[1] A. Bensoussan, G. Da Prato, M. C. Delfour and S. K. Mitter, *Representation and Control of Infinite Dimensional Systems*, Vol. II, Birkhäuser, Boston, **1993**.

[2] F. Bucci, *Absolute stability of feedback systems in Hilbert spaces*, in : Optimal Control: Theory, Algorithms, and Applications, 24–39, W. W. Hager and P. M. Pardalos, Editors, Kluwer Academic Publishers B.V., **1998**.

[3] F. Bucci, *Frequency domain stability of feedback systems with unbounded input operator*, Dynamics of Continuous, Discrete and Impulsive Systems, to appear.

[4] C. Corduneanu, *Integral Equations and Stability of Feedback Systems*, Academic Press, New York, London, **1973**.

[5] R. E. Kalman, *Lyapunov functions for the problem of Lur'e in automatic control*, Proc. Nat. Acad. Sci. USA, **49** (1963), 201–205.

[6] I. Lasiecka and R. Triggiani, *Differential and Algebraic Riccati Equations with Application to Boundary/Point Control Problems: Continuous Theory and Approximation Theory*, Lecture Notes in Control and Information Sci. 164, Springer Verlag, Berlin, **1991**.

[7] A. Pazy, *Semigroups of Linear Operators and Applications to Partial Differential Equations*, Springer Verlag, Berlin, **1983**.

[8] V. M. Popov, *On absolute stability of non-linear automatic control systems*, Automat. i Telemekh., **22** (1961), 961–979 (in Russian). English transl. in Automat. Remote Control, **22** (1962), 857–875.

[9] M. Vidyasagar, *Nonlinear Systems Analysis*, Prentice-Hall Int., **1993**.

[10] D. Wexler, *Frequency domain stability for a class of equations arising in reactor dynamics*, SIAM J. Math. Anal., **10** (1979), 118–138.

[11] D. Wexler, *On frequency domain stability for evolution equations in Hilbert spaces via the algebraic Riccati equation*, SIAM J. Math. Anal., **11** (1980), 969–983.

[12] V. A. Yakubovich, *Solution of certain matrix inequalities occurring in the theory of automatic controls*, Dokl. Akad. Nauk USSR, **143** (1962), 1304–1307 (in Russian). English transl. in Soviet Math. Dokl., **4** (1963), 620–623.

Dipartimento di Matematica Applicata,
Università di Firenze,
Via S. Marta, 3
I-50139 Firenze, Italy
E-mail address: fbucci@dma.unifi.it

International Series of Numerical Mathematics
Vol. 133, © 1999 Birkhäuser Verlag Basel/Switzerland

Shape Control in Hyperbolic Problems

John Cagnol and Jean-Paul Zolésio

Abstract. The vibration of a shell constrained to be in a specific configuration and the vibration of a shell with that shape at its natural reference position are known to be different by physicists. We consider a shell which is under a large displacement and a small deformation, that gives a constrained shell Ω in a static equilibrium. We consider a small vibration of Ω and we are interested in the equation of that vibration. We first investigate a new exact model for constrained shell that is $p(d,\infty)$ which is the counterpart of the model developed by Delfour and Zolésio for shells in the reference configuration. Then we study the regularity of the solution in the interior domain as well as the the boundary regularity. Both regularities were proven to be interesting for shape differentiability (and thus shape control) in hyperbolic equations, we will recall the shape differentiability results for the wave equation.

1. Introduction

1.1. The Problem

We consider an elastic thin shell made of an homogeneous, isotropic and elastic material. We note $\Omega^0 \subset \mathbb{R}^3$ that shell in its reference configuration. Let T_0 be a large displacement with small deformation (*i.e.* T_0 is a mapping from $\mathbb{R}^3$ onto $\mathbb{R}^3$, the norm of the associated Green-St Venant strain tensor $\frac{1}{2}({}^*D\mathrm{T}_0 D\mathrm{T}_0 - I)$ is small but no assumption is made of $\|\mathrm{T}_0\|$). When T_0 is applied to Ω^0, the shell goes in a constrained configuration also called preconstrained or prestressed configuration (the stress field does not vanish). We note $\Omega = \mathrm{T}_0(\Omega^0)$, the shell in that position. The mapping T_0 is called the *static displacement*. We assume T_0 is such that the coefficients of the matrix $E = D\mathrm{T}_0 \circ \mathrm{T}_0{}^{-1}$ belong to $W^{1,\infty}(\mathbb{R}^3)$, which is the case when, for instance, $\mathrm{T}_0 \in W^{2,\infty}(\Omega^0, \mathbb{R}^3)$. For a mapping T_0 arising from the aeronautic industry, see [3].

We suppose the prestressed shell Ω is under a vibration. Let τ be the final time and $t < \tau$, we note $\Omega(t)$ the shell at the time t and $T(t)$ the mapping such that $T(t)(\Omega) = \Omega(t)$. We assume T belongs to $L^2([0,\tau], H^1(\Omega,\mathbb{R}^3)) \cap H^1([0,\tau], L^2(\Omega,\mathbb{R}^3))$ and is compatible with a clamped boundary condition on Γ_D. That is $T_{|\Gamma_D} = I$. The mapping T is a perturbation of I, we have $T = I + u$ where u is small. From now, u will be the parameter to be considered. We note $H^1_{\Gamma_D}(\Omega;\mathbb{R}^3)$

the set of functions of $H^1(\Omega, \mathbb{R}^3)$ that vanish on Γ_D. The parameter u belongs to

$$H(\Omega) = L^2([0,\tau], H^1_{\Gamma_D}(\Omega, \mathbb{R}^3)) \cap H^1([0,\tau], L^2(\Omega, \mathbb{R}^3))$$

We note $\mathrm{T} = T \circ \mathrm{T}_0$. Let $\tilde{u} = u \circ \mathrm{T}_0$, we have $\mathrm{T}(t) = \mathrm{T}_0 + \tilde{u}(t)$. We note C the 4th-order elastic tensor and

$$\varepsilon(u) = \frac{1}{2}({}^*Du + Du) \quad , \quad \Sigma(u) = \frac{2}{|\det(E)|} E(C..({}^*E\varepsilon(u)E))\,{}^*E$$

where E is a matrix depending on the transformation T_0.

Let ϕ and ψ be the value of u and $\partial_t u$ at $t = 0$. We suppose $\phi \in H^1(\Omega, \mathbb{R}^3)$ and $\psi \in L^2(\Omega, \mathbb{R}^3)$. The equation of the vibration around the natural shape of the shell Ω is given by the hyperbolic equation

$$(1) \qquad \begin{cases} \rho\partial_{tt} u - \mathrm{div}\,(\Sigma(u)) = 0 \text{ on } [0;\tau[\times\Omega \\ \Sigma(u).n = 0 \text{ on } \Gamma \;,\; u(0) = \phi \text{ on } \Omega \;,\; \partial_t u(0) = \psi \text{ on } \Omega \end{cases}$$

we will note $P(u) = \rho\partial_{tt} u - \mathrm{div}\,(\Sigma(u))$

Theorem 1.1. *System (1) has a unique solution u that belong to $C([0;\tau]; H^1_{\Gamma_D}(\Omega)) \cap C^1([0;\tau]; L^2(\Omega))$ moreover for all $\nu \in C^1(\Gamma_D)$ with $\nu \geq 0$ on Γ_D and* $\mathrm{supp}\,\nu \in \Gamma_D$ *we have $\nu|\varepsilon(u)| \in L^2(\sigma_D)$.*

1.2. Conventions and First Lemmae Concerning the Tensors

We refer to [10] for the conventions concerning the tensors.

Lemma 1.2. *Let A, B, X and Y be four matrices and C be a third-order tensor then $(XAY)..C..B = A..({}^*X(C..B)\,{}^*Y)$.*

Lemma 1.3. *Let A and Z be two 3rd-order tensors and U and V be two vectors then $(A.V)..(U.Z) = (Z..A)..(U \otimes V)$.*

Let $X = (x_{i,j,k})_{i,j,k}$ be a 3rd-order tensor then $X^{\triangleleft} = (x_{k,i,j})_{i,j,k}$ and $X^{\triangleleft\triangleleft} = (x_{j,k,i})_{i,j,k}$. For a matrix X and a third-order tensor $Y = (Y_{i,j,k})_{i,j,k}$ we note $X \times Y$ (*resp.* $Y \times X$) the third-order tensor $(X \times Y_i)_i$ (*resp.* $(Y_i \times X)_i$) where Y_i is a matrix and $\times$ the multiplication of matrices. For a third-order tensor X let us note $\mathrm{tr}(X)$ the vector whose coefficients are the traces of the matrices X_i. For a matrix X we note $\mathrm{sym}\,(X) = \frac{1}{2}({}^*X + X)$.

2. The Equation

2.1. The Elastic Energy

Let C be the 4th-order elastic tensor. It satisfies $\forall (i,j,k,l) \in \{1,2,3\}^4$, $C_{ijkl} = C_{jikl} = C_{klij}$ and it is constant since the material is homogeneous. Moreover we assume $\exists \alpha > 0$, $\forall \xi$, $C_{ijkl}\xi_{ij}\xi_{kl} > \alpha\xi^2_{i,j}$ this allows the definition of norm $|\cdot|$ such that $|\Xi|^2 = \Xi..C..\Xi$. We consider the Green-St Venant strain tensor

$$\bar{\varepsilon}(u) = \frac{1}{2}({}^*D\mathrm{T}D\mathrm{T} - I)$$

the elements of this matrix belong to $H^1([0,\tau], L^2(\Omega,\mathbb{R}^3)) \cap L^2([0,\tau], H^{-1}(\Omega,\mathbb{R}^3))$. The elastic energy is given by $E_p = \int_{\Omega^0} \bar{\varepsilon}..C..\bar{\varepsilon}$

Lemma 2.1. *At the first order* $2\bar{\varepsilon} = {}^*D\mathrm{T}_0\left((I+2\varepsilon)\circ \mathrm{T}_0\right) D\mathrm{T}_0 - I$

Proposition 2.2. *The elastic energy of* Ω *is given by*

$$E_p = \frac{1}{4}\int_\Omega ({}^*E(I+2\varepsilon(u))E - I)..C..({}^*E(I+2\varepsilon(u))E - I)\frac{1}{|\det(E)|}$$

2.2. The Kinetic Energy

Let ρ be the density and v be the speed vector field of the vibrating body. It is defined on $\Omega(t)$ by $v(t,x) = \partial_t \mathrm{T}\circ \mathrm{T}^{-1}$. Since $DT = I + Du$ the real $\det(DT)$ may be approximated by 1, it follows

Proposition 2.3. *The kinetic energy of* $\Omega(t)$ *is given by*

$$E_k(t) = \frac{1}{2}\rho\int_\Omega (\partial_t u)^2$$

2.3. Equation of the Vibration

Let $\tau < +\infty$ be the final time and let A be defined by $A = \int_0^\tau (E_p - E_k(t))\,dt$. From propositions 2.2 and 2.3 we get

$$A(u) = \frac{1}{4}\int_0^\tau\int_\Omega ({}^*E(I+2\varepsilon(u))E - I)..C..({}^*E(I+2\varepsilon(u))E - I)\frac{1}{|\det(E)|} - 2\rho(\partial_t u)^2$$

Let θ be a real and w a function from $[0,\tau[\times\mathbb{R}^3$ to $\mathbb{R}^3$, we note $A'(u;w) = \frac{\partial}{\partial\theta}A(u+\theta w)|_{\theta=0}$. The physical evolutions of the structures are characterized by the functions u, defined on $[0;\tau[\times\Omega$, such that $A'(u;w) = 0$ for all tests w such that $w(0,\cdot) = 0$ and $w(\tau,\cdot) = 0$. Using the linearity of ϵ and lemma 1.2, we get

$$A'(u;w) = \int_0^\tau\int_\Omega \Sigma(u)..\varepsilon(w) + ({}^*EE - I)..C..({}^*E\varepsilon(w)E)\frac{1}{|\det(E)|} - \rho\partial_t u\partial_t w$$

When $u = 0$ we have $\mathrm{T} = \mathrm{T}_0$ which gives a minimum for the energy, thus we get $A'(0;w) = 0$. Hence

$$\text{(2)}\qquad \text{forall tests } w,\ \int_0^\tau\int_\Omega ({}^*EE - I)..C..({}^*E\varepsilon(w)E)\frac{1}{|\det(E)|}\,dx\,dt = 0$$

Using this identity in the general expression of $A'(u;w) = 0$, we get

$$\text{(3)}\qquad \int_0^\tau\int_\Omega \Sigma(u)..\varepsilon(w) - \rho\partial_t u\partial_t w = 0$$

from the Green's formula we obtain

$$\int_0^\tau\int_\Omega -\mathrm{div}\,(\Sigma(u))w + \rho\partial_{tt}uw + \int_0^\tau\int_\Gamma \Sigma(u)nw - \int_\Omega \partial_t u(0)w(0) + \partial_t u(\tau)w(\tau) = 0$$

that equality holds for all tests w therefore

$$\text{(4)}\qquad \begin{cases} \rho\partial_{tt}u - \mathrm{div}\,(\Sigma(u)) = 0 & \text{on } [0;\tau[\times\Omega \\ \Sigma(u).n = 0 & \text{on } \Gamma \end{cases}$$

2.4. Equation Satisfied by E

The matrix E is $D\mathrm{T}_0 \circ {\mathrm{T}_0}^{-1}$ where T_0 is a local minimum for the elastic energy. The necessary condition (2) leads to the strong formulation

$$\mathrm{div}\left(\frac{1}{|\det E|} E(({}^*EE - I)..C)\,{}^*E\right) = 0$$

that is $\mathrm{div}\,((\frac{1}{\det(D\mathrm{T}_0)} D\mathrm{T}_0(\bar{\varepsilon}(0)..C)\,{}^*D\mathrm{T}_0) \circ {\mathrm{T}_0}^{-1}) = 0$. Moreover T_0 is assumed to be a small deformation hence $\|\bar{\varepsilon}(0)\|_{L^\infty}$ is small as compared to $\|I\|_{L^\infty}$. On the other hand from that last assertion comes the existence of a real $c > -1$ such that

$$\forall v \in \mathbb{R}^3,\ \|{}^*EEv\|_{L^2} \geq (1+c)\|v\|_{L^2} \tag{5}$$

the same inequality holds for $E\,{}^*E$.

3. The Shell Theory

3.1. The Geometry

Let b is the oriented distance function, p is the orthogonal projection from $\mathbb{R}^3$ onto Ω. Those definitions are standard in intrinsic differential calculus, they are recalled in the article of Michel Delfour of those proceedings ([7]) page 98 and detailed in [8]. We will note b^0 and p^0 the corresponding function for Ω^0.

Assumption 3.1 (Shell Form Assumption). *There exists $O \subset \mathbb{R}^3$ such that ∂O is a manifold and $\omega \subset \partial O$ such that $\Omega = \{x \in \mathbb{R}^3$ s.t. $|b_O(x)| < h,\ p(x) \in \omega\}$.*

We shall use, as well, the standard notations of intrinsic geometry for shells. Let $\Gamma_z = \{x \in \Omega \mid b_O(x) = z\}$, we define Γ, S_h, T_z, κ_1 and κ_2 as on page 98 of those proceedings ([7]). Let

$$U_h^d = \left\{U \in H(S_h) \mid \exists u_i \in H(\Gamma),\ U = \sum_{i=0}^{d} b^i\ u_i \circ p\right\} \quad \text{and} \quad U_h = \cup_{d\in\mathbb{N}}\ U_h^d$$

Assumption 3.2. $T_z^{-1} \circ \mathrm{T}_0 = \mathrm{T}_0 \circ T_z^0$

3.2. The Equation

The Federer's measure decomposition on (3) yields

$$\int_0^\tau \int_{-h}^{h} \int_{\Gamma_z} (\Sigma(u)..\varepsilon(w) - \rho\partial_t u \partial_t w)\, d\Gamma_z\, dz\, dt = 0 \tag{6}$$

In order to compute that integral, let us consider $\int_{\Gamma_z} \Sigma(u)..\varepsilon(w)\, d\Gamma_z$, we perform a change of variable and obtain

$$\int_{\Gamma_z} \Sigma(u)..\varepsilon(w)\, d\Gamma_z = \int_\Gamma (\Sigma(u) \circ T_z)..(\varepsilon(w) \circ T_z) \det(DT_z)\, d\Gamma$$

Lemma 3.3. *We have* $\det(DT_z) = 1 + \kappa_1 z + \kappa_2 z^2$. *We will note $j(z)$ that expression.*

Lemma 3.4. *We have* $E \circ T_z = (I + zD^2b).E.((I + zD^2b^0)^{-1} \circ {\mathrm{T}_0}^{-1} \circ T_z)$.

Definition 3.5. *Let us define f and g by $f(z) = z\frac{1+\kappa_1 z}{j(z)}$, and $g(z) = z^2\frac{1}{j(z)}$.*

Lemma 3.6. *with this notation we have $(I + zD^2b)^{-1} = I - f(z)D^2b + g(z)(D^2b)^2$.*

For the sake of simplicity, from now on, we will suppose that Ω^0 is a plate.

Lemma 3.7. *$E \circ T_z = (I + zD^2b).E$ and ${}^*E \circ T_z = {}^*E.(I + zD^2b)$.*

Let X be a matrix. As a consequence of the isotropy of the material and [10], one has $C..X = \frac{\lambda}{2}\mathrm{tr}(X)I + \mu X$ therefore

$$\begin{aligned}\Sigma \circ T_z &= \frac{\lambda}{\det(E \circ T_z)}\mathrm{tr}(({}^*E \circ T_z)(\varepsilon \circ T_z)(E \circ T_z))\ (E\,{}^*E) \circ T_z \\ &\quad +2\mu\frac{1}{\det(E \circ T_z)}((E\,{}^*E) \circ T_z)(\varepsilon \circ T_z)((E\,{}^*E) \circ T_z)\end{aligned}$$

Following [1] we consider the differential operators $\mathrm{D_N}$ and $\mathrm{D_{SHELL}}$ defined by

$$\mathrm{D_N} = \begin{pmatrix} \partial_z \\ 1 \\ f \\ g \end{pmatrix} \quad \text{and} \quad \mathrm{D_{SHELL}}v = \begin{pmatrix} (v \otimes n) \\ (D_\Gamma v) \\ (D_\Gamma v.D^2b) \\ (D_\Gamma v.(D^2b)^2) \end{pmatrix}$$

If $\theta : z \mapsto \theta(z)$ is a scalar function and v is a vector then $\mathrm{D_N}\theta(z)$ is a vector and $\mathrm{D_{SHELL}}v$ is a 3rd-order tensor. Let $u \in U_h^d$ then

$$(Du) \circ T_z = \sum_{i=0}^{d} \mathrm{D_N}z^i.\mathrm{D_{SHELL}}u_i \tag{7}$$

therefore $({}^*E \circ T_z)(\varepsilon \circ T_z)(E \circ T_z)$ is equal to

$$\mathrm{sym}\left({}^*E(I + zD^2b)\left(\sum_{i=0}^{d} \mathrm{D_N}z^i.(\mathrm{D_{SHELL}}u_i)\right)(I + zD^2b)E\right)$$

which is equal to

$$\mathrm{tr}(({}^*E \circ T_z)(\varepsilon \circ T_z)(E \circ T_z)) = \sum_{i=0}^{d} \mathrm{D_N}z^i.({}^*E\mathrm{tr}(\mathrm{D_{SHELL}}u_i)E)$$

$$+\sum_{i=0}^{d} z\mathrm{D_N}z^i.\mathrm{tr}({}^*ED^2b(\mathrm{D_{SHELL}}u_i)E) + \sum_{i=0}^{d} z^2\mathrm{D_N}z^i.\mathrm{tr}({}^*ED^2b(\mathrm{D_{SHELL}}u_i)D^2bE)$$

Since $\mathrm{D_N}z^i = {}^*(iz^{i-1}, z^i, z^i f(z), z^i g(z))$ the real $\mathrm{tr}(({}^*E \circ T_z)(\varepsilon \circ T_z)(E \circ T_z))$ may be written as a function of 10 terms depending on z with coefficients independent of z. With the convention $iz^{i-1} = 0$ for $i = 0$, we get

$$\begin{array}{lll} & i\mathrm{tr}({}^*E(\mathrm{D}_{\text{SHELL}}u_i)_1 E) & z^{i-1} \\ + & 2i\mathrm{tr}({}^*E(\mathrm{D}_{\text{SHELL}}u_i)_1 D^2bE) + \mathrm{tr}({}^*E(\mathrm{D}_{\text{SHELL}}u_i)_2 E) & z^i \\ + & (2\mathrm{tr}({}^*E(\mathrm{D}_{\text{SHELL}}u_i)_2 D^2bE) + i\mathrm{tr}({}^*ED^2b(\mathrm{D}_{\text{SHELL}}u_i)_1 D^2bE)) & z^{i+1} \\ + & \mathrm{tr}({}^*ED^2b(\mathrm{D}_{\text{SHELL}}u_i)_2 D^2bE) & z^{i+2} \\ - & \mathrm{tr}({}^*E(\mathrm{D}_{\text{SHELL}}u_i)_3 E) & z^i f(z) \\ + & 2\mathrm{tr}({}^*E(\mathrm{D}_{\text{SHELL}}u_i)_3 D^2bE) & z^{i+1} f(z) \\ - & \mathrm{tr}({}^*ED^2b(\mathrm{D}_{\text{SHELL}}u_i)_3 D^2bE) & z^{i+2} f(z) \\ + & \mathrm{tr}({}^*E(\mathrm{D}_{\text{SHELL}}u_i)_4 E) & z^i g(z) \\ + & 2\mathrm{tr}({}^*E(\mathrm{D}_{\text{SHELL}}u_i)_4 D^2bE) & z^{i+1} g(z) \\ + & \mathrm{tr}({}^*ED^2b(\mathrm{D}_{\text{SHELL}}u_i)_4 D^2bE) & z^{i+2} g(z) \end{array}$$

Moreover $(E\,{}^*E) \circ T_z = (I + zD^2b)E\,{}^*E(I + zD^2b)$ hence

$$(E\,{}^*E) \circ T_z = E\,{}^*E + z(D^2bE\,{}^*E + E\,{}^*ED^2b) + z^2 D^2bE\,{}^*ED^2b$$

therefore the first term of $\Sigma \circ T_z$ may be written as function of 16 terms depending on z with coefficients being matrices independent of z. The same method applies for the second term of $\Sigma \circ T_z$, it follows

$$\Sigma \circ T_z = \sum_{i=0}^{d} \sum_{\alpha=-1}^{4} a_{\alpha,i} z^{\alpha+i} \frac{1}{j(z)} + b_{\alpha,i} z^{\alpha+i} \frac{f(z)}{j(z)} + c_{\alpha,i} z^{\alpha+i} \frac{g(z)}{j(z)} \tag{8}$$

the coefficients $a_{\alpha,i}$, $b_{\alpha,i}$ and $c_{\alpha,i}$ have been computed with a symbolic computation system and are presented in [6]. From those coefficients we define a third-order tensor $\mathrm{A}^i_{\text{SHELL}}$ which is $3 \times 3 \times 3$ and such that (8) may be rewritten

$$\Sigma \circ T_z = \sum_{i=0}^{d+4} \frac{1}{j(z)} \mathrm{A}^{i\,\triangleleft\triangleleft}_{\text{SHELL}}.V_i(z) \quad , \quad V_i(z) = z^i\,{}^*(1, f(x), g(z))$$

We suppose $w = b^k.(w_k \circ p)$ then $\varepsilon(w) \circ T_z = \mathrm{D_N} z^k.\mathrm{D}_{\text{SHELL}} w$ therefore

$$(\Sigma \circ T_z)..(\varepsilon(w) \circ T_z) \det(DT_z) = \sum_{i=0}^{d+4} \left(\frac{1}{j(z)} \mathrm{A}^{i\,\triangleleft\triangleleft}_{\text{SHELL}}.V_i(z) \right) .. \left(\mathrm{D_N} z^k.\mathrm{D}_{\text{SHELL}} w_k \right) j(z)$$

then, lemma 1.3 yields

$$(\Sigma \circ T_z)..(\varepsilon(w) \circ T_z) \det(DT_z) = \sum_{i=0}^{d+4} ((\mathrm{D}_{\text{SHELL}} w_k)..\mathrm{A}^{i\,\triangleleft\triangleleft}_{\text{SHELL}})..(\mathrm{D_N} z^k \otimes V_i(z))$$

Let us define the matrix L^k_i by $L^k_i = (\mathrm{D}_{\text{SHELL}} w_k)..\mathrm{A}^{i\,\triangleleft\triangleleft}_{\text{SHELL}}$ and the matrix $q^k_i(z)$ by $\mathrm{D_N} z^k \otimes V_i(z)$, that is

$$q^k_i(z) = \begin{pmatrix} kz^{i+k-1} & kz^{i+k-1} f(z) & kz^{i+k-1} g(z) \\ z^{i+k} & z^{i+k} f(z) & z^{i+k} g(z) \\ z^{i+k} f(z) & z^{i+k} f(z)^2 & z^{i+k} f(z) g(z) \\ z^{i+k} g(z) & z^{i+k} f(z) g(z) & z^{i+k} g(z)^2 \end{pmatrix}$$

then $(\Sigma \circ T_z)..(\varepsilon(w) \circ T_z) \det(DT_z) = \sum_{i=0}^{d+4} L_i^k..q_i^k(z)$ hence

$$\int_0^\tau \int_{-h}^h \int_{\Gamma_z} \Sigma(u)..\varepsilon(w)\, d\Gamma_z\, dz\, dt = \sum_{i=0}^{d+4} \int_0^\tau \int_\Gamma L_i^k..(Q_i^k(h) - Q_i^k(-h))\, dz\, dt$$

where Q_i^k is the matrix whose terms are the anti-derivative of the terms of q_i^k. The terms of that matrix can be computed explicitly as the functions to be integrated are rational fractions. On the other hand

$$\partial_t u \circ T_z = \sum_{i=0}^{d} b^i \circ T_z.(u_i \circ p \circ T_z) = \sum_{i=0}^{d} z^i u_i$$

similarly $\partial_t w = z^k w_k$ therefore equation (6) yields for all $k \in \mathbb{N}$

$$\sum_{i=0}^{d+4} \int_0^\tau \int_\Gamma (-\frac{1+(-1)^{i+k}}{i+k+1} \chi_{i \leq d} h^{i+k+1} \rho \partial_t u_i \partial_t w_k + L_i^k..(Q_i^k(h) - Q_i^k(-h))) d\Gamma dt = 0$$

where $\chi_{i \leq d} = 1$ if $i \leq d$ and 0 otherwise.

4. Shape Differentiability in the Wave Equation

We recall the results of shape differentiability for the wave equation, the Dirichlet case has been presented in [5]. Let (H_m) be

$$f \in L^1([0,\tau[, H^m(D)),\ f^{(m)} \in L^1([0,\tau[, H^{m-1}(D))\phi \in H^{m+1}(D),\ \psi \in H^m(D)$$

We consider

$$(9) \qquad \begin{cases} \frac{\partial^2 u}{\partial t^2} - \operatorname{div}(K\nabla u) = f \text{ on } Q \\ u = 0 \text{ on } \Sigma \ ,\quad u(0) = \phi \text{ on } \Omega \ ,\quad \partial_t u(0) = \psi \text{ on } \Omega \end{cases}$$

When $m > 0$, the solution u is shape differentiable in $L^2([0,\tau[\times D)$ and $u' \in C([0,\tau[, H^m(\Omega)) \cap \ldots \cap C^m([0,\tau[, L^2(\Omega))$ is solution to

$$(10) \qquad \begin{cases} \frac{\partial^2 u'}{\partial t^2} - \operatorname{div}(K\nabla u') = 0 \text{ on } Q \\ u' = -\frac{\partial u}{\partial n} \langle V(0), n \rangle \text{ on } \sigma \ ,\ u'(0) = 0 \text{ on } \Omega \ ,\ \partial_t u'(0) = 0 \text{ on } \Omega \end{cases}$$

Although the proof does not work for $m = 0$, the problem u' is solution to, survives because of the Hidden Regularity (I. Lasiecka, J.L. Lions, R. Triggiani, see [11], [12]). We prove that under (H_0) the solution u is weakly shape differentiable in $L^2([0,\tau[\times D)$ and u' belongs to $C([0,\tau[, L^2(\Omega)) \cap C^1([0,\tau[, H^{-1}(\Omega))$. This result is classical for elliptic and parabolic equations but is not for hyperbolic equations with regularity (H_0) (see [5].)

Although the proof of the hidden derivative is absolutely different, once that hidden regularity is obtained, the situation concerning the shape differentiability of the regular case is analogous in the Neumann case. Let u be the solution to

$$\begin{cases} \frac{\partial^2 u}{\partial t^2} - \operatorname{div}(K\nabla u) = f \quad \text{on } Q \\ \frac{\partial u}{\partial n} = g \quad \text{on } \Sigma \ , \quad u(0) = \phi \quad \text{on } \Omega \ , \quad \partial_t u(0) = \psi \quad \text{on } \Omega \end{cases}$$

then the shape derivative of u and is solution to

$$\begin{cases} \partial_{tt} u' - \operatorname{div}(K\nabla u') = 0 \quad \text{on } Q \\ \frac{\partial u'}{\partial n} = (f + \kappa_1 g - \partial_t^2 u)\langle V(0), n\rangle + \operatorname{div}_\Gamma(K\nabla_\Gamma u \langle V(0), n\rangle) + g'_\Gamma \ \text{on } \sigma \\ y'(0) = 0 \quad \text{on } \Omega \ , \quad \partial_t y'(0) = 0 \quad \text{on } \Omega \end{cases}$$

In the next section we obtain the regularity needed to extend such results of differentiability to the equation of the vibration of a prestressed shell (1).

5. Existence and Regularity of the Solutions

5.1. Interior Regularity

The existence and regularity of the solutions will be derived from the theory of semigroups. Let Λ be the linear operator in $L^2(\Omega)$ defined by

$$D(\Lambda) = H^2(\Omega; \mathbb{R}^3) \cap H^1_{\Gamma_D}(\Omega; \mathbb{R}^3) \qquad \Lambda = -\frac{1}{\rho}\operatorname{div}(\Sigma)$$

Lemma 5.1. *The operator Λ is self-adjoint.*

Proof. Let $v \in H(\Omega)$.

$$\int_\Omega \Lambda(u)v = \frac{1}{\rho}\int_\Omega \Sigma(u)..\varepsilon(v)$$

from the symmetry of Σ and lemma 1.2 follows $\Sigma(u)..\varepsilon(v) = \varepsilon(u)..\Sigma(v)$, therefore Λ is self-adjoint. □

Lemma 5.2. *The operator Λ is coercive in $H^1_{\Gamma_D}(\Omega)$.*

Proof. One has

$$\int_\Omega \Lambda(u)u = \frac{1}{\rho}\int_\Omega \frac{2}{|\det(E)|}({}^*E\varepsilon(u)E)..C..({}^*E\varepsilon(u)E)$$

from [10] and the isotropy of the material, it follows

$$\int_\Omega \Lambda(u)u = \frac{1}{\rho}\int_\Omega \frac{1}{|\det(E)|}(\lambda({}^*E\varepsilon(u)E)..I + 2\mu({}^*E\varepsilon(u)E)..({}^*E\varepsilon(u)E))$$

since $\lambda \geq 0$ the following inequality holds

$$\int_\Omega \Lambda(u)u \geq \frac{2\mu}{\rho}\int_\Omega \frac{1}{|\det(E)|}({}^*E\varepsilon(u)E)..({}^*E\varepsilon(u)E)$$

We have $({}^*E\varepsilon(u)E)..({}^*E\varepsilon(u)E) = \varepsilon(u)..(E\,{}^*E\varepsilon(u)E\,{}^*E)$ therefore, using (5), there exists a non negative constant k such that

$$\int_\Omega \Lambda(u)u \geq \frac{2k\mu}{\rho}\int_\Omega \frac{1}{|\det(E)|}\varepsilon(u)..\varepsilon(u)$$

$u_{|\Gamma_D} = 0$ and Korn's inequality yield the existence of a non negative real C such that

$$\int_\Omega \Lambda(u)u \geq \left(\frac{2kC\mu}{\rho}\frac{1}{\max_\Omega|\det(E)|}\right)\|u\|_{H^1_{\Gamma_D}(\Omega)}$$

hence Λ is coercive. □

We note

$$U = \begin{pmatrix} u \\ \partial_t u \end{pmatrix}, \quad U_0 = \begin{pmatrix} \phi \\ \psi \end{pmatrix} \quad \text{and} \quad A = \begin{pmatrix} 0 & 1 \\ -\Lambda & 0 \end{pmatrix}$$

then (1) is equivalent to

$$\begin{cases} \partial_t U = AU \\ U(0) = U_0 \end{cases} \tag{11}$$

Proposition 5.3. *Following* [2, prop. 2.12], *the operator A is the infinitesimal generator of a strongly continuous semi-group of contraction S on $H(\Omega)$. Moreover $^*A = -A$.*

Proposition 5.4. $\exists! u \in C([0;\tau]; H^1_{\Gamma_D}(\Omega)) \cap C^1([0;\tau]; L^2(\Omega))$ *solution to (1)*

Proof. This derives from [2, prop 3.3] and proposition 5.3. □

5.2. Boundary Regularity

Let φ be a solution to (1). We consider a flow mapping T_s and the associated vector field V (cf. [13] or [4]). Let us compute the derivative with respect to s of

$$\frac{1}{2}\int_{Q_s}\frac{1}{|\det(E)|}({}^*E\varepsilon(\varphi\circ T_s^{-1})E)..C..({}^*E\varepsilon(\varphi\circ T_s^{-1})E)$$

at $s=0$, via two different ways. That derivative will be denoted $\mathfrak{E}$ and is call the *extractor.* We want the distributed integral to be defined for $\varphi \in C([0;\tau]; H^1_{\Gamma_D}(\Omega)) \cap C^1([0;\tau]; L^2(\Omega))$ such that $P(\varphi) \in L^2(\Omega)$.

Lemma 5.5. *One has*

$$\mathfrak{E} = \int_Q \frac{1}{|\det(E)|}\left(\operatorname{div}(V(0))|\partial_t\varphi|^2 + \langle D\varphi.\partial_t V(0), \partial_t\varphi\rangle\right)\, dx\, dt -$$

$$\int_\sigma \frac{1}{|\det(E)|}(\langle D\varphi.V(0), \Sigma(\varphi)n\rangle + \frac{1}{2}({}^*E\varepsilon(\varphi)E)..C..({}^*E\varepsilon(\varphi)E - |\partial_t\varphi|^2)\langle V(0), n\rangle)$$

Proof. From [4, lemma 8], the left-hand side of the equality is equal to

$$\int_Q \frac{1}{|\det(E)|}\left({}^*E\frac{\partial}{\partial s}(\varepsilon(\varphi\circ T_s^{-1}))_{s=0}E\right)..C..({}^*E\varepsilon(\varphi)E)\, dx\, dt$$
$$+\frac{1}{2}\int_\sigma \frac{1}{|\det(E)|}({}^*E\varepsilon(\varphi)E)..C..({}^*E\varepsilon(\varphi)E)\langle V(0), n\rangle\, d\Gamma\, dt$$

that is

$$\rho\int_Q \frac{1}{|\det(E)|}\langle D\varphi.V(0), \partial_{tt}\varphi\rangle\, dx\, dt - \int_Q \frac{1}{|\det(E)|}\langle D\varphi.V(0), P(\varphi)\rangle\, dx\, dt$$

$$-\int_{\sigma}\frac{1}{|\det(E)|}\left(\langle D\varphi.V(0),\Sigma(\varphi)n\rangle+\frac{1}{2}({}^*E\varepsilon(\varphi)E)..C..({}^*E\varepsilon(\varphi)E)\,\langle V(0),n\rangle\right)\,d\Gamma\,dt$$

On the other hand

$$\int_{Q}\frac{1}{|\det(E)|}\,\langle D\varphi.V(0),\partial_{tt}\varphi\rangle\,\,dx\,dt=$$

$$-\int_{Q}\frac{1}{|\det(E)|}\,\langle\partial_t D\varphi.V(0),\partial_t\varphi\rangle\,\,dx\,dt+\left[\int_{\Gamma}\frac{1}{|\det(E)|}\,\langle D\varphi.V(0),\partial_t\varphi\rangle\,\,d\Gamma\right]_0^T$$

since T is arbitrary, $V(0)(T)$ may be assumed to vanish moreover $\partial_t\varphi(0)=0$ therefore the first term of the right-hand side of the previous equality may be rewritten as

$$-\int_{Q}\frac{1}{|\det(E)|}\,(\langle V(0),\partial_t\,{}^*D\varphi\partial_t\varphi\rangle+\langle D\varphi.\partial_t V(0),\partial_t\varphi\rangle)\,\,dx\,dt$$

which is equal to

$$\int_{Q}\frac{1}{|\det(E)|}\left(\operatorname{div}(V(0))|\partial_t\varphi|^2+\langle D\varphi.\partial_t V(0),\partial_t\varphi\rangle\right)\,dx\,dt+\int_{\sigma}|\partial_t\varphi|^2\,\langle V(0),n\rangle$$

the identity derives. □

Lemma 5.6. *One has*

$$\mathfrak{E}=\int_{Q}\frac{1}{|\det(E)|}(\frac{1}{2}tr(E^{-1}.DE.V(0))({}^*E\varepsilon(\varphi)E)..C..({}^*E\varepsilon(\varphi)E)$$

$$+({}^*DE.V(0)\varepsilon(\varphi)E)..C..({}^*E\varepsilon(\varphi)E)+\frac{1}{2}\,{}^*E\varepsilon(D\varphi.DV(0))E)..C..({}^*E\varepsilon(\varphi)E)$$

$$+({}^*E\varepsilon(\varphi)DE.V(0))..C..({}^*E\varepsilon(\varphi)E)+\frac{1}{2}\operatorname{div}(V(0))({}^*E\varepsilon(\varphi)E)..C..({}^*E\varepsilon(\varphi)E))\,dx\,dt$$

Proof. The identity derives from the change of variable T_s in the left-hand side and the computation of the derivatives with respect to s. □

Lemma 5.7. *On* Γ_D

$$\langle\Sigma(\varphi).n,D\varphi.V(0)\rangle=({}^*E\varepsilon(\varphi)E)..C..({}^*E\varepsilon(\varphi)E)\,\langle V(0),n\rangle$$
$$=|\,{}^*E(D\varphi.n)\,{}^*nE|\,\langle V(0),n\rangle$$

Proof. Since $\varphi|_\sigma=0$ we have $D\varphi=(D\varphi.n).{}^*n$. Property of C gives the result. □

Proposition 5.8. *Let B is defined by (12), then*

$$\int_{\sigma_D}\frac{1}{|\det(E)|}|\,{}^*E(D\varphi.n)\,{}^*nE|^2\,\langle V(0),n\rangle=B(\varphi)$$

Proof. We use 5.5 and 5.6 to compute $\mathfrak{E}$ via two different ways. We have $\langle V(0), n\rangle =$ 0 on Γ_N moreover $\partial_t\varphi = 0$ on Γ_D therefore the integral over σ is equal to

$$-\int_{\sigma_D} \frac{1}{|\det(E)|}\Big(\langle D\varphi.V(0), \Sigma(\varphi)n\rangle + \frac{1}{2}({}^*E\varepsilon(\varphi)E)..C..({}^*E\varepsilon(\varphi)E)\,\langle V(0), n\rangle\Big)$$

Lemma 5.7 proves the term below is equal to

$$\frac{1}{2}\int_{\sigma_D} \frac{1}{|\det(E)|}|\,{}^*E(D\varphi.n)\,{}^*nE|\,\langle V(0), n\rangle$$

Let

$$\begin{aligned} B(\varphi) \;=\; & \int_Q \frac{2}{|\det(E)|}(\operatorname{div}(V(0))|\partial_t\varphi|^2 + \langle D\varphi.\partial_t V(0), \partial_t\varphi\rangle + \langle D\varphi.V(0), P(\varphi)\rangle \\ & -\frac{1}{2}\operatorname{tr}(E^{-1}.DE.V(0))({}^*E\varepsilon(\varphi)E)..C..({}^*E\varepsilon(\varphi)E) \\ & -({}^*DE.V(0)\varepsilon(\varphi)E + {}^*E\varepsilon(\varphi)DE.V(0))..C..({}^*E\varepsilon(\varphi)E) \\ & -\frac{1}{2(}{}^*E\varepsilon(D\varphi.DV(0))E)..C..({}^*E\varepsilon(\varphi)E) \\ & -\frac{1}{2}\operatorname{div}(V(0))({}^*E\varepsilon(\varphi)E)..C..({}^*E\varepsilon(\varphi)E)) \end{aligned} \tag{12}$$

then

$$\int_{\sigma_D} \frac{1}{|\det(E)|}|\,{}^*E(D\varphi.n)\,{}^*nE|\,\langle V(0), n\rangle = B(\varphi)$$

□

Remark 5.9. *The real $B(\varphi)$ is defined when $\varphi \in C([0;\tau]; H^1_{\Gamma_D}) \cap C^1([0;\tau]; L^2(\Omega))$ and $P(\varphi) \in L^1([0,\tau[, L^2(\Omega))$,*

Let $(\varphi^m)_m$ be a sequence of functions of $C^\infty(Q)$ that vanish on Γ_D and such that $\varphi^m \to u$ in $H^1(Q)$ and $P\varphi^m \to Pu$ in $L^2(Q)$. For the sake of shortness we shall not prove the density in those proceedings, however the proof is similar to the one given for the wave equation in [9]. Since $B(u)$ exists, $\||\varepsilon(\varphi^m)|\sqrt{\langle V(0), n\rangle}\|_{L^2(\sigma_D)}$ is bounded, hence there exists a function ξ in $L^2(\sigma_D)$ and a subsequence such that

$$\varepsilon(\varphi^m)\sqrt{\langle V(0), n\rangle} \rightharpoonup \xi \quad \text{weakly in } L^2(\sigma_D) \quad \text{as } m_k \to +\infty$$

On σ_D the Green's theorem proves $\xi = \Sigma(u)..\varepsilon(u)\sqrt{\langle V(0), n\rangle}$

Proposition 5.10. *For all $\nu \in C^1(\Gamma_D)$ with $\nu \geq 0$ on Γ_D and* supp $\nu \in \Gamma_D$ *we have $\nu|\varepsilon(u)| \in L^2(\sigma_D)$.*

The authors wish to thank Sani Makaï and Bruno Vergnes (Alcatel Space Company) for helpful discussions.

References

[1] Jean-Christophe Aguilar and Jean-Paul Zolésio. Coque fluide intrinsèque dans approximation géométrique (intrinsic fluid shell without geometrical approximation). *Comptes Rendus de l'Académie des Sciences, Paris, series I, Partial Differential Equations*, 326(11):1341–1346, 1998.

[2] Alain Bensoussan, Giuseppe Da Prato, Michel C. Delfour, and S. K. Mitter. *Representation and control of infinite dimensional systems*, volume 1. Birkhäuser, 1993.

[3] John Cagnol and Jean-Paul Marmorat. Static equilibrium of hyperelastic thin shell: Symbolic and numerical computation. *Mathematics and Computers in Simulation*, 46(2):103–115, 1998.

[4] John Cagnol and Jean-Paul Zolésio. Hidden shape derivative in the wave equation. In P. Kall, I. Lasiecka, and M. Polis, editors, *System Modelling and Optimization*, volume 396 of *Chapman & Hall/CRC Research Notes in Mathematics*, pages 42–52. IFIP, CRC Press LLC, Boca Raton, 1998. To appear.

[5] John Cagnol and Jean-Paul Zolésio. Hidden shape derivative in the wave equation with dirichlet boundary condition. *Comptes Rendus de l'Académie des Sciences, Paris, série I*, 326(9):1079–1084, 1998. Partial Differential Equations.

[6] John Cagnol and Jean-Paul Zolésio. The tensor $\mathrm{A}_{\mathrm{SHELL}}$ for a constrained shell: Explicit computation with a symbolic computation system. Technical report, CMA, Ecole des Mines de Paris, France, 1998.

[7] Michel C. Delfour. Intrinsic $p(2,1)$ thin shell model and naghdi's models without *a priori* assumption on the stress tensor. In *those proceedings*. Birkhäuser, 1998.

[8] Michel C. Delfour and Jean-Paul Zolésio. Differential equations for linear shells: comparison between intrinsic and classical models. In Luc Vinet, editor, *Advances in Mathematical Sciences, CRM's 25 years*, pages 42–124, 1997.

[9] Michel C. Delfour and Jean-Paul Zolésio. Hidden boundary smoothness in hyperbolic tangential problems of nonsmooth domains. In P. Kall, I. Lasiecka, and M. Polis, editors, *System Modelling and Optimization*, volume 396 of *Chapman & Hall/CRC Research Notes in Math.*, pages 53–61. IFIP, CRC Press LLC, 1998. To appear.

[10] Paul Germain. *Mécanique*, volume I. Ellipses, Ecole Polytechnique, 1986.

[11] Irena Lasiecka, Jacques-Louis Lions, and Roberto Triggiani. Non homogeneous boundary value problems for second order hyperbolic operators. *Journal de Mathématiques pures et Appliquées*, 65(2):149–192, 1986.

[12] Irena Lasiecka and Roberto Triggiani. Recent advances in regularity of second-order hyperbolic mixed problems, and applications. In C. Jones, editor, *Dynamics reported. Expositions in dynamical systems*, pages 104–162. Springer-Verlag, 1994.

[13] Jan Sokolowski and Jean-Paul Zolésio. *Introduction to Shape Optimization*. SCM 16. Springer-Verlag, 1991.

Centre de Mathématiques Appliquées
Ecole des Mines de Paris – INRIA
2004 route des Lucioles, B.P. 93
06902 Sophia Antipolis Cedex, France
E-mail address: `John.Cagnol@sophia.inria.fr`, `zolesio@inln.cnrs.fr`

International Series of Numerical Mathematics
Vol. 133,

Second Order Optimality Conditions for Some Control Problems of Semilinear Elliptic Equations with Integral State Constraints

Eduardo Casas, Fredi Tröltzsch and Andreas Unger [1]

Abstract. In this paper we discuss necessary and sufficient second order optimality conditions for control problems of semilinear elliptic equations with pointwise control constraints as well as integral state constraints. We consider the case of a boundary control problem, the state equation being described by a Neumann problem. The presence of state constraints along with pointwise control constraints causes some known difficulties in the proof of optimality conditions. Due to the specific integral form of the state-constraints, here we are able to considerably tighten the gap between second order necessary and sufficient condition. The analysis is based on some constraint qualification.

1. Introduction

In the past years, the theory of second order conditions for the optimal control of distributed parameter systems has received a good deal of attention. We refer, for instance, to [1], [2], [3], [4], [6]. Taking advantage of the special integral type of the state-constraints, in this paper we improve our results obtained in [4] for a more general class of state constraints. In contrast to [4], where second order conditions were developed in the spirit of [7], [8], we are able to shrink the critical cone, where coercivity is required in the second order condition. Roughly speaking, the cone used in the sufficient condition is arbitrarily close to the one in the necessary condition.

We first introduce the optimal control problem. Let Ω be a bounded open subset of $\mathbb{R}^n$ with a Lipschitz boundary Γ. Given a function $u \in L^\infty(\Gamma)$, we consider the following boundary value problem

$$(1.1)\qquad \begin{cases} -\Delta y(x) + y(x) = 0 & \text{in } \Omega \\ \partial_\nu y(x) = b(x, y(x), u(x)) & \text{on } \Gamma, \end{cases}$$

[1]This research was partially supported by European Union, under HCM Project number ERBCHRXCT940471, and by Deutsche Forschungsgemeinschaft, under Project number Tr 302/1-2. The first author was also supported by Dirección General de Investigación Científica y Técnica (Spain)

where $\partial_\nu y$ denotes the normal derivative of y and $b : \Gamma \times \mathbb{R} \times \mathbb{R} \longrightarrow \mathbb{R}$ is a function measurable w.r.t. the first variable and of class C^2 w.r.t. the others and satisfying

$$(1.2) \quad \begin{cases} \dfrac{\partial b}{\partial y}(x, y, u) \le 0 \quad \text{a.e. } x \in \Gamma,\ \forall (y, u) \in \mathbb{R}^2; \\ \forall M > 0\ \exists \psi_M \in L^p(\Gamma)\ (p > n-1) \text{ and } C_M > 0 \text{ such that} \\ |b(x, 0, u)| \le \psi_M(x) \quad \text{a.e. } x \in \Gamma \text{ and } |u| \le M; \\ \displaystyle\sum_{1 \le i+j \le 2} \left| \frac{\partial^{i+j} b}{\partial y^i \partial u^j}(x, y, u) \right| \le C_M \quad \text{a.e. } x \in \Gamma,\ |y| \le M \text{ and } |u| \le M. \end{cases}$$

Under these assumptions the existence and uniqueness of a solution of (1.1) can be proved; see Casas and Tröltzsch [2].

The state equation of our control problem is given by (1.1). The cost functional $J : L^\infty(\Omega) \longrightarrow \mathbb{R}$ is defined by

$$J(u) = \int_\Omega f(x, y_u(x))dx + \int_\Gamma g(x, y_u(x), u(x))dS(x),$$

where $y_u = G(u)$ is the solution of (1.1) corresponding to u, $f : \Omega \times \mathbb{R} \longrightarrow \mathbb{R}$ is measurable w.r.t. the first variable and of class C^2 with respect to the second and $g : \Gamma \times \mathbb{R} \times \mathbb{R} \longrightarrow \mathbb{R}$ is also measurable in the first variable and of class C^2 w.r.t. the other two. Moreover we assume that

$$(1.3) \quad \begin{cases} f(\cdot, 0) \in L^1(\Omega); \\ \forall M > 0\ \exists \psi^1_M \in L^1(\Omega) \text{ such that} \\ \left| \dfrac{\partial f}{\partial y}(x, y) \right| + \left| \dfrac{\partial^2 f}{\partial y^2}(x, y) \right| \le \psi^1_M(x) \quad \text{for } |y| \le M \text{ and a.e. } x \in \Omega; \end{cases}$$

$$(1.4) \quad \begin{cases} g(\cdot, 0, 0) \in L^1(\Gamma); \\ \forall M > 0\ \exists \psi^2_M \in L^1(\Gamma) \text{ such that} \\ \displaystyle\sum_{1 \le i+j \le 2} \left| \frac{\partial^{i+j} g}{\partial y^i \partial u^j}(x, y, u) \right| \le \psi^2_M(x) \text{ for } |y| \le M, |u| \le M \text{ and a.e. } x \in \Gamma. \end{cases}$$

Let us consider some functionals $F_j : C(\bar{\Omega}) \longrightarrow \mathbb{R}$ of class C^2, $1 \le j \le m$, and functions $u_a, u_b \in L^\infty(\Gamma)$, with $u_a(x) \le u_b(x)$ a.e. $x \in \Gamma$. The control problem is formulated as follows

$$
\text{(P)} \begin{cases} \text{Minimize } J(u) \\ u_a(x) \leq u(x) \leq u_b(x) \;\; \text{a.e. } x \in \Gamma \\ F_j(y_u) = 0, \;\; 1 \leq j \leq m_1 \\ F_j(y_u) \leq 0, \;\; m_1 + 1 \leq j \leq m. \end{cases}
$$

Let us show some examples of state constraints that fall into the previous abstract framework.

Example 1. For every $1 \leq j \leq m$ let $f_j : \Omega \times \mathbb{R} \longrightarrow \mathbb{R}$ be a measurable function of class C^2 with respect to the second variable such that for each $M > 0$ there exists a function $\eta_M^j \in L^1(\Omega)$ satisfying

$$
|f_j(x,0)| + \left| \frac{\partial f_j}{\partial y}(x,y) \right| + \left| \frac{\partial^2 f_j}{\partial y^2}(x,y) \right| \leq \eta_M^j(x) \;\; a.e. \; x \in \Omega, \; \forall |y| \leq M.
$$

Then the equality and inequality constraints defined by the functions

$$
F_j(y_u) = \int_\Omega f_j(x, y_u(x))dx
$$

are included in the formulation of (P).

Example 2. For every $1 \leq j \leq m$ let $f_j : \Gamma \times \mathbb{R} \longrightarrow \mathbb{R}$ be a measurable function of class C^2 with respect to the second variable such that for each $M > 0$ there exists a function $\eta_M^j \in L^1(\Gamma)$ satisfying

$$
|f_j(x,0)| + \left| \frac{\partial f_j}{\partial y}(x,y) \right| + \left| \frac{\partial^2 f_j}{\partial y^2}(x,y) \right| \leq \eta_M^j(x) \;\; a.e. \; x \in \Gamma, \; \forall |y| \leq M.
$$

Then the functionals

$$
F_j(y_u) = \int_\Gamma f_j(x, y_u(x))dS(x)
$$

define some integral constraints included in the formulation of (P).

Example 3. Given m functions $f_i : \mathbb{R} \longrightarrow \mathbb{R}$ of class C^2, a set of points $\{x_i\}_{i=1}^m \subset \bar{\Omega}$, and some integer m_1, $1 \leq m_1 \leq m$, the constraints

$$
f_i(y_u(x_i)) = 0, \; 1 \leq i \leq m_1 \;\; \text{and} \;\; f_i(y_u(x_i)) \leq 0, \; m_1 + 1 \leq i \leq m
$$

can be written in the above framework by putting $F_i(y) = f_i(y(x_i))$.

In the sequel we will denote by $G : L^\infty(\Gamma) \longrightarrow H^1(\Omega) \cap C(\bar{\Omega})$ the mapping associating to every function u the solution of (2.1) and G_j, $1 \leq j \leq m$, will be the composite functionals $G_j : L^\infty(\Omega) \longrightarrow \mathbb{R}$ defined by $G_j(u) = F_j(G(u))$. The next theorems provide formulas for the derivatives of G, J and G_j; see Casas and Tröltzsch [2] for the corresponding proofs. We begin with the first order derivative of the control-state mapping G.

Theorem 1. *G is of class C^2, i.e. twice continuously Fréchet differentiable. If $u, v \in L^\infty(\Gamma)$, $y = G(u)$ and $z_v = G'(u)v$, then z_v is the solution of*

$$\left\{ \begin{array}{l} -\Delta z + z = 0 \quad in\ \Omega \\ \\ \partial_\nu z = \dfrac{\partial b}{\partial y}(\cdot, y, u)z + \dfrac{\partial b}{\partial u}(\cdot, y, u)v \quad on\ \Gamma. \end{array} \right. \tag{1.5}$$

If $v_1, v_2 \in L^\infty(\Gamma)$ and $z_{v_1 v_2} = G''(u)[v_1, v_2]$, then $z_{v_1 v_2}$ is the solution of

$$\left\{ \begin{array}{l} -\Delta z + z = 0 \quad in\ \Omega \\ \\ \partial_\nu z = \dfrac{\partial b}{\partial y}(\cdot, y, u)z + \dfrac{\partial^2 b}{\partial y^2}(\cdot, y, u)z_{v_1} z_{v_2} + \\ \\ \dfrac{\partial^2 b}{\partial y \partial u}(\cdot, y, u)(z_{v_1} v_2 + z_{v_2} v_1) + \dfrac{\partial^2 b}{\partial u^2}(\cdot, y, u)v_1 v_2 \quad on\ \Gamma. \end{array} \right. \tag{1.6}$$

The next result concerns the first and second order derivative of the objective functional J. Notice that differentiability is considered in the space $L^\infty(\Gamma)$.

Theorem 2. *The functional J is of class C^2 and for every $\bar{u}, v \in L^\infty(\Gamma)$ we have*

$$J'(\bar{u})v = \int_\Gamma \left\{ \bar{\varphi}_0 \frac{\partial b}{\partial u}(\cdot, \bar{y}, \bar{u}) + \frac{\partial g}{\partial u}(\cdot, \bar{y}, \bar{u}) \right\} v dS(x) \tag{1.7}$$

and

$$\begin{aligned} J''(\bar{u})v^2 = & \int_\Omega \frac{\partial^2 f}{\partial y^2}(\cdot, \bar{y}) z_v^2 dx + \int_\Gamma \left[\bar{\varphi}_0 \frac{\partial^2 b}{\partial y^2}(\cdot, \bar{y}, \bar{u}) + \frac{\partial^2 g}{\partial y^2}(\cdot, \bar{y}, \bar{u}) \right] z_v^2 dS(x) + \\ & 2 \int_\Gamma \left[\bar{\varphi}_0 \frac{\partial^2 b}{\partial y \partial u}(\cdot, \bar{y}, \bar{u}) + \frac{\partial^2 g}{\partial y \partial u}(\cdot, \bar{y}, \bar{u}) \right] z_v v dS(x) + \\ & \int_\Gamma \left[\bar{\varphi}_0 \frac{\partial^2 b}{\partial u^2}(\cdot, \bar{y}, \bar{u}) + \frac{\partial^2 g}{\partial u^2}(\cdot, \bar{y}, \bar{u}) \right] v^2 dS(x), \end{aligned} \tag{1.8}$$

where $\bar{y} = G(\bar{u})$, $z_v \in H^1(\Omega) \cap C(\bar{\Omega})$ is the solution of (1.5) *corresponding to $(\bar{y}, \bar{u})$, i.e. $z_v = G'(\bar{u})v$, and $\bar{\varphi}_0 \in W^{1,s}(\Omega)$ for every $s < n/(n-1)$ is the solution of*

$$\left\{ \begin{array}{l} -\Delta \bar{\varphi}_0 + \bar{\varphi}_0 = \dfrac{\partial f}{\partial y}(\cdot, \bar{y}) \quad in\ \Omega \\ \partial_\nu \bar{\varphi}_0 = \dfrac{\partial b}{\partial y}(\cdot, \bar{y}, \bar{u})\bar{\varphi}_0 + \dfrac{\partial g}{\partial y}(\cdot, \bar{y}, \bar{u}) \quad on\ \Gamma. \end{array} \right. \tag{1.9}$$

Now, the functionals defining the state constraints are handled analogously.

Theorem 3. *The functionals G_j, $1 \le j \le m$, are of class C^2 and for every $\bar{u}, v \in L^\infty(\Gamma)$ we have*

$$G_j'(\bar{u})v = \int_\Gamma \bar{\varphi}_j \frac{\partial b}{\partial u}(\cdot, \bar{y}, \bar{u}) v \, dS(x), \tag{1.10}$$

where $\bar{y} = G(\bar{u})$ and $\{\bar{\varphi}_j\}_{j=1}^m \subset W^{1,s}(\Omega)$ for every $s < n/(n-1)$ satisfy

$$\begin{cases} -\Delta \bar{\varphi}_j + \bar{\varphi}_j = F_j'(\bar{y})|_\Omega & \text{in } \Omega \\ \partial_\nu \bar{\varphi}_j = \dfrac{\partial b}{\partial y}(\cdot, \bar{y}, \bar{u})\bar{\varphi}_j + F_j'(\bar{y})|_\Gamma & \text{on } \Gamma. \end{cases} \tag{1.11}$$

Moreover

$$G_j''(\bar{u})v^2 = F_j''(\bar{y})z_v^2 +$$

$$\int_\Gamma \bar{\varphi}_j \left\{ \frac{\partial^2 b}{\partial y^2}(\cdot, \bar{y}, \bar{u}) z_v^2 + 2\frac{\partial^2 b}{\partial y \partial u}(\cdot, \bar{y}, \bar{u}) z_v v + \frac{\partial^2 b}{\partial u^2}(\cdot, \bar{y}, \bar{u}) v^2 \right\} dS(x), \tag{1.12}$$

where $z_v = G'(\bar{u})v$.

2. First and Second Order Necessary Optimality Conditions

Second order *necessary* conditions have already been derived in our paper [2]. We recall them along with the known first order necessary conditions for convenience. In this way, we are able to compare the second order sufficient conditions with their necessary counterpart.

In this section we will assume that $\bar{u}$ is a local solution for problem (P). We introduce by $I_0 = \{j \le m \mid F_j(\bar{y}) = 0\}$ and $I_- = \{j \le m \mid F_j(\bar{y}) < 0\}$ the sets of indices of active and inactive inequality constraints, respectively, where $\bar{y} = G(\bar{u})$ is the state associated to $\bar{u}$. It is obvious that $\{1, \dots, m_1\} \subset I_0$. Define the set of "$\epsilon$-inactive control constraints"

$$\Gamma_\epsilon = \{x \in \Gamma : u_a(x) + \epsilon \le \bar{u}(x) \le u_b(x) - \epsilon\} \quad \text{for } \epsilon \ge 0.$$

We rely on the following regularity assumption

$$\begin{cases} \exists \epsilon_{\bar{u}} > 0 \text{ and } \{h_j\}_{j \in I_0} \subset L^\infty(\Gamma), \text{ with } \operatorname{supp} h_j \subset \Gamma_{\epsilon_{\bar{u}}}, \text{ such that} \\ G_i'(\bar{u})h_j = \delta_{ij}, \quad i, j \in I_0. \end{cases} \tag{2.1}$$

Obviously, our assumption is equivalent to the independence of the gradients $\{G_j'(\bar{u})\}_{j \in I_0}$ in $L^\infty(\Gamma_{\epsilon_{\bar{u}}})$. Using Theorem 3 we can write the previous assumption in the following way

$$\int_\Gamma \bar{\varphi}_i \frac{\partial b}{\partial u}(\cdot, \bar{y}, \bar{u}) h_j \, dS(x) = \delta_{ij}, \; i, j \in I_0. \tag{2.2}$$

Now we establish the first order necessary conditions for optimality satisfied by $\bar{u}$.

Theorem 4. *Let us assume that* (2.1) *holds. Then there exist real numbers* $\{\bar\lambda_j\}_{j=1}^m \subset \mathbb{R}$ *and functions* $\bar y \in H^1(\Omega) \cap C^\alpha(\bar\Omega)$, *for some* $\alpha \in (0,1)$, *and* $\bar\varphi \in W^{1,s}(\Omega)$ *for all* $s < n/(n-1)$ *such that*

$$\bar\lambda_j \geq 0, \quad m_1 \leq j \leq m, \ \bar\lambda_j = 0 \text{ if } j \in I_-; \tag{2.3}$$

$$\begin{cases} -\Delta \bar y + \bar y = 0 & \text{in } \Omega \\ \partial_\nu \bar y = b(\cdot, \bar y, \bar u) & \text{on } \Gamma, \end{cases} \tag{2.4}$$

$$\begin{cases} -\Delta\bar\varphi + \bar\varphi = \dfrac{\partial f}{\partial y}(\cdot,\bar y) + \displaystyle\sum_{j=1}^m \bar\lambda_j F_j'(\bar y)|_\Omega & \text{in } \Omega \\ \partial_\nu\bar\varphi = \dfrac{\partial b}{\partial y}(\cdot,\bar y,\bar u)\bar\varphi + \dfrac{\partial g}{\partial y}(\cdot,\bar y,\bar u) + \displaystyle\sum_{j=1}^m \bar\lambda_j F_j'(\bar y)|_\Gamma & \text{on } \Gamma. \end{cases} \tag{2.5}$$

$$\int_\Gamma \left[\bar\varphi\frac{\partial b}{\partial u}(\cdot,\bar y,\bar u) + \frac{\partial g}{\partial u}(\cdot,\bar y,\bar u)\right](u-\bar u)\,dS(x) \geq 0 \quad \text{for all } u_a \leq u \leq u_b. \tag{2.6}$$

Moreover, if $\bar\varphi_0$ *is the solution of* (1.9) *and* $\bar\varphi_j$ *is the solution of* (1.11), $1 \leq j \leq m$, *then*

$$\bar\varphi = \bar\varphi_0 + \sum_{j=1}^m \bar\lambda_j\bar\varphi_j. \tag{2.7}$$

The second order necessary optimality conditions are stated in the following theorem:

Theorem 5. *Let* $\bar u$ *be a local solution of* (P) *and* $\bar y$, $\bar\varphi$, $\{\bar\varphi_j\}_{j=0}^m$ *and* $\{\bar\lambda_j\}_{j=1}^m$ *given by Theorems 2, 3 and 4. Let us assume that the regularity hypothesis* (2.1) *holds. Suppose that* $h \in L^\infty(\Gamma)$ *is any direction satisfying*

(2.8)

$$\int_\Gamma \bar\varphi_j\frac{\partial b}{\partial u}(\cdot,\bar y,\bar u)h\,dS(x) = 0 \text{ if } (j \leq m_1) \text{ or } (j > m_1,\ F_j(\bar y) = 0 \text{ and } \bar\lambda_j > 0);$$

$$\int_\Gamma \bar\varphi_j\frac{\partial b}{\partial u}(\cdot,\bar y,\bar u)h\,dS(x) \leq 0 \quad \text{if } j > m_1,\ F_j(\bar y) = 0 \text{ and } \bar\lambda_j = 0;$$

$$h(x) = \begin{cases} \geq 0 & \text{if } \bar u(x) = u_a(x); \\ \leq 0 & \text{if } \bar u(x) = u_b(x) \end{cases}$$

and put

$$d(x) = \bar\varphi(x)\frac{\partial b}{\partial u}(x,\bar y(x),\bar u(x)) + \frac{\partial g}{\partial u}(x,\bar y(x),\bar u(x)).$$

Then the following inequality is satisfied

$$\int_\Omega \frac{\partial^2 f}{\partial y^2}(\cdot,\bar y)z_h^2\,dx + \int_\Gamma\left[\bar\varphi\frac{\partial^2 b}{\partial y^2}(\cdot,\bar y,\bar u) + \frac{\partial^2 g}{\partial y^2}(\cdot,\bar y,\bar u)\right]z_h^2\,dS(x) +$$

(2.9)
$$2\int_\Gamma \left[\bar\varphi \frac{\partial^2 b}{\partial y \partial u}(\cdot,\bar y,\bar u) + \frac{\partial^2 g}{\partial y \partial u}(\cdot,\bar y,\bar u)\right] z_h h dS(x) +$$
$$\int_\Gamma \left[\bar\varphi \frac{\partial^2 b}{\partial u^2}(\cdot,\bar y,\bar u) + \frac{\partial^2 g}{\partial u^2}(\cdot,\bar y,\bar u)\right] h^2 dS(x) +$$
$$\sum_{j=1}^m \bar\lambda_j F_j''(\bar y) z_h^2 \geq -2\|d\|_{L^2(\Gamma)}\|h\|^2_{L^2(\Gamma_d)},$$

where

$$\Gamma_d = \{x \in \Gamma : |d(x)| > 0\}$$

and $z_h \in C(\bar\Omega) \cap H^1(\Omega)$ *is the solution of*

(2.10)
$$\begin{cases} -\Delta z_h + z_h = 0 & \text{in } \Omega \\ \partial_\nu z_h = \dfrac{\partial b}{\partial y}(\cdot,\bar y,\bar u) z_h + \dfrac{\partial b}{\partial u}(\cdot,\bar y,\bar u) h & \text{on } \Gamma. \end{cases}$$

For the proof of the previous two theorems the reader is referred to [2].

3. Sufficient Optimality Conditions

The theory of second order *sufficient* condition has to deal with two essential difficulties. First, we are faced with the well-known two-norm discrepancy: The L_2-norm appearing in the second order coercivity assumption cannot be used to define the Fréchet derivatives of the nonlinearities in the control problem. Second, it is difficult to select those control- or state-constraints, which are said to be strongly active in some sense.

The two-norm discrepancy leads to additional assumptions on the regularity of the given functionals. Therefore, we exclude point functionals having a support up to the boundary Γ. In this case, we are able to deal with domains Ω of arbitrary dimension n, while otherwise we would have the restriction to $n = 2$. Following [5], the second difficulty is overcome by introducing a set Γ^τ of strongly active control constraints.

On the other hand, here we can take advantage of the simple structure of the state-constraints, while [4] dealt with a quite general type of pointwise state-constraints. It is natural that all inequality state-constraints are strongly active, which have positive Lagrange multiplier. In the critical cone of the second order necessary condition, these constraints appeared as homogeneous equalities. This should be the same for the *sufficient* condition. In this section, we are able to confirm this expectation, while in [4] we had to work with associated inequalities. This essential improvement is based on a different method of proof. Moreover, the splitting technique of [4] is avoided at the expense of a slightly modified coercivity condition.

We need the following additional assumptions. The functions ψ_M^1 and ψ_M^2 introduced in (1.3) and (1.4) satisfy

$$\psi_M^1 \in L^\infty(\Omega) \text{ and } \psi_M^2 \in L^\infty(\Gamma). \tag{3.1}$$

There exist an open set $\Omega_0 \subset \bar{\Omega}_0 \subset \Omega$ and functions $\phi_j \in L^p(\Omega \setminus \Omega_0)$, $p > n/2$, and $\psi_j \in L^q(\Gamma)$, $q > n-1$, $1 \le j \le m$, such that

$$\langle F_j'(y)|_{\bar{\Omega}\setminus\Omega_0}, \varphi\rangle = \int_{\Omega\setminus\Omega_0} \phi_j(x)\varphi(x)dx + \int_\Gamma \psi_j(x)\varphi(x)dS(x) \ \forall y, \varphi \in C(\bar{\Omega}). \tag{3.2}$$

Notice that the functionals are required to be more regular close to the boundary. In this way, we avoid point functionals with support up to the boundary. Assumption (3.2) is satisfied in our examples if

Example 1: $f_j(\cdot, 0) \in L^1(\Omega)$ and $\eta_M^j \in L^p(\Omega)$.

Example 2: $f_j(\cdot, 0) \in L^1(\Gamma)$ and $\eta_M^j \in L^q(\Gamma)$.

Example 3: $\{x_j\}_{j=1}^m \subset \Omega$.

Finally we state the second order sufficient optimality conditions

Theorem 6. *Let $\bar{u}$ be a feasible point of* (P) *satisfiying together with the functions $\bar{\varphi} \in W^{1,s}(\Omega)$, for all $s < n/(n-1)$, $\bar{y} \in H^1(\Omega) \cap C^\alpha(\bar{\Omega})$, for some $\alpha \in (0,1)$, and Lagrange multipliers $\{\bar{\lambda}_j\}_{j=1}^m \subset \mathbb{R}$ the first order necessary conditions* (2.3)–(2.7). *Assume that the regularity assumption* (2.1) *holds and define the set*

$$\Gamma^\tau = \{x \in \Gamma : |d(x)| \ge \tau\}.$$

Let the inequality

$$\begin{aligned} &\int_\Omega \frac{\partial^2 f}{\partial y^2}(\cdot, \bar{y}) z_h^2 dx + \int_\Gamma \left[\bar{\varphi}\frac{\partial^2 b}{\partial y^2}(\cdot, \bar{y}, \bar{u}) + \frac{\partial^2 g}{\partial y^2}(\cdot, \bar{y}, \bar{u})\right] z_h^2 dS(x) + \\ &2\int_\Gamma \left[\bar{\varphi}\frac{\partial^2 b}{\partial y \partial u}(\cdot, \bar{y}, \bar{u}) + \frac{\partial^2 g}{\partial y \partial u}(\cdot, \bar{y}, \bar{u})\right] z_h h dS(x) + \\ &\int_\Gamma \left[\bar{\varphi}\frac{\partial^2 b}{\partial u^2}(\cdot, \bar{y}, \bar{u}) + \frac{\partial^2 g}{\partial u^2}(\cdot, \bar{y}, \bar{u})\right] h^2 dS(x) + \sum_{j=1}^m \bar{\lambda}_j F_j''(\bar{y}) z_h^2 \ge \\ &\delta_1 \|h\|^2_{L^2(\Gamma\setminus\Gamma^\tau)} - \delta_2 \|h\|^2_{L^2(\Gamma^\tau)}, \end{aligned} \tag{3.3}$$

hold for every h satisfying (2.8), *some $\delta_1 > 0$, $\delta_2 \ge 0$, and some $\tau > 0$, where $z_h \in H^1(\Omega) \cap C^\alpha(\bar{\Omega})$ is the solution of* (2.10). *Then there exist $\epsilon > 0$ and $\delta > 0$ such that*

$$J(u) \ge J(\bar{u}) + \delta \|u - \bar{u}\|^2_{L^2(\Gamma)}$$

holds for every feasible control u such that

$$\|u - \bar{u}\|_{L^\infty(\Gamma)} < \epsilon.$$

The (direct) proof of this result is quite long. It will be published elsewhere.

References

[1] F. Bonnans. Second order analysis for control constrained optimal control problems of semilinear elliptic systems. To appear in *Appl. Math. Optimization.*

[2] E. Casas and F. Tröltzsch. Second order necessary optimality conditions for some state-constrained control problems of semilinear elliptic equations. To appear in *Appl. Math. Optimization.*

[3] E. Casas, F. Tröltzsch, and A. Unger. Second order sufficient optimality conditions for a nonlinear elliptic control problem. *J. for Analysis and its Appl.* 15 (1996), pp. 687–707.

[4] E. Casas, F. Tröltzsch, and A. Unger. Second order sufficient optimality conditions for some state-constrained control problems of semilinear elliptic equations. To appear in *SIAM J. Control Optimization.*

[5] A.L. Dontchev, W.W. Hager, A.B. Poore, and B. Yang. Optimality, stability, and convergence in nonlinear control. *Appl. Math. Optimization* 31 (1995), No. 3, pp. 297–326.

[6] H. Goldberg, and F. Tröltzsch. Second order sufficient optimality conditions for a class of non–linear parabolic boundary control problems. *SIAM J. Control Optimization* 31 (1993), pp. 1007–1027.

[7] H. Maurer. First and second order sufficient optimality conditions in mathematical programming and optimal control. *Math. Programming Study* 14 (1981), pp. 163–177.

[8] H. Maurer, and J. Zowe. First– and second–order conditions in infinite–dimensional programming problems. *Math. Programming* 16 (1979), pp. 98–110.

Eduardo Casas
Departamento de Matemática Aplicada y Ciencias de la Computación
E.T.S.I. Industriales y de Telecomunicación
Universidad de Cantabria
39071 Santander, Spain

Fredi Tröltzsch, Andreas Unger
Fakultät für Mathematik
Technische Universität Chemnitz-Zwickau
D-09107 Chemnitz, Germany

International Series of Numerical Mathematics
Vol. 133, © 1999 Birkhäuser Verlag Basel/Switzerland

Intrinsic $P(2,1)$ Thin Shell Model and Naghdi's Models without A Priori Assumption on the Stress Tensor

Michel C. Delfour

Abstract. In earlier papers (cf. [6, 7, 8, 9, 5]) a completely intrinsic differential calculus on $C^{1,1}$ submanifolds of codimension one in $\mathbf{R}^N$ has been developed. Its potential has been illustrated by investigating some linear models of thin shells based on truncated series expansions with respect to the variable normal to the midsurface. In this paper we characterize the solution space of the $P(2,1)$ model for an arbitrary constitutive law and a midsurface with Lipschitzian boundary in a $C^{1,1}$ submanifold of $\mathbf{R}^N$. We further obtain Naghdi's thin shell models by elimination of variables in the $P(2,1)$ model without using the a priori assumption $\sigma_{33}=0$ on the stress tensor σ.

1. Introduction

In recent papers (cf. [10, 11, 5]) it was established that the polynomial $P(2,1)$ model is both pertinent and basic in the theory of *thin shells*. It was shown in [5] that its solution converges to the solution of a coupled system of variational equations which yields (as the thickness $2h$ goes to zero) the *membrane shell equation* and the *asymptotic bending equation* for the plate or the bending dominated shell. Moreover it yields by variable elimination and approximation *Naghdi's modified model* without using the a priori assumption $\sigma_{33}=0$ on the stress tensor σ.

The object of this paper is to further investigate the $P(2,1)$ model and its relationship to *Naghdi's model*. We first give a complete characterization of the space E^{01} of solutions of the $P(2,1)$ model which was introduced in the form of a completion in [5]. Out of the three vectors (v_h^0, v_h^1, v_h^2) which specify its solution only v_h^0 and the tangential component $v_{h\Gamma}^1$ of v_h^1 belong to the Sobolev space H^1. The normal components v_{hn}^1 and v_{hn}^2 are L^2-functions and the tangential component $v_{h\Gamma}^2$ of v_h^2 belongs to H^{-1}. This is due to the boundary layer phenomenon developing in the high-order terms of the solution as the thickness goes to zero.

We first eliminate the *rough variable* v_h^2 and get the *reduced $P(2,1)$ model* for (v_h^0, v_h^1). There is no approximation involved. The substitution process shifts quadratic terms involving v_h^2 to the linear right-hand side, but the total strain energy remains unchanged. The two-dimensional *effective constitutive law* associated with a general three-dimensional constitutive law is naturally introduced and an explicit formula is given as a by-product of the substitution process.

By a similar process the normal component v^1_{hn} of v^1_h is eliminated and a new variational equation is obtained for $(v^0_h, v^1_{h\Gamma})$ in $H^1 \times H^1$ with a second effective constitutive law. Again no approximation is made. Quadratic terms involving v^1_{hn} are shifted to the linear right-hand side and the total strain energy is not affected. This model contains many terms. It can be simplified by only retaining terms of order 1 or h^2 and absorbing other terms into dominating ones. The maximal simplification yields *Naghdi's modified model* without the a priori assumption $\sigma_{33} = 0$ on the stress tensor σ. The analogue of Naghdi's model and modified model are discussed in [1] (cf. eq. (7.1) p. 35) for Koiter's model which is a special case of Naghdi's model under the Love-Kirchhoff condition. The two models differ by two related tensors. The analogue of the tensor occuring in Naghdi's modified model was introduced in [12, 13] and [14]. Its relative advantages are discussed in [4].

Naghdi's model and modified model both converge to the right asymptotic model for plates and bending dominated shells. However the general asymptotic model contains a coupling term involving the mean curvature and the strain tensor associated with the membrane energy (cf [5]). Naghdi's models can be enriched by a more careful approximation and selective elimination of the quadratic terms in order to preserve the general asymptotic model. This new model will be referred to as *Naghdi's enriched model*

NOTATION AND BACKGROUND MATERIAL. For a detailed account of the intrinsic differential calculus on a $C^{1,1}$-submanifold, the reader is referred to the now available lecture notes [9, 5]. The inner product in $\mathbf{R}^N$ and the double inner product in $\mathcal{L}(\mathbf{R}^N;\mathbf{R}^N)$, the space of $N \times N$ matrices or tensors, are denoted as

$$x \cdot y = \sum_{i=1}^{N} x_i\, y_i, \quad A \cdot\cdot B = \sum_{i=1}^{N}\sum_{j=1}^{N} A_{ij}\, B_{ij}.$$

${}^*A$ denotes the transpose of A. Given Ω in $\mathbf{R}^N$, Ω not empty (resp. $\Gamma \overset{\text{def}}{=} \partial\Omega$ not empty) the *distance function* (resp. *oriented distance function*) is defined as

$$d_\Omega(x) \overset{\text{def}}{=} \inf_{y\in\Omega} |y - x| \text{ (resp. } b_\Omega(x) = d_\Omega(x) - d_{\mathbf{R}^N - \Omega}(x)).$$

When Ω is a domain of class $C^{1,1}$ in $\mathbf{R}^N$, $b = b_\Omega$ is $C^{1,1}$ in a neighborhood of every point of Γ and the converse is true. Its gradient ∇b coincides with the exterior unit normal n to the boundary on Γ. The *projection* p onto Γ and the *orthogonal projection* P onto the *tangent plane* $T_x\Gamma$ are given by

$$p(x) \overset{\text{def}}{=} x - b(x)\,\nabla b(x), \quad P(x) \overset{\text{def}}{=} I - \nabla b(x)\,{}^*\nabla b(x),$$

where ${}^*V$ denotes the transpose of a column vector V in $\mathbf{R}^N$. Given $h > 0$ and an open domain ω in Γ, a *shell* is the open domain

$$S_h(\omega) \overset{\text{def}}{=} \left\{x \in \mathbf{R}^N : \left|b_\Omega(x)\right| < h, p(x) \in \omega\right\}$$

in $\mathbf{R}^{\mathrm{N}}$. When $\omega = \Gamma$ the shell has *no boundary*; otherwise we denote by γ the (relative) *boundary* of ω in Γ and by

$$\Sigma_h(\gamma) \stackrel{\text{def}}{=} \{x \in \mathbf{R}^{\mathrm{N}} : |b_\Omega(x)| < h, p(x) \in \gamma\}$$

its *lateral boundary.* For $b \in C^{1,1}(S_{2h}(\omega))$, $|z| < h$, $X \in \omega$, define

$$T_z(X) \stackrel{\text{def}}{=} X + z\,\nabla b(X),\ j_z \stackrel{\text{def}}{=} \det DT_z(X) = \det\big[I + z\,D^2 b(X)\big] = \sum_{i=0}^{N-1} \kappa_i(X)\,z^i,$$

where the κ_i's are the coefficients of the polynomial j_z of degree $(N-1)$ in z. For $n \geq 0$ also define the following functions of $X \in \omega$ and their averages

$$\boxed{\alpha_n(h) \stackrel{\text{def}}{=} \int_{-h}^{h} j_z\, z^n\, dz = \sum_{j=n}^{n+N-1} \left[1 - (-1)^{j+1}\right] \frac{h^{j+1}}{j+1} \kappa_{j-n},\ \overline{\alpha}_n(h) \stackrel{\text{def}}{=} \frac{\alpha_n(h)}{2h}} \tag{1}$$

The $\alpha_n(h)$'s are polynomials in odd powers of h. As h goes to zero

$$\frac{\overline{\alpha}_0(h) - 1}{h^2} \to \frac{1}{3}\kappa_2, \quad \frac{\overline{\alpha}_1(h)}{h^2} \to \frac{1}{3}\kappa_1, \quad \frac{\overline{\alpha}_2(h)}{h^2} \to \frac{1}{3}, \quad \frac{\overline{\alpha}_3(h)}{h^4} \to \frac{1}{5}\kappa_1.$$

For $N = 3$, $\kappa_1 = \Delta b$ is twice the *mean curvature* and κ_2 the *Gauss curvature.* It will be convenient to introduce the following notation for the decompositions of an $N \times N$ matrix τ into its tangential and normal parts along ω

$$\tau^P \stackrel{\text{def}}{=} P\tau P, \quad \tau_{nn} \stackrel{\text{def}}{=} \tau n \cdot n, \quad \tau^n \stackrel{\text{def}}{=} \tau - \tau_{nn}\, n^* n$$

$$\tau = \tau^P + (P\tau n)^* n + n^*(P\tau n) + \tau_{nn}\, n^* n = \tau^n + \tau_{nn}\, n^* n$$

and the spaces of symmetric matrices

$$\mathrm{Sym}_{\mathrm{N}} \stackrel{\text{def}}{=} \{\tau \in \mathcal{L}(\mathbf{R}^{\mathrm{N}}; \mathbf{R}^{\mathrm{N}}) : {}^*\tau = \tau\}, \quad \mathrm{Sym}_{\mathrm{N}}^{\mathrm{n}} \stackrel{\text{def}}{=} \{\tau \in \mathrm{Sym}_{\mathrm{N}} : \tau_{nn} = 0\}$$

$$\mathrm{Sym}_{\mathrm{N}}^{\mathrm{P}} \stackrel{\text{def}}{=} \{\tau \in \mathrm{Sym}_{\mathrm{N}} : \tau n = 0\} \implies \forall \tau \in \mathrm{Sym}_{\mathrm{N}},\ \tau^P \in \mathrm{Sym}_{\mathrm{N}}^{\mathrm{P}} \text{ and } \tau^n \in \mathrm{Sym}_{\mathrm{N}}^{\mathrm{n}}.$$

2. The $P(2,1)$ Thin Shell Model

2.1. Equation of linear elasticity

We first deal with a shell without boundary or with homogeneous Neumann boundary conditions. For a shell with homogeneous Dirichlet boundary conditions on a part γ_0 of the boundary γ the constructions, proofs and results are similar.

Assumption 2.1. *The* compliance C *(or the* constitutive law C^{-1}*) is a linear bijective and symmetrical transformation* $C\colon \mathrm{Sym}_{\mathrm{N}} \to \mathrm{Sym}_{\mathrm{N}}$ *for which there exists a constant* $\alpha > 0$ *such that* $C^{-1}\tau \cdot\cdot\, \tau \geq \alpha\, \tau \cdot\cdot\, \tau$ *for all* $\tau \in \mathrm{Sym}_{\mathrm{N}}$.

For instance for the Lamé constants $\mu > 0$ and $\lambda \geq 0$, the special constitutive law $C^{-1}\tau = 2\mu\,\tau + \lambda\, \mathrm{tr}\,\tau\, I$ verifies Assumption 2.1 with $\alpha = 2\mu$.

Assumption 2.2. *Assume that* $\exists \bar{h} > 0$ *such that* $b \in C^{1,1}(S_{\bar{h}}(\omega))$ *and*

$$\exists \beta,\ 0 < \beta < 1, \forall X \in \omega, \quad \bar{h}\,\|D^2 b(X)\| \leq \beta.$$

Given such a C, the N-dimensional variational equation of linear elasticity

$$\boxed{\begin{aligned}&\exists V(h) \in H^1\big(S_h(\omega)\big)^N \text{ such that } \forall V \in H^1\big(S_h(\omega)\big)^N\\ &\int_{S_h(\omega)} C^{-1}\varepsilon\big(V(h)\big)\cdot\cdot\,\varepsilon(V) - F\cdot V - G\cdot DV\nabla b\,dx = 0\end{aligned}} \tag{2}$$

makes sense as a variational equation in $H^1\big(S_h(\omega)\big)^N/\ker\varepsilon$ for vector functions F and G in $L^2\big(S_h(\omega)\big)^N$ verifying the condition

$$\forall V \in \ker\varepsilon,\quad \int_{S_h(\omega)} F\cdot V + G\cdot DV\nabla b\,dx = 0,\quad \varepsilon(V) \stackrel{\text{def}}{=} \frac{1}{2}(DV + {}^*DV). \tag{3}$$

From this condition there exists a constant $c(h) = c\big(S_h(\omega)\big) > 0$ such that

$$\boxed{\forall V \in H^1\big(S_h(\omega)\big)^N,\ \left|\int_{S_h(\omega)} F\cdot V + G\cdot DV\nabla b\,dx\right| \le c(h)\,\big\|\varepsilon(V)\big\|_{L^2(S_h(\omega))}.} \tag{4}$$

Assume that ω is a bounded open connected domain in Γ for which Assumptions 2.2 is verified. Let γ_0 be a subset of the boundary γ with non-zero $(N-1)$-capacity. For the N-dimensional Dirichlet boundary conditions the underlying space can be chosen as

$$H^1_{\gamma_0}\big(S_h(\omega)\big)^N \stackrel{\text{def}}{=} \big\{V \in H^1\big(S_h(\omega)\big)^N : V|_{\Sigma_h(\gamma_0)} = 0\big\}, \tag{5}$$

where $\Sigma_h(\gamma_0)$ is the piece of the lateral boundary whose projection onto Γ is equal to γ_0. With $H^1_{\gamma_0}\big(S_h(\omega)\big)^N$ in place of $H^1\big(S_h(\omega)\big)^N$ in (2) and (3), the variational equation (2) has a unique solution in $H^1_{\gamma_0}\big(S_h(\omega)\big)^N$ under condition (3).

2.2. The $P(2,1)$ model and characterization of the space E^{01} of solutions

We now proceed by polynomial approximation $P(k,\ell)$ of order k for the *displacement vector* $v = V\circ T_z$ and of order ℓ for the *linear strain tensor* $\varepsilon(V)\circ T_z$. For the $P(2,1)$ model

$$V\circ T_z \simeq v^0 + z\,v^1 + z^2\,v^2 \text{ and } \varepsilon(V)\circ T_z \simeq \varepsilon^0(v^0,v^1) + z\,\varepsilon^1(v^0,v^1,v^2)$$

where the $\varepsilon^i(v)$'s are given by the expressions

$$\begin{aligned}\varepsilon^0(v^0,v^1) &\stackrel{\text{def}}{=} \varepsilon^0(v) = \frac{1}{2}\,(v^1\,{}^*n + n\,{}^*v^1) + \varepsilon_\Gamma(v^0)\\ \varepsilon^1(v^0,v^1,v^2) &\stackrel{\text{def}}{=} [v^2\,{}^*n + n\,{}^*v^2] + \varepsilon_\Gamma(v^1) - \frac{1}{2}\big[D_\Gamma(v^0)\,D^2b + D^2b\,{}^*D_\Gamma(v^0)\big].\end{aligned} \tag{6}$$

It will be convenient to introduce the tensor

$$\boxed{e^1(v^0,v^1) \stackrel{\text{def}}{=} \varepsilon^1(v^0,v^1,0).} \tag{7}$$

and use the compact notation $v = (v^0,v^1,v^2)$, $e^1(v) = e^1(v^0,v^1)$, $\varepsilon^0(v) = \varepsilon^0(v^0,v^1)$, and $\varepsilon^1(v) = \varepsilon^1(v^0,v^1,v^2)$.

The natural space of solution associated with the tensors $\varepsilon^0(v)$ and $\varepsilon^1(v)$ will involve the so called *approximate rigid displacements*. Under Assumptions 2.2 the set of *approximate rigid displacements* associated with a bounded open domain ω with a Lipschitzian boundary γ in Γ is the finite dimensional subspace (cf. [9, 5])

$$\boxed{\begin{aligned} K \stackrel{\text{def}}{=} \big\{(v^0, v^1) \colon & (v^0, v^1) \in H^1(\omega)^N \times H^1(\omega)^N \\ & \varepsilon^0(v^0, v^1) = 0 \text{ and } \varepsilon^1(v^0, v^1, 0) = 0\big\} \\ = \big\{(v^0, v^1) \colon & v^0(X) = a + AX, \quad v^1(X) = An(X) \\ & \forall a \in \mathbf{R}^{\mathrm{N}} \text{ and all matrices } A \text{ such that } A + {}^*A = 0\big\} \end{aligned}}$$

$$\boxed{\begin{aligned} \ker \varepsilon^0 \cap \ker \varepsilon^1 = \big\{(v^0, v^1, v^2) \colon & (v^0, v^1, v^2) \in H^1(\omega)^N \times H^1(\omega)^N \times L^2(\omega)^N \\ & \varepsilon^0(v^0, v^1) = 0 \text{ and } \varepsilon^1(v^0, v^1, v^2) = 0\big\} \\ = \big\{(v^0, v^1, 0) \colon & (v^0, v^1) \in K\big\}. \end{aligned}}$$

The natural *space of solution* is then the completion E^{01} of the quotient space

$$\frac{H^1(\omega)^N \times H^1(\omega)^N \times L^2(\omega)^N}{\ker \varepsilon^0 \cap \ker \varepsilon^1} = \frac{H^1(\omega)^N \times H^1(\omega)^N}{K} \times L^2(\omega)^N$$

with respect to the norm

$$\big\{\|\varepsilon^0(v^0, v^1)\|^2_{L^2(\omega)} + \|\varepsilon^1(v^0, v^1, v^2)\|^2_{L^2(\omega)}\big\}^{1/2}. \tag{8}$$

Denote by $H^1_{\gamma_0}(\omega)$ the space $\big\{v \in H^1(\omega) \colon v|_{\gamma_0} = 0\big\}$. The space of solutions associated with the Dirichlet case will be the completion $E^{01}_{\gamma_0}$ of $H^1_{\gamma_0}(\omega) \times H^1_{\gamma_0}(\omega) \times L^2(\omega)^N$ with respect to the norm (8). With this notation the $P(2,1)$ model is characterized by the following variational equation: to find $v_h = (v^0_h, v^1_h, v^2_h)$ in E^{01} (resp. $E^{01}_{\gamma_0}$) such that for all $v = (v^0, v^1, v^2)$ in E^{01} (resp. $E^{01}_{\gamma_0}$)

$$\boxed{\begin{aligned} \int_\omega & \overline{\alpha}_0(h)\, C^{-1}\varepsilon^0_h \cdot\cdot\, \varepsilon^0 + \overline{\alpha}_1(h)\, \big[C^{-1}\varepsilon^1_h \cdot\cdot\, \varepsilon^0 + C^{-1}\varepsilon^0_h \cdot\cdot\, \varepsilon^1\big] \\ & + \overline{\alpha}_2(h)\, C^{-1}\varepsilon^1_h \cdot\cdot\, \varepsilon^1 \, d\Gamma = \ell_h(v^0, v^1, v^2) \end{aligned}} \tag{9}$$

where $\varepsilon^i_h \stackrel{\text{def}}{=} \varepsilon^i(v_h)$, $\varepsilon^i \stackrel{\text{def}}{=} \varepsilon^i(v)$, and $\ell_h(v^0, v^1, v^2)$ is defined as

$$\stackrel{\text{def}}{=} \frac{1}{2h} \int_{S_h(\omega)} F \cdot (v^0 \circ p + bv^1 \circ p + b^2 v^2 \circ p) + G \cdot (v^1 \circ p + 2\, b\, v^2 \circ p)\, dx \tag{10}$$

Further assume that F and G in $L^2\big(S_h(\omega)\big)^N$ verify the condition: there exists $c_h > 0$ such that for all v in E^{01} (resp. $E^{01}_{\gamma_0}$)

$$\big|\ell_h(v^0, v^1, v^2)\big| \le c_h \big\{\|\varepsilon^0(v^0, v^1)\|^2_{L^2(\omega)} + \|\varepsilon^1(v^0, v^1, v^2)\|^2_{L^2(\omega)}\big\}^{1/2} \tag{11}$$

Then for h sufficiently small the variational equation (9) has a unique solution $v_h = (v^0_h, v^1_h, v^2_h)$ in E^{01} (resp. $E^{01}_{\gamma_0}$).

The structure of the space of solutions E^{01} can be further specified.

Theorem 2.3. *Under Assumptions* 2.1 *and* 2.2, *let* ω *be a bounded open domain with a Lipschitzian boundary* γ *in* Γ.

(i) ($P(1,1)$ model)

$$\left\{\left\|\varepsilon^0(v^0,v^1)\right\|^2_{L^2(\omega)} + \left\|\varepsilon^1(v^0,v^1,0)\right\|^2_{L^2(\omega)} + \|v^0\|^2_{L^2(\omega)} + \|v^1_\Gamma\|^2_{L^2(\omega)}\right\}^{1/2} \tag{12}$$

is a norm equivalent to the standard norm on $H^1(\omega)^N \times H^1(\omega)^N$ *and*

$$\left\{\left\|\varepsilon^0(v^0,v^1)\right\|^2_{L^2(\omega)} + \left\|\varepsilon^1(v^0,v^1,0)\right\|^2_{L^2(\omega)}\right\}^{1/2} \tag{13}$$

is a norm equivalent to the canonical quotient norm on

$$\boxed{Q \overset{\text{def}}{=} \frac{H^1(\omega)^N \times H^1(\omega)^N}{K}}$$

The linear subspace

$$\boxed{Q^{n0} \overset{\text{def}}{=} \left\{(v^0,v^1) \in H^1(\omega)^N \times H^1(\omega)^N \,:\, v^1_n = 0\right\}/K} \tag{14}$$

of Q *is closed with respect to the norm*

$$\left\{\|\varepsilon^0(v^0,v^1_\Gamma)\|^2 + \|e^1(v^0,v^1_\Gamma)\|^2\right\}^{1/2}. \tag{15}$$

For any $h>0$ *such that*

$$h\|D^2 b\|_{L^\infty(\omega)} \le 1/\sqrt{2}$$

$$\frac{1}{2}\left[\|\varepsilon^0(v^0,v^1_\Gamma)\|^2 + h^2\|e^1(v^0,v^1_\Gamma)\|^2\right] \le \|\varepsilon^0(v^0,v^1_\Gamma)\|^2 + h^2\|e^{1P}(v^0,v^1_\Gamma)\|^2$$

$$\|\varepsilon^0(v^0,v^1_\Gamma)\|^2 + h^2\|e^{1P}(v^0,v^1_\Gamma)\|^2 \le \|\varepsilon^0(v^0,v^1_\Gamma)\|^2 + h^2\|e^1(v^0,v^1_\Gamma)\|^2$$

and the norm (15) *is equivalent to the following norm on* Q^{n0}

$$\left\{\|\varepsilon^0(v^0,v^1_\Gamma)\|^2 + \|e^{1P}(v^0,v^1_\Gamma)\|^2\right\}^{1/2} \tag{16}$$

(ii) ($P(2,1)$ model)

$$\left\{\left\|\varepsilon^0(v^0,v^1)\right\|^2_{L^2(\omega)} + \left\|\varepsilon^1(v^0,v^1,v^2)\right\|^2_{L^2(\omega)} + \|v^0\|^2_{L^2(\omega)} + \|v^1_\Gamma\|^2_{L^2(\omega)} + \|v^2_\Gamma\|^2_{L^2(\omega)}\right\}^{1/2}$$

is a norm equivalent to the standard norm on $H^1(\omega)^N \times H^1(\omega)^N \times L^2(\omega)^N$. *The linear space*

$$\boxed{Q^n \overset{\text{def}}{=} \left\{(v^0,v^1) \in H^1(\omega)^N \times L^2(\omega)^N \,:\, v^1_\Gamma \in H^1(\omega)^N\right\}/K} \tag{17}$$

is closed for the norm

$$\left\{\|\varepsilon^0(v^0,v^1_\Gamma)\|^2 + \|v^1_n\|^2 + \|e^{1P}(v^0,v^1_\Gamma)\|^2\right\}^{1/2} \tag{18}$$

which is equivalent to either of the norms

$$\left\{\|\varepsilon^0(v^0,v^1)\|^2 + \|e^{1P}(v^0,v^1_\Gamma)\|^2\right\}^{1/2} \quad \boxed{\left\{\|\varepsilon^0(v^0,v^1)\|^2 + \|e^{1P}(v^0,v^1)\|^2\right\}^{1/2}} \tag{19}$$

The map

$$(v^0,v^1,v^2) \mapsto (v^0,v^1,P\varepsilon^1(v^0,v^1,v^2_\Gamma)n + v^2_n n) \,:\, E^{01} \to Q^n \times L^2(\omega)^N \tag{20}$$

is a continuous linear bijection and

$$E^{01} = \left\{ (v^0, v^1, v^2) \,:\, \begin{array}{c} (v^0, v^1) \in Q^n,\ v_n^2 \in L^2(\omega)\, and \\ v_\Gamma^2 \in H^{-1}(\omega)^N \ such\ that \\ P\varepsilon^1(0, v_n^1 n, v_\Gamma^2) \in L^2(\omega)^N \end{array} \right\} \tag{21}$$

In particular the closed linear subspace

$$\{(v^0, v^1, v^2) \in E^{01} \,:\, \varepsilon^1(v^0, v^1, v^2) n = 0\} \tag{22}$$

is isomorphic to Q^n *which is closed for the second norm* (19) *and the closed linear subspace*

$$\{(v^0, v^1, v^2) \in E^{01} \,:\, \varepsilon^1(v^0, v^1, v^2) n = 0 \ and\ \varepsilon^0(v^0, v^1)_{nn} = 0\} \tag{23}$$

is isomorphic to Q^{n0} *which is closed for the norm* (16) *or* (15).

The same considerations apply to the homogeneous Dirichlet boundary conditions on a part γ_0 of the boundary γ with the corresponding spaces $E^{01}_{\gamma_0}$,

$$Q_{\gamma_0} \stackrel{\text{def}}{=} \{(v^0, v^1) \in H^1_{\gamma_0}(\omega)^N \times H^1(\omega)^N : v^1_\Gamma|_{\gamma_0} = 0\} \tag{24}$$

$$Q^{n0}_{\gamma_0} \stackrel{\text{def}}{=} \{(v^0, v^1) \in H^1_{\gamma_0}(\omega)^N \times H^1_{\gamma_0}(\omega)^N \,:\, v^1_n = 0\} \tag{25}$$

$$Q^{n}_{\gamma_0} \stackrel{\text{def}}{=} \{(v^0, v^1) \in H^1_{\gamma_0}(\omega)^N \times L^2(\omega)^N \,:\, v^1_\Gamma \in H^1_{\gamma_0}(\omega)^N\} \tag{26}$$

3. Reduced $\boldsymbol{P(2,1)}$ Model: Elimination of the Variable $\boldsymbol{v_h^2}$

From condition (11) on $\ell_h(v^0, v^1, v^2)$, there exists $q_h \in L^2(\omega)^N$ such that

$$\forall v^2 \in L^2(\omega)^N, \quad \int_\Omega q_h \cdot v^2 \, d\Gamma \stackrel{\text{def}}{=} \ell_h(0, 0, v^2). \tag{27}$$

It is now possible to eliminate the variable v_h^2. In order to minimize the volume of the computations, first rewrite the variational equation in the form

$$\int_\omega \overline{\alpha}(h)\, C^{-1}\varepsilon_h^0 \cdot\cdot\, \varepsilon^0 + \frac{\overline{\alpha}_1(h)}{\overline{\alpha}_2(h)} C^{-1} \left(\overline{\alpha}_1(h)\varepsilon_h^0 + \overline{\alpha}_2(h)\varepsilon_h^1\right) \cdot\cdot\, \varepsilon^0 + C^{-1} \left(\overline{\alpha}_1(h)\varepsilon_h^0 + \overline{\alpha}_2(h)\varepsilon_h^1\right) \cdot\cdot\, \varepsilon^1 \, d\Gamma = \ell_h(v^0, v^1, v^2)$$

for all $v \in E^{01}$, where

$$\overline{\alpha}(h) \stackrel{\text{def}}{=} \overline{\alpha}_0(h) - \frac{\overline{\alpha}_1(h)^2}{\overline{\alpha}_2(h)} \tag{28}$$

Recall that $\varepsilon^1(v^0, v^1, v^2) = e^1(v^0, v^1) + v^2\, {}^*n + n\, {}^*v^2$. By choosing test functions of the form $(0, 0, v^2)$, we get

$$2\left[C^{-1}(\overline{\alpha}_1(h)\varepsilon_h^0 + \overline{\alpha}_2(h)\varepsilon_h^1)\right] n = q_h.$$

For smooth test functions $(v^0, v^1, 0)$ the variational equation reduces to

$$\int_\omega \overline{\alpha}(h)\, C^{-1}\varepsilon_h^0 \cdot\cdot\, \varepsilon^0 + \frac{\overline{\alpha}_1(h)}{\overline{\alpha}_2(h)} C^{-1}\left(\overline{\alpha}_1(h)\varepsilon_h^0 + \overline{\alpha}_2(h)\varepsilon_h^1\right) \cdot\cdot\, \varepsilon^0 \\ + C^{-1}\left(\overline{\alpha}_1(h)\varepsilon_h^0 + \overline{\alpha}_2(h)\varepsilon_h^1\right) \cdot\cdot\, e^1 \, d\Gamma = \ell_h(v^0, v^1, 0)$$

It turns out that the first equation uniquely determines v_h^2 as a function of (v_h^0, v_h^1). By introducing an effective constitutive law, we now eliminate v_h^2 from the second equation and get a variational equation for (v_h^0, v_h^1) in Q^n.

Theorem 3.1. [5] *Let* C *verify Assumption* 2.1. *The transformation of* $\mathbf{R}^{\mathrm{N}}$

$$\boxed{N(u) \overset{\text{def}}{=} [C^{-1}(u\,{}^*n + n\,{}^*u)]n, \quad u \in \mathbf{R}^{\mathrm{N}}} \tag{29}$$

and the effective constitutive law C_{eP}: $\mathrm{Sym}_{\mathrm{N}}^{\mathrm{P}} \to \mathrm{Sym}_{\mathrm{N}}^{\mathrm{P}}$ *defined as*

$$\boxed{C_{eP}^{-1}\tau \overset{\text{def}}{=} C^{-1}\{\tau - \{N^{-1}([C^{-1}\tau]n)\,{}^*n + n\,{}^*N^{-1}([C^{-1}\tau]n)\}\},\ \tau \in \mathrm{Sym}_{\mathrm{N}}^{\mathrm{P}}} \tag{30}$$

are bijective, symmetrical and coercive. For any $\sigma \in \mathrm{Sym}_{\mathrm{N}}$

$$\boxed{C^{-1}\tau \cdot\cdot\, \sigma = C_{eP}^{-1}\tau^P \cdot\cdot\, \sigma^P + 2[C^{-1}\tau]n \cdot N^{-1}([C^{-1}\sigma]n).} \tag{31}$$

In our case, recalling that $\varepsilon^{1P}(v^0, v^1, v^2) = e^{1P}(v^0, v^1)$, we can eliminate the variable v_h^2 from the original variational equation:

$$\boxed{\begin{aligned} &\int_\omega \overline{\alpha}(h)\, C^{-1}\varepsilon^0(v_h^0, v_h^1) \cdot\cdot\, \varepsilon^0(v^0, v^1) + \frac{\overline{\alpha}_1(h)^2}{\overline{\alpha}_2(h)} C_{eP}^{-1}\varepsilon_\Gamma^P(v_h^0) \cdot\cdot\, \varepsilon_\Gamma^P(v^0) \\ &\quad + \overline{\alpha}_1(h)\left[C_{eP}^{-1}e^{1P}(v_h^0, v_h^1) \cdot\cdot\, \varepsilon_\Gamma^P(v^0) + C_{eP}^{-1}\varepsilon_\Gamma^P(v_h^0) \cdot\cdot\, e^{1P}(v^0, v^1)\right] \\ &\quad + \overline{\alpha}_2(h) C_{eP}^{-1}e^{1P}(v_h^0, v_h^1) \cdot\cdot\, e^{1P}(v^0, v^1)\, d\Gamma \\ &= \ell_h\left(v^0, v^1, v^2 - N^{-1}\left(\left[C^{-1}\left(\varepsilon^1(v^0, v^1, v^2) + \frac{\overline{\alpha}_1(h)}{\overline{\alpha}_2(h)}\varepsilon^0(v^0, v^1)\right)\right]n\right)\right) \end{aligned}} \tag{32}$$

In the substitution process to eliminate the variable v_h^2, the new quadratic part is continuous and coercive for the space Q^n (cf. Theorem 2.3 (iii)). The terms eliminated from the quadratic part now combine with those of the linear right-hand side where the test variable v^2 also disappears. It is easy to show that

$$\ell_h\left(v^0, v^1, v^2 - N^{-1}\left(\left[C^{-1}\left(\varepsilon^1(v^0, v^1, v^2) + \frac{\overline{\alpha}_1(h)}{\overline{\alpha}_2(h)}\varepsilon^0(v^0, v^1)\right)\right]n\right)\right) = \tilde{\ell}_h(v^0, v^1)$$

where $\tilde{\ell}_h(v^0, v^1)$ is the resulting reduced right-hand side

$$\boxed{\tilde{\ell}_h(v^0, v^1) \overset{\text{def}}{=} \ell_h\left(v^0, v^1, -N^{-1}\left(\left[C^{-1}\left(e^1(v^0, v^1) + \frac{\overline{\alpha}_1(h)}{\overline{\alpha}_2(h)}\varepsilon^0(v^0, v^1)\right)\right]n\right)\right)} \tag{33}$$

By assumption (11) there exists a constant $c_h' > 0$ such that

$$\forall (v^0, v^1) \in Q^n, \quad |\tilde{\ell}_h(v^0, v^1)| \leq c_h' \left(\|\varepsilon^0(v^0, v^1)\| + \|e^{1P}(v^0, v^1)\|\right).$$

In the substitution process, the right-hand side is modified is such a way that it becomes continuous with respect to the topology of Q^n associated with the new quadratic part of the equation. As a result the variational equation (32) becomes an equation in Q^n. By continuity and coercivity of the bilinear form and continuity of the linear form with respect to the norm (19) as h goes to zero, there exists a unique solution in Q^n. For the special constitutive law $C^{-1}\varepsilon = 2\mu\varepsilon + \lambda \operatorname{tr}\varepsilon I$, it is easy to check that the associated *effective constitutive law* C_{eP}^{-1} is

$$\boxed{C_{eP}^{-1}\tau = 2\mu\tau + \frac{2\mu\lambda}{2\mu+\lambda}(\operatorname{tr}\tau)P, \quad \tau\in \mathrm{Sym}_{\mathrm{N}}^{\mathrm{P}}.} \tag{34}$$

4. Naghdi's Models: Elimination of the Variable v^1_{hn}

In this section as in §3 we only cover the cases of shells without boundary and shells with homogeneous Neumann boundary conditions. However the same considerations apply to and the same results and variational equations hold for homogeneous Dirichlet boundary conditions on a part γ_0 of the boundary γ with the corresponding spaces $E^{01}_{\gamma_0}$, Q_{γ_0}, $Q^{n0}_{\gamma_0}$, and $Q^n_{\gamma_0}$ as specified by (24), (25) and (26).

4.1. Naghdi's model and modified model

The mathematical existence and uniqueness theory of Naghdi's model can be found in [2, 1]. It has been shown in [9] that for the special constitutive law $C^{-1}\varepsilon = 2\mu\varepsilon + \lambda \operatorname{tr}\varepsilon I$, the quadratic part of Naghdi's model can be written in the form

$$\boxed{\int_\omega C_e^{-1}\varepsilon^0(v_h^0, v_{h\Gamma}^1)\cdot\cdot\,\varepsilon^0(v^0,v_\Gamma^1) + \frac{h^2}{3}C_e^{-1}\overline{e}^{1P}(v_h^0, v_{h\Gamma}^1)\cdot\cdot\,\overline{e}^{1P}(v^0,v_\Gamma^1)\,d\Gamma} \tag{35}$$

where $C_e^{-1}\varepsilon = 2\mu\varepsilon + (2\mu\lambda)/(2\mu+\lambda)\operatorname{tr}\varepsilon I$ and the tensor

$$\boxed{\overline{e}^{1P}(v^0,v_\Gamma^1) \overset{\text{def}}{=} \varepsilon_\Gamma^P(v^1) + \frac{1}{2}\left[D^2b\,D_\Gamma^P(v^0) + {}^*D_\Gamma^P(v^0)\,D^2b\right]} \tag{36}$$

is associated with the tensor

$$\begin{aligned}
&\overline{\varepsilon}^1(v^0,v^1,v^2) \overset{\text{def}}{=} v^2\,{}^*n + n\,{}^*v^2 + \varepsilon_\Gamma(v^1) + \frac{1}{2}\left[D^2b\,D_\Gamma(v^0) + {}^*D_\Gamma(v^0)\,D^2b\right]\\
\Rightarrow\ &\overline{\varepsilon}^{1P}(v^0,v^1,v^2) = \varepsilon_\Gamma^P(v^1) + \frac{1}{2}\left[D^2b\,D_\Gamma^P(v^0) + {}^*D_\Gamma^P(v^0)\,D^2b\right] = \overline{e}^{1P}(v^0,v^1).
\end{aligned}$$

This is compatible with the notation $\overline{e}^1(v^0,v^1) \overset{\text{def}}{=} \overline{\varepsilon}^1(v^0,v^1,0)$, and it is readily seen that the tensors $\overline{\varepsilon}^1(v^0,v^1,v^2)$ and $\varepsilon^1(v^0,v^1,v^2)$ are related as follows

$$\begin{aligned}
\overline{\varepsilon}^1(v^0,v^1,v^2) &= \varepsilon^1(v^0,v^1,v^2) + \frac{1}{2}\left[D^2b\,\varepsilon_\Gamma(v^0) + \varepsilon_\Gamma(v^0)\,D^2b\right]\\
\overline{\varepsilon}^{1P}(v^0,v^1,v^2) &= \varepsilon^{1P}(v^0,v^1,v^2) + \frac{1}{2}\left[D^2b\,\varepsilon_\Gamma^P(v^0) + \varepsilon_\Gamma^P(v^0)\,D^2b\right]
\end{aligned}$$

$$\Rightarrow \boxed{\overline{e}^{1P}(v^0,v^1) = e^{1P}(v^0,v^1) + \frac{1}{2}\left[D^2b\,\varepsilon_\Gamma^P(v^0) + \varepsilon_\Gamma^P(v^0)\,D^2b\right] = \overline{\varepsilon}^{1P}(v^0,v^1,0).}$$

In our analysis the tensors ε^1, e^1 and $e^{1P}(v^0, v^1_\Gamma)$ will naturally occur and we shall consider *Naghdi's modified model* which is characterized by the quadratic form

$$\boxed{\int_\omega C_e^{-1}\varepsilon^0(v_h^0, v_{h\Gamma}^1)\cdot\cdot\, \varepsilon^0(v^0, v_\Gamma^1) + \frac{h^2}{3} C_e^{-1} e^{1P}(v_h^0, v_{h\Gamma}^1)\cdot\cdot\, e^{1P}(v^0, v_\Gamma^1)\, d\Gamma} \tag{37}$$

The two models (35) and (37) enjoy the same mathematical properties and have the same space of solutions. The two related tensors are discussed for Koiter's model in [1] (cf. equation (7.1) p. 35) who points out that the analogue of the tensor e^{1P} was introduced in [12, 13] and [14]. Its advantages over $\overline{e}^{1P}$ are discussed in [4].

4.2. From the reduced $P(2,1)$ model to Naghdi's modified model

When compared to the reduced $P(2,1)$ model characterized by the variational equation (32), the main difference with Naghdi's modified model is the absence of the variable v_{hn}^1. By choosing test functions of the form $(0, v_n^1\, n)$ in (32) we get an equation for v_{hn}^1 as a function of $(v_h^0, v_{h\Gamma}^1)$. This follows from the fact that, for the reduced $P(2,1)$ model, v_n^1 can be isolated in the two tensors ε^0 and e^{1P}

$$\varepsilon^0(v^0, v^1) = \varepsilon^0(v^0, v_\Gamma^1) + v_n^1\, n\, {}^*n \quad \Rightarrow \|\varepsilon^0(v^0, v^1)\|^2 = \|\varepsilon^0(v^0, v_\Gamma^1)\|^2 + \|v_n^1\|^2$$
$$e^{1P}(v^0, v^1) = e^{1P}(v^0, v_\Gamma^1) + v_n^1\, D^2 b.$$

Therefore

$$\frac{1}{2}\left\{\|\varepsilon^0(v^0, v_\Gamma^1)\| + \|v_n^1\|\right\} \le \|\varepsilon^0(v^0, v^1)\| \le \|\varepsilon^0(v^0, v_\Gamma^1)\| + \|v_n^1\|$$
$$\|e^{1P}(v^0, v_\Gamma^1)\| - h\|D^2 b\|_{L^\infty}\|v_n^1\| \le \|e^{1P}(v^0, v^1)\| \le \|e^{1P}(v^0, v_\Gamma^1)\| + h\|D^2 b\|_{L^\infty}\|v_n^1\|$$
$$\begin{aligned}\Rightarrow\ & \frac{1}{2}\|\varepsilon^0(v^0, v_\Gamma^1)\| + h\|e^{1P}(v^0, v_\Gamma^1)\| + \left(\frac{1}{2} - h\|D^2 b\|_{L^\infty}\right)\|v_n^1\| \\ & \le \|\varepsilon^0(v^0, v^1)\| + h\|e^{1P}(v^0, v^1)\| \\ & \le \|\varepsilon^0(v^0, v_\Gamma^1)\| + h\|e^{1P}(v^0, v_\Gamma^1)\| + \left(1 + h\|D^2 b\|_{L^\infty}\right)\|v_n^1\|.\end{aligned}$$

Thus for h sufficiently small we have the equivalence of norms and the continuity of the right-hand side (33) with respect to the new norm. In particular we have the $L^2(\omega)$-continuity with respect to the variable v_n^1 and

$$\boxed{\exists f_h \in L^2(\omega) \text{ such that } \forall v_n^1 \in L^2(\omega) \quad \int_\omega f_h\, v_n^1\, d\Gamma \overset{\text{def}}{=} \tilde{\ell}_h(0, v_n^1 n)} \tag{38}$$

We now introduce another effective constitutive law to eliminate v_{hn}^1 and get a variational equation for $(v_h^0, v_{h\Gamma}^1)$.

Theorem 4.1. [5] *Let C be a constitutive law verifying Assumption* 2.1. *Then*

$$\nu \overset{\text{def}}{=} 2[C^{-1} n\, {}^*n]_{nn} = N(n)\cdot n > 0. \tag{39}$$

The effective constitutive law $C_{en}\colon \mathrm{Sym}_N^n \to \mathrm{Sym}_N^n$ *defined as*

$$\boxed{C_{en}^{-1}\tau \overset{\text{def}}{=} C^{-1}\tau - 2\nu^{-1}[C^{-1}\tau]_{nn}\, C^{-1}(n\, {}^*n), \quad \tau \in \mathrm{Sym}_N^n} \tag{40}$$

is bijective, symmetrical and coercive. For all $\sigma \in \mathrm{Sym_N}$

$$\boxed{C^{-1}\tau \cdot\cdot\, \sigma = C_{en}^{-1}\tau^n \cdot\cdot\, \sigma^n + 2\nu^{-1}[C^{-1}\tau]_{nn}\,[C^{-1}\sigma]_{nn}.} \tag{41}$$

For the constitutive law $C^{-1}\varepsilon = 2\mu\,\varepsilon + \lambda\,\mathrm{tr}\,\varepsilon\,I$ and $\tau \in \mathrm{Sym_N^n}$

$$\boxed{C_{en}^{-1}\tau = 2\mu\,\tau + \frac{2\mu\lambda}{2\mu+\lambda}\,\mathrm{tr}\,\tau\,P} \tag{42}$$

and C_{en}^{-1} coincides with C_{eP}^{-1} given by (34).

Recall the variational equation (32) for the reduced $P(2,1)$ model

$$\begin{aligned}\int_\omega \overline{\alpha}(h)\,C^{-1}\varepsilon^0(v_h^0,v_h^1)\cdot\cdot\,\varepsilon^0(v^0,v^1) + \frac{\overline{\alpha}_1(h)^2}{\overline{\alpha}_2(h)}C_{eP}^{-1}\varepsilon_\Gamma^P(v_h^0)\cdot\cdot\,\varepsilon_\Gamma^P(v^0)&\\ +\,\overline{\alpha}_1(h)\left[C_{eP}^{-1}e^{1P}(v_h^0,v_h^1)\cdot\cdot\,\varepsilon_\Gamma^P(v^0) + C_{eP}^{-1}\varepsilon_\Gamma^P(v_h^0)\cdot\cdot\,e^{1P}(v^0,v^1)\right]&\\ +\,\overline{\alpha}_2(h)C_{eP}^{-1}e^{1P}(v_h^0,v_h^1)\cdot\cdot\,e^{1P}(v^0,v^1)\,d\Gamma = \tilde{\ell}_h(v^0,v^1)&\end{aligned}$$

where $\tilde{\ell}_h(v^0,v^1)$ is given by (33). For test functions $(0, v_n^1 n)$, $v_n^1 \in L^2(\omega)$, we get

$$\overline{\alpha}(h)\,[C^{-1}\varepsilon^0(v_h^0,v_h^1)]_{nn} + C_{eP}^{-1}(\overline{\alpha}_1(h)\varepsilon_\Gamma^P(v_h^0) + \overline{\alpha}_2(h)e^{1P}(v_h^0,v_h^1))\cdot\cdot\,D^2b = f_h$$

and

$$\begin{aligned}\left\{[C^{-1}n\,{}^*n]_{nn} + \frac{\overline{\alpha}_2(h)}{\overline{\alpha}(h)}C_{eP}^{-1}D^2b\cdot\cdot\,D^2b\right\}v_{hn}^1 + [C^{-1}\varepsilon^0(v_h^0,v_{h\Gamma}^1)]_{nn}&\\ +\,C_{eP}^{-1}\left(\frac{\overline{\alpha}_1(h)}{\overline{\alpha}(h)}\varepsilon_\Gamma^P(v_h^0) + \frac{\overline{\alpha}_2(h)}{\overline{\alpha}(h)}e^{1P}(v_h^0,v_{h\Gamma}^1)\right)\cdot\cdot\,D^2b = \frac{1}{\overline{\alpha}(h)}f_h&\end{aligned}$$

Recalling that $\nu = 2[C^{-1}n\,{}^*n]_{nn}$, define

$$\nu_2 \stackrel{\mathrm{def}}{=} 2\,C_{eP}^{-1}D^2b\cdot\cdot\,D^2b \text{ and } \nu_h \stackrel{\mathrm{def}}{=} \nu + \frac{\overline{\alpha}_2(h)}{\overline{\alpha}(h)}\nu_2. \tag{43}$$

Then v_{hn}^1 is given by the following expression

$$\boxed{\begin{aligned}v_{hn}^1 = \frac{2}{\nu_h}\frac{1}{\overline{\alpha}(h)}f_h - \frac{2}{\nu_h}\Big\{&[C^{-1}\varepsilon^0(v_h^0,v_{h\Gamma}^1)]_{nn}\\ &+ C_{eP}^{-1}\left(\frac{\overline{\alpha}_1(h)}{\overline{\alpha}(h)}\varepsilon_\Gamma^P(v_h^0) + \frac{\overline{\alpha}_2(h)}{\overline{\alpha}(h)}e^{1P}(v_h^0,v_{h\Gamma}^1)\right)\cdot\cdot\,D^2b\Big\}\end{aligned}}$$

Using new test function of the form (v^0, v_Γ^1) in (32) we get the second equation

$$\boxed{\begin{aligned}&\int_\omega \overline{\alpha}(h)\,\{C^{-1}\varepsilon^0(v_h^0,v_{h\Gamma}^1)\cdot\cdot\,\varepsilon^0(v^0,v_\Gamma^1) + v_{hn}^1\,[C^{-1}\varepsilon^0(v^0,v_\Gamma^1)]_{nn}\}\\ &\quad+\frac{\overline{\alpha}_1(h)^2}{\overline{\alpha}_2(h)}C_{eP}^{-1}\varepsilon_\Gamma^P(v_h^0)\cdot\cdot\,\varepsilon_\Gamma^P(v^0)\\ &\quad+\overline{\alpha}_1(h)\left\{\begin{aligned}&C_{eP}^{-1}e^{1P}(v_h^0,v_{h\Gamma}^1)\cdot\cdot\,\varepsilon_\Gamma^P(v^0) + C_{eP}^{-1}\varepsilon_\Gamma^P(v_h^0)\cdot\cdot\,e^{1P}(v^0,v_\Gamma^1)\\ &+ v_{hn}^1\,C_{eP}^{-1}D^2b\cdot\cdot\,\varepsilon_\Gamma^P(v^0)\end{aligned}\right\}\\ &\quad+\overline{\alpha}_2(h)\,\{C_{eP}^{-1}e^{1P}(v_h^0,v_{h\Gamma}^1)\cdot\cdot\,e^{1P}(v^0,v_\Gamma^1) + v_{hn}^1\,C_{eP}^{-1}D^2b\cdot\cdot\,e^{1P}(v^0,v_\Gamma^1)\}\,d\Gamma\\ &= \tilde{\ell}_h(v^0,v_\Gamma^1)\end{aligned}}$$

This is a new coupled system for $(v_h^0, v_{h\Gamma}^1)$ and v_{hn}^1 and we can substitute for v_{hn}^1 in the second equation to get a single variational equation for $(v_h^0, v_{h\Gamma}^1) \in Q^n$.

However the substitution yields a large number of terms and it is preferable to introduce some approximation. So we substitute for v_{hn}^1 and at the same time only retain terms of order 1 or h^2 in the second equation. The term involving f_h in the integrand is given by

$$f_h \frac{2}{\nu_h} \left\{ \left[C^{-1}\varepsilon^0(v^0, v_\Gamma^1)\right]_{nn} + C_{eP}^{-1} \left(\frac{\overline{\alpha}_1(h)}{\overline{\alpha}(h)} \varepsilon_\Gamma^P(v^0) + \frac{\overline{\alpha}_2(h)}{\overline{\alpha}(h)} e^{1P}(v^0, v_\Gamma^1) \right) \cdot\cdot\, D^2 b \right\}.$$

When combined with $\tilde{\ell}_h$ the new right-hand side becomes

$$\tilde{\ell}_h(v^0, v_\Gamma^1 - \frac{2}{\nu_h} \left\{ \left[C^{-1}\varepsilon^0(v^0, v_\Gamma^1)\right]_{nn} \right.$$
$$\left. + C_{eP}^{-1} \left(\frac{\overline{\alpha}_1(h)}{\overline{\alpha}(h)} \varepsilon_\Gamma^P(v^0) + \frac{\overline{\alpha}_2(h)}{\overline{\alpha}(h)} e^{1P}(v^0, v_\Gamma^1) \right) \cdot\cdot\, D^2 b \right\} n).$$

We again only retain terms of order 1 or h^2. By definition of ν_h, we get

$$\tilde{\ell}_h(v^0, v_\Gamma^1 - \frac{2}{\nu} \left\{ \left(1 - \frac{\overline{\alpha}_2(h)}{\overline{\alpha}(h)} \frac{\nu_2}{\nu}\right) \left[C^{-1}\varepsilon^0(v^0, v_\Gamma^1)\right]_{nn} \right.$$
$$\left. + C_{eP}^{-1} \left(\frac{\overline{\alpha}_1(h)}{\overline{\alpha}(h)} \varepsilon_\Gamma^P(v^0) + \frac{\overline{\alpha}_2(h)}{\overline{\alpha}(h)} e^{1P}(v^0, v_\Gamma^1) \right) \cdot\cdot\, D^2 b \right\} n).$$

We now turn to the quadratic terms in the integrand. First the term in $\overline{\alpha}(h)$

$$\begin{aligned}
&\overline{\alpha}(h)\ \{C^{-1}\varepsilon^0(v_h^0, v_{h\Gamma}^1)\cdot\cdot\, \varepsilon^0(v^0, v_\Gamma^1) + v_{hn}^1\, [C^{-1}\varepsilon^0(v^0, v_\Gamma^1)]_{nn}\} \\
\simeq\, &\overline{\alpha}(h) \left\{ C^{-1}\varepsilon_h^{0n}\cdot\cdot\, \varepsilon^{0n} - \frac{2}{\nu}[C^{-1}\varepsilon_h^{0n}]_{nn}\, [C^{-1}\varepsilon^{0n}]_{nn} \right. \\
&\quad + \frac{2}{\nu} \frac{\overline{\alpha}_2(h)}{\overline{\alpha}(h)} \frac{\nu_2}{\nu} [C^{-1}\varepsilon_h^{0n}]_{nn}\, [C^{-1}\varepsilon^{0n}]_{nn} \\
&\quad \left. - \frac{2}{\nu} C_{eP}^{-1} \left(\frac{\overline{\alpha}_1(h)}{\overline{\alpha}(h)} \varepsilon_\Gamma^P(v_h^0) + \frac{\overline{\alpha}_2(h)}{\overline{\alpha}(h)} e^{1P}(v_h^0, v_{h\Gamma}^1) \right) \cdot\cdot\, D^2 b\, [C^{-1}\varepsilon^{0n}]_{nn} \right\} \\
=\, &\overline{\alpha}(h) \left\{ C_{en}^{-1}\varepsilon_h^{0n}\cdot\cdot\, \varepsilon^{0n} + \frac{2}{\nu} \frac{\overline{\alpha}_2(h)}{\overline{\alpha}(h)} \frac{\nu_2}{\nu} [C^{-1}\varepsilon_h^{0n}]_{nn}\, [C^{-1}\varepsilon^{0n}]_{nn} \right. \\
&\quad \left. - \frac{2}{\nu} C_{eP}^{-1} \left(\frac{\overline{\alpha}_1(h)}{\overline{\alpha}(h)} \varepsilon_\Gamma^P(v_h^0) + \frac{\overline{\alpha}_2(h)}{\overline{\alpha}(h)} e^{1P}(v_h^0, v_{h\Gamma}^1) \right) \cdot\cdot\, D^2 b\, [C^{-1}\varepsilon^{0n}]_{nn} \right\}
\end{aligned}$$

For the term in $\overline{\alpha}_1(h)$

$$\begin{aligned}
&\overline{\alpha}_1(h) \left\{ \begin{array}{l} C_{eP}^{-1} e^{1P}(v_h^0, v_{h\Gamma}^1)\cdot\cdot\, \varepsilon_\Gamma^P(v^0) + C_{eP}^{-1}\varepsilon_\Gamma^P(v_h^0)\cdot\cdot\, e^{1P}(v^0, v_\Gamma^1) \\ + v_{hn}^1\, C_{eP}^{-1} D^2 b\cdot\cdot\, \varepsilon_\Gamma^P(v^0) \end{array} \right\} \\
\simeq\, &\overline{\alpha}_1(h) \left\{ \begin{array}{l} C_{eP}^{-1} e^{1P}(v_h^0, v_{h\Gamma}^1)\cdot\cdot\, \varepsilon_\Gamma^P(v^0) + C_{eP}^{-1}\varepsilon_\Gamma^P(v_h^0)\cdot\cdot\, e^{1P}(v^0, v_\Gamma^1) \\ - \frac{2}{\nu}[C^{-1}\varepsilon_h^{0n}]_{nn}\, C_{eP}^{-1} D^2 b\cdot\cdot\, \varepsilon_\Gamma^P(v^0) \end{array} \right\}
\end{aligned}$$

Finally for the term in $\overline{\alpha}_2(h)$

$$\begin{aligned}
&\overline{\alpha}_2(h)\, \{C_{eP}^{-1} e^{1P}(v_h^0, v_{h\Gamma}^1)\cdot\cdot\, e^{1P}(v^0, v_\Gamma^1) + v_{hn}^1\, C_{eP}^{-1} D^2 b\cdot\cdot\, e^{1P}(v^0, v_\Gamma^1)\} \\
\simeq\, &\overline{\alpha}_2(h) \left\{ C_{eP}^{-1} e^{1P}(v_h^0, v_{h\Gamma}^1)\cdot\cdot\, e^{1P}(v^0, v_\Gamma^1) - \frac{2}{\nu}[C^{-1}\varepsilon_h^{0n}]_{nn}\, C_{eP}^{-1} D^2 b\cdot\cdot\, e^{1P}(v^0, v_\Gamma^1) \right\}
\end{aligned}$$

Putting everything together we get a new reduced variational equation

$$
\boxed{
\begin{aligned}
&\int_\omega \overline{\alpha}(h)\, C_{en}^{-1}\varepsilon^0(v_h^0, v_{h\Gamma}^1)\cdot\cdot\, \varepsilon^0(v^0, v_\Gamma^1) + \frac{\overline{\alpha}_1(h)^2}{\overline{\alpha}_2(h)} C_{eP}^{-1}\varepsilon_\Gamma^P(v_h^0)\cdot\cdot\, \varepsilon_\Gamma^P(v^0) \\
&\quad + \overline{\alpha}_1(h) \left\{ \begin{aligned} & C_{eP}^{-1} e^{1P}(v_h^0, v_{h\Gamma}^1)\cdot\cdot\, \varepsilon_\Gamma^P(v^0) + C_{eP}^{-1}\varepsilon_\Gamma^P(v_h^0)\cdot\cdot\, e^{1P}(v^0, v_\Gamma^1) \\ & - \frac{2}{\nu}\,[C^{-1}\varepsilon^0(v_h^0, v_{h\Gamma}^1)]_{nn}\, C_{eP}^{-1} D^2 b\cdot\cdot\, \varepsilon_\Gamma^P(v^0) \\ & - \frac{2}{\nu}\, C_{eP}^{-1}\varepsilon_\Gamma^P(v_h^0)\cdot\cdot\, D^2 b\,[C^{-1}\varepsilon^0(v^0, v_\Gamma^1)]_{nn} \end{aligned} \right\} \\
&\quad + \overline{\alpha}_2(h) \left\{ \begin{aligned} & C_{eP}^{-1} e^{1P}(v_h^0, v_{h\Gamma}^1)\cdot\cdot\, e^{1P}(v^0, v_\Gamma^1) \\ & + \frac{2}{\nu}\frac{\nu_2}{\nu}\,[C^{-1}\varepsilon^0(v_h^0, v_{h\Gamma}^1)]_{nn}\,[C^{-1}\varepsilon^0(v^0, v_\Gamma^1)]_{nn} \\ & - \frac{2}{\nu}[C^{-1}\varepsilon^0(v_h^0, v_{h\Gamma}^1)]_{nn}\, C_{eP}^{-1} D^2 b\cdot\cdot\, e^{1P}(v^0, v_\Gamma^1) \\ & - \frac{2}{\nu}\, C_{eP}^{-1} e^{1P}(v_h^0, v_{h\Gamma}^1)\cdot\cdot\, D^2 b\,[C^{-1}\varepsilon^0(v^0, v_\Gamma^1)]_{nn} \end{aligned} \right\} d\Gamma \\
&= \tilde{\ell}_h(v^0, v_\Gamma^1 - \frac{2}{\nu}\Big\{ \Big(1 - \frac{\overline{\alpha}_2(h)}{\overline{\alpha}(h)}\frac{\nu_2}{\nu}\Big)\,\big[C^{-1}\varepsilon^0(v^0, v_\Gamma^1)\big]_{nn} \\
&\qquad + C_{eP}^{-1}\Big(\frac{\overline{\alpha}_1(h)}{\overline{\alpha}(h)}\varepsilon_\Gamma^P(v^0) + \frac{\overline{\alpha}_2(h)}{\overline{\alpha}(h)} e^{1P}(v^0, v_\Gamma^1)\Big)\cdot\cdot\, D^2 b\Big\} n)
\end{aligned}
}
\tag{44}
$$

It is easy to check that as h goes to zero the quadratic part is coercive and continuous with respect to the norm

$$
\boxed{\left\{\left\|\varepsilon^0(v^0, v_\Gamma^1)\right\|^2 + \left\|e^{1P}(v^0, v_\Gamma^1)\right\|^2\right\}^{1/2}.}
\tag{45}
$$

and that the linear part is continuous with respect to that same norm. Hence from Theorem 2.3 (i) we have existence and uniqueness of solution in Q^{n0}.

It is important to observe that we have kept all terms of order 1 and h^2 in order to preserve the global asymptotic behavior of the model as h goes to zero (cf. [5]). But several terms can be *mathematically absorbed* by the sum

$$
C_{en}^{-1}\varepsilon^0(v_h^0, v_{h\Gamma}^1)\cdot\cdot\, \varepsilon^0(v^0, v_\Gamma^1) + h^2\, C_{eP}^{-1} e^{1P}(v_h^0, v_{h\Gamma}^1)\cdot\cdot\, e^{1P}(v^0, v_\Gamma^1)
\tag{46}
$$

of the two dominating terms of (44). However in some cases doing this in a systematic way changes the resulting asymptotic model as h goes to zero. The *maximal mathematical reduction* of model (44) is Naghdi's modified model. Starting from the observation that as h goes to zero $\overline{\alpha}_0(h) \to 1$, $\overline{\alpha}_1(h) \to \kappa_1 h^2/3$, $\overline{\alpha}_2(h) \to h^2/3$, $\overline{\alpha}(h) \to 1$, then

$$
\frac{\overline{\alpha}_1(h)^2}{\overline{\alpha}_2(h)} C_{eP}^{-1}\varepsilon_\Gamma^P(v_h^0)\cdot\cdot\, \varepsilon_\Gamma^P(v^0) \simeq \kappa_1^2 \frac{h^2}{3} C_{eP}^{-1}\varepsilon_\Gamma^P(v_h^0)\cdot\cdot\, \varepsilon_\Gamma^P(v^0)
$$

is absorbed in the first term in the sum (46). The group of terms multiplied by $\overline{\alpha}_1(h) \simeq \kappa_1 h^2/3$ in (44) is also absorbed by (46). Furthermore in the last group of terms multiplied by $\overline{\alpha}_2(h) \simeq h^2/3$ in (44) the last three terms are also absorbed by (46). Finally model (44) reduces to

$$\boxed{\begin{aligned}&\int_\omega C_{en}^{-1}\varepsilon^0(v_h^0, v_{h\Gamma}^1)\cdot\cdot\,\varepsilon^0(v^0, v_\Gamma^1) + \frac{h^2}{3} C_{eP}^{-1} e^{1P}(v_h^0, v_{h\Gamma}^1)\cdot\cdot\, e^{1P}(v^0, v_\Gamma^1)\, d\Gamma \\ &= \tilde{\ell}_h(v^0, v_\Gamma^1 - \frac{2}{\nu}\Big\{\Big(1 - \frac{h^2}{3}\frac{\nu_2}{\nu}\Big)\big[C^{-1}\varepsilon^0(v^0, v_\Gamma^1)\big]_{nn} \\ &\qquad + \frac{h^2}{3} C_{eP}^{-1}\left(\kappa_1 \varepsilon_\Gamma^P(v^0) + e^{1P}(v^0, v_\Gamma^1)\right)\cdot\cdot\, D^2 b\Big\} n)\end{aligned}} \tag{47}$$

This maximal mathematical reduction has not affected the underlying space of solutions. For the special constitutive law $C^{-1}\varepsilon = 2\mu\,\varepsilon + \lambda\,\mathrm{tr}\,\varepsilon I$, $C_{eP}^{-1} = C_{en}^{-1}$, and (47) is exactly the *quadratic part of Naghdi's modified model.* The reader will also notice that we have not used the *a priori assumption* $\sigma_{33} = 0$ on the stress tensor σ, that is $[C^{-1}\varepsilon_h^0]_{nn} = 0$ and $[C^{-1}\varepsilon_h^1]_{nn} = 0$ in our notation. This results from the fact that the substitution process has appropriately modified the linear right-hand side of the equation. The existence and uniqueness of solutions to Naghdi's linear model is usually given for C^2 midsurfaces (cf. [2, 1]). In our framework everything has been done for $C^{1,1}$ midsurfaces. A recent note by [3] gives this extension by different methods within the covariant/contravariant framework for the special constitutive law $C^{-1}\tau = 2\mu\,\tau + \lambda\ \mathrm{tr}\,\tau\, I$.

4.3. Naghdi's enriched model

Finally we know from [5] that there is a missing coupling term in the system of variational equations which characterizes the asymptotic model obtained from Naghdi's (resp. Naghdi's modified) model. This can be fixed by only dropping the terms of order h^2 which involve both ε_h^0 and ε^0 in (44). The *Naghdi's enriched model* which yields the correct asymptotic model in all cases is given by the following variational equation

$$\boxed{\begin{aligned}&\int_\omega C_{en}^{-1}\varepsilon^0(v_h^0, v_{h\Gamma}^1)\cdot\cdot\,\varepsilon^0(v^0, v_\Gamma^1) \\ &\quad + \kappa_1\frac{h^2}{3}\left\{C_{eP}^{-1} e^{1P}(v_h^0, v_{h\Gamma}^1)\cdot\cdot\,\varepsilon_\Gamma^P(v^0) + C_{eP}^{-1}\varepsilon_\Gamma^P(v_h^0)\cdot\cdot\, e^{1P}(v^0, v_\Gamma^1)\right\} \\ &\quad + \frac{h^2}{3}\left\{\begin{aligned}&C_{eP}^{-1} e^{1P}(v_h^0, v_{h\Gamma}^1)\cdot\cdot\, e^{1P}(v^0, v_\Gamma^1) \\ &- \frac{2}{\nu}[C^{-1}\varepsilon^0(v_h^0, v_{h\Gamma}^1)]_{nn}\, C_{eP}^{-1} D^2 b\cdot\cdot\, e^{1P}(v^0, v_\Gamma^1) \\ &- \frac{2}{\nu} C_{eP}^{-1} e^{1P}(v_h^0, v_{h\Gamma}^1)\cdot\cdot\, D^2 b\,[C^{-1}\varepsilon^0(v^0, v_\Gamma^1)]_{nn}\end{aligned}\right\} d\Gamma \\ &= \tilde{\ell}_h(v^0, v_\Gamma^1 - \frac{2}{\nu}\Big\{\Big(1 - \frac{h^2}{3}\frac{\nu_2}{\nu}\Big)\big[C^{-1}\varepsilon^0(v^0, v_\Gamma^1)\big]_{nn} \\ &\qquad + \frac{h^2}{3} C_{eP}^{-1}\left(\kappa_1 \varepsilon_\Gamma^P(v^0) + e^{1P}(v^0, v_\Gamma^1)\right)\cdot\cdot\, D^2 b\Big\} n)\end{aligned}} \tag{48}$$

References

[1] M. Bernadou, *Méthodes d'éléments finis pour les problèmes de coques minces*, Masson, Paris, Milan, Barcelone, 1994.

[2] M. Bernadou, Ph.G. Ciarlet, and B. Miara, *Existence theorems for two-dimensional linear shell theories*, J. Elasticity **34** (1994), 111–138.

[3] A. Blouza, *Existence et unicité pour le modèle de Naghdi pour une coque peu régulière*, C. R. Acad. Sci. Paris Sér. I Math. **324** (1997), 839–844.

[4] B. Budiansky and J.L. Sanders, *On the "best" first-order linear shell theory*, Progr. in Appl. Mech. (W. Prager Anniversary Volume), pp. 129-140, Macmillan, New York 1967.

[5] M.C. Delfour, *Intrinsic differential geometric methods in the asymptotic analysis of linear thin shells*, Boundaries, interfaces and transitions (M.C. Delfour, ed.), CRM Proc. Lecture Notes, Amer. Math. Soc., Providence, RI, 1998, pp. 19-90.

[6] M.C. Delfour and J.-P. Zolésio, *On a variational equation for thin shells*, Control and Optimal Design of Distributed Parameter Systems (J. Lagnese, D. L. Russell, and L. White, eds.), Springer-Verlag, Berlin, New York 1994, pp. 25–37.

[7] ———, *A boundary differential equation for thin shells*, J. Differential Equations **119** (1995), 426–449.

[8] ———, *Tangential differential equations for dynamical thin/shallow shells*, J. Differential Equations **128** (1996), 125–167.

[9] ———, *Differential equations for linear shells: comparison between intrinsic and classical models*, Advances in Mathematical Sciences–CRM's 25 years (Luc Vinet, ed.), CRM Proc. Lect. Notes, Amer. Math. Soc., Providence, RI, 1997, pp. 42–124.

[10] ———, *Convergence to the asymptotic model for linear thin shells*, Optimization Methods in Partial Differential Equations (S. Cox and I. Lasiecka, eds.), Contemp. Math., vol. 209, Amer. Math. Soc., Providence, RI, 1977.

[11] ———, *Convergence of the linear $P(1,1)$ and $P(2,1)$ thin shells to asymptotic shells*, Proc. Plates and Shells: from theory to practice (M. Fortin, ed.), CRM Proc. Lect. Notes ser., AMS Publications, Providence, RI, 1997 (1996 CMS Annual Seminar) (to appear).

[12] W.T. Koiter, *On the nonlinear theory of thin elastic shells*, in Proc. Kon. Nederl. Akad. Wetensch. B59 (1966), 1–54.

[13] ———, *On the nonlinear theory of thin elastic shells*, in Proc. Kon. Nederl. Akad. Wetensch. B73 (1970), 169–195.

[14] J.L. Sanders, *An improved first approximation theory of thin shells*, NASA Report 24, 1959.

Centre de recherches mathématiques et
Département de Mathématiques et de statistique,
Université de Montréal,
C. P. 6128, succ. Centre-ville,
Montréal QC, Canada H3C 3J7
E-mail address: delfour@CRM.UMontreal.CA

International Series of Numerical Mathematics
Vol. 133,

On the Approximate Controllability for some Explosive Parabolic Problems

J.I. Díaz and J.L. Lions

Abstract. We consider in this paper distributed systems governed by parabolic evolution equations which can blow up in finite time and which are controlled by initial conditions. We study here the following question : Can one choose the initial condition in such a way that the solution does not blow up before a given time T and which is, at time T, as close as we wish from a given state ? Some general results along these lines are presented here for semilinear second order parabolic equations as well as for a non local nonlinear problem. We also give some results proving that "the more the system will blow up" the "cheaper" it will be the control.

1. Introduction

We consider *distributed systems* of evolution, i.e. systems whose state (denoted by y) is given by the solution (or by a solution) of a Partial Differential Equation (PDE) of evolution. In this paper we consider distributed systems which, *if not controlled, can blow up in finite time.* We *conjecture* that these systems are *approximately controllable* (and even *exactly* controllable). In other words, by a "*suitable set of actions*" (the control), the system can be driven, in a finite time T, from an initial state y^0 to a neighborhood of the target y^T (or to reach exactly y^T). We also conjecture that "the more the system will blow-up", the *"cheaper"* it will be the control.

Of course, all this has to be made precise!

This is what we intend to do in the present paper, *when the control is the initial state.*

Before proceeding, let us notice that considering the initial state as a control is a standard point of view in the *assimilation of data* in Meteorology or in Climatology. Cf. Blayo, Blum and Verron [2], Le Dimet and Charpentier [10].

We consider first semilinear parabolic problems of the type

$$\begin{cases} y_t + Ay &= f(y) \quad \text{in } \Omega \times (0,T), \\ y &= 0 \quad \text{on } \Gamma_D \times (0,T), \\ \frac{\partial y}{\partial \nu} &= 0 \quad \text{on } \Gamma_N \times (0,T), \\ y(0,x) &= u(x) \quad \text{on } \Omega, \end{cases} \tag{1}$$

where Ω is a bounded open regular set of R^N, $\partial\Omega = \Gamma_D \cup \Gamma_N$, A is a linear second order elliptic operator of the form

$$Ay = -\sum_{i,j=1}^{N} \frac{\partial}{\partial x_i}(a_{ij}\frac{\partial y}{\partial x_j}) + a_0 y,$$

where $a_{ij} \in C^1(\overline{\Omega}), a_0 \in C^0(\overline{\Omega})$, $a_0 \geq 0$ and there exists $\alpha > 0$ such that

$$\sum_{i,j=1}^{N} a_{ij}(x)\xi_i\xi_j \geq \alpha\left|\xi\right|^2, \ \forall \xi = \{\xi_i\}_{i=1}^N \in R^N, \text{ a.e. } x \in \Omega.$$

In (1), vector ν denotes the associated *conormal vector,* $\nu = \{\nu_i\}_{i=1}^N$,

$$\nu_i = \sum_{j=1}^{N} a_{ij}(x) n_j,$$

with $\mathbf{n} = \{n_i\}_{i=1}^N$ the unit outward normal to Γ_N. Function f is assumed to be locally Lipschitz and u denotes the control function.

In (1) the function f is not necessarily decreasing (nor sublinear at infinity), so that the associate solutions can *blow up in a finite time.* Let us recall that by well-known results (see, e.g., Cazenave and Haraux [3]) given $u \in L^q(\Omega)$, for some $q \in [1, +\infty]$ there is a unique (local in time) solution, defined on a *maximal interval* $[0, T_m[$, $T_m = T_m(u)$ and $\|y(t)\|_{L^q(\Omega)} \nearrow +\infty$ if $t \nearrow T_m$ when $T_m(u) < +\infty$.

The problem of approximate controllability alluded to above can now be stated in the following fashion: let $T > 0$ be given and let $\mathcal{E}(T)$ be the set of elements $u \in L^q(\Omega)$ for which $T_m(u) > T$; is the set $\{y(T : u), u \in \mathcal{E}(T)\}$ dense in $L^q(\Omega)$?

Some positive answers will be presented in what follows for equations of type (1) and also for other *non local* models as we explain below.

Of course the problem of approximate (resp. exact) controllability can be raised for systems *not necessarily blowing up in finite time,* i.e. for systems such that $T_m(u) = +\infty$.

For the case of $A = -\Delta$ and Dirichlet boundary conditions (i.e., $\partial\Omega = \Gamma_D$), the linear case ($f(s) = as + b$) was already considered by Lattes and Lions [9] and Lions [11]. Concerning the nonlinear case, it seems that the first result in the literature was due to Bardos and Tartar [1]. They proved a negative answer: if $f(s) = -\left|s\right|^{p-1} s$, and $p > 1$ then any solution of (1) satisfies the "universal" estimate

$$|y(x,t)| \leq Ct^{-1/(p-1)}, \text{ for any } (x,t) \in \Omega \times (0,T). \tag{2}$$

for some positive constant C independent of u. So, in this case, the approximate controllability fails if $y^T \in L^q(\Omega)$ is such that

$$\left|y^T(x)\right| > CT^{-1/(p-1)}, \text{ on a positively measured subset of } \Omega.$$

A first positive result for nonlinear problems was due to Henry [8] who proved the L^2 approximate controllability when f is assumed to be globally Lipschitz. More recently his result was improved in Fabre, Puel and Zuazua [7] where the authors obtain the mentioned property in $L^q(\Omega)$,with q arbitrary $1 \leq q \leq \infty$, under the condition f locally Lipschitz and *sublinear at the infinity*, i.e. such that

$$|f(s)| \leq C(1+|s|), \text{ for } |s| \text{ large enough.} \tag{3}$$

The above results can be easily extended to more general elliptic operators A and when Γ_N is not empty.

We shall assume that

$$sf(s) \geq 0 \text{ for any } s \text{ with } |s| > s_0, \text{ for some } s_0 \geq 0 \tag{4}$$

so that solutions may blow up. A first result concerns the case of small time T

Theorem 1.1. *Assume (4) and*

$$\max\left\{\left|\int_{-\infty}^{-s_1} \frac{ds}{f(s)}\right|, \int_{s_1}^{+\infty} \frac{ds}{f(s)}\right\} < \infty \tag{5}$$

for some $s_1 > s_0$. *Let* $y^T \in L^q(\Omega)$, *for some* $q \in [1,+\infty]$, *and let* $\varepsilon \in (0,1)$. *Then, there exists* $u \in C^2(\overline{\Omega})$ *and* $\tau_0 > 0$ *such that* $T_m(u) > \tau_0$ *and*

$$\left\|y(\tau_0;u) - y^T\right\|_{L^q(\Omega)} \leq \varepsilon.$$

The proof of this result will be given in Section 2. Although several remarks and generalizations will be also given in that section, we point out that condition (5) requires a *superlinear* growth on f (when $|s|$ is large). In fact, it is easy to see that (5) is fulfilled in the cases of $f(s) = \lambda |s|^{p-1} s$, with $p > 1$, and, for instance, $|f(s)| \leq \lambda e^{|s|}$, for any $\lambda > 0$, for which *blow-up phenomena* may arise.

The proof of Theorem 1.1 shows that time τ_0 must be (in general) small enough. A natural question is to find conditions on the data in order to have $\tau_0 = T$ arbitrary. Two different results can be obtained in that direction. The first one concerns the case of the "pure" Neumann problem and y^T "near a constant":

Theorem 1.2. *Let* $\varepsilon > 0$ *be given. Assume (4), (5),* $\partial\Omega = \Gamma_N$ *and assume that* $y^T \in L^q(\Omega)$, *for some* $q \in [1,+\infty]$, *is such that*

$$\left\|y^T - M\right\|_{L^q(\Omega)} \leq \varepsilon/2, \text{ for some constant } M. \tag{6}$$

Then, for any $T > 0$ *and* $\varepsilon \in (0,1)$ *there exists* $u \in C^2\left(\overline{\Omega}\right)$ *with* $T_m(u) > T$ *and*

$$\left\|y(T;u) - y^T\right\|_{L^q(\Omega)} \leq \varepsilon.$$

If y^T is near a stationary state (even if it is an *unstable one*) we shall prove

Theorem 1.3. *Assume (4), (5) and let $y^T \in L^q(\Omega)$, for some $q \in [1, +\infty]$, be such that there exists $g^* \in L^\infty(\Omega)$ verifying that*

$$\begin{cases} Ag^* &= f(g^*) & \text{in } \Omega, \\ g^* &= 0 & \text{on } \Gamma_D, \\ \frac{\partial g^*}{\partial \nu} &= 0 & \text{on } \Gamma_N. \end{cases}$$

and

$$\| g^* - y^T \|_{L^q(\Omega)} \leq \varepsilon/2.$$

Then, for any $T > 0$ and $\varepsilon \in (0,1)$ there exists $u \in C^2\left(\overline{\Omega}\right)$, different of g^*, with $T_m(u) > T$ and

$$\|y(T;u) - y^T\|_{L^q(\Omega)} \leq \varepsilon.$$

The proofs of these results are contained in Section 3 where, again, some generalizations and remarks will be given.

As a final question, it seems interesting to study the optimality of the control u. This is done in Section 4. A partial result in this direction is the following

Theorem 1.4. *Assume that the conditions of Theorem 1.2 or 1.3 are satisfied with $q = +\infty$. Then the set*

$$\begin{aligned} K = \ & \Big\{ v : v \in L^\infty(\Omega) : \|y(T;v) - y^T\|_{L^\infty(\Omega)} \leq \varepsilon, \\ & \|y(t;v)\|_{L^\infty(\Omega)} \leq \|y^T\|_{L^\infty(\Omega)} + 1, \forall t \in [0,T] \Big\} \end{aligned}$$

is not empty. Moreover there exists $v_0 \in K$ such that

$$\|v_0\|_{L^\infty(\Omega)} = \inf\{\|v\|_{L^\infty(\Omega)}, v \in K\}.$$

We indicated above that we conjecture that the "more it blows up", the cheaper it will to (approximately) control the system. Such a result is provided by the following

Theorem 1.5. *Assume that the conditions of Theorem 1.2 holds true with $q = +\infty$ and $f(s) = \lambda F(s)$, $\lambda > 0$, $F(0) = 0$. Then*

$$\|v_0(\lambda)\|_{L^\infty(\Omega)} \searrow 0 \text{ when } \lambda \nearrow +\infty,$$

where $v_0(\lambda)$ denotes the control obtained in Theorem 1.4.

All the above results are improvements of the paper Díaz and Lions [6]. Our approach is based on a suitable use of the solution $Y(t)$ of the associated *backward* Cauchy problem

$$(CP : Y_d) \begin{cases} \frac{dY}{dt}(t) = f(Y(t)) - f(0), \ t < 0, \\ Y(0) = Y_d. \end{cases}$$

(Notice that in the above formulation we replaced the *final time* T by $t = 0$, which is possible since the ODE is autonomous, i.e., f is time independent).

We conclude this paper by some few remarks on another type of *non linear non local* system. Namely we consider the problem

$$\begin{cases} y_t + Ay = cy \int_\Omega y^2 dx & \text{in } \Omega \times (0,T), \\ y = 0 & \text{on } \partial\Omega \times (0,T), \\ y(0,x) = u(x) & \text{on } \Omega, \end{cases} \tag{7}$$

when A is as above but *symmetric*

$$a_{ij} = a_{ji}$$

and where $c > 0$.

We shall prove (by different methods) the following

Theorem 1.6. *System (7) is approximately controllable in* $L^2(\Omega)$.

Moreover we shall prove

Theorem 1.7. *The "cost of approximate controllability" decreases to* 0 *as* c *increases to* $+\infty$ *in (7).*

2. On the approximate controllability for small time

Proof of Theorem 1.1. Let $g^* \in C^2\left(\overline{\Omega}\right)$ be such that

$$g^* = 0 \text{ on } \Gamma_D, \; \frac{\partial g^*}{\partial \nu} = 0 \text{ on } \Gamma_N \text{ and } \left\| g^* - y^T \right\|_{L^q(\Omega)} \leq \varepsilon/2.$$

Assume, for the moment, that

$$f \in C^2(R).$$

First step. Let $x \in \overline{\Omega}$ arbitrary and let $Y_d = g^*(x)$. Since the ODE of the $(CP : Y_d)$ is an equation of "separable variables", it is easy to see that there exists $T(f, g^*) > 0$ such that the solution $Y(t : Y_d)$ of $(CP : Y_d)$ can be continued backwards to $[-T(f,g^*), 0]$. In particular, for any $\tau \in [0, T(f,g^*)]$ we can define the function

$$u_\tau(x) := Y(-\tau : g^*(x)), x \in \overline{\Omega}. \tag{8}$$

Clearly $Y(t : g^*(x))$ depends continuously on the initial data and therefore on $x \in \overline{\Omega}$. Thus, $u_\tau \in C(\overline{\Omega}.)$. Moreover, if $x \in \Gamma_D$, then $g^*(x) = 0$ and so $u_\tau(x) = 0$ since $Y_\infty = 0$ is an *equilibrium point* of the ODE

$$\frac{dY}{dt}(t) = f(Y(t)) - f(0).$$

On the other hand, we have

$$\nabla u_\tau(x) = \frac{\partial Y}{\partial \alpha}(-\tau : g^*(x)) \nabla g^*(x).$$

Thus $\frac{\partial g^*}{\partial \nu} = 0$ on Γ_N . Moreover

$$\frac{\partial^2 u_\tau}{\partial x_i \partial x_j}(x) = \frac{\partial^2 Y}{\partial \alpha^2}(-\tau : g^*(x))\frac{\partial g^*}{\partial x_i}(x)\frac{\partial g^*}{\partial x_j}(x) + \frac{\partial Y}{\partial \alpha}(-\tau : g^*(x))\frac{\partial^2 g^*}{\partial x_i \partial x_j}(x).$$

But from the Peano Theorem we know that $\frac{\partial Y}{\partial \alpha}(t : g^*(x))$ is given as the solution of the linear Cauchy problem

$$\begin{cases} \frac{dZ}{dt}(t) = f'(Y(t : g^*(x)))Z(t) \\ Z(0) = 1, \end{cases}$$

and so, $\frac{\partial Y}{\partial \alpha}(t : g^*(x))$ depends continuously on $x \in \overline{\Omega}$. Since $f \in C^2(R)$, applying again the Peano Theorem we get that $\frac{\partial^2 Y}{\partial \alpha^2}(t : g^*(x))$ also depends continuously on $x \in \overline{\Omega}$. In conclusion, $u_\tau \in C^2(\overline{\Omega})$. Let

$$M_1(T(f, g^*)) := \sup\{\|u_\tau\|_{C^2(\overline{\Omega})} : \tau \in [0, T(f, g^*)]\}.$$

Is is clear that $M_1(T(f, g^*)) < \infty$. Consider, now, the function $U \in C^2(\overline{\Omega} \times [0, \tau])$ given by

$$U(x, t) := Y(t - \tau : g^*(x)).$$

(notice that, in fact, U is defined on $\overline{\Omega} \times [-(T(f, g^*) - \tau), \tau]$). Then

$$U(x, 0) = Y(-\tau : g^*(x)) = u_\tau(x), \ x \in \overline{\Omega},$$

$$U(x, t) = 0, \text{if } x \in \Gamma_D \text{ and } \frac{\partial U}{\partial \nu}(x, t) = 0 \text{ if } x \in \Gamma_N. \ , \ t \in [0, T(f, g^*)],$$

$$U_t + AU - f(U) = h(x, t), \ x \in \Omega, \ t \in (0, T(f, g^*)),$$

where

$$h(x, t) := -f(0) \ + AU.$$

(notice that from the above arguments $h \in C(\overline{\Omega} \times [0, T(f, g^*)])$). Define

$$M_2(T(f, g^*)) := \sup\{\|h\|_{L^\infty(\Omega \times (0,\tau))} : \tau \in [0, T(f, g^*)]\}$$

We point out that $M_2(T(f, g^*)) < \infty$ and that

$$U(x, \tau) = Y(0 : g^*(x)) = g^*(x), \ x \in \overline{\Omega}.$$

Second step. Let us show that the solution $y(x, t : u_\tau)$, with u_τ given by (8), is a global solution on $\overline{\Omega} \times [0, \tau]$, i.e. that $T_m(u_\tau) > \tau$. More precisely, let us show that

$$\sup\{\|y(x, t : u_\tau)\|_{L^\infty(\Omega \times (0,\tau))} : \tau \in [0, T(f, g^*)]\} \leq M_3(T(f, g^*)), \tag{9}$$

for some $M_3(T(f, g^*)) < \infty$. Let us start by assuming that $f(0) \leq 0$. Define $m^+(t)$ by

$$m^+(t) = Y(t - \tau : \|[u_\tau]_+\|_{L^\infty(\Omega)}) \ ,$$

where, in general, $[u]_+(x) = \max(u(x),0)$. From assumption (5) we know that $m_+(t)$ is defined at least on $[-(T(f,g^*)-\tau),\tau]$. Moreover we have

$$\begin{cases} y_t + Ay - f(y) &= 0 \leq m_t^+ + Am^+ - f(m^+) & \text{in } \Omega\times(0,\tau),\\ y(t,x) &= 0 \leq m^+(t) & \text{on } \Gamma_D\times(0,\tau),\\ \frac{\partial y}{\partial\nu} = \frac{\partial m^+}{\partial\nu} &= 0 & \text{on } \Gamma_N\times(0,\tau),\\ y(0,x) &= u_\tau(x) \leq m^+(0) = \|[u_\tau]+\|_{L^\infty(\Omega)} & \text{on } \Omega. \end{cases}$$

Then, by the comparison principle (which holds for problem (1)), we conclude that $y(x,t) \leq m^+(t)$, a.e. $x\in\Omega$, for any $t\in[0,\tau]$. If $f(0)>0$ we replace the barrier function $m^+(t)$ by the solution of

$$\begin{cases} M_t^+(t) &= f(M^+(t))\\ M^+(0) &= \|[u_\tau]_+\|_{L^\infty(\Omega)} \end{cases}$$

and again we get that $y(x,t)\leq M^+(t)$, a.e. $x\in\Omega$, for any $t\in[0,\tau]$ (notice that $m^+(t)\leq M^+(t)$). In a similar way we can construct negative barrier functions $m^-(t)$ and $M^-(t)$ and the conclusion holds.
Third step. Given $\varepsilon\in(0,1)$ let $\tau_0\in(0,T(f,g^*)]$ be such that

$$\tau_0 e^{K(T(f,g^*))\tau_0} M_2(T(f,g^*)) \leq \varepsilon/2 \tag{10}$$

where

$$K(T(f,g^*)) := \max\{|f'(s)| : s\in[-M_3(T(f,g^*)), M_3(T(f,g^*))]\}. \tag{11}$$

Then, as f is globally Lipschitz on the interval $[-M_3(T(f,g^*)), M_3(T(f,g^*))]$, using the $L^\infty(\overline{\Omega}\times[0,\tau_0])$-estimates on functions $y(x,t:u_{\tau_0})$ and $U(x,t)$, and since the functional operator $\mathbf{A}: D(\mathbf{A})\to L^s(\Omega)$ given by

$$\begin{aligned} D(\mathbf{A}) &= \{w\in W^{1,q}(\Omega) : w=0 \text{ on } \Gamma_D,\ \tfrac{\partial w}{\partial\nu}=0 \text{ on } \Gamma_N, Ay\in L^q(\Omega)\} \text{ and}\\ \mathbf{A}y &= Ay, \text{ if } y\in D(\mathbf{A}), \end{aligned} \tag{12}$$

generates a semigroup of contractions on $L^q(\Omega)$, using Gronwall's inequality, we conclude that, for any $t\in[0,\ T(f,g^*)]$, we have

$$\|y(.,t:u_{\tau_0}) - U(.,t)\|_{L^q(\Omega)} \leq e^{Kt}\int_0^t \|h(.,s)\|_{L^q(\Omega)}\,ds \leq te^{Kt}M_2(T(f,g^*)) \leq \varepsilon/2 \tag{13}$$

In particular, making $t=\tau_0$ we get that

$$\left\|y(.,\tau_0:u_{\tau_0}) - y^T\right\|_{L^q(\Omega)} \leq \|y(.,\tau_0:u_{\tau_0}) - g^*\|_{L^q(\Omega)} + \left\|\ g^* - y^T\right\|_{L^q(\Omega)} \leq \varepsilon.$$

Fourth step. If $f\notin C^2(R)$ we approximate f by f_ε with $f_\varepsilon\in C^2(R)$ such that

$$|f(s)-f_\varepsilon(s)| \leq \varepsilon/2, \text{ for any } s\in R.$$

We modify the definition of function $U(x,t)$ by replacing f by f_ε in the definition of $Y(t)$. The rest of the proof follows with obvious modifications.

Remark 2.1. *It is possible to obtain some expressions for the solution $Y(t)$ of $(CP : Y_d)$. We start by pointing out that, for the arguments of the proof of Theorem 1.1, we can assume without loss of generality that f is strictly increasing. Indeed, once that we have the estimate (9), we can replace the equation of (1) by*

$$y_t + \widetilde{A}y = \widetilde{f}(y) \quad in\ \Omega \times (0,T),$$

with

$$\widetilde{A}y = Ay + (K+1)y,\ \widetilde{f}(y) = f(y) + (K+1)y, \\ K = K(T(f,g^*))$$

and now $\widetilde{f}(y)$ is strictly increasing. Coming back to $(CP : Y_d)$ we have

$$\int_{Y(t)}^{Y_d} \frac{ds}{f(s)-f(0)} = -t \tag{14}$$

Then, if we define the (strictly decreasing) function

$$\Psi(r) := \int_r^\infty \frac{ds}{f(s)-f(0)}$$

and its inverse function

$$\eta = \Psi^{-1}, \quad then\ (14)\ says\ that \quad \Psi(Y(t)) - \Psi(Y_d) = -t.$$

In fact, it is clear that defining

$$Y(t) = \eta(\Psi(Y_d) - t)$$

we get the (unique) solution of $(CP : Y_d)$. In this way, the function $u_\tau(x)$ of the proof of Theorem 1.1 is given by

$$u_\tau(x) := \eta(\Psi(g^*(x)) + \tau), x \in \overline{\Omega},$$

and

$$U(x,t) := Y(t-\tau : g^*(x)) = \eta(\Psi(g^*(x)) + \tau - t).$$

If, for instance,

$$f(s) = s^3$$

then

$$\Psi(r) := \int_r^\infty \frac{ds}{f(s)-f(0)} = \frac{1}{2r^2},$$

$$\eta(s) = \Psi^{-1}(s) = \frac{1}{\sqrt{2s}},$$

$$u_\tau(x) := \eta(\Psi(g^*(x)) + \tau) = \frac{g^*(x)}{\sqrt{1+2\tau g^*(x)^2}}$$

$$U(x,t) := Y(t-\tau : g^*(x)) = \eta(\Psi(g^*(x)) + \tau - t) = \frac{g^*(x)}{\sqrt{1+2(\tau-t)g^*(x)^2}}.$$

In this special case it is also possible to check directly that if, for instance, $A = -\Delta$, then

$$\Delta U(x,t) = -\frac{6(\tau - t)g^*(x)\left|\nabla g^*(x)\right|^2}{2[1+2(\tau - t)g^*(x)^2]^{5/2}} + \frac{\Delta g^*(x)}{[1+2(\tau - t)g^*(x)^2]^{3/2}}.$$

Remark 2.2. *We point out that if (4) holds true with $s_0 = 0$ and $q = +\infty$, estimate (9) implies that*

$$\|y(t:u_\tau)\|_{L^\infty(\Omega)} \leq \left\|y^T\right\|_{L^\infty(\Omega)}, \text{ for any } t \in [0,\tau]$$

(it suffices to see that $Y(t)$ and $M(t)$ are increasing functions and that we can assume that $\|g^*\|_{L^\infty(\Omega)} \leq \left\|y^T\right\|_{L^\infty(\Omega)}$).

Remark 2.3. *The proof of Theorem 1.1 allows to see that assumptions (4), and (5) are merely used to be sure that the Cauchy problem $(CP: Y_d)$ has a solution backward continuable when we take $Y_d = g^*(x)$ with $x \in \overline{\Omega}$ arbitrary. In fact, assumption (5) implies that the continuation interval is "universal" (i.e. independent of the values of $g^*(x)$). More generally, the same proof applies to the case in which (4), and (5) are replaced by the following general condition on f and y^T*

$$H(f,y^T) \equiv \begin{cases} \exists\, T(f,y^T) \in (0,T] \text{ such that the solution of } (CP: Y_d) \\ \text{is continuable to } [-T(f,y^T),0] \text{ for any } Y_d = y^T(x), x \in \overline{\Omega}. \end{cases}$$

as well as that $g^* = y^T$, i.e.,

$$y^T \in C^2\left(\overline{\Omega}\right) \quad \text{and } y^T = 0 \text{ on } \Gamma_D, \frac{\partial y^T}{\partial \nu} = 0 \text{ on } \Gamma_N.$$

It is easy to see that assumption $H(f,y^T)$ is verified if f is sublinear (in that case $T(f,y^T) = +\infty$), as well as in the case in which f is a decreasing function, as , for instance, $f(s) = -\left|s\right|^{p-1}s$, with $p > 1$. In this last case $T(f,y^T) < +\infty$ and $T(f,y^T)$ strongly depends on the concrete values of $Y_d = y^T(x)$. So, if $y^T \in L^\infty(\Omega)$ we get the approximate controllability at least for some small $\tau_0 \in (0, T(f,y^T)]$. In some sense, this generalization of Theorem 1.1 covers the gap open by the negative results of Bardos and Tartar [1] and gives an answer to a conjecture posed in Fabre, Puel and Zuazua [7] concerning this special function f.

Remark 2.4. *It is also possible to get other type of generalizations, this time concerning the elliptic operator A. In the proof of Theorem 1.1 we merely applied the L^s-continuous dependence and the comparison principle for the associated functional operator $\mathbf{A} : D(\mathbf{A}) \to L^s(\Omega)$ and the linearity of $\mathbf{A}$ was not used. So the results remain valid for other diffusion operators (some quasilinear operators as, for instance, the p-Laplacian, the minimal surface operator and some fully nonlinear operators).*

Remark 2.5. *The constructive nature of the proof of Theorem 1.1 supplies additional qualitative informations on the constructed control u. So, for instance, u vanishes (resp. is strictly positive, resp. strictly negative) on the same subset of Ω*

where g^* vanishes (resp. is strictly positive, resp. strictly negative). This type of qualitative informations were obtained in Díaz [4] and Díaz, Henry and Ramos [5] for different nonexplosive semilinear problems.

Remark 2.6. *It is easy to see that there is not uniqueness of the control u (in fact of the pair $\{u, \tau_0\}$).*

Remark 2.7. *A general comment concerns a different type of arguments in order to prove conclusions for short time such as the one presented in Theorem 1.1. In fact, we can consider the abstract semilinear Cauchy problem*

$$\begin{cases} \frac{dy}{dt}(t) + \mathbf{A}y(t) = f(y(t)) & t \in (0,T), \text{ in } X, \\ y(0) = u & \text{in } X, \end{cases}$$

where X denotes a Banach space. Under very general conditions (see, e.g., the exposition made in Vrabie [14] it is well known that the solution depends continuously on the initial data and therefore the conclusion of Theorem 1.1 holds trivially by choosing $u = y^T$ (or $u = g^*$ with g^* a regularization of y^T). Although this kind of arguments would lead to very general results (even of a greater generality than Theorem 1.1) we point out that the proof of Theorem 1.1 has a constructive character which is very useful in order to study the approximate controllability for large time and other qualitative properties (see Sections 3 and 4 below).

3. On the approximate controllability for large time

Proof of Theorem 1.2. Arguing as in the proof of Theorem 1.1 with $g^* = M$ we construct the function

$$v(x) := Y^*(-T : M), x \in \overline{\Omega},$$

where $Y^*(t : M)$ denotes the solution of the ordinary Cauchy problem

$$\begin{cases} \frac{dY^*}{dt}(t) = f(Y^*(t)), \ t < 0, \\ Y^*(0) = M. \end{cases}$$

Defining

$$U^*(x,t) := Y^*(t - T : M)$$

we have that

$$\frac{\partial U^*}{\partial \nu}(x,t) = 0 \text{ if } x \in \Gamma_N \ , \ t \in [0,T]$$

(notice that since Γ_D is empty we do not need to have U^* vanishing in any part of $\partial\Omega$: this is the reason why we replace $Y(t : M)$ by $Y^*(t : M)$). The rest of the arguments of the proof of Theorem 1.1 can be repeated but now with

$$h(x,t) := AU^*(x,t) = 0$$

and the conclusion holds for the initial control $u = v$.

Remark 3.1. *As in Remark 2.1, if we assume f strictly increasing (or more generally (4) with $s_0 = 0$) then we have that*

$$U^*(x,t) := Y^*(t-T : M) = \eta^*(\Psi^*(M) + T - t)$$

with

$$\Psi^*(r) := \int_r^\infty \frac{ds}{f(s)}, \text{ if } r > 0,$$

and

$$\eta^* = (\Psi^*)^{-1}.$$

Proof of Theorem 1.3. We introduce the positive constants

$$\omega = \max\{|f'(s)| : s \in [-\|g^*\|_{L^\infty(\Omega)} - 1, \|g^*\|_{L^\infty(\Omega)} + 1]\}$$

and

$$\widetilde{\varepsilon} = (\varepsilon/(2|\Omega|^{1/q}))\exp(-\omega T).$$

By Theorem 1.1 there exists $\{v_0, \tau_0\}$ such that

$$\|y(\tau_0; v_0) - g^*\|_{L^\infty(\Omega)} \leq \widetilde{\varepsilon}.$$

We can assume that $\tau_0 < T$ (otherwise the proof ends). For $t < T_m(v_0)$ and $x \in \overline{\Omega}$ we introduce

$$z(x,t) := y(x,t : v_0) - g^*(x).$$

We also define

$$T^* = \sup\{t \mid t \geq \tau_0, \|z(\widetilde{t})\|_{L^\infty(\Omega)} < 1 \text{ for any } \widetilde{t} \in (\tau_0, t)\}.$$

It is clear that $T^* < T_m(v_0)$. So, by construction, $\|z(t)\|_{L^\infty(\Omega)} < 1$ for any $t \in (\tau_0, T^*)$. Then, if we define

$$\phi(x,s) = \begin{cases} f(s + g^*(x)) - f(g^*(x)) & \text{if } |s| \leq 1, \\ f(1 + g^*(x)) - f(g^*(x)) & \text{if } s > 1, \\ f(-1 + g^*(x)) - f(g^*(x)) & \text{if } s < -1, \end{cases}$$

we get that $f(z(x,t) + g^*(x)) - f(g^*(x)) = \phi(x, z(x,t))$. We point out that $\phi(x,s)$ is a Lipschitz function of constant smaller or equal than ω. Since

$$\begin{cases} z_t + Az &= \phi(x, z(x,t)) & \text{in } \Omega \times (\tau_0, T^*), \\ z(t,x) &= 0 & \text{on } \Gamma_D \times (0,\tau), \\ \frac{\partial z}{\partial \nu} &= 0 & \text{on } \Gamma_N \times (0,\tau), \\ \|z(\tau_0)\|_{L^\infty(\Omega)} &\leq \widetilde{\varepsilon}, \end{cases}$$

we deduce that

$$\|z(t)\|_{L^\infty(\Omega)} \leq e^{\omega t} \|z(\tau_0)\|_{L^\infty(\Omega)} \text{ for any } t \in (\tau_0, T^*).$$

In particular,

$$\|z(T^*)\|_{L^\infty(\Omega)} \leq (\varepsilon/(2|\Omega|^{1/q}))e^{\omega(T^*-T)}$$

and so $T \leq T^*$ (otherwise we get a contradiction with the definition of T^*). Then

$$\|z(T)\|_{L^\infty(\Omega)} \leq \varepsilon/(2\,|\Omega|^{1/q})$$

and the proof is completed since, again,

$$\left\|y(.,T:v_0) - y^T\right\|_{L^q(\Omega)} \leq |\Omega|^{1/q}\,\|y(.,T:v_0) - g^*\|_{L^\infty(\Omega)} + \left\|\,g^* - y^T\right\|_{L^q(\Omega)} \leq \varepsilon.$$

Remark 3.2. *The same proof applies for more general functions $y^T \in L^q(\Omega)$, for some $q \in [1,+\infty]$, such that*

$$\begin{cases} \textit{there exists } g^* \in L^\infty(\Omega) \textit{ verifying that } \left\|\,g^* - y^T\right\|_{L^q(\Omega)} \leq \varepsilon/2 \textit{ and} \\ \|Ag^* - f(g^*)\|_{L^\infty(\Omega)} \leq (\varepsilon/(2\,|\Omega|^{1/q}\,T))\exp(-\omega T) \textit{ with} \\ \omega = \max\{|f'(s)| : s \in [-\|g^*\|_{L^\infty(\Omega)} - 1, \|g^*\|_{L^\infty(\Omega)} + 1]\}\;, \\ g^* = 0 \textit{ on } \Gamma_D,\ \frac{\partial g^*}{\partial \nu} = 0 \textit{ on } \Gamma_N. \end{cases}$$

4. On the "cost" of the controls

Proof of Theorem 1.4. By applying Theorems 1.2 or 1.3, with $q = +\infty$, and Remark 2.2 we have that the set

$$K = \begin{array}{l}\Big\{v : v \in L^\infty(\Omega) : \left\|y(T;v) - y^T\right\|_{L^\infty(\Omega)} \leq \varepsilon, \\ \|y(t;v)\|_{L^\infty(\Omega)} \leq \left\|y^T\right\|_{L^\infty(\Omega)} + 1, \forall t \in [0,T]\Big\}\end{array}$$

is not empty. Let us start by assuming that

$$f \text{ is strictly decreasing on } [-\left\|y^T\right\|_{L^\infty(\Omega)} - 1, \left\|y^T\right\|_{L^\infty(\Omega)} + 1]. \tag{15}$$

Then any solution of problem (1) with $u \in K$ coincides with the solution of a similar problem in which we replace the function f by the truncated function

$$\widetilde{f}(s) = \begin{cases} f(s) & \text{if } \ |s| \leq \left\|y^T\right\|_{L^\infty(\Omega)} + 1, \\ f(\left\|y^T\right\|_{L^\infty(\Omega)} + 1) & \text{if } \ s > \left\|y^T\right\|_{L^\infty(\Omega)} + 1, \\ f(-\left\|y^T\right\|_{L^\infty(\Omega)} - 1) & \text{if } \ s < -\left\|y^T\right\|_{L^\infty(\Omega)} - 1. \end{cases}$$

Now, let $\{v_n\}$ be a minimizing sequence. Obviously

$$\|v_n\|_{L^\infty(\Omega)} \leq \ \|v_0\|_{L^\infty(\Omega)}\,,$$

with v_0 the control given in the statement of Theorems 1.2 or 1.3. So, $v_n \rightharpoonup v$ weakly-star in $L^\infty(\Omega)$. By applying the theory of accretive operators generating compact semigroups (see, e.g. Vrabie [14]) it is easy to see that $y(t;v_n) \to y(t;v)$ strongly in $L^q(\Omega)$, for any $q \in (1,+\infty)$, and that $\|y(t;v)\|_{L^\infty(\Omega)} < \left\|y^T\right\|_{L^\infty(\Omega)} + 1$. Thus the conclusion follows. If (15) fails, we argue as in Remark 2.1, i.e. we replace the equation of (1) by

$$y_t + \widehat{A}y = \widehat{f}(y) \quad \text{in } \Omega \times (0,T),$$

with

$$\widetilde{A}y = Ay - (K+1)y, \ \widetilde{f}(y) = f(y) - (K+1)y,$$

for some $K > 0$ large enough and now $\widetilde{f}(y)$ becomes strictly increasing (notice that by the change of variables $y = e^{\lambda t}y^*$, we can always assume that $a_0 > K+1$).

Remark 4.1. *It would be interesting to know if the conclusion of Theorem 1.4 remains true after replacing the set K by the more general set*

$$\widetilde{K} = \{v \mid \ v \in L^\infty(\Omega), \left\|y(T;v) - y^T\right\|_{L^\infty(\Omega)} \le \varepsilon\}$$

and if we also replace the exponent $q = +\infty$ by $q \in [1,+\infty)$ arbitrary (see some related results for $q = 2$ and $f(s) = s^3$ in Lions [12] *(Section 1.12)).*

Proof of Theorem 1.5. We can assume without loss of generality that F is a strictly increasing function. Using the notation of Remark 3.1 we have that

$$u(x) = U^*(x,0) = \eta_F^*(\Psi_F^*(M) + \lambda T)$$

where

$$\Psi_F^*(r) := \int_r^\infty \frac{ds}{F(s)}, \text{ if } r > 0,$$

and

$$\eta_F^* = (\Psi_F^*)^{-1}$$

(notice that $\Psi^*(r) = \Psi_F^*(r)/\lambda$). On the other hand, since F is a Lipschitz function near $r = 0$ and $F(0) = 0$ we have that

$$\Psi_F^*(r) \nearrow +\infty \text{ if } r \searrow 0$$

(and also $\Psi_F^*(r) \searrow 0$ if $r \nearrow +\infty$). Analogous properties hold if $r < 0$. Then $\|u\|_{L^\infty(\Omega)} \searrow 0$ if $\lambda \nearrow +\infty$ and, since by construction

$$\|v_0(\lambda)\|_{L^\infty(\Omega)} \le \|u\|_{L^\infty(\Omega)},$$

we get the result.

Remark 4.2. *We conjecture that the conclusion of Theorem 1.5 remains true under more general conditions (more general functions y^T, more general boundary conditions, etc.).*

Remark 4.3. *For previous results along the lines of Theorem 1.5 with distributed (or boundary) control, cf. Lions and Zuazua (*[13]*) (cf. also Theorem 1.7 below).*

5. Approximate controllability of non local systems

We consider now system (7) of the Introduction and we are going to prove Theorems 1.6 and 1.7.

Let us first notice that given $u \in L^2(\Omega)$ the existence of a solution (local in time) defined on a *maximal interval* $[0, T_m[$ ($T_m = T_m(u)$ and $\|y(t)\|_{L^2(\Omega)} \nearrow +\infty$ if $t \nearrow T_m$ when $T_m(u) < +\infty$) can be proved by using compactness methods (see, e.g. Theorem 4.3.1. of Vrabie [14]). Notice also that problem (7) has a *superlinear* nature since if we denote

$$\mathbf{F}(y) = cy \int_\Omega y^2 dx$$

then $\mathbf{F} : L^2(\Omega) \to L^2(\Omega)$ and

$$\|\mathbf{F}(y)\|_{L^2(\Omega)} = c \|y\|^3_{L^2(\Omega)} .$$

On the other hand, if we set

$$\|y(t)\|^2 = \int_\Omega y(x,t)^2 dx,$$

we obtain, after multiplying (7) by y, that

$$\frac{1}{2}\frac{d}{dt} \|y(t)\|^2 + a(y(t)) - c\|y(t)\|^4 = 0 \tag{16}$$

where

$$a(y(t)) = a(y(t), y(t)) = \int_\Omega (\sum_{i,j=1}^N a_{ij} \frac{\partial y}{\partial x_i}\frac{\partial y}{\partial x_j} + a_0 y^2) dx.$$

The estimate (16) shows the "unboundedness" of y as $t \to +\infty$ and the *"increase in instability"* as $c \nearrow +\infty$. This is made more precise below.

Before proving Theorems 1.6 and 1.7 let us show some examples in which $T_m(u) < +\infty$. For this and later purposes we are going to use the eigenfunctions

$$\begin{cases} Aw_i = \lambda_i w_i, & 0 < \lambda_1 \leq \lambda_2 \leq ... \\ w_i = 0 & \text{on } \partial\Omega, \\ \int_\Omega w_i w_j dx = \delta_{ij}. \end{cases} \tag{17}$$

We recall that $w_1(x) > 0$ a.e. $x \in \Omega$.

Proposition 5.1. *Let $u \in L^2(\Omega)$, $u \geq 0$ on Ω, be such that*

$$\left(\int_\Omega u w_1 dx\right)^2 \geq \frac{\lambda_1}{c}.$$

Then $T_m(u) < +\infty$.

Proof. By multiplying by $y_-(t) = \min(y(t), 0)$ and applying Gronwall's inequality we get that $y(t) \geq 0$ on Ω for any $t \in [0, T_m[$. We introduce the function

$$H(t) = \int_\Omega y(t) w_1 dx \text{ for } t \in [0, T_m[.$$

Then

$$H'(t) = \int_\Omega y_t(t) w_1 dx = -\lambda_1 H(t) + c(\int_\Omega y(t)^2 dx) H(t).$$

But

$$H(t)^2 \leq \int_\Omega y(t)^2 w_1 dx \leq \int_\Omega y(t)^2 dx.$$

Therefore

$$H'(t) \geq H(t)(-\lambda_1 + cH(t)^2), \text{ on }]0, T_m[$$

and the result follows in a standard way (see, e.g. Proposition 5.4.1 of Cazenave and Haraux [3]).

In order to prove Theorems 1.6 and 1.7 we introduce

$$E_m = \text{space generated by } w_i,\ 1 \leq i \leq m. \tag{18}$$

Let T be given. We shall have *approximate controllability* if we know that

$$\left.\begin{array}{l} \text{given } y^T \in E_m, \text{ there exists } u \in L^2(\Omega) \text{ such that (7) has} \\ \text{a solution defined in } \Omega \times (0,T) \text{ and which is such that } y(T) = y^T. \end{array}\right\} \tag{19}$$

Actually, we are going to verify (19) by choosing $u \in E_m$, i.e.

$$u = \sum_{j=1}^m u_j w_j. \tag{20}$$

The controls are now $\{u_j\} \in R^m$.

For the initial condition (20), the solutions of (7) can be computed as follows. The solution $y(t) \in E_m$, so that

$$y(t) = \sum_{j=1}^m y_j(t) w_j,$$

where

$$\left.\begin{array}{l} y_j'(t) + \lambda_j y_j(t) = c y_j(t) Y(t), \\ y_j(0) = u_j, \\ Y(t) = \sum_{j=1}^m y_j(t)^2. \end{array}\right\} \tag{21}$$

One has to find (if possible) u_j such that $T < T_m(u)$ and

$$y_j(T) = y_j^T \text{ (where } y^T = \sum_{j=1}^m y_j^T w_j). \tag{22}$$

The solution of (21) is given by

$$\left.\begin{array}{l} y_j(t) = u_j \exp(-\lambda_j t + cZ(t)), \text{ where} \\ Z(t) = \int_0^t Y(s) ds. \end{array}\right\} \tag{23}$$

But (23) is still *implicit.* We make it explicit in the following way. Let us set

$$U(t) = \sum_{j=1}^{m} {u_j}^2 e^{-2\lambda_j t}.$$

Then (23) leads to

$$Y(t) = U(t)e^{2cZ(t)}$$

and since $Y(t) = Z'(t)$, it follows that

$$\begin{aligned} Z'(t)e^{-2cZ(t)} &= U(t), \text{ i.e.} \\ e^{-2cZ(t)} - 1 &= -2c\int_0^t U(s)ds = -2c\sum_{j=1}^{m}(\frac{1-e^{-2\lambda_j t}}{2\lambda_j})u_j^2. \end{aligned} \tag{24}$$

Formulae (23) (24) define "explicitly" y_j in function of the terms u_j. One has to find, if possible, u_j such that (22) holds true, i.e.

$$u_j e^{-\lambda_j T} = e^{-cZ(T)} y_j^T$$

i.e. finally

$$u_j = e^{\lambda_j T} y_j^T e^{-cZ(T)}.$$

Let us set

$$\mu_j(T) = \frac{1-e^{-2\lambda_j T}}{\lambda_j}, \; z_j^T = e^{\lambda_j T} y_j^T.$$

Then one has to find, if possible, u_j such that

$$u_j = z_j^T (1 - c\sum_{j=1}^{m} \mu_j(T)u_j^2)^{1/2}. \tag{25}$$

It follows from (25) that

$$\sum_{j=1}^{m} \mu_j(T)u_j^2 = (\sum_{j=1}^{m} \mu_j(T)(z_j^T)^2)(1 - c\sum_{j=1}^{m} \mu_j(T)u_j^2).$$

If we set

$$\|z^T\| = \sum_{j=1}^{m} \mu_j(T)(z_j^T)^2,$$

it follows that

$$\sum_{j=1}^{m} \mu_j(T)u_j^2 = \frac{\|z^T\|^2}{1 + c\|z^T\|^2}, \tag{26}$$

hence

$$u_j = \frac{z_j^T}{(1 + c\|z^T\|^2)^{1/2}}.$$

This proves Theorem 1.6.

Moreover we can define as the *cost* the quantity $\sum_{j=1}^{m} \mu_j(T)u_j^2$. It is given by (26), which shows Theorem 1.7.

Remark 5.2. *By using (23) and (24) we get that the solution of (21) is given by*

$$y_j(t) = e^{-\lambda_j t} \frac{u_j}{(1 - c\sum_{j=1}^{m}(\frac{1-e^{-2\lambda_j t}}{\lambda_j})u_j^2)^{1/2}}.$$

It is easy to see that the associated function $y(t)$ blows up in a finite time (i.e. $T_m(u) < +\infty$) if the initial datum u is such that

$$\sum_{j=1}^{m} \frac{u_j^2}{\lambda_j} > \frac{1}{c}.$$

Remark 5.3. *In the above proof, the fact that $c > 0$ is essential. Things become different, for instance in (26), if $c < 0$.*

References

[1] Bardos, C. and Tartar, L., 1973, Sur l'unicité rétrograde des équations paraboliques et quelques questions voisines, *Arch. Ration. Mech. Analysis,* **50**, pp. 10–25

[2] Blayo, E., Blu m, J. and Verron, J., 1998, Assimilation variationelle de données en Océanographie et réduction de la dimension de l'espace de contrôle. In *Équations aux dérivées partielles et applicatons: Articles dédiés à Jacques-Louis Lions,* Gauthier-Villars, Paris, pp. 199–220.

[3] Cazenave, Th. and Haraux, A., 1990, *Introduction aux problèmes d'évolution semi-linéaires,* Ellipses, Paris.

[4] Díaz, J.I., 1991, Sur la contrôlabilité approchée de inéquations variationelles et d'autres problèmes paraboliques non linéaire, *C. R. Acad. Scie. de Paris,* **1312**, Série I, pp. 519–522.

[5] Díaz, J.I., Henry, J. and Ramos, A.M., 1998, On the Approximate Conrollability of Some Semilinear Parabolic Boundary-Value Problems, *Appl. Math. Optim.,* **37**, pp. 71–97.

[6] Díaz, J.I. and Lions, J.-L., 1998, Sur la contrôlabilité de problèmes paraboliques avec phénomènes d'explosion, *C. R. Acad. Scie. de Paris.* To appear.

[7] Fabre, C., Puel, J.P., and Zuazua, E., 1995, On the density of the range of the semi-group for semilinear heat equations. In *Control and Optimal Design of Distributed Parameter Systems.* Springer-Verlag, New York, IMA Volumes #70, pp. 73–92.

[8] Henry, J., 1978, *Etude de la contrôlabilité approchée de certaines équations paraboliques,* Thèse d'Etat, Université de Paris VI.

[9] Lattes, R. and Lions, J.L., 1967, *Méthode de quasiréversibilité et applications,* Dunod, Paris.

[10] Le Dimet, F.X. and Charpentier, I., 1998, Méthodes du second ordre en assimilation de données. In *Équations aux dérivées partielles et applicatons: Articles dédié Jacques-Louis Lions,* Gauthier-Villars, Paris, pp. 623–640.

[11] Lions, J.L. 1968, *Contrôle optimal de systèmes gouvernés par des équations aux derivées partielles*, Dunod, Paris.

[12] Lions, J.L. 1983, *Contrôle des systêmes distribués singuliers*, Gauthier-Villars, Bordas, Paris.

[13] Lions, J.L. and Zuazua, E., 1997, The cost of controlling unstable systems: time irreversible systems, *Revista Matemática de la Univ. Complutense de Madrid,* **10**, pp. 481–523.

[14] Vrabie, I.I., 1995, *Compactness Methods for Nonlinear Evolutions,* Pitman Monographs, Longman, Harlow.

J.I.D.
Departamento de Matemática Aplicada,
Facultad de Matemáticas,
Universidad Complutense de Madrid,
E-28040 Madrid, Spain
J.L.L.
Collège de France,
3, rue d'Ulm,
F-75005 Paris, France
E-mail address: jidiaz@sunma4.mat.ucm.es

International Series of Numerical Mathematics
Vol. 133,

Fréchet-Differentiability and Sufficient Optimality Conditions for Shape Functionals

Karsten Eppler

Abstract. For a heuristic approach of the boundary variation in shape optimization, applicable for a special class of domains, the computation of second derivatives of domain and boundary integral functionals are discussed. Moreover, for this approach the functionals are Fréchet-differentiable, because an embedding into a Banach space problem is possible. This implies symmetry and allows the discussion of sufficient condition in terms of a coercivity assumption on the second Fréchet-derivative. The theory is illustrated by a discussion of the Dido problem.

1. Introduction

Throughout the last 25–30 years shape optimization problems have been intensively studied in the literature with respect to various directions of investigation. A lot of methods for the description of the domain variation are developed and derivatives of functionals and solutions of state equations with respect to these domain or boundary variations can be computed. Moreover, necessary optimality conditions are given and numerical algorithms for a wide variety of problems are applied (cf. the surveys in [11] and [13]). Nevertheless, due to some difficulties arising from theoretical as well as from technical point of view, the study of sufficient conditions seems to be not very well developed at the moment. Only a few number of papers are concerned with related investigations [6], [1]. Therefore, we discuss the most easiest case of shape functionals only for a special class of domains.

More precisely, we shall study 2-dimensional simply connected bounded domains $\Omega \subset D$, where D is given and the domains under consideration satisfying a condition of starshapeness with respect to a neighbourhood $U_\delta(x_0) = \{y \in \mathbb{R}^2 \mid |y - x_0| < \delta\}$, with some fixed $\delta > 0$. Without loss of generality we assume in the sequel $x_0 = \mathbf{0}$. The main advantage of this assumption is that the boundary $\Gamma = \partial\Omega$ of such domains can be described by a Lipschitz continuous function $r = r(\phi)$ of the polar angle ϕ (i.e., $\Gamma := \{\gamma(\phi) = \begin{pmatrix} r(\phi)\cos\phi \\ r(\phi)\sin\phi \end{pmatrix} \mid \phi \in [0, 2\pi]\}$ and, vice verca, each domain (boundary) can be identified with this describing function.

Remark 1: Due to a result of Mazja [10], the boundary function of a domain Ω, which is starshaped to an open subset U_δ, is Lipschitz continuous with a constant, depending only on δ an on

$$d_\Omega := \sup\{|x| \mid x \in \Omega\}.$$

Consequently, if we assume that all domains under consideration are **uniformly** bounded (i.e., there exists a bounded outer "security set" D), then they have **uniform** Lipschitz continuous boundaries.
Moreover, the assumption $\Gamma \in C^k, \quad (k \in \mathbb{N})$ is equivalent to

$$r(\cdot) \in C_p^k[0, 2\pi] := \{r(\cdot) \in C^k[0, 2\pi] \mid r^{(i)}(0) = r^{(i)}(2\pi), \ i = 0, \dots, k\}. \tag{1}$$

Based on this, formulae for first and second order derivative of shape functionals can be obtained very easily. Furthermore, second order sufficient optimality conditions are investigated. Finally, these conditions were applied to the Dido problem.

2. First and second order derivatives

For transformations into polar coordinates we recall well-known formulas for the curvature $\kappa(\cdot)$ (for $\Gamma \in C^2$) and arclength $l(\cdot)$

$$\kappa(\phi) = \frac{2r'^2(\phi) + r^2(\phi) - r(\phi)r''(\phi)}{\sqrt{r^2(\phi) + r'^2(\phi)}^3}, \text{ and } l(\phi) = \sqrt{r^2(\phi) + r'^2(\phi)}, \tag{2}$$

and unscaled and scaled outer normal of the boundary

$$\vec{a}(\phi) = \begin{pmatrix} r(\phi)\cos\phi + r'(\phi)\sin\phi \\ r(\phi)\sin\phi - r'(\phi)\cos\phi \end{pmatrix} \Rightarrow \vec{n}(\phi) = \frac{1}{\sqrt{r^2(\phi) + r'^2(\phi)}} \vec{a}(\phi). \tag{3}$$

In the following a reference domain $\Omega \in C^1$ is given, where the boundary Γ is associated with the describing function $r \in C_p^1[0, 2\pi]$. In this way, the "optimization variables" (the admissible domains) are identified with elements of an (open) subset of the Banach space $C_p^1[0, 2\pi]$, and differential calculus in Banach spaces can be applied for the study of the problem.
Remark 2: Because of

$$\vec{e}_r \cdot \vec{n} = \frac{r}{\sqrt{r^2 + r'^2}} > 0, \ \vec{e}_r = \begin{pmatrix} \cos\phi \\ \sin\phi \end{pmatrix} \text{ – the radial unit vector,}$$

the above perturbations are always regular, i.e., the perturbation field is a tangential field if and only if $r_1(\cdot) \equiv 0$.

Lemma 2.1. *Let $h \in C(D)$ and $g \in C^1(D)$ be given. Then the functionals* $J_1 = \int\limits_\Omega h\,dx$ *and* $J_2 = \int\limits_\Gamma g\,dS_\Gamma$ *are Fréchet-differentiable with respect to* $C_p^1[0, 2\pi]$ *at* *every admissible* Ω *with the derivatives*

$$\nabla J_1(r)[r_1] = \int_0^{2\pi} r(\phi) r_1(\phi) h(r(\phi), \phi)\, d\phi, \tag{4}$$

and

$$\nabla J_2(r)[r_1] = \int_0^{2\pi} r_1 \sqrt{r^2 + r'^2}\, \frac{\partial g}{\partial r}(r(\phi), \phi) + g(r(\phi), \phi)\, \frac{r r_1 + r' r_1'}{\sqrt{r^2 + r'^2}}\, d\phi. \tag{5}$$

Proof. In order to calculate the directional derivative in the direction $r_1 \in C_p^1[0, 2\pi]$, we define perturbed domains Ω_ε by the connection

$$\Gamma_\varepsilon = \{ r_\varepsilon(\phi) \vec{e}_r(\phi) \mid \phi \in [0, 2\pi] \}, \quad r_\varepsilon(\phi) = r(\phi) + \varepsilon r_1(\phi),$$

and $\varepsilon > 0$ is taken sufficiently small, provided that $r_\varepsilon(\phi) > \delta$, $\phi \in [0, 2\pi]$ is satisfied. Using Fubini's Theorem for transformation into polar coordinates we get

$$\begin{aligned} J_1(\Omega_\varepsilon) - J_1(\Omega) &= \int_0^{2\pi} \int_{r(\phi)}^{r_\varepsilon(\phi)} h(r, \phi) r dr\, d\phi \\ &= \int_0^{2\pi} r_1(\phi) \int_0^{\varepsilon} h(r_\tau(\phi), \phi) r_\tau(\phi)\, d\tau\, d\phi, \quad \text{with } r_\tau = r + \tau r_1 \\ &= \varepsilon \cdot \int_0^{2\pi} r(\phi) r_1(\phi) h(r(\phi), \phi)\, d\phi + o(\varepsilon). \end{aligned}$$

Taking $\varepsilon \to 0$, we obtain that J_1 is directionally differentiable at every admissible r and the derivative is given by (4). Moreover, this represents a Gateaux-derivative $G(r; \cdot)$, because it is obviously linear and continuous with respect to $r_1 \in C^1$. Additionally we deduce the Gateaux-differentiability of J_1 in the same way in a neighbourhood of an arbitrary r_0 (or Ω_0), where the norm of $G(r; \cdot)$ depends continuously on $r \in U_\eta(r_0)$ ($\eta > 0$ sufficiently small). Hence, by a standard argument from functional analysis (cf. [2], [9]), the functional J_1 is continuously Fréchet-differentiable at r_0. For J_2 the transformation

$$J_2(\Omega_\varepsilon) = \int_{\Gamma_\varepsilon} g\, dS_\Gamma = \int_0^{2\pi} g(r_\varepsilon(\phi), \phi) \sqrt{r_\varepsilon^2 + {r'}_\varepsilon^2}\, d\phi$$

and differentiation w. r. t. ε leads to (5) (for the directional derivative). By $r_0(\phi) \geq c_0 > 0$ we have an upper bound of $\dfrac{1}{\sqrt{r^2 + r'^2}}$ uniformly for $r \in U_\eta(r_0)$. Therefore, the same arguments as for J_1 imply the Fréchet-differentiability. □

Remark 3: It is well known ([13], [11]) that the directional derivative of J_1 for other approaches is given by a boundary integral with the trace of h multiplied by the normal component of the domain variation at the boundary. Because of $r_1(\vec{e}_r \cdot \vec{n})dS_\Gamma = r(\phi)r_1(\phi)d\phi$ in our situation, this is (4). Also for J_2 formulae for directional derivatives similar to (5) are known for other approaches ([11], [13], [5]).

Remark 4: The shape derivative of the unit normal and the curvature are given by

$$\frac{d}{d\varepsilon}\vec{n}_\varepsilon(\phi)|_{\varepsilon=0} = \frac{rr_1' - r'r_1}{r^2 + r'^2}(\phi) \cdot \vec{\tau}(\phi) \perp \vec{n}(\phi),$$

where $\vec{\tau}(\phi)$ denotes the unit tangential vector on Γ directed to increasing ϕ, and

$$\frac{d}{d\varepsilon}\kappa_\varepsilon|_{\varepsilon=0} = \frac{2rr_1 + 4r'r_1' - rr_1'' - r''r_1}{\sqrt{r^2 + r'^2}^3} - 3\kappa \cdot \frac{rr_1 - r'r_1'}{r^2 + r'^2},$$

respectively. The relation $\frac{d}{d\varepsilon}\vec{n}_\varepsilon(t)|_{\varepsilon=0} \perp \vec{n}$ is also known for more general cases (see [13]).

Second Fréchet-derivatives. As we had already announced, second shape derivatives for starshaped domains can be computed "straightforward", if the data fields are smooth enough.

Theorem 2.2. *Let $h \in C^1(D)$ and $g \in C^2(D)$ be given. Then the functionals $J_1 = \int_\Omega h\,dx$ and $J_2 = \int_\Gamma g\,dS_\Gamma$ are twice Fréchet-differentiable with respect to $C_p^1[0, 2\pi]$ at Ω with the second derivatives*

$$\nabla^2 J_1(r)[r_1; r_2] = \int_0^{2\pi} r_1(\phi)r_2(\phi)h(r,\phi) + r(\phi)r_1(\phi)r_2(\phi)\frac{\partial h}{\partial r}(r,\phi)\,d\phi, \tag{6}$$

and

$$\begin{aligned}\nabla^2 J_2(r)[r_1; r_2] &= \int_0^{2\pi} \left\{ r_1 r_2 \sqrt{r^2 + r'^2}\frac{\partial^2 g}{\partial r^2} + \frac{\partial g}{\partial r}\left[r_1 \frac{rr_2 + r'r_2'}{\sqrt{r^2 + r'^2}} + r_2 \frac{rr_1 + r'r_1'}{\sqrt{r^2 + r'^2}} \right] \right. \\ &+ \left. g\frac{(r_1 r_2 + r_1' r_2')(r^2 + r'^2) - (rr_1 + r'r_1')(rr_2 + r'r_2')}{\sqrt{r^2 + r'^2}^3} \right\} d\phi. \end{aligned} \tag{7}$$

Remark 5: Due to the Banach space embedding, the boundary variation r_1 on perturbed boundaries $\Gamma_{\delta r_2}$ and on Γ is defined in the same way without any additional problem. Therefore, differentiation can be carried out in the sense of

$$d^2 J_i(\Omega)[r_1; r_2] = \lim_{\delta \to 0} \frac{dJ_i(\Omega_{\delta r_2})[r_1] - dJ_i(\Omega)[r_1]}{\delta}, \quad i = 1, 2,$$

and leads obviously to symmetry with respect to r_1 and r_2. Moreover, we need no additional regularity of the boundary for the definition of higher order derivatives of shape functionals.

Some examples. For the volume $J_1 = \int\limits_\Omega dx$ of a domain we have

- $dJ_1(\Omega)[r_1] = \nabla J_1(r)[r_1] = \int\limits_0^{2\pi} r(\phi) r_1(\phi)\, d\phi,$
- $d^2 J_1(r)[r_1; r_2] = \int\limits_0^{2\pi} r_1(\phi) r_2(\phi)\, d\phi.$

The second derivative of the volume does not depend on the reference domain, hence, third derivatives will vanish (for 2D-domains).

Similarly for the perimeter $J_2 = \int\limits_\Gamma dS_\Gamma$ we obtain

- $dJ_2(\Omega)[r_1] = \nabla J_2(r)[r_1] = \int\limits_0^{2\pi} \frac{r r_1 + r' r_1'}{\sqrt{r^2 + r'^2}}\, d\phi,$
- $d^2 J_2(r)[r_1; r_2] = \int\limits_0^{2\pi} \frac{(r_1 r_2 + r_1' r_2')(r^2 + r'^2) - (r r_1 + r' r_1')(r r_2 + r' r_2')}{\sqrt{r^2 + r'^2}^3}\, d\phi.$

3. Optimality conditions

Whereas necessary optimality conditions can be easily obtained by using directional derivatives of first and second order, the situation for sufficient conditions is more complicated in general in shape optimization. Due to the special approach for starshaped domains, standard methods are applicable. In this paper we shall study only the case of free (local) minima, where the "distance" between domains is understood as the distance of boundary describing functions in related Banach spaces.

Optimality conditions for volume functionals. Let Ω_0 be an optimal domain with boundary Γ_0. From the standard necessary condition it follows immediately ("all $r_1 \in C_p^1$ are admissible")

$$dJ_1(\Omega_0)[r_1] = \nabla J_1(r_0)[r_1] = \int\limits_0^{2\pi} r_0(\phi) r_1(\phi) h(r_0, \phi)\, d\phi = 0 \quad \Rightarrow h|_{\Gamma_0} \equiv 0. \tag{8}$$

Moreover, according to (6) we get for a domain, satisfying the necessary condition

$$\nabla^2 J_1(r_0)[r; r] = \int\limits_0^{2\pi} r^2(\phi) r_0(\phi) \frac{\partial h}{\partial r}|_0(\phi) d\phi. \tag{9}$$

Optimality can be guaranteed often by some coercivity of the second Fréchet-derivative. However, it is impossible to have coercivity with respect to C^1 (the "space of differentiation"), we can only expect an estimate

$$\nabla^2 J_1(r_0)[r,r] \geq c_0 \|r\|^2_{L_2}, \quad \text{(where } c_0 > 0 \text{ is ensured by } \frac{\partial h}{\partial r}|_0(\phi) > 0, \ \forall \phi).$$

This is known from other control problems as the "two-norm-discrepancy".

Remark 6: The conditions $\frac{\partial h}{\partial r}|_0 > 0$ and $\frac{\partial h}{\partial n}|_0 > 0$ are equivalent for starshaped domains (we have $(\vec{e}_r, \vec{n}) > 0 \ \forall \phi$ and $\frac{\partial h}{\partial \vec{\tau}}|_0 = 0 \Rightarrow \frac{\partial h}{\partial r}|_0 = \frac{\partial h}{\partial n}|_0(\vec{e}_r, \vec{n})$).

Theorem 3.1. *For $\Omega_0 \in C^1$ and $h \in C^2$ the conditions $h|_{\Gamma_0} \equiv 0$ and $\frac{\partial h}{\partial r}|_0 > 0$ are sufficient for optimality.*

Proof. We have (from differential calculus):

$$J_1(r_0 + r) - J_1(r_0) = \frac{1}{2}\left[d^2 J_1(r_0)[r,r] + R_2(r),\right], \ \left(\frac{|R_2(r)|}{\|r\|^2_{C^1}} \to 0 \text{ for } \|r\|_{C^1} \to 0\right),$$

but this is not enough to ensure optimality. Nevertheless, by a more careful estimate of the remainder $R_2(r) = d^2 J_1(r_\nu)[r,r] - d^2 J_1(r_0)[r,r]$ (where $r_\nu := r_0 + \nu r$) it follows

$$|R_2(r)| = |\int_0^{2\pi} r^2 \left[h(r_\nu, \phi) - 0 + (r_\nu)\frac{\partial h}{\partial r}|_\nu - r_0 \frac{\partial h}{\partial r}|_0\right] d\phi|$$

$$\leq \max |r(\phi)| \int_0^{2\pi} r^2 [c_1(h,\eta) + c_2(h,\eta) + c_3(h,\eta)]\, d\phi$$

$$\leq c(h,\eta) \|r\|_C \|r\|^2_{L_2}, \text{ with } \|r\|_C < \eta.$$

We arrive at (for η sufficiently small)

$$J_1(r_0 + r) - J_1(r_0) \geq \frac{c_0}{2} \|r\|^2_{L_2}, \text{ if } \|r\|_{C^2} < \eta,$$

which ensures the optimality of Ω_0. □

Remark 7: The easy situation allows an interpretation as follows:
From the necessary and sufficient condition we have for the data field h

$$(i)\ h|_{\Gamma_0} = 0, \ (ii)\ h(x) > 0, \forall x \in U_\delta(\Gamma_0) \setminus \bar{\Omega}_0, \ (iii)\ h < 0, \forall x \in U_\delta(\Gamma_0) \cap \Omega_0.$$

Therefore, each perturbation of the boundary increases the functional value.

Optimality conditions for boundary functionals. Here we additionally assume $\Omega_0 \in C^2$. The necessary condition follows after an integration by parts

$$\nabla J_2(r_0)[r_1] = \int_0^{2\pi} r_1 \sqrt{r_0^2 + r'^2_0} \frac{\partial g}{\partial r}|_0 + g_0 \frac{r_0 r_1 + r'_0 r'_1}{\sqrt{r_0^2 + r'^2_0}}\, d\phi =$$

$$= \int_0^{2\pi} r_1 \left\{ \left[\frac{r_0^2}{\sqrt{r_0^2 + r'^2_0}} \frac{\partial g}{\partial r}|_0 - \frac{r'_0}{\sqrt{r_0^2 + r'^2_0}} \frac{\partial g}{\partial \phi}|_0 \right] + g_0 r_0 \frac{r_0^2 + 2r'^2_0 - r_0 r''_0}{\sqrt{r_0^2 + r'^2_0}^3} \right\} dS_\Gamma,$$

$$= \int_0^{2\pi} r_0 r_1 \left(\frac{\partial g}{\partial n}|_0 + g\kappa_0 \right) dS_\Gamma = 0 \quad \Rightarrow \quad \left[\frac{\partial g}{\partial n} + g\kappa \right] |_0 \equiv 0.$$

For the study of sufficient condition we need the second derivative $d^2 J_2(\Omega_0)[r; r]$.

$$\nabla^2 J_2(r_0)[r; r] = \int_0^{2\pi} r^2 \sqrt{r_0^2 + r'^2_0} \frac{\partial^2 g}{\partial r^2} + 2r \frac{\partial g}{\partial r} \frac{r_0 r + r'_0 r'}{\sqrt{r_0^2 + r'^2_0}} + g \frac{(r'_0 r - r_0 r')^2}{\sqrt{r_0^2 + r'^2_0}^3} \, d\phi.$$

By integration by parts of the "mixed terms" $r' r \cdot f(\phi)$ we arrive at

$$\nabla^2 J_2(r_0)[r; r] = \int_0^{2\pi} r^2 \cdot f_1(\nabla^2 g, \nabla g, g, r_0) + r'^2 \cdot f_2(g, r_0) \, d\phi, \tag{10}$$

where f_1 and f_2 are given by

$$f_1(\nabla^2 g, \nabla g, g, r_0)(\phi) = r_0 \frac{\partial}{\partial n} \left(\frac{\partial g}{\partial r} \right) + \frac{\partial g}{\partial \phi} \frac{r_0 r'_0}{\sqrt{r_0^2 + {r'_0}^2}^3}$$

$$+ \frac{\partial g}{\partial r} \frac{2r_0^3 + 4{r'_0}^2 r_0 - r_0^2 r''_0}{\sqrt{r_0^2 + {r'_0}^2}^3} + g \frac{2{r'_0}^4 + r_0^3 r''_0 - 2r_0 {r'_0}^2 r''_0 - r_0^2 r'^2_0}{\sqrt{r_0^2 + {r'_0}^2}^5}$$

$$\text{and} \qquad f_2(g, r_0)(\phi) = \frac{r_0^2 g}{\sqrt{r_0^2 + {r'_0}^2}^3}.$$

Remark 8: Here, only a H^1-estimate is possible

$$\nabla^2 J_2(r_0)[r; r] \ \geq \ c_0 \|r\|^2_{H^1}, \text{ with some } c_0 > 0. \tag{11}$$

For the verification a Riccati equation technique may be used.

Theorem 3.2. *For $\Omega_0 \in C^2$ and $g \in C^3$ the condition* $\left[\frac{\partial g}{\partial n} + g\kappa \right] |_0 \equiv 0$ *and estimate (11) are sufficient for optimality.*

Proof. Similar to the volume case we have to estimate $R_2(r) = d^2 J_2(r_\nu)[r, r] - d^2 J_2(r_0)[r, r]$. From (10) it follows

$$|R_2(r)| \leq \int_0^{2\pi} r^2 |f_1^\nu(\phi) - f_1^0(\phi)| + r'^2 |f_2^\nu(\phi) - f_2^0(\phi)| \, d\phi,$$

where $f_1^\nu(\phi) = f_1(\nabla^2 g, \nabla g, g, r_\nu)(\phi)$ and $f_2^\nu(\phi) = f_2(g, r_\nu)(\phi)$, respectively.
Moreover, with $g \in C^3$ and (10) we get

(because of $|f_1^\nu(\phi) - f_1^0(\phi)| \le |\frac{r_\nu^2}{\sqrt{r_\nu^2 + r'^2_\nu}} \frac{\partial^2 g}{\partial r^2}|_\nu - \frac{r_0^2}{\sqrt{r_0^2 + r'^2_0}} \frac{\partial^2 g}{\partial r^2}|_0| + \quad \ldots +$

$$+|\frac{2r_\nu^3 + 4r'^2_\nu - r_\nu^2 r''_\nu}{\sqrt{r_\nu^2 + r'^2_\nu}^3} \frac{\partial g}{\partial r}|_\nu - \frac{2r_0^3 + 4r'^2_0 - r_0^2 r''_0}{\sqrt{r_0^2 + r'^2_0}^3} \frac{\partial g}{\partial r}|_0| + \quad \ldots \quad \text{and so on)}$$

$$R_2(r)| \le \int_0^{2\pi} r^2 \{c_1|r| + c_2|r'| + c_3|r''|\} + r'^2 c_4|r| \, d\phi,$$

with $\quad c_i = c_i(g, r_0, \eta)$, $i = 1(1)4$.

$$\le \tilde{c}(g, r_0, \eta) \cdot \|r\|_{C^2} \cdot \|r\|_{H^1}^2, \text{ for } \|r\|_{C^2} < \eta.$$

Summarizing up, we are able to estimate (for sufficiently small $\eta > 0$)

$$J_2(r_0 + r) - J_2(r_0) \ge \frac{c_0}{2} \|r\|_{H^1}^2, \text{ for } \|r\|_{C^2} < \eta.$$

□

Remark 9: If $g < 0$ holds somewhere on Γ, the domain Ω cannot be optimal.
Remark 10: The assumptions on the data h and g can be weakened, because a Lipschitz-estimate for the remainder is not necessary.

4. The Dido problem

As an illustrating example we want to apply the foregoing investigations to the Dido problem of maximizing the volume (area) of a domain subject to a given length of the perimeter. There are two elementary proofs known for the optimality of the circle (see, for example [14]). One of them is mainly based on investigations of Zenodorus in the ancient greece. The second proof was developed by Steiner in the 19th century. Moreover, several formulations of the problem are given in the calculus of variation (cf. [9]). If we restrict our considerations to starshaped domains only, the problem seems to become

$$\text{(P)} \quad J_1(r_0) = \int_{\Omega_0} -1 dx \to \inf, \text{ subject to } J_2(r_0) = \int_{\Gamma_0} 1 \, dS_\Gamma = l_0.$$

However, the problem is invariant with respect to parallel shifting. Hence, for the investigation of **sufficient** condition we additionally fix the baricentre, for convenience at the origin, which "forbids" the parallel shifting and does not influence the original problem otherwise. We arrive at the following modified problem

$$
\text{(PM)} \begin{cases} J_1(r_0) &= \int\limits_{\Omega_0} -1\,dx = \int\limits_0^{2\pi} -\frac{1}{2} r_0^2(\phi) d\phi \;\rightarrow\; \inf, \\ \text{subject} & \text{to} \\ J_2(r_0) &= \int\limits_{\Gamma_0} 1\,dS_\Gamma - l_0 = \int\limits_0^{2\pi} \sqrt{r_0^2(\phi) + r'^2_0(\phi)} d\phi - l_0 = 0, \\ J_3(r_0) &= \int\limits_{\Omega_0} x_1\,dx = \int\limits_0^{2\pi} \cos\phi \int\limits_0^{r_0(\phi)} \rho^2 d\rho d\phi = 0, \\ J_4(r_0) &= \int\limits_{\Omega_0} x_2\,dx = \int\limits_0^{2\pi} \sin\phi \int\limits_0^{r_0(\phi)} \rho^2 d\rho d\phi = 0. \end{cases}
$$

Whereas the discussion of necessary conditions is known from calculus of variation, we repeat it in terms of shape functionals. We define the Lagrangian

$$
L(r_0;\lambda) = J_1(r_0) - \sum_{k=2}^{4} \lambda_k J_k(r_0), \qquad \text{and obtain for } r_0 \in C_p^2
$$

$$
\begin{aligned} dL(r_0;\lambda)[r_1] &= \int\limits_0^{2\pi} -r_0 r_1 [1 + \lambda_3 r_0 \cos\phi + \lambda_4 r_0 \sin\phi] - \lambda_2 \frac{r_0 r_1 + r'_0 r'_1}{\sqrt{r_0^2 + r'^2_0}}\, d\phi \\ &= \int\limits_0^{2\pi} -r_0 r_1 \left(1 + \lambda_2 \cdot \kappa_0 + \lambda_3 \cos\phi r_0 + \lambda_4 \sin\phi r_0\right) d\phi \stackrel{!}{=} 0, \end{aligned}
$$

$$
\Rightarrow \qquad 1 + \lambda_2 \cdot \kappa_0(\phi) + \lambda_3 \cos\phi r_0(\phi) + \lambda_4 \sin\phi r_0(\phi) = 0, \quad \phi \in [0, 2\pi].
$$

With $\lambda_3^0 = \lambda_4^0 = 0$ and according to our constraints, we get

$$
\kappa_0 \equiv \text{ const.} \neq 0 \;\Rightarrow\; r_0(\phi) \equiv r_0,\; \lambda_2^0 = -\kappa_0^{-1} = -r_0 = -\frac{l_0}{2\pi}.
$$

Remark 11: The assertion $\lambda_3^0 = \lambda_4^0 = 0$ makes sense, because the optimal value function is obviously constant with respect to a variation of the value of the second and third constraint. Moreover, a vanishing Lagrange multiplier of the objective (i.e., $\lambda_1 = 0$) implying $\lambda_2 = 0$ or $\kappa_0 \equiv 0$. Therefore, regularity of the Lagrangian can be assumed.

Remark 12: The additional constraints are formally not needed for the necessary condition. Also for Problem (P) we obtain

$$
\kappa_0 \equiv \text{ const.} \neq 0 \;\text{ and }\; \lambda_2^0 = -\kappa_0^{-1} = -\frac{l_0}{2\pi}.
$$

However, we cannot conclude uniquely $r_0(\phi) \equiv r_0$, because all "shifted" circle with centre at $\vec{\varepsilon} = (\varepsilon_1, \varepsilon_2)^T$ (for $\varepsilon_1^2 + \varepsilon_2^2 < r_0^2$) satisfies the necessary condition.

For the validity of a sufficient second order condition we need

$$
\nabla^2 L(r_0, \lambda^0)[r; r] \;\geq\; c_0 \|r\|_{H^1}^2,
$$

for all r from the tangent cone T_c^0 at Ω_0 of the constraints. Due to the regularity, the tangent cone coincides with the linearizing cone, i.e., according to the derivatives of J_k,

$$T_c(\Omega_0) = \left\{ r \in C^2 \mid \int_0^{2\pi} r(\phi) d\phi = 0, \int_0^{2\pi} r(\phi) \cos\phi d\phi = 0, \int_0^{2\pi} r(\phi) \sin\phi d\phi = 0 \right\}.$$

Lemma 4.1. *It holds*

$$\nabla^2 L(r_0, \lambda^0)[r;r] \geq \frac{3}{5} \|r\|^2_{H^1},$$

for all $r \in T_c^0$, ensuring that a sufficient second order condition is satisfied for the circle.

Proof. An easy calculation yields

$$\nabla^2 L(r_0, \lambda^0)[r;r] = \int_0^{2\pi} r'^2(\phi) - r^2(\phi)\, d\phi.$$

Moreover, the system of trigonometric functions $\{\mathbf{1}, \cos n\phi, \sin n\phi, n \geq 1\}$ is complete in C^2 and an orthogonal basis in H^1, hence,

$$\|r\|^2_{H^1} = \int_0^{2\pi} r'^2(\phi) + r^2(\phi)\, d\phi = \mu_0^2(r) + (1+n^2) \sum_{n=1}^{\infty} \mu_n^2(r) + \nu_n^2(r),$$

where the Fourier-coefficients of r are related to the normalized functions. Furthermore, the tangent cone is contained in the closure of the linear hull of $\{\cos n\phi, \sin n\phi, n \geq 2\}$. Therefore, we are able to estimate as follows for $r \in T_c^0$

$$\int_0^{2\pi} r'^2(\phi) - r^2(\phi)\, d\phi = \sum_{n=2}^{\infty} (n^2 - 1)[\mu_n^2(r) + \nu_n^2(r)] \geq \sum_{n=2}^{\infty} \frac{3(n^2+1)}{5} [\mu_n^2(r) + \nu_n^2(r)].$$

Hence, we have the desired coercivity of $\nabla^2 L(r_0, \lambda^0)[r;r]$. □

Remark 13: From calculus of variation the validity of

$$\nabla^2 L(r_0, \lambda^0)[r;r] \geq 0, \ \forall r \in T_{c2}^0 = \{r \in C^2 \mid \int_0^{2\pi} r(\phi) d\phi = 0\}$$

is known. However, this is directly clear from the discussion above. Moreover, the functions $r_1(\phi) = \cos\phi$ and $r_2(\phi) = \sin\phi$ are associated with the "linearized directions of parallel shifting" at Ω_0 with respect to x_1 and x_2, respectively.

Concluding remarks. Similar investigations for second derivatives and sufficient conditions are possible for some other native approaches, using only boundary perturbations for the description of the domain variation ([4]). Because some of the results are obviously or intuitively clear, optimality conditions for shape functionals are not to much important. Nevertheless, it can be a first step for the study of more interesting shape optimization problems. For example, it seems to be possible to combine the presented technique with BIE- or potential methods ([8]) for the calculation of shape derivatives for solutions of elliptic equations ([12], [7], [3]), also related to investigations of Fuji ([5], [6], [1]). This will be discussed in a forthcoming paper.

References

[1] S. Belov and N. Fuji: Symmetry and sufficient condition of optimality in a domain optimization problem, Control and Cybernetics vol. **26** (1997), No. 1, 45–56.

[2] Bögel, K. and M. Tasche: Analysis in normierten Räumen, Akademie-Verlag, Berlin, 1974.

[3] Eppler, K.: Optimal shape design for elliptic equations via BIE-methods, preprint TU Chemnitz (1998).

[4] Eppler, K.: On the symmetry of second directional derivatives in optimal shape design and sufficient optimality conditions for shape functionals, preprint TU Chemnitz (1998).

[5] Fuji, N.: Second order necessary conditions in a domain optimization problem, Journal of Optimization Theory and Applications vol. **65** (1990), No. 2, 223–245.

[6] Fuji, N.: Sufficient conditions for optimality in shape optimizations, Control and Cybernetics vol. **23** (1994), No. 3, 393–406.

[7] Fuji, N. and Y. Goto: A potential method for shape optimization problems, Control and Cybernetics vol. **23** (1994), No. 3, 383–392.

[8] Guenther, N. M.: Die Potentialtheorie und ihre Anwendungen auf Grundaufgaben der mathematischen Physik, Teubner, Leipzig, 1957.

[9] Ioffe, A. D. and V. M. Tichomirow: Theorie der Extremalaufgaben, Deutscher Verlag der Wissenschaften, Berlin 1979.

[10] Mazja, W.: Einbettungssätze für Sobolewsche Räume I, Teubner, Leipzig, 1979.

[11] Pironneau, O.: Optimal Shape Design for Elliptic Systems, Springer, New York, 1983.

[12] Potthast, R.: Fréchet differentiability of boundary integral operators in inverse acoustic scattering, Inverse Problems vol. **10** (1994), 431–447.

[13] Sokolowski, J. and J.-P. Zolesio: Introduction to Shape Optimization, Springer, Berlin, 1992.

[14] V. M. Tichomirow: Stories about Maxima and Minima, Mathematical World **1**, American Mathematical Society, 1990.

Technical University of Chemnitz
Faculty of Mathematics
PSF 964
D-09107 Chemnitz, Germany
E-mail address: k.eppler@mathematik.tu-chemnitz.de

International Series of Numerical Mathematics
Vol. 133, © 1999 Birkhäuser Verlag Basel/Switzerland

State Constrained Optimal Control for some Quasilinear Parabolic Equations

Luis A. Fernández [1]

Abstract. We study optimal control problems where the system is governed by some quasilinear parabolic equations and there are restrictions on the control as well as on the state. The distributed control can appear in all the coefficients of the operator. State constraints of integral type and also pointwise in time are considered. Our main interest is the derivation of the first order optimality conditions. Finally, an application to exact controllability in finite dimensional subspaces is given.

1. Introduction

This paper deals with state constrained optimal control problems of systems governed by some quasilinear parabolic equations in divergence form. Our framework is more general than those corresponding to previous studies (see [1], [11], [4] and [6]) in the sense that all the coefficients of the operator (not only the secondary part) can depend on the control u, probably in a nonlinear fashion. With respect to the state restrictions, two different situations are examined. In the first one, equality and inequality state constraints in an integral form are included, depending both on the state and its spatial gradient. In the second problem, pointwise (in time) state constraints are considered. These cases have been extensively studied for linear equations (see, among others, [2], [8] and [9]) and semilinear problems (see for instance [12] and [10]). In the quasilinear case it is difficult to give precise references.

Our main interest is the derivation of the first order optimality conditions. This is accomplished by proving that the relation control-state is differentiable.

In spite of its general appearance, it should be mentioned that studies [1] and [11] can not be applied even to the Heat operator: one of the abstract conditions assumed there implies (in practice) the finite-dimensional character of the functional spaces involved (see condition ii) of Lemma 3.3 in [1] and hypothesis $H(A)_4 - (6)$ in [11]). Here, the hypotheses are established in terms of polynomial growth conditions to make the verification easier. Moreover, we provide concrete examples within our framework and out of the scope of previous known works.

[1] This work was partially supported by DGICYT (Spain). Proyect PB 94 – 1067.

Finally, we present an application of our results to the exact controllability in finite dimensional subspaces.

2. State Equation

Let Ω be a bounded open subset of $\mathbb{R}^n$ with Lipschitz continuous boundary $\partial\Omega$, $n \in \mathbb{N}$ and $0 < T < +\infty$. The following parabolic initial boundary value problem will be the state equation for our optimal control problems:

$$\begin{cases} y_t(x,t) + A(u)y(x,t) = 0 & \text{in } Q_T = \Omega \times (0,T), \\ y(x,t) = 0 & \text{on } \Sigma = \partial\Omega \times (0,T), \\ y(x,0) = y_0(x) & \text{in } \Omega, \end{cases} \tag{1}$$

with $x = (x_1, \dots, x_n) \in \Omega \subset \mathbb{R}^n$, $y_t = \dfrac{\partial y}{\partial t}$, $\nabla y = \left(\dfrac{\partial y}{\partial x_1}, \dots, \dfrac{\partial y}{\partial x_n}\right)$ and

$$A(u)y = -div(a(y,\nabla y,u)) + a_0(y,\nabla y,u) = -\sum_{i=1}^{n} \frac{\partial}{\partial x_i}(a_i(y,\nabla y,u)) + a_0(y,\nabla y,u),$$

where $y_0 \in L^2(\Omega)$ and the coefficients $a(y,\eta,u) = (a_1(y,\eta,u), \dots, a_n(y,\eta,u))$ and $a_0(y,\eta,u)$ are C^1−functions that satisfy the following conditions:

$$\sum_{i,j=1}^{n} \frac{\partial a_i}{\partial \eta_j}(y,\eta,u)\xi_i\xi_j \geq \mu_1 \|\xi\|^2 \tag{2}$$

$$\sum_{i=1}^{n} \left(\sum_{j=1}^{n} \left| \frac{\partial a_i}{\partial \eta_j}(y,\eta,u) \right| + \left| \frac{\partial a_i}{\partial u}(y,\eta,u) \right| \right) \leq \mu_2 \tag{3}$$

$$\sum_{i=1}^{n} \left| \frac{\partial a_i}{\partial y}(y,\eta,u) \right| + \sum_{j=1}^{n} \left| \frac{\partial a_0}{\partial \eta_j}(y,\eta,u) \right| + \left| \frac{\partial a_0}{\partial u}(y,\eta,u) \right|$$
$$\leq \mu_2(\|\eta\|^\alpha + |u|^\alpha + |y|^\beta + 1) \tag{4}$$

$$\left| \frac{\partial a_0}{\partial y}(y,\eta,u) \right| \leq \mu_2(\|\eta\|^{2\alpha} + |u|^{2\alpha} + |y|^{2\beta} + 1) \tag{5}$$

$$a(y,\eta,u) \cdot \eta + a_0(y,\eta,u)y \geq \mu_1\|\eta\|^2 - \mu_2(y^2 + |u|^2 + 1) \tag{6}$$

for every $y \in \mathbb{R}$, $\eta, \xi \in \mathbb{R}^n$, $u \in K_1 \subset \mathbb{R}$, some positive constants μ_1, μ_2 and some exponents α, β satisfying

$$\alpha \in [0, 2/(n+2)) \quad \text{and} \quad \beta \in [0, 2/n). \tag{7}$$

When $\alpha = \beta = 0$, hypotheses (2) - (5) imply condition (6) and for each control $u \in L^2(Q_T)$ and initial datum $y_0 \in L^2(\Omega)$, problem (1) has a unique solution in the space

$$\widetilde{W} = \{y \in L^2(0,T;H_0^1(\Omega)) : y_t \in L^2(0,T;H^{-1}(\Omega))\},$$

where $H_0^1(\Omega)$ is the usual Hilbert Sobolev space and $H^{-1}(\Omega)$ its dual. In the general case, if hypothesis (6) is omitted, the solution of (1) does not belong to $\widetilde{W}$, even it can blow up at some point of interval $[0,T]$ (for instance if $Ay = -\triangle y - |y|^{\gamma-1}y$ with $\gamma > 1$). Let us introduce the functional framework where the state equation (1) is well posed:

Definition 2.1. *Given $n \in \mathbb{N}$, we consider the space*

$$W = \{y \in L^2(0,T;H_0^1(\Omega)) \cap L^\rho(Q_T) : y_t \in L^2(0,T;H^{-1}(\Omega)) + L^{\rho'}(Q_T)\}$$

equipped with the norm

$$\|y\|_W = \|y\|_{L^2(0,T;H_0^1(\Omega))} + \|y\|_{L^\rho(Q_T)} + \|y_t\|_{L^2(0,T;H^{-1}(\Omega))+L^{\rho'}(Q_T)}$$

where $\rho = 2 + 4/n$ and $\rho' = (2n+4)/(n+4)$ are conjugated exponents.

By using classical results (see [7, pp.74-5]) it can be proved that $\widetilde{W}$ is a dense subspace of W; so, W is contained in $C([0,T];L^2(\Omega))$ with continuous inclusion and the formula of integration by parts holds. These are basic ingredients in order to prove the main theorems of this work. First, let us show that (1) is a well-posed problem in W:

Theorem 2.2. *Let us suppose (2)–(7) and $y_0 \in L^2(\Omega)$. Therefore,*

a) *For each $u \in L^2(Q_T)$, the problem (1) has a unique solution $y_u \in W$.*
b) *Given $\{u_m\}_{m\in\mathbb{N}} \subset L^2(Q_T)$ such that*

$$u_m \to u \quad in \quad L^2(Q_T) \quad when \quad m \to +\infty,$$

then

$$y_{u_m} \to y_u \quad in \quad W \quad when \quad m \to +\infty.$$

Proof. Given a control $u \in L^2(Q_T)$, problem (1) has (at least) one solution y_u in $L^2(0,T;H_0^1(\Omega)) \cap L^\infty(0,T;L^2(\Omega))$: this is a consequence of [7, Theorem 6.7, pp. 466-7]. At this level, the restrictions assumed on the exponents α and β are sharp. It is well known that the space $L^2(0,T;H_0^1(\Omega)) \cap L^\infty(0,T;L^2(\Omega))$ is contained in $L^\rho(Q_T)$ with continuous inclusion for $\rho = 2 + 4/n$, see for instance [7, pp. 74-5]. Moreover, conditions (3)–(5) imply that if $y_u \in L^2(0,T;H_0^1(\Omega)) \cap L^\rho(Q_T)$, hence $a(y_u, \nabla y_u, u) \in (L^2(Q_T))^n$ and $a_0(y_u, \nabla y_u, u) \in L^{p'}(Q_T)$, for certain $p' > \rho'$. Therefore, $y_u \in W$.

Uniqueness of solution in W for problem (1) can be deduced as follows: by using the Mean Value Theorem, given two solutions $y, \hat{y}$ of (1), the difference $z = y - \hat{y}$ can be viewed as a solution of a linear problem that satisfies the requirements of [7, Theorem 3.1, p. 145] with $q = r = 1 + \frac{n}{2}$. Uniqueness of solution for this kind of linear problems with null data implies that $z(x,t) = 0$. Here again, restriction (7) on the exponents α and β is needed. This proves *a)*.

Given a sequence $\{u_m\}_{m\in\mathbb{N}} \subset L^2(Q_T)$ and applying the corresponding equation to $y_m = y_{u_m}$ to the proper y_m we deduce that $\{y_m\}_{m\in\mathbb{N}}$ is a bounded sequence in $L^2(0,T;H_0^1(\Omega)) \cap C([0,T];L^2(\Omega))$ and therefore in W. In particular, since $W \subset L^2(0,T;H_0^1(\Omega)) \cap W^{1,\rho'}([0,T];H^{-1}(\Omega))$, the classical embedding results imply that $\{y_m\}_{m\in\mathbb{N}}$ is a relatively compact sequence in $L^2(Q_T)$. Hence, we can extract a subsequence (denoted the same) verifying

$$y_m \to \tilde{y} \ \text{ weakly in } W \text{ and strongly in } L^p(Q_T), \tag{8}$$

for some element $\tilde{y} \in W$ and every $p \in [1,\rho)$. Moreover, $\tilde{y}_t = \tilde{y}_1 - \tilde{y}_2$, where

$$div(a(y_m, \nabla y_m, u_m)) \to \tilde{y}_1 \ \text{ weakly in } L^2(0,T;H^{-1}(\Omega)) \ \text{ and} \tag{9}$$

$$a_0(y_m, \nabla y_m, u_m) \to \tilde{y}_2 \ \text{ weakly in } L^{p'}(Q_T) \tag{10}$$

for some $p' > \rho'$. We finish the proof of *b)* by showing that $\tilde{y} = y_u$ and convergence (8) is valid in the strong topology of W. To this end, we apply integration by parts to $y_m - \tilde{y}$ as follows:

$$\begin{aligned} &\int_{Q_T} (a(y_m, \nabla y_m, u_m) - a(y_m, \nabla \tilde{y}, u_m)) \cdot (\nabla y_m(x,s) - \nabla \tilde{y}(x,s)) dxds \\ &+ \frac{1}{2}\|y_m(T) - \tilde{y}(T)\|^2_{L^2(\Omega)} = \int_{Q_T} (\tilde{y}_2(x,s) - a_0(y_m, \nabla y_m, u_m))(y_m - \tilde{y})(x,s) dxds \\ &- \int_0^T < \tilde{y}_1(s), y_m(s) - \tilde{y}(s) > ds - \int_{Q_T} a(y_m, \nabla \tilde{y}, u_m) \cdot (\nabla y_m - \nabla \tilde{y})(x,s) dxds. \end{aligned} \tag{11}$$

It is easy to see (selecting subsequences, if necessary) that

$$a(y_m, \nabla \tilde{y}, u_m) \to a(\tilde{y}, \nabla \tilde{y}, u) \quad \text{in} \quad L^2(Q_T).$$

By taking the limit as m goes to $+\infty$ in relation (11) and using all the convergences obtained before together with the hypothesis, we have

$$\lim_m \int_{Q_T} (a(y_m, \nabla y_m, u_m) - a(y_m, \nabla \tilde{y}, u_m)) \cdot (\nabla y_m(x,s) - \nabla \tilde{y}(x,s)) dxds = 0. \tag{12}$$

Condition (2) together with the Mean Value Theorem implies that the convergence (8) is valid in the strong topology of $L^2(0,T;H_0^1(\Omega))$. Now, it is straightforward to pass to the limit in the equation of y_m when m tends to $+\infty$ and to conclude that $\tilde{y} = y_u$. Integrating by parts as in (11) in $(0,t)$ and taking sup in t, we deduce

$$\|y_m - y_u\|_{C([0,T];L^2(\Omega))} \to 0 \ \text{ as } m \to \infty.$$

Now, convergence (8) in the strong topology of W follows from previous convergences and the equation. □

Let us present some concrete examples of operators A satisfying all the assumptions and not considered in previous studies:

Example 2.3.

- $A(u)y = -\triangle y + ye^{-y^2}(1+|u|^2)^{\alpha}$. *Here, u acts in a nonlinear way in the secondary term of the operator.*
- $A(u)y = -div\left(\nabla y + \dfrac{\nabla y}{\sqrt{1+u^2+\|\nabla y\|^2}}\right) + y(1+y^2)^{\beta}$. *In this case, control u appears nonlinearly in the main term of the operator.*
- $A(u)y = -div\left(\nabla y + \dfrac{\nabla y\|\nabla y\|^u}{\|\nabla y\|^u+1}\right)$. *Now, the control u intervenes in the principal part as exponent. This type of operator appears in the modelling of electromagnetic potentials, see for instance [5]. In order to satisfy the hypotheses we assume $1<\delta_1\le u(x,t)\le\delta_2$ for a.e. $(x,t)\in Q_T$.*

3. Control Problems

In this work we are concerned with two different optimal control problems:

$$(P_1)\begin{cases} \text{Minimize } G_0(u) \\ \text{subject to } \ u\in\mathbb{K} \ \text{ and} \\ G_1(u)=0, \\ G_2(u)\le 0, \end{cases}$$

and

$$(P_2)\begin{cases} \text{Minimize } G_0(u) \\ \text{subject to } \ u\in\mathbb{K} \ \text{ and} \\ y_u(t)\in B \ \ \forall t\in[0,T] \end{cases}$$

where $\mathbb{K}$ is a non-empty convex subset of $L^2(Q_T)$ and B is a convex subset of $L^2(\Omega)$ with non empty interior.

The cost functional G_0 is defined as follows

$$G_0(u)=\int_{Q_T} L_0(y_u(x,t),\nabla y_u(x,t),u(x,t))dxdt$$

and for each $j=1,2$, the functional G_j is given by

$$G_j(u)=\int_{Q_T} L_j(y_u(x,t),\nabla y_u(x,t))dxdt,$$

where $L_0 : \mathbb{R} \times \mathbb{R}^n \times \mathbb{R} \to \mathbb{R}$, $L_j : \mathbb{R} \times \mathbb{R}^n \to \mathbb{R}$ are C^1-functions. The following hypotheses will be assumed:

$$\left| \frac{\partial L_0}{\partial y}(y, \eta, u) \right| + \sum_{j=1}^{2} \left| \frac{\partial L_j}{\partial y}(y, \eta) \right| \leq \mu_2 (1 + |y|^{\rho/\rho'} + \|\eta\|^{2/\rho'} + |u|^{2/\rho'}), \tag{13}$$

$$\sum_{i=1}^{n} \left| \frac{\partial L_0}{\partial \eta_i}(y, \eta, u) \right| + \left| \frac{\partial L_0}{\partial u}(y, \eta, u) \right| + \sum_{j=1}^{2} \sum_{i=1}^{n} \left| \frac{\partial L_j}{\partial \eta_i}(y, \eta) \right|$$

$$\leq \mu_2 (1 + |y|^{\rho/2} + \|\eta\| + |u|), \tag{14}$$

for every $(y, \eta, u) \in \mathbb{R} \times \mathbb{R}^n \times \mathbb{R}$.

Existence of solutions (optimal controls) for this type of problems can be obtained under classical hypotheses (coercivity + convexity w.r.t. u) at least when control u only acts in the secondary term. In the general case (i.e. u appears in the principal part of the operator) it is well known that to obtain existence results becomes more complicated, but it is still possible in some particular situations, for instance when $\mathbb{K}$ is a compact subset of $L^2(Q_T)$.

In Theorem 2.2, we have proved that the relation control-state is continuous. In order to deal with first order optimality conditions we are going to show that, under our assumptions, this relation is, in fact, Gâteaux differentiable:

Theorem 3.1. *Let us assume (2)–(7) and consider the functional*

$$F : L^2(Q_T) \to W$$

defined by $F(u) = y_u$, *where* y_u *denotes the unique solution of state equation (1) corresponding to* u.

Then, F *is Gâteaux differentiable. Moreover, given elements* $u, \tilde{u} \in L^2(Q_T)$, $z = DF(u)(\tilde{u})$ *is the unique solution in* W *of the linearized problem*

$$\begin{cases} z_t(x,t) - div \left\{ \dfrac{\partial a}{\partial \eta}(y_u, \nabla y_u, u) \cdot \nabla z + \dfrac{\partial a}{\partial y}(y_u, \nabla y_u, u) z \right\} \\ + \dfrac{\partial a_0}{\partial \eta}(y_u, \nabla y_u, u) \cdot \nabla z + \dfrac{\partial a_0}{\partial y}(y_u, \nabla y_u, u) z \\ = div \left\{ \dfrac{\partial a}{\partial u}(y_u, \nabla y_u, u) \tilde{u} \right\} - \dfrac{\partial a_0}{\partial u}(y_u, \nabla y_u, u) \tilde{u} & \text{in } Q_T, \\ z(x,t) = 0 & \text{on } \Sigma, \\ z(x,0) = 0 & \text{in } \Omega. \end{cases} \tag{15}$$

Proof. Take $u, \tilde{u} \in L^2(Q_T)$. First of all, let us point out that the linearized problem (15) has a unique solution in W thanks to hypotheses (2)–(7) and [7, Theorem 4.1, p. 153] with $q = r = 1 + \frac{n}{2}$ and $q_1 = r_1 = \rho'$. Given $\lambda \in (0,1)$ we can obtain the Gâteaux differentiability of F by showing that $z^\lambda = \frac{1}{\lambda}(y_{u+\lambda\tilde{u}} - y_u)$ converges towards z in W as λ tends to 0: to this end, each z^λ can be viewed as the unique solution of a linear problem. Thanks to Theorem 2.2-*b*) we know that

$$y_{u+\lambda\tilde{u}} \to y_u \quad \text{in } W \quad \text{as } \lambda \to 0.$$

Now, the convergence of z^λ towards z in $L^2(0,T;H_0^1(\Omega)) \cap C([0,T];L^2(\Omega))$ follows from [7, Theorem 4.5, p. 166], by using the Dominated Convergence Theorem and the hypotheses. As a consequence of this fact, it is easy to conclude that z_t^λ converges towards z_t in $L^2(0,T;H^{-1}(\Omega)) + L^{\rho'}(Q_T)$. □

Once we have proved the differentiable character of relation control-state, the derivation of optimality conditions can be done by classical methods, using an abstract Lagrange Multipliers Rule and showing that the functionals involved satisfy the required regularity assumptions. Here we have the main result for our problem (P_1):

Theorem 3.2. *Let $\bar{u}$ be an optimal control for problem (P_1) and $\bar{y} \in W$ the associated optimal state.*

Then, there exist $\lambda = (\lambda_0, \lambda_1, \lambda_2) \in \mathbb{R}^3$ and a unique $\bar{p} \in W$ satisfying the following conditions:

$$\lambda_0 \geq 0, \quad \lambda_2 \geq 0 \quad \textit{and} \quad \|\lambda\| > 0, \tag{16}$$

$$\begin{cases} -\bar{p}_t(x,t) - div\left\{\left[\dfrac{\partial a}{\partial \eta}(\bar{y},\nabla\bar{y},\bar{u})\right]^* \cdot \nabla\bar{p} + \dfrac{\partial a_0}{\partial \eta}(\bar{y},\nabla\bar{y},\bar{u})\bar{p}\right\} & \\ +\left[\dfrac{\partial a}{\partial y}(\bar{y},\nabla\bar{y},\bar{u})\right]^* \cdot \nabla\bar{p} + \dfrac{\partial a_0}{\partial y}(\bar{y},\nabla\bar{y},\bar{u})\bar{p} = \lambda_0\dfrac{\partial L_0}{\partial y}(\bar{y},\nabla\bar{y},\bar{u}) & \\ +\displaystyle\sum_{j=1}^{2}\lambda_j\frac{\partial L_j}{\partial y}(\bar{y},\nabla\bar{y}) - div\left(\lambda_0\frac{\partial L_0}{\partial \eta}(\bar{y},\nabla\bar{y},\bar{u}) + \sum_{j=1}^{2}\lambda_j\frac{\partial L_j}{\partial \eta}(\bar{y},\nabla\bar{y})\right) & \textit{in } Q_T, \\ \bar{p}(x,t) = 0 & \textit{on } \Sigma, \\ \bar{p}(x,T) = 0 & \textit{in } \Omega, \end{cases} \tag{17}$$

where * denotes the transposition operation,

$$\lambda_2 \cdot G_2(\bar{u}) = 0, \tag{18}$$

$$\int_{Q_T} f(\bar{y},\bar{p},\bar{u})(x,t)(u-\bar{u})(x,t)\,dxdt \geq 0 \tag{19}$$

for every $u \in \mathbb{K}$, *where*

$$f(\bar{y}, \bar{p}, \bar{u}) = \lambda_0 \frac{\partial L_0}{\partial u}(\bar{y}, \nabla \bar{y}, \bar{u}) - \left(\frac{\partial a_0}{\partial u}(\bar{y}, \nabla \bar{y}, \bar{u})\bar{p} + \left[\frac{\partial a}{\partial u}(\bar{y}, \nabla \bar{y}, \bar{u}) \right]^* \cdot \nabla \bar{p} \right)$$

Proof. We are going to prove that each functional G_j is of class C^1. Let us argue in the case $j = 0$, the rest being analogous. It is straightforward to deduce that G_0 is Gâteaux differentiable with

$$DG_0(u) \cdot (\tilde{u}) =$$

$$= \int_{Q_T} \left(\frac{\partial L_0}{\partial y}(y_u, \nabla y_u, u)z + \frac{\partial L_0}{\partial \eta}(y_u, \nabla y_u, u) \cdot \nabla z + \frac{\partial L_0}{\partial u}(y_u, \nabla y_u, u)\tilde{u} \right) dxdt, \tag{20}$$

where $z = DF(u)(\tilde{u}) \in W$ with the notation of Theorem 3.1.

Now, introducing the adjoint state p_u as the unique solution in W of the adjoint problem (17) with $\lambda_0 = 1, \lambda_1 = \lambda_2 = 0$ and (y_u, u) instead of $(\bar{y}, \bar{u})$ (recall that right-hand term of the equation belongs to $L^2(0, T; H^{-1}(\Omega)) + L^{\rho'}(Q_T)$ thanks to (13)–(14)), we obtain that

$$DG_0(u) = f(y_u, p_u, u) \in L^2(Q_T),$$

with $\lambda_0 = 1$. From [7, Theorem 4.5, p. 166] and Theorem 2.2, it follows that

$$DG_0(u_m) \longrightarrow DG_0(u) \quad \text{in} \quad L^2(Q_T),$$

assumed that

$$u_m \longrightarrow u \quad \text{in} \quad L^2(Q_T)$$

when $m \to +\infty$. This implies that G_0 is of class C^1.

Since $\mathbb{K}$ is a closed convex set, we can apply the classical Lagrange Multipliers Rule to deduce that there exists $\lambda = (\lambda_0, \lambda_1, \lambda_2) \in \mathbb{R}^3$ verifying (16), (18) and

$$\sum_{j=0}^{2} \lambda_j DG_j(\bar{u}) \cdot (u - \bar{u}) \geq 0 \quad \forall u \in \mathbb{K} \tag{21}$$

Finally, the inequality (21) can be transformed into (19) with the help of the adjoint state $\bar{p}$. □

As it is classical in Control Theory, from expression (19) we can obtain qualitative information about the optimal control $\bar{u}$: in particular, their behaviour is of "bang-bang" type when the structure of $\mathbb{K}$ is as follows:

Corollary 3.3. *Under the hypotheses of Theorem 3.2, let us assume that*

$$\mathbb{K} = \{u \in L^2(Q_T) : u(x,t) \in [\theta_0(x,t), \theta_1(x,t)] \quad a.e. \ (x,t) \in Q_T\}$$

where $\theta_i \in L^\infty(Q_T)$ *for* $i = 0, 1$.

Then,

$$\bar{u}(x,t) = \begin{cases} \theta_0(x,t) & \textit{where } f(\bar{y},\bar{p},\bar{u})(x,t) > 0 \\ \theta_1(x,t) & \textit{where } f(\bar{y},\bar{p},\bar{u})(x,t) < 0 \end{cases}$$

In a similar way we can obtain the optimality system for problem (P_2):

Theorem 3.4. *Let $\bar{u}$ be an optimal control for problem (P_2) and $\bar{y} \in W$ the associated optimal state.*

Then, there exist $\lambda_0 \in \mathbb{R}$, $\bar{p}_1 \in W$, $\bar{p}_2 \in L^2(0,T;H_0^1(\Omega)) \cap L^\rho(Q_T)$ and $\bar{\nu} \in BV([0,T];L^2(\Omega))$ satisfying the following conditions:

$$\lambda_0 \geq 0 \quad \textit{and} \quad \lambda_0 + \|\bar{\nu}\|_{BV([0,T];L^2(\Omega))} > 0, \tag{22}$$

$$\int_0^T (y(t) - \bar{y}(t))d\bar{\nu}(t) \leq 0, \tag{23}$$

for every $y \in C([0,T];L^2(\Omega))$ such that $y(t) \in B$ for all $t \in [0,T]$ and

$$\int_{Q_T} f(\bar{y},\bar{p}_1 + \bar{p}_2,\bar{u})(x,t)(u - \bar{u})(x,t)dxdt \geq 0 \tag{24}$$

for every $u \in \mathbb{K}$, where f is given in Theorem 3.2 and $\bar{p}_1$ is the unique solution of the adjoint state equation (17) with $\lambda_1 = \lambda_2 = 0$ and $BV([0,T];L^2(\Omega))$ denotes the space of bounded variation functions with values in $L^2(\Omega)$.

Proof. Consider $G : L^2(Q_T) \to C([0,T];L^2(\Omega))$ given by $G(u) = y_u$ and

$$C = \{y \in C([0,T];L^2(\Omega)) : y(t) \in B \ \ \forall t \in [0,T]\}.$$

Thanks to Theorem 3.1 and the continuous inclusion $W \subset C([0,T];L^2(\Omega))$, it is clear that the cost functional G_0 and operator G are Gâteaux differentiable. Moreover, C is a convex subset of $C([0,T];L^2(\Omega))$ with non empty interior, because B is a convex subset of $L^2(\Omega)$ with non empty interior. By virtue of the abstract Lagrange Multiplier Rule proved in [3], we deduce the existence of $\lambda_0 \in \mathbb{R}$ and $\bar{\nu} \in BV([0,T];L^2(\Omega))$ verifying (22)–(23) and

$$\lambda_0 DG_0(\bar{u})(u - \bar{u}) + \int_0^T DF(\bar{u})(u - \bar{u})(t)d\bar{\nu}(t) \geq 0 \tag{25}$$

for every $u \in \mathbb{K}$, using once more the notation of Theorem 3.1. In order to interpret (25) we use the transposition method: let us introduce $W_0 = \{z \in W : z(0) = 0\}$ and the operator $T : W_0 \to L^2(0,T;H^{-1}(\Omega)) + L^{\rho'}(Q_T)$ where $T(z)$ is given by the left-hand side of the equation in (15) with $u = \bar{u}$. It is easy to verify that T is an isomorphism and, therefore, the adjoint operator $T^* : L^2(0,T;H_0^1(\Omega)) \cap L^\rho(Q_T) \to (W_0)'$ is an isomorphism too. We conclude by taking $\bar{p}_2 = (T^*)^{-1}(\bar{\nu})$ and noticing that relation (25) can be written as (24). □

Under supplementary assumptions it is possible to assert that $\lambda_0 \neq 0$ in Theorems 3.2 and 3.4 (i.e. the corresponding control problem is normal). As it is usual in the literature, this is the case when some type of Slater condition is satisfied, see for instance [3] where some other situations are mentioned.

4. Finite Dimensional Exact Controllability

It is well known the difficulty for obtaining controllability results in the context of quasilinear equations, when the control is acting in a (possible small) subdomain of Q_T. Here we can apply the previous theory to deduce one interesting result in this direction:

Theorem 4.1. *Let us consider the initial boundary problem*

$$\begin{cases} y_t - div\,(a(y,\nabla y)) + a_0(y,\nabla y) = u(x,t)\chi_{q_T}(x,t) & in\ Q_T, \\ y(x,t) = 0 & on\ \Sigma, \\ y(x,0) = y_0(x) & in\ \Omega, \end{cases}$$

where the coefficients satisfy the hypotheses (2)–(5) with $\alpha = \beta = 0$, $y_0 \in L^2(\Omega)$, ω is a non empty open subset of Ω, $q_T = \omega \times (0,T)$ and χ_{q_T} denotes the characteristic function of q_T.

Then, for each $y_d \in L^2(\Omega)$ and E a finite dimensional subspace of $L^2(\Omega)$, there exists $u \in L^2(q_T)$ such that

$$\Pi_E(y_u(T)) = \Pi_E(y_d),$$

where Π_E denotes the orthogonal projection from $L^2(\Omega)$ onto E.

Proof. First, let us consider the case of a linear equation with bounded coefficients (here, the difficulty comes from the lack of regularity of the coefficients). Since the equation is linear, we can suppose (without loss of generality) that $y_0 = 0$. Given E, we can find a hilbertian basis of E, say $\{\phi_1, \dots, \phi_m\}$, where each ϕ_i is an analytic function in Ω (for instance, we can use some eigenfunctions associated to the Laplace operator). Let us define $p_i \in \widetilde{W}$ as the unique solution of the adjoint problem with final value ϕ_i at time T. It is easy to show that the family $\{p_i\}_{i=1}^m$ is linearly independent in q_T; hence, the $m \times m$ matrix M whose $(i,j)-$ element is given by

$$M(i,j) = \int_{q_T} p_i(x,t)p_j(x,t)dxdt,$$

has non zero determinant. Given $y_d \in L^2(\Omega)$, there exists $\delta = (\delta_1, \dots, \delta_m) \in \mathbb{R}^m$ such that $\Pi_E(y_d) = \sum_{i=1}^m \delta_i\phi_i$. Taking $\tilde{\delta} = M^{-1}\cdot\delta$ and $u = \sum_{i=1}^m \tilde{\delta}_i p_i|_{q_T} \in L^2(q_T)$, it follows that

$$\int_\Omega y_u(T)(x)\phi_j(x)dx = \int_{q_T} u(x,t)p_j(x,t)dxdt = \delta_j,$$

or, equivalently, $\Pi_E(y_u(T)) = \Pi_E(y_d)$. Now, to deal with the quasilinear case, we introduce a family of optimal control problems: for each $k \in \mathbb{N}$ let us consider

$$(P_k)\begin{cases} \text{Minimize} \quad J_k(u) = \dfrac{1}{2}\displaystyle\int_{q_T} u^2(x,t)dxdt + k\|\Pi_E(y_u(T) - y_d)\|_{L^2(\Omega)} \\ \text{over} \quad u \in L^2(q_T). \end{cases}$$

Since (P_k) is a well-defined problem, J_k is weakly continuous and coercive in $L^2(q_T)$, there exists at least one solution u_k.

We claim that there exists some $k_0 \in \mathbb{N}$ such that $\Pi_E(y_{k_0}(T)) = \Pi_E(y_d)$, where $y_k = y_{u_k}$.

Let us suppose that this is not the case: i.e., $\Pi_E(y_k(T) - y_d) \neq 0$ for every $k \in \mathbb{N}$. Thanks to Theorem 3.1, the optimality system for each problem gives

$$J_k'(u_k)v = \int_{q_T} u_k(x,t)v(x,t)dxdt +$$

$$+k\|\Pi_E(y_k(T) - y_d)\|_{L^2(\Omega)}^{-1}(\Pi_E(y_k(T) - y_d), \Pi_E(z_v^k(T)))_{L^2(\Omega)} = 0 \quad \forall\, v \in L^2(q_T) \tag{26}$$

where z_v^k is the unique solution of the initial boundary value problem

$$\begin{cases} z_t(x,t) - div\left\{\dfrac{\partial a}{\partial \eta}(y_k, \nabla y_k)\cdot \nabla z + \dfrac{\partial a}{\partial y}(y_k, \nabla y_k) z\right\} + \\ \\ +\dfrac{\partial a_0}{\partial \eta}(y_k, \nabla y_k)\cdot \nabla z + \dfrac{\partial a_0}{\partial y}(y_k, \nabla y_k) z = v(x,t) & \text{in } Q_T, \\ \\ z(x,t) = 0 & \text{on } \Sigma, \\ \\ z(x,0) = 0 & \text{in } \Omega. \end{cases} \tag{27}$$

On the other hand, there exist an element $\psi \in E$ and a subsequence (still labeled in the same way) such that

$$\|\Pi_E(y_k(T) - y_d)\|_{L^2(\Omega)}^{-1}\Pi_E(y_k(T) - y_d) \longrightarrow \psi \text{ in } L^2(\Omega) \quad \text{as } k \to +\infty.$$

Let us point out that this convergence is strong because E is a finite dimensional subspace; hence, $\|\psi\|_{L^2(\Omega)} = 1$.

Taking into account that $J_k(u_k) \leq J_k(0)$, it follows that $\{u_k/\sqrt{k}\}_{k\in\mathbb{N}}$ is a bounded sequence in $L^2(q_T)$. Moreover, thanks to the hypothesis $\alpha = \beta = 0$, all the coefficients of the equation in (27) remain bounded and it is standard (selecting a subsequence, if necessary) to show that $\{z_v^k\}$ converges towards an element z_v in the weak topology of $\widetilde{W}$.

Dividing the expression (26) by k and making k tend to $+\infty$, we have

$$(\psi, \Pi_E(z_v(T)))_{L^2(\Omega)} = 0 \quad \forall\, v \in L^2(q_T).$$

Finally, we can choose $v \in L^2(q_T)$ such that $\Pi_E(z_v(T)) = \psi$ (remember that z_v is the solution of a linear problem), arriving at the contradiction $\|\psi\|_{L^2(\Omega)} = 0$. □

References

[1] N. U. Ahmed. Optimal control of a class of strongly nonlinear parabolic systems. *J. of Math. Anal. & Appl.*, 61:188–207, 1977.

[2] V. Barbu and Th. Precupanu. *Convexity and Optimization in Banach Spaces.* Editura Academiei, Sijthoff & Noordhoff, Bucharest, 1978.

[3] E. Casas. Boundary control of semilinear elliptic equations with pointwise state constraints. *SIAM J. on Control & Optimiz.*, 31(4):993–1006, 1993.

[4] E. Casas, L.A. Fernández, and J. Yong. Optimal control of quasilinear parabolic equations. *Proc. Royal Soc. Edinburgh*, 125A:545–565, 1995.

[5] J. Haslinger and P. Neittaanmäki. *Finite Element Approximation for Optimal Shape Design.* John Wiley & Sons, Chichester, 1988.

[6] B. Hu and J. Yong. Pontryagin maximum principle for semilinear and quasilinear parabolic equations with pointwise state constraints. *SIAM J. on Control & Optimiz.*, 33(6):1857–1880, 1995.

[7] O. A. Ladyzhenskaya, V. A. Solonnikov, and N. N. Ural'tseva. *Linear and Quasilinear Equations of Parabolic Type.* Am. Math. Soc, Providence, R. I., 1968.

[8] I. Lasiecka. State constrained control problems for parabolic systems: regularity of optimal solutions. *Appl. Math. Optim.*, 6:1–29, 1980.

[9] U. Mackenroth. On parabolic distributed optimal control problems with restrictions on the gradient. *Appl. Math. Optim.*, 10:69–95, 1983.

[10] P. Neittaanmäki and D. Tiba. *Optimal Control of Nonlinear Parabolic Systems.* Marcel Dekker, Inc., New York, 1994.

[11] N. S. Papageorgiou. On the optimal control of strongly nonlinear evolution equations. *J. of Math. Anal. & Appl.*, 164:83–103, 1992.

[12] F. Tröltzsch. *Optimality Conditions for Parabolic Control Problems and Applications.* Teubner-Texte, Leipzig, 1984.

Departamento de Matemáticas, Estadística y Computación
Facultad de Ciencias,
Universidad de Cantabria,
Avda. de los Castros, s/n
39071 Santander, Spain
E-mail address: lafernandez@besaya.unican.es

International Series of Numerical Mathematics
Vol. 133, © 1999 Birkhäuser Verlag Basel/Switzerland

Controllability Property for the Navier-Stokes Equations

Andrei V. Fursikov

Abstract. Different statements of a controllability problem for the Navier-Stokes equations are given. Theorems on exact, on local exact and on approximate controllability of the 3D Navier-Stokes equations are obtained when the Navier-Stokes equations are supplied with periodic boundary conditions (i.e. these equations are defined on torus Π), a control is distributed and it is concentrated in a subdomain of Π.

In this paper we consider different statements of a controllability problem for the 3D Navier-Stokes equations. We study the case when the Navier-Stokes system is equipped with periodic boundary conditions (i.e. it is defined on torus Π) and it is controlled by a local distributed control i.e. when the control vector field is the right-hand side of the Navier-Stokes equations concentrated in an arbitrary open subset of Π.

Zero controllability of the Navier-Stokes equations was established for the 2D case in [1], and for the 3D case in [2]. Local exact controllability of the 2D Navier-Stokes equations was proved in [3]. The same property for the 3D Navier-Stokes and Boussinesq equations was established in [4],[5],[6]. Approximate controllability for the 2D Navier-Stokes equations was proved in [7]. Exact controllability for the 2D Navier-Stokes equations was established in [8].

In this paper we establish local exact controllability by a more constructive method than in [4],[5],[6]. In addition we establish here the approximate controllability and exact controllability properties and give drafts of proofs for corresponding theorems. The results of this paper were obtained by the author together with O.Yu. Imanuvilov. All complete proofs will be published in [9].

1. Statement of problems and formulation of the main results

Let $L > 0$ be a magnitude, $\Pi = R^3/LZ^3$ be the torus of dimension 3 with the length of each generatrix equal to L. Let $\omega \subset \Pi$ be an open subset of the torus Π, $Q = (0,T)\times\Pi$, $Q^\omega = (0,T)\times\omega$. On the cylinder Q we consider the Navier-Stokes system

$$\partial_t v(t,x) - \Delta v + (v,\nabla)v + \nabla p = f(t,x) + u(t,x), \qquad \operatorname{div} v = 0 \tag{1.1}$$

supplied with the initial condition

$$v(t,x)|_{t=0} = v_0(x). \tag{1.2}$$

Here f, v_0 are given vector fields, and $u(t,x)$ is a control concentrated on the cylinder Q^ω. Note that to define (1.1), (1.2) on torus Π means to assume that (1.1), (1.2) are determined for an arbitrary $x = (x_1, x_2, x_3) \in R^3$ and all vector fields from (1.1), (1.2) are periodic with respect to each x_i with period L.

The exact controllability problem for the Navier-Stokes system is as follows: given a solution $(\hat{v}(t,x), \hat{p}(t,x))$ of the system

$$\partial_t \hat{v}(t,x) - \Delta \hat{v} + (\hat{v}, \nabla)\hat{v} + \nabla \hat{p} = f(t,x), \quad \operatorname{div} \hat{v} = 0, \tag{1.3}$$

to find a control $u(t,x)$, $\operatorname{supp} u \subset Q^\omega$, such that the solution $v(t,x)$ of problem (1.1), (1.2) coincides at the instant $t = T$ with $\hat{v}(T,x)$:

$$v(t,x)|_{t=T} \equiv \hat{v}(T,x). \tag{1.4}$$

To make this statement of the problem more precise and to formulate the main result we introduce some necessary functions spaces. We set for $1 \le p \le \infty, k \ge 0$,

$$V_p^k(\Omega) = \left\{ v(x) = (v_1, v_2, v_3) \in (W_p^k(\Pi))^3 : \quad \operatorname{div} v = 0 \right\}, \tag{1.5}$$

$$V_p^{1,2(k)}(Q) = \left\{ v \in Ł_2(0,T; V_p^{2+k}(\Pi)) : \partial_t v \in Ł_2(0,T; V_p^k(\Omega)) \right\}, \tag{1.6}$$

where $W_p^k(\Omega)$ is the Sobolev space with the summable power index p and smoothness index k,

$$V^{1,2}(Q) = V_2^{1,2(0)}(Q), \quad V^k(\Pi) = V_2^k(\Pi), \quad H^k(\Pi) = W_2^k(\Pi). \tag{1.7}$$

We define the control space

$$U(\omega) = \left\{ u(t,x) \in (L_2(Q))^3 : \operatorname{supp} u \in Q^\omega \right\}. \tag{1.8}$$

The following theorem on exact controllability of Navier-Stokes system (1.1), (1.2) by a local distributed control is the main result of this paper.

Theorem 1.1 *Let $f \in L_2(0,T; V^2(\Pi))$, $v_0 \in V^4(\Pi)$ and a solution $(\hat{v}, \hat{p}) \in C^1(0,T; V^4(\Pi)) \times L_2(0,T; H^1(\Pi))$ of system (1.3), defined on the cylinder Q be given. Then there exists a solution $(v,p,u) \in V^{1,2}(Q) \times L_2(0,T; H^1(\Pi)) \times U(\omega)$ of problem (1.1), (1.2), (1.4).*

To prove Theorem 1.1 we first of all weaken the notion of exact controllability to two directions by introducing the notions of local exact controllability and of approximate controllability.

Let a pair $(\hat{v}, \hat{p})$ satisfy Navier-Stokes equations (1.3) on Q. Navier-Stokes system (1.1) supplied with (1.2) is called *locally exact controllable* with respect to control space (1.8), if there exists $\varepsilon > 0$ such that for an arbitrary initial condition v_0 satisfying the inequality

$$\|\hat{v}(0,\cdot) - v_0\|_{V^1(\Pi)} \le \varepsilon \tag{1.9}$$

there exists a control $u \in U(\omega)$ such that the solution (v,p) of problem (1.1), (1.2) exists in the space $V^{1,2}(Q) \times L_2(0,T;H^1(\Pi))$ and satisfies condition (1.4).

Theorem 1.2 *The Navier-Stokes system is locally exact controllable with respect to control space (1.8). Moreover, the parameter ε depends on the magnitude $\|\hat{v}\|_{C^1(0,T;V^2_\infty(\Pi))} + 1/T$ continuously and is monotonically decreasing.*

Let an initial condition $v_0 \in V^4(\Pi)$ and a right side $f \in L_2(0,T;V^2(\Pi))$ be given. Consider the set of vector fields $v_i \in V^4(\Pi)$ satisfying

$$\sum_{i=0}^{1} \|v_i\|_{V^4(\Pi)} \le R. \tag{1.10}$$

Navier-Stokes system (1.1), (1.2) is called *approximately controllable* with respect to control space (1.8), if for any $\varepsilon > 0, R > 0$ and for an arbitrary v_1 belonging to set (1.10) there exist an instant $T = T_{\varepsilon,R}$ and a solution

$$(v,p,u) \in V^{1,2}(Q_{T_{\varepsilon,R}}) \times L_2(0,T_{\varepsilon,R};H^1(\Pi)) \times U(\omega,Q_{T_{\varepsilon,R}}),$$

of problem (1.1),(1.2) such that

$$\|v(T_{\varepsilon,R}) - v_1\|_{V^1(\Pi)} \le \varepsilon. \tag{1.11}$$

Theorem 1.3 *Navier-Stokes system (1.1), (1.2) is approximately controllable with respect to control space (1.8). Moreover, for each $\varepsilon > 0$ one can choose the time $T_{\varepsilon,R}$ such that*

$$T_{\varepsilon,R} \to 0 \quad as \quad \varepsilon \to 0 \tag{1.12}$$

for an arbitrary $R > 0$.

Note that one can easily derive Theorem 1.1 from Theorems 1.2, 1.3. We will discuss the proof of Theorems 1.2, 1.3 in sections 2,3.

2. Solvability of the local exact controllability problem for the Navier-Stokes system

First of all, we linearize Navier-Stokes system (1.1) at the point $(\hat{v},\hat{p})$:

$$L(y) \equiv \partial_t y - \Delta y + (\hat{v},\nabla)y + (y,\nabla)\hat{v} = -\nabla p + f + u, \quad \operatorname{div} y = 0. \tag{2.1}$$

We set for (2.1) the initial condition and final condition:

$$y(t,x)|_{t=0} = y_0(x), \quad y(T,x) \equiv 0. \tag{2.2}$$

For a precise formulation of the controllability problem we introduce function spaces for the data and for a solution of problem (2.1),(2.2).

Let, just as earlier, $\omega \in \Pi$ be a subdomain of torus Π, $\xi_\omega(x)$ be the characteristic function of the set ω. We introduce the function

$$\eta = \eta_\lambda(t,x) = \frac{e^{\frac{4\lambda}{3}\|\psi\|_{C(\Pi)}} - e^{\lambda\psi(x)}}{T-t}, \tag{2.3}$$

where $\lambda > 1$ is a parameter, $\psi \geq 1$ is a function defined on Π such that $\nabla\psi > 0$ for $x \in \Pi \setminus \omega$. Existence of such a function ψ is easily established.

We introduce the weight functions

$$\rho(t,x) = e^{\eta_{\lambda-1}(t,x)}\xi_\omega(x) + e^{\eta_\lambda(t,x)}(1-\xi_\omega(x)), \quad \rho_1(t,x) = \frac{e^{\eta_\lambda(t,x)}}{(T-t)},$$

where $\eta(t,x)$ is function (2.3). We set also $\hat{\eta}_\lambda(t) = \min_{x\in\Pi} \eta_\lambda(t,x)$.

Let

$$L_2(\rho,Q) = \left\{ u(t,x) \in L_2(Q) : \|y\|_{L_2(\rho,Q)} = \int_Q \rho y^2 dxdt < \infty \right\}.$$

The space $L_2(\rho_1, Q)$ is defined in the same way. We set

$$Y^{1,2}(Q) = \big\{ y(t,x) \in V^{1,2}(Q) : e^{\frac{2\hat{\eta}_\lambda}{3}} y \in V^{1,2}(Q),\ y \in L_2(\rho,Q),$$
$$(1-\xi_\omega)(\partial_t y - \Delta y + (\hat{v},\nabla)y + (y,\nabla)\hat{v}) \in L_2(\rho,Q) \big\}.$$

Let us introduce the space of right sides for equation (2.1):

$$F(Q) = \big\{ f \in L_2(0,T;(L_2(\Pi))^n) : \exists f_1 \in (L_2(\rho_1,Q))^n \quad \exists f_2 \in L_2(0,T;H^1(\Pi)) :$$
$$f = f_1 + \nabla f_2; \quad \|f\|_{F(Q)} = \inf_{f=f_1+\nabla f_2} (\|f_1\|^2_{L_2(\rho_1,Q)} + \|f_2\|_{L_2(0,T;H^1(\Pi))})^{\frac{1}{2}} \big\}.$$

The space of controls is defined as follows:

$$U(\rho_1,\omega) = \{ u \in (L_2(\rho_1,Q))^n, \mathrm{supp} u \in Q^\omega = (0,T)\times\omega \}.$$

Theorem 2.1 *Let $\hat{v} \in W^1_\infty(0,T;V^2_\infty(\Pi))$. Then for arbitrary $y_0 \in V^1(\Pi)$, $f \in F(Q)$, there exists a solution*

$$(y,p,u) \in V^{1,2}(\rho,Q) \times L_2(0,T;H^1(\Pi)) \times U(\rho_1,Q)$$

of problem (2.1),(2.2).

Proof. To prove Theorem 2.1 we apply a variant of the penalty method. We consider the following problem of minimization:

$$J_k(y,u) = \frac{1}{2}\int_Q \rho(t,x)\,|y|^2\,dxdt + \frac{1}{2}\int_Q m_k(t,x)\,|u|^2\,dxdt \to inf \tag{2.4}$$

when $(y,u,\nabla p)$ satisfy the restrictions

$$L(y) = \nabla p + f + u, \quad y(t,x)|_{t=0} = y_0(x), \tag{2.5}$$

where $L(y)$ is operator (2.1). Here $m_k(t,x)$ is defined by the formula

$$m_k(t,x) = \xi_\omega(x)\min(k, e^{\eta_{\lambda-1}(t,x)}) + (1-\xi_\omega(x))k,$$

where $\xi_\omega(x)$ is the characteristic function of the set ω, $\eta(t,x)$ is the function from (2.3) with $\lambda \gg 1$, (the magnitude of λ will be specified below), and k is a natural number.

Lemma 2.1. *Let $f \in (L_2(\rho, Q))^3, y_0 \in V^1(\Pi)$. For an arbitrary natural number k there exists the unique solution*

$$(y, \nabla p, u) \in V^{1,2}(Q) \times L_2(0,T;(L_2(\Pi))^3) \times (L_2(Q))^3$$

of problem (2.4),(2.5). Moreover, $y \in (L_2(\rho, Q))^3$.

Let us derive the optimality system for problem (2.4), (2.5).

Lemma 2.2. *Let conditions of Lemma 2.1 be fulfilled and*

$$(y_k, \nabla p_k, u_k) \in (V^{1,2}(Q) \cap (L_2(\rho, Q))^3) \times (L_2(Q))^3 \times (L_2(Q))^3$$

be a solution of problem (2.4), (2.5). Then there exists a pair

$$(z_k, \nabla q_k) \in L_2(0,T;V^0(\Pi)) \times (L_2(Q))^3$$

such that

$$L(y_k) = -\nabla p_k + f + u_k, \qquad y_k(0,x) = y_0(x) \tag{2.6}$$

$$L^*(z_k) = \nabla q_k - \rho y_k \quad \text{in} \quad Q \tag{2.7}$$

$$z_k - m_k u_k = 0 \qquad \text{in} \quad Q, \tag{2.8}$$

where

$$L^*(z_k) = N^*(z_k) = -\partial_t z_k - \Delta z_k - (\hat{v}, \nabla) z_k + ((z_k, \nabla)\hat{v})^*, \quad \operatorname{div} z_k = 0. \tag{2.9}$$

Moreover, z_k satisfies the inequality

$$\int_Q (T-t)|z_k|^2 e^{-\eta_\lambda} dxdt + \|z_k(0)\|^2_{V^0(\Pi)} \tag{2.10}$$

$$\leq c_2 \left(\int_Q \rho^2 |y_k|^2 e^{-\eta_\lambda} dxdt + \int_{Q^\omega} (m_k^2 |u_k|^2 + \rho^2 |y_k|^2) e^{-\eta_{\lambda-1}} dxdt \right),$$

where C_2 does not depend on (y_k, u_k).

Proof. Optimality system (2.6)-(2.8) is simply derived with the help of the Lagrange principle. In order to prove inequality (2.10) we use the following

Theorem 2.2. *Let conditions of Theorem 2.1 and Lemma 2.1 be fulfilled. Then there exists $\hat{\lambda} > 0$, such that for $\lambda > \hat{\lambda}$ a solution of equation (2.7) satisfies the estimate*

$$\int_Q (T-t)|z_k(t,x)|^2 e^{-\eta_\lambda(t,x)} dxdt + \|z_k(0,\cdot)\|^2_{V^0(\Pi)}$$

$$\leq c_\lambda \left(\int_Q |\rho y_k(t,x)|^2 e^{-\eta_\lambda(t,x)} dxdt + \int_{Q^\omega} (|\rho y_k(t,x)|^2 + |y_k(t,x))|^2) e^{-\eta_{\lambda-1}(t,x)} dxdt \right), \tag{2.11}$$

where c_λ depends on λ, but does not depend on ρy_k.

To derive this bound we develop the proof of analogous estimates from [2], [3], [6], [7].

Applying to (2.7) inequality (2.11) and after that substituting (2.8) into the integral over Q^ω from the right side of (2.11), we obtain inequality (2.10). □

Lemma 2.3. *Let $f \in (L_2(\rho, Q))^3$, $f|_{Q^\omega \equiv 0}$, and $(y_k, \nabla p_k, u_k)$ be a solution of problem (2.4), (2.5), constructed in Lemma 2.1. Then there exists a constant $c > 0$, which does not depend on k, such that*

$$J_k(y_k, u_k) \le c\left(\|f\|^2_{(L_2(\rho,Q))^n} + \|y_0\|^2_{V^0(\Pi)}\right). \tag{2.12}$$

Proof. Let z_k be the functions constructed in Lemma 2.2 and satisfying estimate (2.10). Taking into account the definitions of the functions $m_k(t, x)$ and $\rho(t, x)$ we obtain that

$$|\xi_\omega(x) m_k(t,x) e^{-\eta_{\lambda-1}(t,x)}| \le 1, \quad \rho^2(t,x) e^{-\eta_{\lambda-1}(t,x)} \xi_\omega(x) = \rho(t,x)\xi_\omega(x),$$
$$\rho^2(t,x) e^{-\eta_\lambda(t,x)}(1-\xi_\omega(x)) = \rho(t,x)(1-\xi_\omega(x)).$$

Applying these relations to the right side of inequality (2.10) and taking into account definition (2.4) of the functional J_k, one can derive from (2.10) the following estimate:

$$\int_Q (T-t)|z_k|^2 e^{-\eta_\lambda} dx dt + \|z_k(0,\cdot)\|^2_{V^0(\Pi)} \le c_1 J_k(y_k, u_k), \tag{2.13}$$

where the constant c_1 does not depend on k. Multiplying equality (2.7) on y_k, integrating the obtained equation over Q, after that integrating by parts with respect to x and t, and taking into account (2.6) we get the equality

$$\int_Q \rho |y_k|^2 e^{-\eta_\lambda} dx dt = -(z_k(0,\cdot), y_0)_{V^0(\Pi)} - \int_Q (f + u_k)\cdot z_k dx dt.$$

This equality, (2.8), and definition (2.4) of the functional J_k yield:

$$2J_k(y_k, u_k) = -(z_k(0,\cdot), y_0)_{V^0(\Pi)} - \int_Q f \cdot z_k dx dt$$

$$\le (\|y_0\|^2_{V^0(\Pi)} + \|f\|^2_{(L_2(\rho_1,Q))^n})^{1/2}(\|z_k(0,\cdot)\|^2_{V^0(\Pi)} + \int_Q (T-t)|z_k|^2 e^{-\eta_\lambda} dx dt)^{1/2}.$$

Applying estimate (2.13) to the right side of the last inequality, we obtain (2.12). □

With help of Lemma 2.3 we pass to the limit in the sequence $(y_k, \nabla p_k, u_k)$ of solutions for problem (2.4)–(2.8) as parameter k tends to infinity. As a result we obtain a solution $(y, p, u) \in V^{1,2}(\rho, Q) \times L_2(0, T; H^1(\Pi)) \times U(\rho_1, Q)$ of problem (2.1), (2.2). This proves Theorem 2.1. □

Now one can derive Theorem 1.3 from Theorem 2.1 with the help of a version of the Implicit Function Theorem.

3. Approximate controllability of the Navier-Stokes system

In this section we present the idea of the proof of approximate controllability for Navier-Stokes system (1.1), (1.2). Let $m(t,x) = (m_1, m_2, m_3) \in C^\infty(Q)$ be a vector field defined on Q such that

$$\operatorname{div} m(t,x) = 0, \quad (t,x) \in Q \quad \text{and} \quad m(t,x) = \nabla\gamma(t,x), \quad (t,x) \in Q \setminus Q^\omega, \quad (3.1)$$

where $\gamma(t,x)$ is a function determined on $Q \setminus Q^\omega$, which by (3.1) is harmonic on this set.

We intend to reduce approximate controllability problem (1.1), (1.2), (1.11) to an exact controllability problem for a system of linear differential equations of the first order with coefficients that we will choose ourselves. These coefficients will be defined by a certain vector field m of the form (3.1). We will look for a solution (v,p) of (1.1) in the form

$$v = z + m, \quad \operatorname{div} z = 0, \quad \operatorname{div} m = 0 \quad (3.2)$$

(what pressure p we will get will become clear later). Substitution of (3.2) into (1.1) yields

$$\partial_t z + (m,\nabla)z + (z,\nabla)m + (z,\nabla)z - \Delta z + \partial_t m - \Delta m + (m,\nabla)m - \nabla p = f + u. \quad (3.3)$$

It is easy to see, by virtue of (3.1), that the equality

$$\partial_t m - \Delta m + (m,\nabla)m = \nabla q_1 \quad (3.4)$$

holds, where q_1 is a certain function.

Below we will make a contraction of time and because of that the terms $(z,\nabla)z - \Delta z$, f in (3.3) will become small. Eliminating these terms in (3.3), we get the system

$$\partial_t z + (m,\nabla)z + (z,\nabla)m + \nabla q = u', \quad \operatorname{div} z = 0, \quad (3.5)$$

where $\nabla q = \nabla p + \nabla q_1$ and ∇q_1 is defined in (3.4). Initial condition (1.2) and approximate controllability condition (1.11) generate the following conditions for (3.5):

$$z|_{t=0} = v_0, \quad z(T,\cdot) = v_1. \quad (3.6)$$

We assume for briefness that $v_i \in V^0(\Pi) \cap (C^\infty(\Pi))^n$, $\quad i = 0,1$. The coefficients of system (3.5) are defined by the vector field m which is constructed in the following Lemma:

Lemma 3.1. *There exists an instant $T > 0$ and a vector field*

$$m(t,x) = (m_1, m_2, m_3) \in (C^\infty(Q))^3$$

satisfying conditions (3.1) such that

$$m(0,x) \equiv m(T,x) \equiv 0, \quad \left.\frac{\partial^k m(t,x)}{\partial t^k}\right|_{t=0} = \left.\frac{\partial^k m(t,x)}{\partial t^k}\right|_{t=T} \equiv 0, \quad k \in N \quad (3.7)$$

and for any $x_0 \in \Pi$ the following relation holds:

$$\{(t, x(t,x_0)), t \in (0,T)\} \cap Q^\omega \neq \emptyset,$$

where $x(t,x_0)$ is a solution of the following Cauchy problem:

$$dx(t,x_0)/dt = m(t,x(t,x_0)), \quad x(t,x_0)|_{t=0} = x_0.$$

Moreover, for any $x_0 \in \Pi$ the equality $x(T,x_0) = x_0$ is true. Besides, there exists a finite covering $\{O_i, \quad i=1,\dots,M\}$ of the torus Π by open sets O_i and there exists a number $\Delta > 0$, such that for each i all curves $x(t,x_0), \quad x_0 \in O_i$ remain within ω simultaneously during a certain time interval of length Δ.

Theorem 3.1. *Let $m(t,x)$ satisfy all conditions of Lemma 3.1 and let*

$$v_i \in (C^\infty(Q))^3 \cap V^1(Q), \quad i=0,1,$$

be given. Then there exists a solution

$$(z,\nabla p,u) \in (C^\infty(Q))^3 \cap V^{1,2}(Q)) \times (C^\infty(Q))^3 \times (U(\omega,Q) \cap C^\infty(Q))^4)$$

of problem (3.5), (3.6) which satisfies

$$\|\partial_t z\|^2_{L_2(0,T;V^{k-1}(\Pi))} + \|z\|^2_{L_2(0,T;V^k(\Pi))} + \|\nabla q\|^2_{L_2(0,T;H^{k-1}(\Pi))}$$
$$+ \|u\|^2_{L_2(0,T;H^{k-1}(\Pi))^3} \le c_k \sum_{j=0}^{1} \|\tilde v_i\|_{V^k(\Pi)} \tag{3.8}$$

for each $k \in N$, where the constant c_k depends only on vector field $m(t,x)$ and its derivatives of order not more than k.

Using $z, \nabla q, m, u$ obtained in Theorem 3.1 we construct the functions

$$z_\delta(t,x) = z(t/\delta,x), \quad m_\delta(t,x) = m(t/\delta,x)/\delta,$$
$$\nabla q_\delta(t,x) = \nabla q(t/\delta,x)/\delta, \quad u_\delta(t,x) = u(t/\delta,x)/\delta. \tag{3.9}$$

Functions (3.9) satisfy system (3.5) and relations (3.6) transformed to the equalities

$$z_\delta\,|_{t=0} = v_0, \quad z_\delta(\delta T,\cdot) = v_1. \tag{3.10}$$

Now we look for the precise solution of problem (1.1), (1.2), (1.11) in the following form:

$$v = z_\delta + m_\delta + y, \quad u = u_\delta, \tag{3.11}$$

We substitute (3.11) into (1.1) and taking into account equality (3.5) for functions (3.9) we write the obtained equalities in the form of equations with respect to y which are defined for $(t,x) \in Q_{\delta T}$:

$$\partial_t y - \Delta y + (y,\nabla)(y + z_\delta + m_\delta) + (z_\delta + m_\delta,\nabla)y - \nabla q_2 = f_1, \quad \operatorname{div} y = 0, \tag{3.12}$$

where

$$f_1 = f + \Delta z_\delta - (z_\delta,\nabla)z_\delta, \qquad \nabla q_2 = \nabla p - \nabla q_1(t\delta,x)/\delta, \tag{3.13}$$

and ∇q_1 is defined in (3.4).

Relations (3.11), (1.2), (3.10), (3.7) lead to the initial condition for y:

$$y|_{t=0} = 0. \tag{3.14}$$

In virtue of (3.11), (3.7), (3.10), to prove (1.11) one has to establish the inequality

$$\|y(\delta T,\cdot)\|^2_{V^1(\Pi)} < \varepsilon^2. \tag{3.15}$$

In virtue of (3.13), (3.9) we get $\|f_1\|_{L_2(Q_{\delta T})} \to 0$ as $\delta \to 0$.

This relation, (3.14) and energy estimates for solution y of (3.12) and for its gradient ∇y yield

$$\|y(t,\cdot)\|^2_{V^0(\Pi)} + \|y(t,\cdot)\|^2_{V^1(\Pi)} + \int_0^t \|y(s,\cdot)\|^2_{V^2(\Pi)}\, ds \to 0 \qquad \text{as} \quad \delta \to 0.$$

This proves (3.15).

References

[1] A.V. Fursikov and O.Yu. Imanuvilov, *On exact boundary zero-controllability of two-dimensional Navier-Stokes equations,* Acta Applicandae Mathematicae **37** (1994), 67–76.

[2] A.V. Fursikov, *Exact boundary zero controllability of three dimensional Navier-Stokes equations,* J. of Dynamical and Control Syst., **1**, No 3 (1995), 325–350.

[3] A.V. Fursikov and O.Yu. Imanuvilov, *Local exact controllability of two dimensional Navier-Stokes system with control on the part of the boundary,* Sbornik. Math., **187**, No 9 (1996), 1355–1390.

[4] A.V. Fursikov and O.Yu. Imanuvilov, *Local exact controllability of the Navier-Stokes equations,* C.R.Acad.Sci.Paris, **323**, Ser.1 (1996), 275–280.

[5] A.V. Fursikov and O.Yu. Imanuvilov, *Local exact boundary controllability of the Navier-Stokes system.* Contemporary Mathematics, **209** (1997), 115–129.

[6] A.V. Fursikov and O.Yu. Imanuvilov, *Local exact boundary controllability of the Boussinesq equation,* SIAM Journal on Control and optimization, **36**, No 2 (1998), 391–421.

[7] J.-M. Coron, *On the controllability of the 2-D incompressible Navier-Stokes equations with the Navier slip boundary conditions,* ESIAM Control, Optimization and Calculus of Variations **1** (1996), 35–75.

[8] J.-M. Coron and A.V. Fursikov, *Global exact controllability of the 2D Navier-Stokes Equations on a manifold without boundary,* Russian J. of Math. Physics, **4**, No 4 (1996), 429–448.

[9] A.V. Fursikov and O.Yu. Imanuvilov, *Exact controllability of the Navier-Stokes and Boussinesq equations,* Russian Math. Surveys (to appear).

Department of Mechanics and Mathematics,
Moscow State University,
119899 Moscow, Russia
E-mail address: fursikov@dial01.msu.ru

International Series of Numerical Mathematics
Vol. 133, © 1999 Birkhäuser Verlag Basel/Switzerland

Shape Sensitivity and Large Deformation of the Domain for Norton-Hoff Flows

Nicolas Gomez and Jean-Paul Zolésio

Abstract. In this paper, we derive new results concerning shape control and analysis for non-newtonian fluids. *Via* the speed-method, we prove the continuity of the weak-solution of Norton-Hoff problem with respect to the domain and the shape-differentiability of the energy functional.

From the so-called Shape Differential Equation, we prove the existence of a virtual large deformation of the domain that increase the energy functional.

1. Norton-Hoff Penalized Problem

The model we study here was first introduced by Norton to modelize the creep of steel at high-temperature, and generalized by Hoff and Friâa who finally gave the formulation we use in [1]. This model is used to describe non-newtonian flows. We will study here a steady penalized formulation with mixed boundary conditions. The convergence of the penalized model towards the incompressible one has been studied in [4].

1.1. Weak Solution

We consider the problem of a visco-plastic fluid occupying a domain Ω in $\mathbb{R}^N$ and undergoing the action of volume forces of density f. The speed of the fluid is subject to:

$$\mathcal{P}(\Omega)\quad\left\{\begin{array}{c} -\operatorname{div}\left[K|\varepsilon(u)|^{p-2}\varepsilon(u)\right]-\eta\nabla\left(|\operatorname{div}u|^{p-2}\operatorname{div}u\right)=f \quad \text{on} \quad \Omega \\ u=0 \quad \text{on} \quad \Gamma_D\subset\partial\Omega \\ K|\varepsilon(u)|^{p-2}\varepsilon(u).n+\eta\nabla\left(|\operatorname{div}u|^{p-2}\operatorname{div}u\right)n=0 \text{ on } \Gamma_N=\partial\Omega\setminus\overline{\Gamma_D} \end{array}\right.$$

where $K>0$ and $1<p\leq 2$ characterize the material, η is a non-negative penalization parameter. When $p=2$, we recover the case of newtonian fluids and when p tends to one, the model converges towards plasticity model of Prandtl-Reuss. We refer to [2, 3] for asymptotic studies.

We assume Ω is a Lipschitz, open and bounded subset of $\mathbb{R}^N$ and Γ_D is an open subset of $\Gamma=\partial\Omega$ with non-zero Hausdorff-measure. It is natural to consider the Banach space $\mathcal{W}(\Omega)=\left\{\, v\in \mathrm{W}^{1,p}(\Omega,\mathbb{R}^N)\;\middle|\; v=0 \text{ on } \Gamma_D\right\}$ equiped with

$\|v\|_{\mathcal{W}(\Omega)} = \left[\int_\Omega \frac{K}{p}|\varepsilon(v)|^p + \frac{\eta}{p}|\mathrm{div}\, v|^p\right]^{\frac{1}{p}}$. Thanks to Poincaré's and Korn's inequalities, this norm is equivalent to the one induced by $\mathrm{W}^{1,p}(\Omega, \mathbb{R}^N)$. We have the following existence and uniqueness result.

Proposition 1.1. *Let $f \in \mathcal{W}(\Omega)'$. The problem $\mathcal{P}(\Omega)$ is the Euler equation for the functional Φ_Ω defined on $\mathcal{W}(\Omega)$ by $\Phi_\Omega(v) = \|v\|^p_{\mathcal{W}(\Omega)} - \langle f, v\rangle_{\mathcal{W}(\Omega)'\times\mathcal{W}(\Omega)}$. Thus, it has a unique weak-solution in $\mathcal{W}(\Omega)$.*

This is obtained by classical techniques of the direct method in the calculus of variations.

1.2. Moving Problem

For general considerations for the speed-method, we use the results and briefly reminds some notations of [10]. We fix $D \subset \mathbb{R}^N$ as a smooth open bounded hold-all. We denote $\mathcal{O}_k$ the set of all N-dimensional $\mathcal{C}_k$-submanifolds of D where k is a fixed nonnegative integer. We also fix $\Omega_0 \in \mathcal{O}_{k+1}$, an initial domain and the part Γ_D of its boundary. The domain Ω_0 is moved by the flow of a non autonomous field V in the space $\mathcal{V}_k = \left\{V \in \mathcal{C}^0\left(\mathbb{R}^+, \mathcal{C}^k(\bar{D}, \mathbb{R}^N)\right) \middle| \langle V, \nu\rangle = 0 \text{ on } \partial D,\ V_{|\Gamma_D} = 0\right\}$ where ν is the unit outward normal on ∂D.

Lemma 1.2. *For any V in $\mathcal{V}_k$, there exists an invertible flow mapping $s \mapsto T_s(V)$ from $\mathbb{R}^+$ to $\mathcal{C}^k(\bar{D}, \bar{D}) \cap \mathrm{L}^\infty_{loc}\left(\mathbb{R}^+, \mathcal{C}^k(\bar{D}, \mathbb{R}^N)\right)$.*

We refer to [10] for instance and classical results on O.D.E for the proof. For any $s \in \mathbb{R}^+$, we name $\Omega_s(V)$ the transported domain $T_s(V)(\Omega_0)$, denote $\Gamma_s = T_s(\Gamma)$, and $\Gamma_{N,s} = T_s(\Gamma_N)$. We emphasize the deformation preserves the part Γ_D of the boundary and the regularity of the initial domain.

In order to formulate the *moving weak problem* $\mathcal{P}(\Omega_s(V))$ and get some results, we need more assumptions on the data. *We assume* $f \in \mathrm{W}^{1,p'}(D, \mathbb{R}^N)$. So, $\mathcal{P}(\Omega_s(V))$ is the Euler equation for the moving functional $\Phi_{\Omega_s(V)}$. By proposition 1.1, for any s in $\mathbb{R}^+$, $\mathcal{P}(\Omega_s(V))$ has an unique solution denoted $u_{\Omega_s(V)}$ in $\mathcal{W}(\Omega_s(V))$.

1.3. Energy Functional and Energy Estimates

We are interested in the *energy functional* $E(\Omega) = \min_{v\in\mathcal{W}(\Omega)} \Phi_\Omega(v) = \Phi_\Omega(u_\Omega)$, which will be proved to be shape-differentiable. This will take us to the shape continuity of the weak solution of Norton-Hoff problem. We enounce here some useful technical lemmas.

Lemma 1.3. *For any Ω in $\mathcal{O}_k$, $E(\Omega) = \frac{1-p}{p}\int_\Omega f u_\Omega = \frac{1-p}{p}\|u_\Omega\|^p_{\mathcal{W}(\Omega)}$.*

Proof. A choice of the state as a test function in the necessary and sufficient condition of optimality for minimizing Φ_Ω provides $\int_\Omega K|\varepsilon(u_\Omega)|^p + \eta|\mathrm{div}\, u_\Omega|^p = \int_\Omega f u_\Omega$. Thus the lemma is obtained by substitution in the expression of the energy $E(\Omega) = \Phi_\Omega(u_\Omega)$. □

Lemma 1.4. *Both* $s \mapsto \|u_{\Omega_s(V)}\|_{\mathcal{W}(\Omega_s(V))}$ *and* $s \mapsto |E(\Omega_s(V))|$ *lay in* $\mathrm{L}^\infty_{loc}(\mathbb{R}^+, \mathbb{R})$.

Proof. Since for any $s \in \mathbb{R}^+$, $\Phi_{\Omega_s(V)}(u_{\Omega_s(V)}) \leq \Phi_{\Omega_s(V)}(0) = 0$ we have

$$\int_{\Omega_s(V)} \frac{K}{p}|\varepsilon(u_{\Omega_s(V)})|^p + \frac{\eta}{p}|\mathrm{div}\, u_{\Omega_s(V)}|^p \leq \int_{\Omega_s(V)} f.u_{\Omega_s(V)}$$

and we come to

$$\left\|u_{\Omega_s(V)}\right\|^p_{\mathcal{W}(\Omega_s(V))} \leq \left\|u_{\Omega_s(V)}\right\|_{\mathcal{W}(\Omega_s(V))} \|f\|_{\mathcal{W}(\Omega_s(V))'} \tag{1}$$

But $\|f\|_{\mathcal{W}(\Omega_s(V))'}$ is locally bounded, due to lemma 1.2. The local boundedness of the energy comes from lemma 1.3. □

2. Shape Analysis

In this section, we derive the shape-continuity of the weak-solution of Norton-Hoff penalized problem from the previous lemmas and the shape-differentiability of the energy functional. The question of the differentiability of the state is, as far as we know, still opened. Usual techniques for linear equations fail, mainly because of a lack of regularity.

We will move the initial domain Ω_0 with the flow of given a field $V \in \mathcal{V}_k$. In this section, we omit the reference to V. Let Ψ be the functional defined on $\mathbb{R}^+ \times \mathcal{W}(\Omega_0)$ by

$$\Psi(s, v) = \Phi_{\Omega_s}(v \circ T_s^{-1}) = \int_{\Omega_0} \frac{K}{p}|\mathfrak{s}(\mathrm{D}v.A_s)|^p + \frac{\eta}{p}|\mathrm{tr}\ (\mathrm{D}v.A_s)|^p - f \circ T_s.v \tag{2}$$

where γ_s is the Jacobian $|\det \mathrm{D}T_s|$, and $A_s = (\mathrm{D}T_s(V))^{-1}$. The operator $\mathfrak{s}$ denotes the symmetry operator for matrices. The moving weak-problem $\mathcal{P}(\Omega_s)$, for $s \in \mathbb{R}^+$ can be settled in a "Lagrangian point of view" *via* Necas' Lemma for the transport of Sobolev spaces. For any $s \in \mathbb{R}^+$, we denote $u^s = u_{\Omega_s} \circ T_s \in \mathcal{W}(\Omega_0)$ the *transported solution* which is the unique minimum of $\Psi(s, .)$

$$\min_{v \in \mathcal{W}(\Omega_s)} \Phi_{\Omega_s}(v) = E(\Omega_s) = \min_{v \in \mathcal{W}(\Omega_0)} \Psi(s, v) = \Psi(s, u^s) \tag{3}$$

2.1. Shape-Differentiability of the Energy Functional

The main idea here is to avoid differentiating the energy functional *via* a differentiation of the state. This is possible because the energy is a minimum, and techniques of differentiation of a minimum with respect to a parameter can be applied.

Proposition 2.1. *For any* v *in* $\mathcal{W}(\Omega_0)$, $s \mapsto \Psi(s, v)$ *is differentiable.*

It is well known that it is sufficient to prove the differentiability at 0. It is quite technical and comes from the Lebesgue's theorem and the expression of Ψ. This differentiability will yield the existence of a shape gradient for the energy functional.

Proposition 2.2. *The energy possesses a shape gradient given by*

$$\mathrm{d}E(\Omega_0;V) = \partial_s\Psi(s,u_0)\Big|_{s=0} =$$
$$\int_{\Omega_0} -K|\varepsilon(u)|^{p-2}\varepsilon(u)..(\mathrm{D}u.\mathrm{D}V(0)) - \eta|\mathrm{div}\, u|^{p-2}\mathrm{div}\, u\, \mathrm{tr}\, (\mathrm{D}u.\mathrm{D}V(0)) \quad (4)$$
$$+\int_{\Omega_0}\left[\frac{K}{p}|\varepsilon(u)|^p + \frac{\eta}{p}|\mathrm{div}\, u|^p - fu\right]\mathrm{div}\, V(0) - \mathrm{D}f.V(0).u$$

Proof. For every v in $\mathcal{W}(\Omega_0)$, and $s>0$, we have

$$\Psi(0,u^0)\leq\Psi(0,v) \quad \text{and} \quad \Psi(s,u^s)\leq\Psi(s,v) \tag{5}$$

which yields

$$\frac{\Psi(s,u^s)-\Psi(0,u^0)}{s}\leq\frac{\Psi(s,u^0)-\Psi(0,u^0)}{s}$$

If we denote $\overline{\partial_s}E(\Omega_0)=\limsup_{s\to 0}[E(\Omega_s)-E(\Omega_0)]s^{-1}$, we can rewrite the previous inequality passing to the limit

$$\overline{\partial_s}E(\Omega_0)\leq\partial_s\Psi(0,u^0) \tag{6}$$

The inequalities (5) also yields $\frac{\Psi(s,u^s)-\Psi(0,u^s)}{s}\leq\frac{\Psi(s,u^s)-\Psi(0,u^0)}{s}$. Since the mapping $s\mapsto\Psi(s,v)$, from $[0,1]$ to $\mathbb{R}$, is differentiable for any v in $\mathcal{W}(\Omega_0)$, there exists $\theta(s,v)$ in $[0,1]$ such that $\Psi(s,v)-\Psi(0,v)=s\partial_s\Psi(\theta(s,v)s,v)$. For $v=u^s$, we come $\partial_s\Psi(\theta(s,u^s)s,u^s)\leq\frac{\Psi(s,u^s)-\Psi(0,u^0)}{s}$. Passing to the limit, the continuity of $\partial_s\Psi$ yields

$$\partial_s\Psi(0,u^0)\leq\underline{\partial_s}E(\Omega_0) \tag{7}$$

where $\underline{\partial_s}E(\Omega_0)=\liminf_{s\to 0}[E(\Omega_s)-E(\Omega_0)]s^{-1}$ which achieve the proof. □

From expression (4), it is clear that $\mathrm{d}E(\Omega_0;V)$ is linear and continuous with respect to $V(0)$. Hence there exists a distribution $\mathcal{G}(\Omega_0)$ in $\mathcal{A}'_k$ where

$$\mathcal{A}_k=\left\{V\in\mathcal{C}^k(\bar{D},\mathbb{R}^N)\Big|\langle V,\nu\rangle=0 \text{ on } \partial D,\ V_{|\Gamma_D}=0\right\}$$

such that $\mathrm{d}E(\Omega_0;V)=\langle\mathcal{G}(\Omega_0)\,,\,V(0)\rangle_{\mathcal{A}'_k\times\mathcal{A}_k}$. Using structure result on the gradient (see [9]), we can prove this distribution has support on the boundary of Ω_0. Due to a lack of regularity for the problem, this boundary expression cannot be used the same way it is for classical cases. It is also well known that the expression of the gradient can be generalized in the expression of the gradient for any value of the parameter s in $\mathbb{R}^+$:

$$\mathrm{d}E(\Omega_s(V);V) = \langle \mathcal{G}(\Omega_s(V))\,,\, V(s)\rangle_{\mathcal{A}'_k\times\mathcal{A}_k} = \tag{8}$$
$$\int_{\Omega_s(V)} \left[\frac{K}{p}|\varepsilon(u_s)|^p + \frac{\eta}{p}|\mathrm{div}\,(u_s)|^p - f.u_s\right] \mathrm{div}\,V(s)$$
$$-\int_{\Omega_s(V)} K\,|\varepsilon(u_s)|^{p-2}\varepsilon(u_t)..\,(\mathsf{D}u_t.\mathsf{D}V(s))$$
$$-\int_{\Omega_s(V)} \left[\eta\,|\mathrm{div}\,u_s|^{p-2}\mathrm{div}\,u_s.\mathrm{tr}\;(\mathsf{D}u_s.\mathsf{D}V(s)) + \mathsf{D}f.V(s).u_s\right]$$

2.2. Continuity of the State with Respect to the Domain

In linear equations, the continuity of the state with respect to the domain often comes from *a priori* estimates which yields a weak-continuity *via* compactness arguments, and from the weak formulation that provides the convergence of the norms. The same technique may be used here since the spaces $\mathcal{W}(\Omega_s)$ are uniformly convex. The weak-continuity will derive from the following lemma and the convergence of the norms deduced of the following results.

Lemma 2.3. *The mapping* $s \mapsto \|u^s\|_{\mathcal{W}(\Omega_0)}$ *belongs to* $\mathrm{L}^\infty_{loc}(\mathbb{R}^+,\mathbb{R})$.

The proof is technical and mainly based on lemma 1.4 and the local boundedness property with respect to s of the mappings $v \mapsto v\circ T_s$ from $\mathcal{W}(\Omega_s(V))$ to $\mathcal{W}(\Omega_0)$, derived from the lemma 1.2.

Proposition 2.4. *When s tends to zero, the following convergences hold*

$$\|u_{\Omega_s}\|_{\mathcal{W}(\Omega_s)} \to \|u_{\Omega_0}\|_{\mathcal{W}(\Omega_0)} \qquad\qquad \|u^s\|_{\mathcal{W}(\Omega_0)} \to \left\|u^0\right\|_{\mathcal{W}(\Omega_0)}$$

Proof. The first convergence is a mere consequence of the continuity of the energy functional and lemma 1.3.

The second convergence comes from the first one and, when s tends to 0

$$\int_{\Omega_s} |\varepsilon(u_{\Omega_s})|^p - \int_{\Omega_0} |\varepsilon(u^s)|^p = \int_{\Omega_0} \gamma_s|\mathfrak{s}(\mathsf{D}u^s.A_s)|^p - |\varepsilon(u^s)|^p \to 0 \tag{9}$$

$$\int_{\Omega_s} |\mathrm{div}\,u_{\Omega_s}|^p - \int_{\Omega_0} |\mathrm{div}\,u^s|^p = \int_{\Omega_0} \gamma_s|\mathrm{tr}\;(\mathsf{D}u^s.A_s)|^p - |\mathrm{div}\,u^s|^p \to 0 \tag{10}$$

To prove (9), it is sufficient to notice that classical calculus yields

$$\int_{\Omega_s} |\varepsilon(u_{\Omega_s})|^p - \int_{\Omega_0} |\varepsilon(u^s)|^p = \int_{\Omega_0} \left(\gamma_s^{\frac{2}{p}}\,\mathfrak{s}(\mathsf{D}u^s.A_s)..(\mathsf{D}u^s.A_s)\right)^{\frac{p}{2}} - \left(\mathfrak{s}(\mathsf{D}u^s)..\mathsf{D}u^s\right)^{\frac{p}{2}}$$
$$\leq \int_{\Omega_0} \left|\gamma_s^{\frac{2}{p}}\,\mathfrak{s}(\mathsf{D}u^s.A_s).A_s^* - \mathfrak{s}(\mathsf{D}u^s)\right|^{\frac{p}{2}} |\mathsf{D}u^s|^{\frac{p}{2}}$$
$$\leq \left\|\gamma_s^{\frac{2}{p}}\,\mathfrak{s}(\mathsf{D}u^s.A_s).A_s^* - \mathfrak{s}(\mathsf{D}u^s)\right\|^{\frac{1}{2}}_{\mathrm{L}^p(\Omega_0,\mathbb{R}^{N^2})} \|\mathsf{D}u^s\|^{\frac{1}{2}}_{\mathrm{L}^p(\Omega_0,\mathbb{R}^{N^2})}$$

Lemma 2.3 provides the boundedness of $\|\mathrm{D}u^s\|^{\frac{1}{2}}_{\mathrm{L}^p(\Omega_0,\mathbb{R}^{N^2})}$. Due to the regularity of the flow mapping, A_s converges towards the identity matrix Id in $\mathrm{L}^\infty(\Omega_0,\mathbb{R}^{N^2})$ and γ_s converges to 1 in $\mathrm{L}^\infty(\Omega_0)$. This provides

$$\left\|\gamma_s^{\frac{2}{p}}\,\mathfrak{s}(\mathrm{D}u^s.A_s).A_s^* - \mathfrak{s}(\mathrm{D}u^s)\right\|_{\mathrm{L}^p(\Omega_0,\mathbb{R}^{N^2})} \to 0$$

From the same arguments we establish (10). □

Theorem 2.5. *The mapping $s \mapsto u^s = u_{\Omega_s} \circ T_s$ is continuous from $\mathbb{R}^+$ to $\mathcal{W}(\Omega_0)$.*

Proof. As usual, it is sufficient to prove the continuity at 0, the continuity elsewhere in $\mathbb{R}^+$ is deduced from semi-group properties of the flow mapping.

Let (s_n) be a sequence in $\mathbb{R}^+$ that converges to 0. It is sufficient to prove that $u^{s_n} \rightharpoonup u^0$ in $\mathcal{W}(\Omega_0)$ and $\|u^{s_n}\|_{\mathcal{W}(\Omega_0)} \to \|u^0\|_{\mathcal{W}(\Omega_0)}$. This last condition is given by proposition 2.4, so we just have to prove the weak convergence.

From lemma 2.3, there exists a u_* in $\mathcal{W}_0$ such that $u_{s_n} \rightharpoonup u^*$ in $\mathcal{W}(\Omega_0)$, up to passing to a subsequence. Since for any v in $\mathcal{W}(\Omega_0)$, $\Psi(s_n, u^{s_n}) \leq \Psi(s_n, v)$, we get $\liminf_{n\to\infty} \Psi(s_n, u^{s_n}) \leq \Psi(0, v)$. Using the weak-semi continuity of Ψ, we come, for any v in $\mathcal{W}(\Omega_0)$, to

$$\Psi(0, u^*) \leq \Psi(0, v)$$

The uniqueness of the solution of $\mathcal{P}(\Omega_0)$ provides $u^* = u_{\Omega_0} = u^0$. Accordingly, the mapping $s \mapsto u^s$ is weakly continuous from $\mathbb{R}^+$ to $\mathcal{W}(\Omega_0)$. □

This result does not provide the continuity of the state, but the continuity of the weak-solution of $\mathcal{P}(\Omega_s)$ transported in Ω_0. We are going to use this continuity to prove the continuity of the state. This has to be precised since the states do not lay in the same space. The use of an extension operator is now usual in this situation.

We consider a continuous extension operator $\Pi_0 \in \mathcal{L}(\mathcal{W}(\Omega_0), \mathcal{W}(D))$. For any s in $\mathbb{R}^+$, the operator $\Pi_s \in \mathcal{L}(\mathcal{W}(\Omega_s), \mathcal{W}(D))$ by $\Pi_s(v) = \Pi_0(u \circ T_s) \circ T_s^{-1}$ is an extension operator. The continuity of the state now makes sense, and we have the following result.

Theorem 2.6. *The mapping $s \mapsto \Pi_s(u_{\Omega_s})$ is continuous from $\mathbb{R}^+$ to $\mathcal{W}(D)$.*

Proof. As we set $u^s = u_{\Omega_s} \circ T_s^{-1}$, we have $\Pi_s(u_{\Omega_s}) = \Pi_0(u^s) \circ T_s^{-1}$. Accordingly, theorem 2.5 provides the result because the mapping $s \mapsto v \circ T_s^{-1}$ is continuous for any v in $\mathcal{W}(D)$ (and even differentiable in L^p, see [10, Sec. 2.14]). □

3. Shape Differential Equation

We are going to develop a gradient-method to increase the energy functional with respect to the domain. This problem is related to finding the most resisting shape, under the flow of a given Norton-Hoff material (undergoing a prescribed action of density and surfaces forces).

The absolute continuity of the energy functional is written

$$\forall s \in \mathbb{R}^+, \quad E(\Omega_s(V)) = E(\Omega_0) + \int_0^s \langle \mathcal{G}(\Omega_t(V)) \, , \, V(t) \rangle_{\mathcal{A}'_k \times \mathcal{A}_k} \, dt \tag{11}$$

Thus, we can govern the evolution of the energy with respect to the domain in a gradient-like way.

From now on, we assume that f is supported in a smooth subset E of D. We denote $\mathbb{A}_k = \{V \in \mathcal{A}_k \big| V = 0 \text{ on } E\}$, $\mathbb{V}_k = \mathcal{C}^0(\mathbb{R}^+, \mathbb{A}_k)$. We also assume $E \subset \Omega_0$ and $\Gamma_D \cap \partial E$ has non-zero Hausdorff-measure. These assumptions will pass the difficulty of a monotony of Korn's constant with respect to the domain (such a monotony is an open problem but for very specific cases,see [6, 5] for instance) and change the local boundedness properties of the previous section in uniform boundedness properties we respect to V. These assumptions make sense for some problems arising in metal forming. Equation (12) can be rewritten for $V \in \mathbb{V}_k$

$$\forall s \in \mathbb{R}^+, \quad E(\Omega_s(V)) = E(\Omega_0) + \int_0^s \langle \mathcal{G}(\Omega_t(V)) \, , \, V(t) \rangle_{\mathbb{A}'_k \times \mathbb{A}_k} \, dt \tag{12}$$

Using the so-called Shape Differential Equation technique, we are going to prove

Proposition 3.1. *There exist a V in $\mathbb{V}_{k+1} \cap \mathrm{L}^2(\mathbb{R}^+, \mathbb{A}_{k+1})$ such that*

$$\forall t \in \mathbb{R}^+, \quad \langle \mathcal{G}(\Omega_t(V)), V(t) \rangle_{\mathbb{A}'_k \times \mathbb{A}_k} = \|\mathcal{G}(\Omega_t(V))\|^2_{\mathbb{A}'_k} = \|V(t)\|^2_{\mathbb{A}_k} \quad \textit{in } D \tag{13}$$

3.1. Equation for the Speed-Field

Let $\ell \in \mathbb{N}$ s.t.$\mathcal{H} = \{V \in \mathrm{H}^\ell_0(D, \mathbb{R}^N) \, \big| \, V_{|\Gamma_D} = 0 \text{ and } V = 0 \text{ on } E\} \subset \mathbb{A}_{k+1}$ and $A : \mathcal{H} \to \mathcal{H}'$ the canonical (linear) duality operator. To prove proposition 3.1, it is sufficient to prove:

Theorem 3.2. *There exists V in $\mathcal{C}^0(\mathbb{R}^+, \mathcal{H})$ such that*

$$\forall s \in \mathbb{R}^+, \quad A^{-1}\mathcal{G}(\Omega_s(V)) - V(s) = 0 \tag{14}$$

This equation is the so-called *Shape Differential Equation* which has been introduced by J.P. Zolésio in [11] and [12]. A solution of this equation is a fixed point of the composite G:

$$\begin{array}{ccccc} \mathcal{C}^0(\mathbb{R}^+, \mathcal{H}) & \stackrel{\bar{G}}{\to} & \mathcal{C}^0(\mathbb{R}^+, \mathcal{H}') & \stackrel{A^{-1}}{\to} & \mathcal{C}^0(\mathbb{R}^+, \mathcal{H}) \\ V & \mapsto & \mathcal{G}(\Omega_.(V)) & \mapsto & A^{-1}\mathcal{G}(\Omega_.(V)) \end{array}$$

We are going to prove the existence of a fixed point by Leray-Schauder's Theorem. The question of the uniqueness remains opened.

3.2. Boundedness Properties

First we prove the existence of a solution of (14) in the interval $I = [0, 1]$. Then we shall derive the unbounded case.

Lemma 3.3. *There exists m_1 s.t. $\forall V \in \mathcal{C}^0(I, \mathbb{A}_k)$, $\forall s \in I$, $\|u_{\Omega_s(V)}\|_{\mathcal{W}(\Omega_s(V))} \leq m_1$.*

Proof. From equation (1), we just have to prove that $\sup_{s\in I} \|f\|_{\mathcal{W}(\Omega_s(V))'}$ is uniformly bounded with respect to V. This holds since $\|f\|_{\mathcal{W}(\Omega_s(V))'} \leq \|f\|_{\mathcal{W}(E)'}$, for any $s \in I$. □

Lemma 3.4. *There exists m_2 s.t. $\forall V \in \mathcal{C}^0(I, \mathbb{A}_k)$, $\forall s \in I$, $\|\mathcal{G}(\Omega_s(V)\|_{\mathbb{A}_k'} \leq m_2$.*

Proof. For any W in $\mathbb{A}_k$ and $\Omega \in \mathcal{O}_k$, we can extend the expression of the gradient distribution:

$$\langle \mathcal{G}(\Omega), W\rangle = \left\{ \begin{array}{l} \displaystyle\int_\Omega \left[\frac{K}{p}|\varepsilon(u_\Omega)|^p + \frac{\eta}{p}|\mathrm{div}\,(u_\Omega)|^p - f.u_\Omega\right] \mathrm{div}\, W \\ \displaystyle\qquad - \int_\Omega K\,|\varepsilon(u_\Omega)|^{p-2}\varepsilon(u_\Omega)..\,(\mathrm{D}u_\Omega.\mathrm{D}W) \\ \displaystyle - \int_\Omega \eta\,|\mathrm{div}\, u_\Omega|^{p-2}\mathrm{div}\, u_\Omega.\mathrm{tr}\;(\mathrm{D}u_\Omega.\mathrm{D}W) + \mathrm{D}f.W.u_\Omega \end{array}\right. \tag{15}$$

$$\leq \left\{ \begin{array}{l} c\|u_\Omega\|^p_{\mathcal{W}(\Omega)}\|\mathrm{div}\, W\|_{\mathrm{L}^\infty(D)} + \|f\|_{\mathcal{W}(\Omega)'}\|u_\Omega\|_{\mathcal{W}(\Omega)}\|\mathrm{div}\, W\|_{\mathrm{L}^\infty(D)} \\ +c\|u_\Omega\|^p_{\mathcal{W}(\Omega)}\|\mathrm{D}W\|_{\mathrm{L}^\infty(D)} + c'\|\mathrm{D}f\|_{\mathrm{L}^{p'}(\Omega)}\|u_\Omega\|_{\mathcal{W}(\Omega)}\|W\|_{\mathrm{L}^\infty(\mathfrak{D})} \end{array}\right.$$

for convenient constants c and c' (which involves Korn's and Poincaré's constants in E). With lemma 3.3 we can find a constant m such that $\forall V \in \mathcal{C}^0(I, \mathbb{A}_k)$, $\forall s \in I$, $\langle \mathcal{G}(\Omega_s(V)), W\rangle \leq m\|W\|_{\mathrm{W}^{1,\infty}(D,\mathbb{R}^N)}$. Eventually, the result comes from the inclusion $\mathbb{A}_k \hookrightarrow \mathrm{W}^{1,\infty}(D, \mathbb{R}^N)$. □

So, it holds that the mapping $V \mapsto \mathcal{G}(\Omega_.(V))$ from $\mathcal{C}^0(I, \mathcal{H})$ to $\mathcal{C}^0(I, \mathcal{H}')$ ranges in a ball of radius m_2 in $\mathcal{C}^0(I, \mathcal{H}')$. Let C be a closed bounded convex containing the ball of radius m_2 in $\mathcal{C}^0(I, \mathcal{H})$. Since the norm of A^{-1} is 1, the mapping G maps C in itself.

3.3. Continuity and Compactness Properties

In order to prove the continuity and compactness of the operator G, we introduce the subset $\mathcal{O}_k(\Omega_0)$ of $\mathcal{O}_k$ containing all the $\mathcal{C}_k$-submanifolds of D which are $\mathcal{C}_k$-diffeomorphic to Ω_0. This space has been studied by Micheletti in [7] who proved it being a complete metric space when endowed with Courant's metric.

$$d_k(\Omega, \Omega') = \inf\left\{\|T - Id\|_{k,\infty} + \|T^{-1} - Id\|_{k,\infty} \middle| T : \Omega \to \Omega'\ \mathcal{C}^k\text{-diffeomorphism}\right\}$$

We consider the following splitting of $G = A^{-1} \circ G_2 \circ G_1$ with,

$$\begin{array}{ccccc} \mathcal{C}^0(I, \mathcal{H}) & \overset{G_1}{\to} & \mathcal{C}^0(I, \mathcal{O}_{k+1}(\Omega_0)) \subset \mathcal{C}^0(I, \mathcal{O}_k(\Omega_0)) & \overset{G_2}{\to} & \mathcal{C}^0(I, \mathcal{H}') \\ V & \mapsto & \Omega_.(V) & \mapsto & \mathcal{G}(\Omega_.(V)) \end{array}$$

The continuity of A^{-1} is immediate. The following (equi-)continuity result is quoted from [12].

Theorem 3.5. *The mapping G_1 is continuous. Moreover, it maps bounded sets on equi-continuous parts.*

The only gap to be filled is the continuity of G_2. From the following lemma, we also have the compacity of G.

Lemma 3.6. *The mapping G_2 is continuous and compact.*

Proof. To prove the continuity, it is sufficient to prove the continuity of $\Omega \mapsto \mathcal{G}(\Omega)$ from $\mathcal{O}_k(\Omega_0)$ to $\mathbb{A}'_k$. For shortness, we just give the main steps of the proof. We first notice that the application $\tilde{\mathcal{G}}$ that maps (χ, v) to

$$W \mapsto \begin{cases} \displaystyle\int_D \chi \left[\frac{K}{p}|\varepsilon(v)|^p + \frac{\eta}{p}|\mathrm{div}\,(v)|^p - f.v\right] \mathrm{div}\, W \\ \displaystyle\qquad - \int_D \chi\, K\, |\varepsilon(v)|^{p-2}\varepsilon(v)..\,(\mathsf{D}v.\mathsf{D}W) \\ \displaystyle - \int_D \chi\, \eta\, |\mathrm{div}\, v|^{p-2}\mathrm{div}\, v.\mathrm{tr}\ (\mathsf{D}v.\mathsf{D}W) + \mathsf{D}f.W.v \end{cases}$$

from $\mathrm{L}^\infty(D) \times \mathcal{W}(D)$ to $\mathbb{A}'_k$ is continuous for the strong topology induced by $\mathrm{L}^2(D) \times \mathcal{W}(D)$. The gradient mapping $\mathcal{G}$ is the composite of $\tilde{\mathcal{G}}$ with the mapping $\Omega \mapsto (\chi_\Omega, \Pi_\Omega(u_\Omega))$ from $\mathcal{O}_k$ to $\mathrm{L}^\infty(D) \times \mathcal{W}(D)$, where $\Pi_\Omega \in \mathcal{L}(\mathcal{W}(\Omega), \mathcal{W}(D))$ is any extension operator. Since the convergence of domains in $\mathcal{O}_k(\Omega_0)$ implies the convergence of characteristic functions in $\mathrm{L}^\infty(D)$ weak-$\star$, the continuity of G_2 is derived from the continuity of the state with respect to the domain.

For the compacity, notice that due to our choice of the space $\mathcal{H}$, G_2 actually ranges in $\mathcal{C}^0(I, \mathcal{O}_{k+1}(\Omega_0)) \subset \mathcal{C}^0(I, \mathcal{O}_k(\Omega_0))$. Murat and Simon have proved in [8] that the inclusion of $\mathcal{O}_{k+1}(\Omega_0)$ in $\mathcal{O}_k(\Omega_0)$ is compact. Ascoli's theorem yields that a subset B of $\mathcal{C}^0(I, \mathcal{O}_{k+1}(\Omega_0))$ is relatively compact in $\mathcal{C}^0(I, \mathcal{O}_k(\Omega_0))$ as soon as B is an equicontinuous family. With theorem 3.5, this holds if B, is the image under G_1 of a bounded subset of $\mathbb{V}_{k+1}$. □

This result justifies the extra-regularity on the speed-field and the domain we imposed: $\mathcal{H} \subset \mathbb{A}_{k+1}$ and $\Omega \in \mathcal{O}_{k+1}$.

3.4. Proof of Theorem 3.2

Eventually, the mapping G is continuous and compact from $\mathcal{C}^0(I, \mathcal{H})$ to itself and maps the convex C in itself. Applying Leray-Schauder's fixed point theorem, we prove there exists V in $C \subset \mathcal{C}^0(I, \mathcal{H})$ such that

$$\forall s \in I, \quad A^{-1}\mathcal{G}(\Omega_s(V)) - V(s) = 0 \tag{16}$$

and we can govern the energy functional in I. Hence we can built a sequence (Ω^n) in $\mathcal{O}_k(\Omega_0)$ and a sequence (V^n) in $\mathcal{C}^0([0,1], \mathcal{H})$ with flow T^n, such that

$$\begin{cases} A^{-1}(\mathcal{G}(\Omega_t(V^n))) + V_t^n &= 0 \text{ on } [0,1] \times D \\ \Omega_0(V^n) &= \Omega_1(V^{n-1}) \\ \Omega_0(V^0) &= \Omega_0 \end{cases} \tag{17}$$

and we denote $\Omega^n = \Omega_0(V^n)$. We consider the field $V \in \mathcal{C}^0(I, \mathcal{H})$ such that $V_{|[n,n+1]} = V^n$. The continuity of the V^n, equations (16) and (17) provide the continuity of V. This field is a solution of the shape differential equation (14).

To complete the proof of theorem 3.2, we just have to prove that $V \in \mathrm{L}^2(\mathbb{R}^+, \mathcal{H})$. This just comes from equation (12), which can be rewritten

$$E(\Omega^n) = E(\Omega_0) + \sum_{i=0}^{n-1} \|V^i\|^2_{\mathrm{L}^2(0,1;\mathcal{H})} \tag{18}$$

Since the energy is bounded (lemmas 3.3,1.4), the serie $\sum_{i\geq 0} \|V^i\|_{\mathrm{L}^2(0,1;\mathcal{H})}$ converges and $V \in \mathrm{L}^2(\mathbb{R}^+, \mathcal{H})$.

References

[1] A. Friaa. La loi de Norton-Hoff généralisée en plasticité et viscoplasticité. Thèse de Doctorat dÉtat, 1979.

[2] A. Bensoussan, J. Frehse Asymptotic behaviour of the time-dependant Norton-Hoff law in plasticity theory and H^1 regularity Com. Math. Univ. Carolinae 37,2 (1996).

[3] R. Temam A generalized Norton-Hoff model and the Prandtl-Reuss law of Plasticity Arch. for Rat. Mech. and Anal. 95,2 (1986).

[4] N. Gomez and J.P. Zolésio. Shape differential equation in norton hoff flows. Preprint INLN 97.41.

[5] I. Hlavacek. Korn's inequality uniform with a class of axisymmetric bodies. *Apl. Mat.*, 34(2):146–154, 1989.

[6] A.P. Maldonado. Korn's inequality in a class of domains satisfying the uniform cone property. *C.R. Acad. Sci. Paris*, Ser. I, 318(4):315–319, 1994.

[7] A.M. Micheletti. Metrica per famiglie di domini limitati e proprietà generiche degli autovalori. *Annali della Scuola Normale Superiore di Pisa*, 26(fasc. 3), 1972.

[8] F. Murat and J. Simon. Sur le contrôle par un domaine géométrique. *Publication de l'Université Pierre et Marie Curie (Paris 6)*, (76015), 1977.

[9] M. Delfour and J.P. Zolésio Structure of shape derivative for non-smooth domains. J. Funct. Anal., vol 104,2 (1995).

[10] J. Sokolowski and J.P. Zolésio. *Introduction to Shape Optimisation: Shape sensitivity analysis*, volume 16 of *Computational Mathematics*. Springer-Verlag, New York, Berlin, Heidelberg, 1992.

[11] J.P. Zolésio. An optimal design procedure for optimal support. In A. Auslander, editor, *Convex Analysis and its Applications*, Lect. Notes in Eco. and Math. Systems, pages 205–219. Springer-Verlag, 1977.

[12] J.P. Zolésio. Identification de Domaines par Déformations. Thèse de Doctorat d'Etat, Nice, 1979.

Centre de Mathématiques Appliquées, I.N.R.I.A,
2004 rte des Lucioles - B.P. 93,
06902 Sophia Antipolis Cedex, France
E-mail address: nicolas.gomez@sophia.inria.fr
E-mail address: jean-paul.zolesio@sophia.inria.fr

International Series of Numerical Mathematics
Vol. 133,

On a Distributed Control Law with an Application to the Control of Unsteady Flow around a Cylinder

Michael Hinze and Andreas Kauffmann

Abstract. A new class of feedback laws for distributed control of dynamical systems is presented. The approach is based on instantaneous control, i.e. control at every time step to the underlying dynamical system. The closed-loop controllers obtained in this way can be proved to be stable in the distributed control case for dynamical systems in finite dimensions, provided that the parameters of the controller are suitably adjusted [4]. As an application a distributed controller for the unsteady flow around a cylinder is numerically investigated. As the numerical results show the approach presented is capable of stabilizing systems with very large dimensions.

1. Introduction

In this work we consider the problem of steering a dynamical system to, or into a small tube around, a given state. This problem may be formulated mathematically as a constraint minimization problem for a cost function of tracking type, where the subsidiary conditions are given by the dynamical system that describes the state. The construction of optimal closed-loop controllers is in the general non-linear case closely connected to the solution of the Hamilton-Jacobi-Bellmann equations. Even in the case of the linear-quadratic regulator problem with quadratic costs and linear constraints the closed-loop controller is characterized by means of a Riccati matrix differential equation whose numerical solution for large scale problems necessitates an enormous amount of computational work.

In this note we present an suboptimal control approach that leads to practically implementable and reliable control laws. It is based on an instantaneous control approach to linear and non-linear dynamical systems. For this purpose the dynamical system is discretized in time. Then, at every time instance a stationary control problem related to the one given in advance is solved either exactly, or approximately using gradient-type methods. It is shown that each of these methods may be interpreted as a special stable time discretization of a class of closed-loop control laws, provided the parameters involved in the approach are adjusted in a proper way. This is of particular importance for the construction of practically relevant controllers for fluid flows, as is pointed out in [3].

The paper is organized as follows. In Section 2 we briefly describe the model problem, which is chosen to be the linear-quadratic regulator problem in its simplest form and state some well-known results concerning solutions and numerical solution techniques. In Section 3 we introduce the concept of instantaneous stabilization. This approach turns out to yield practically realizable feedback laws for which we provide some theoretical results in the finite dimensional case. In Section 4 we discuss the approach at hand of the well-known inverted pendulum equations. It turns out that the qualitative properties of the instantaneous controller are comparable to to that of the optimal stationary Riccati controller. We close with Section 5, where we present the numerical results of instantaneous stabilization applied to the control of unsteady flow around a cylinder.

2. The model problem

In this section we line out the derivation of control laws at hand of the linear-quadratic regulator problem (LQR problem) in its simplest form. To make the ideas of the approach as transparent as possible we do not concern about the questions of controllability and observability. These topics in connection with the approach presented here are discussed in [11]. Thus, the model problem will be formulated as follows: Given the desired state vector $\bar{x} \in \mathbb{R}^n$ and an initial state $x_0 \in \mathbb{R}^n$, find a control vector $u(t) \in \mathbb{R}^n$ such that the objective function

$$J(u) = \frac{1}{2} \int_0^T \gamma |u(t)|^2 + |x(t) - \bar{x}|^2 \, dt \tag{1}$$

is minimal compared to the value of J at all input vectors $v(t) \in \mathbb{R}^n$ satisfying

$$\dot{x}(t) + Ax(t) = b(t) + v(t), \qquad x(0) = x_0 \tag{2}$$

Here, $\gamma > 0$ weights the costs for the control input, the matrix $A \in \mathbb{R}^{n \times n}$ is assumed to be regular and the inhomogeneity is denoted by b. It is well known that this problem admits a unique solution

$$u^* = \frac{1}{\gamma} \mu, \tag{3}$$

where the adjoint state $\mu \in \mathbb{R}^n$ is connected to the state x by the linear forward-backward in time Hamilton equations

$$\begin{array}{rclrcl} \dot{x} + Ax & = & b + \frac{1}{\gamma}\mu, & -\dot{\mu} + A^*\mu & = & -(x - \bar{x}), \\ x(0) & = & x_0, & \mu(T) & = & 0. \end{array} \tag{4}$$

Thus, the control u^* in (3) for $\bar{x}$ independent of time may be computed with the help of the solution of the two-point boundary value problem

$$\begin{aligned} \ddot{\mu} - (A^* - A)\,\dot{\mu} - \left(AA^* + \tfrac{1}{\gamma} A\right) \mu &= b - A\bar{x} \\ A^*\mu(0) - \dot{\mu}(0) &= \bar{x} - x_0 \\ \mu(T) &= 0. \end{aligned} \tag{5}$$

Choosing for the control the Ansatz

$$u(t) = \frac{1}{\gamma}\{-P(t)(x - \bar{x}) + p(t)\} \tag{6}$$

leads to the time-variant feedback matrix P and a time-dependent vector field p satisfying the Riccati matrix differential equations

$$\begin{aligned} \dot{P} - A^*P - PA + I - \tfrac{1}{\gamma}P^2 &= 0 \\ \dot{p} - \tfrac{1}{\gamma}Pp - A^*p &= -PA\bar{x} \\ P(T) &= 0 \\ p(T) &= 0. \end{aligned} \tag{7}$$

In practical applications the use of an numerical approximation of the optimal control law (6) amounts to the storage of an $n \times n$-matrix and an n-vector at every time instance. Especially for large time intervals $T \gg 0$ or/and large system dimension the application of this law is not workable, not to mention the enormous amount of computational work necessary to provide the numerical solution of (7). Concerning storage the problems arising for an forward-backward iterative solution approach to Eq.(4) (compare [5]) or for a finite-difference method applied to (5) are similar.

For infinite time horizon $T = \infty$ the optimal control law is also given by (6), where now the feedback matrix P and the vector p are the suitable time-invariant stationary solution of Eqs. (7), i.e. P and p satisfy the quadratic system

$$\begin{aligned} A^*P - PA + I - \tfrac{1}{\gamma}P^2 &= 0 \\ -\tfrac{1}{\gamma}Pp - A^*p &= -PA\bar{x}. \end{aligned} \tag{8}$$

Investigations on algorithms for the numerical solution of this problem are discussed, for example, by Rauter and Sachs in [6]. However, for large system dimensions as they arise from the discretization of parabolic equations, say the effort to solve Eqs. 8 numerically is enormous. It is one goal of this paper to present an approach which provides numerically computable control laws even for systems of very large dimension.

In the next section we introduce an approach to the design of time invariant control laws that is completely different to the method sketched above as it relies on consistent time integration of the underlying dynamical system.

3. Instantaneous control strategy

As pointed out in the previous section the construction and numerical implementation of optimal control laws suffer from practical implementability, at least when the problems under consideration have large system dimension. In order to overcome these difficulties, we now suggest a different way of constructing closed-loop controllers for dynamical systems.

The approach that we take is based on a time discretization of the state equation given in (2) with $v = 0$. For this purpose let $0 = t_0 < t_1 < \cdots < t_m = T$

denote a equidistant grid on the time interval $[0,T]$ with step size $h = \frac{T}{m}$. At each discrete time level t_i a stationary control problem is solved approximately for an approximate optimal control u_i^* and this control is used to steer the system from t_i to t_{i+1}, where a new approximate optimal control is determined. It cannot be claimed that the controls obtained in this manner approximate the optimal control for (1),(2) as the discretization parameter tends to zero. However, this procedure will be justified by the effectiveness that it exhibits for numerical examples and the interpretation that it allows as closed-loop control law.

As time discretization method we choose the implicit Euler method and as cost function an instantaneous version of (1) is taken. The optimization problem in every time step then has the form

$$(P_i)\begin{cases} \min J(u^{j+1}, x^{j+1}) = \frac{\gamma}{2}|u^{j+1}|^2 + \frac{1}{2}|x^{j+1} - \bar{x}^j|^2 \\ \text{s.t.} \\ (I + hA)x^{j+1} \quad = \quad x^j + hb^j + u^{j+1}, \end{cases} \tag{9}$$

which, due to the quadratic character of the cost function, admits a unique solution (x^{j+1}, u^{j+1}). This tuple together with the uniquely determined Lagrange multiplier μ^{j+1} solves the corresponding first order optimality conditions given by

$$\left.\begin{array}{rcl} (I + hA)x^{j+1} & = & x^j + hb^j + u^{j+1} \\ (I + hA^*)\mu^{j+1} & = & -(x^{j+1} - \bar{x}^j) \\ \gamma u^{j+1} - \mu^{j+1} & = & 0, \end{array}\right\} \tag{10}$$

see [4]. Having in mind to treat cost functions involving more general non-linearities than quadratic ones suitable methods to solve for (x^{j+1}, μ^{j+1}) in (9) are gradient-type methods. The approach presented in the Algorithm following next uses exactly one gradient step at each time level t_i to approximate the solution of problem (9). As is shown later this leads to a control law which is capable of steering the state x of the dynamical system to the desired state $\bar{x}$.

Algorithm 3.1. One-gradient-step stabilization (ogss)

1. Given initial values x^0, set $j = 0$, $t_0 = 0$
2. Given u_0^j, solve

$$\begin{array}{rcl} (I + hA)x & = & x^j + hb^j + u_0^j \\ (I + hA^*)\mu & = & -(x - \bar{x}^j) \end{array}$$

3. Set $\nabla J(u_0^j) = \gamma u_0^j - \mu$
4. Given $\rho > 0$, set $u^{j+1} = u_0^j - \rho \nabla J(u_0^j)$
5. Solve

$$(I + hA)x^{j+1} = x^j + hb^j + u^{j+1}$$

6. Set $t_{j+1} = t_j + h$, $j = j + 1$
7. If $t_j < T$ goto 2.

Here we note, that the gradient of J at u in direction v is given by

$$\nabla J(u)(v) = (\gamma u - \mu, v), \tag{11}$$

so that the additional computational effort at every time step in comparison with the uncontrolled approach consists in the two solves in step 2. The optimal control problem (4), (2) has thus been replaced by a sequence of stationary problems (P_i) which can be realized on workstations even for systems with large dimensions.

Remark 3.1. One may also use the exact solution of problem (9) to obtain an algorithm similar to Algorithm 3.1, see [4, Algorithm 3.3]. In numerical applications promising results are obtained with a near to optimal gradient descent $\rho* \approx \operatorname{argmin}_{\rho \geq 0} J(u_0^j - \rho \nabla J(u_0^j))$ in step 4. of the above algorithm, compare also the numerical results presented in Section 5 and [3].

The major ingredient in the convergence proof for Algorithm 3.1 is its interpretation as conditionally stable semi-implicit discretization scheme of a dynamical system related to the one given in eq. (2). For this purpose we abbreviate

$$B^* := (I + hA)^{-1}, \quad B := (I + hA^*)^{-1},$$

recall $\bar{x} \in \mathbb{R}^n$ to be the desired state, set $f := A\bar{x}$ and require $h \max\{\|A\|, \|A^*\|\} < 1$, so that the matrices B and B^* are well defined.

Theorem 3.2. Algorithm 3.1 is equivalent to the semi-implicit time discretization

$$(I + hA)x^{j+1} = x^j + hb^j - \rho BB^*(x^j - \bar{x}^j) - h\rho BB^*(b^j - f^j), \quad x^0 := x_0, \tag{12}$$

of the dynamical system

$$\dot{x} + Ax = b - \frac{\rho}{h} BB^*(x - \bar{x}) - \rho BB^*(b - f), \quad x(0) = x_0, \tag{13}$$

provided $u_0^j := 0$ for every $j \in \mathbb{N}$.

Proof. $u_0^j = 0$ in step 4. gives $u^j = \rho\mu$. Solving first for x and then for μ in step 2 gives, under consideration of $B\bar{x}^j = BB^*\bar{x}^j + hBB^*f^j$ (since $f^j = A\bar{x}^j$) and step 5, the desired result. □

Remark 3.3. The differential equation in (13) admits a unique smooth solution.

This, as well as the result of Theorem 3.2 holds true also in the infinite dimensional case, provided that the operator $A : X \to X$ is linear with real-part of the spectra bounded from below, say, where X denotes a Banach space.

As the next step we prove that the difference scheme (12) is stable for certain configurations of the parameters involved. For this purpose it is sufficient to ensure that the matrix $C_I := (I + hA)^{-1}(I - \rho BB^*)$ has spectral radius lower than or equal to 1. A sufficient condition for this to hold is given in

Theorem 3.4. Let the parameters h and $0 < \rho < 2$ satisfy the relation

$$\frac{1}{h} \geq \frac{(14\rho + 2)\max\{\|A\|, \|A^*\|\}}{1 - |1 - \rho|}. \tag{14}$$

Then the discretization (12) of (13) is stable.

Proof. A detailed proof of the theorem is given in [4, Theorem 5.1]. For the convenience of the reader we present a sketch. To begin with note, that the matrix C_I may be written as

$$C_I = I \underbrace{-\rho BB^* - hA}_{=:hC_A} + \rho hABB^* + \left(\sum_{k=2}^{\infty} (-h)^k A^k \right) (I - \rho BB^*) . \qquad (15)$$

Using the Neumann expansions of B and B^* and restricting h to satisfy $2h \max\{\|A\|, \|A^*\|\} < 1$, rough estimation yields the desired condition, see [4]. □

Here and in the following we use the item stability as it is defined in the book [8, Definition 5.1.1].

As an immediate consequence of the proof of this theorem we obtain a condition for exponential stability of the differential equation (13).

Theorem 3.5. Let the assumptions of Theorem 3.4 be satisfied. Furthermore, let condition (14) hold with strict inequality sign. Then, the differential equation (13) is exponentially stable.

Proof. Again, we present a sketch. The matrix C_A is the system matrix of the differential equation (13). Following the lines of the proof of the previous theorem in [4] one concludes that the spectral radius r_σ of the matrix $I + hC_A$ satisfies

$$r_\sigma (I + hC_A) \leq \|I + hC_A\| < 1 \quad \text{(for all } h \text{ satisfying (14))}.$$

This implies $\mathrm{Re}(\lambda(C_A)) < 0$ for all eigenvalues $\lambda(C_A)$ of C_A. □

In view of what follows we denote the set of m-times continuously differentiable functions with bounded derivatives up to order m on the infinite time interval $[0, \infty)$ and values in $\mathbb{R}^n$ by $C_b^m([0, \infty), \mathbb{R}^n)$.

In the next theorem we prove decay of the solution of the differential equation (13) and of the iterates $\{x^j\}_{j\in\mathbb{N}}$ of Algorithm 3.1 into a tube around $\bar{x}$ of size proportional to the parameter h.

Theorem 3.6. Let $\bar{x}$ be independent of time, let the assumptions of Theorem 3.2, 3.4 and 3.5 be satisfied. Furthermore, let the inequality in Eq. 14 be strict and assume that $f - b \in C_b^0([0, \infty), \mathbb{R}^n)$. Then, the iterates $\{x^j\}_{j\in\mathbb{N}}$ and the controls $\{u^j\}_{j\in\mathbb{N}}$ of Algorithm 3.1 satisfy

$$\sup_{j\in\mathbb{N}, j\geq j_0} |x^j - \bar{x}| \leq C h, \qquad \sup_{j\in\mathbb{N}, j\geq j_0} |u^j| \leq C h \qquad (16)$$

with some fixed $j_0(h, \rho, x^0)$ and some positive constant C independent of the parameter h.

If, in addition $\rho = 1$, there exists a finite time t_0 such that

$$\sup_{t\geq t_0} |x(t) - \bar{x}| \leq C h. \qquad (17)$$

Proof. We start with proving the second claim. To begin with, set $\Phi : x - \bar{x}$ and observe that Φ solves the stable differential equation

$$\dot{\Phi} = \underbrace{-(A + \frac{\rho}{h} BB^*)}_{=:C_A} \Phi \underbrace{-(\rho BB^* - I)}_{=:C_B}(b - f), \tag{18}$$

which admits the unique solution

$$\Phi(t) = e^{C_A t}\Phi(0) + \int_0^t e^{C_A(t-s)} C_B (b - f)(s)\, ds.$$

Using the expansion $BB^* = I - h(A + A^*) + O(h^2)$ $(h \to 0)$, the regularity requirements on the difference $b - f$, $\rho = 1$ and the exponential stability of the differential equation (18) one obtains for some $t_0 \geq 0$

$$|\Phi(t)| \leq Ch \quad \forall t \geq t_0,$$

where the positive constant can be chosen independent of h.

In order to prove the first claim we exploit the stability properties of the semi-implicit Euler discretization of (18) given by

$$(I + hA)\Phi^{j+1} = \Phi^j - \rho BB^*\Phi^j - h(\rho BB^* - I)(b^j - f^j). \tag{19}$$

Since strict inequality in condition (14) implies $\|C_I\| < 1$ we can conclude

$$|\Phi^j| \leq \|C_I\|^j\, |x^0 - \bar{x}| + \frac{h}{1 - \|C_I\|} \|b - f\|_\infty \leq C\, h \ (j \geq j_0).$$

This is the desired result for the iterates x^j. For the controls u^j the result immediately follows from the representation

$$u^{j+1} = -\rho BB^*(x^j - \bar{x}^j) - h\rho BB^*(b^j - f^j),$$

the regularity requirements on the difference $b - f$ and the the estimate for the iterates x_j. □

Remark 3.7. Note, that if b is chosen as to satisfy $b = A\bar{x}$, the choice $\rho = 1$ in Theorem 3.6 is not necessary to obtain the results in (17).

4. An example

As example we investigate the linearization of the pendulum equation

$$\ddot{\varphi} + \sin\varphi = 0$$

around the unstable stationary solution $\varphi = \pi, \dot{\varphi} = 0$:

$$\dot{x} + \underbrace{\begin{bmatrix} 0 & -1 \\ -1 & 0 \end{bmatrix}}_{=:A} x = \underbrace{\begin{bmatrix} 0 \\ -\pi \end{bmatrix}}_{=:f}.$$

The state $\bar{x}$ to be observed is given by $\bar{x} = [\pi, 0]^t$. Further we set $b = [0, -\pi]$, so that $A\bar{x} = b$ holds true.

We compute the stationary Riccati solution in order to compare it to the results obtained by the control law given by Algorithm 3.1. Using the stationary Riccati solutions P and p from (8) to stabilize the linearized pendulum equations leads to the feedback law

$$\dot{x} = -\left(A + \tfrac{1}{\gamma}P\right)x + \left(f - \tfrac{1}{\gamma}p\right),$$

where the positive definite matrix P and the vector p are given by

$$P = \begin{bmatrix} \sqrt{\gamma^2+\gamma} & \gamma \\ \gamma & \sqrt{\gamma^2+\gamma} \end{bmatrix}, \quad p = \begin{bmatrix} -\sqrt{\gamma^2+\gamma}\pi \\ -\gamma\pi \end{bmatrix}.$$

Using this, a short calculation gives the optimal state for the infinite horizon problem,

$$x = \begin{bmatrix} \pi \\ 0 \end{bmatrix} + e^{-t\cdot\sqrt{1+\frac{1}{\gamma}}} \begin{bmatrix} a \\ b \end{bmatrix},$$

where a and b depend on the initial state of the differential equation. The control has the form $u = -\frac{1}{\gamma}(Px + p)$ and is given by

$$u = -\tfrac{1}{\gamma} \begin{bmatrix} a\sqrt{\gamma^2+\gamma} + b\gamma \\ a\gamma + b\sqrt{\gamma^2+\gamma} \end{bmatrix} e^{-t\cdot\sqrt{1+\frac{1}{\gamma}}}.$$

Two more lines of computation finally yield that the cost $J = \displaystyle\int_0^\infty \|x - \bar{x}\|^2 + \gamma\|u\|^2\, dt$ is given by

$$J = \sqrt{\gamma^2+\gamma}\left(a^2 + 2ab\sqrt{\tfrac{\gamma}{\gamma+1}} + b^2\right) = O(\gamma)\ (\gamma \to 0).$$

Now we proceed similar for the instantaneous control law. To begin with, let $\frac{1}{h} > \|A\| = 1$. Then, there holds $B = B^* = 1 - hA + O(h^2)$ $(h \to 0)$ and $BB^* = 1 - 2hA + O(h^2)$ $(h \to 0)$. For the matrix $C_A = -(A + \frac{\rho}{h}BB^*)$ (see (18)) we obtain the approximation

$$\tilde{C}_A \doteq \begin{bmatrix} -\frac{\rho}{h} & 1 - 2\rho \\ 1 - 2\rho & -\frac{\rho}{h} \end{bmatrix},$$

which is of order $O(h)$ $(h \to 0)$. Using this approximations it is sufficient to investigate the feedback control given by

$$\dot{x} = \tilde{C}_A x + \tfrac{\rho}{h}(I - 2hA)\bar{x} + b - \rho(I - 2hA)(b - f)$$

since its exact solution

$$x = ae^{\lambda_1 t}\begin{bmatrix} 1 \\ 1 \end{bmatrix} + be^{\lambda_2 t}\begin{bmatrix} 1 \\ -1 \end{bmatrix} + \begin{bmatrix} \pi \\ 0 \end{bmatrix}$$

deviates from that of the original control law (13) only by a term of size $O(h)$ $(h \to 0)$. The eigenvalues $\lambda_{1/2}$ of the matrix $\tilde{C}_A$ are given by

$$\lambda_{1/2} \;=\; -\frac{\rho}{h} \pm \operatorname{sign}(1-2\rho)|1-2\rho|,$$

so that negative eigenvalues are guaranteed, provided the parameter h is sufficiently small. Finally, we compute the control and the costs. For the control we obtain

$$u = -\left(a e^{\lambda_1 t} \left(\tfrac{\rho}{h} + 2\rho\right) \begin{bmatrix} 1 \\ 1 \end{bmatrix} + b e^{\lambda_2 t} \left(\tfrac{\rho}{h} + 2\rho\right) \begin{bmatrix} 1 \\ -1 \end{bmatrix} \right) \longrightarrow 0 \ (t \to \infty),$$

the value of the costs J are given by

$$J = \frac{a^2}{\lambda_1} \left(1 + \gamma \left(\frac{\rho}{h} + 2\rho\right)^2\right) + \frac{b^2}{\lambda_2} \left(1 + \gamma \left(\frac{\rho}{h} + 2\rho\right)^2\right) \le C \left(\frac{h}{\rho} + \frac{\gamma\rho}{h}\right)$$

with some positive constant C independent of ρ, γ and h. As a result we obtain that both, the optimal Riccati controller and the controller of Algorithm 3.1 have the same qualitative behaviour.

We conclude with presenting the numerical results for the distributed control law when it is applied to the regulation of the flow around a cylinder.

5. Distributed control of flow around a cylinder

Here we present numerical results related to the control of laminar unsteady flow around a cylinder in a two-dimensional spatial domain Ω. The instantaneous stabilization Algorithm 3.1 is applied to reconstruct the Stokes flow in an observation volume Ω_s behind the cylinder. Using this observations a volume force is computed in every time step in order to minimize the difference in the L^2-norm between the flow and the Stokes flow.

5.1. The distributed control problem

We now put this problem into the framework developed in Section 3. For this purpose we consider the control problem

$$(P) \quad \begin{cases} \min J(f) = \frac{\gamma}{2} \int_0^T \left[\int_{\Omega_s} |u - \bar{u}|^2 \, d\Omega \; + \; \frac{\gamma}{2} \int_{\Omega} |f|^2 \, d\Omega \right] dt \\ \text{s.t.} \\ \quad u_t - \frac{1}{\mathrm{Re}} \Delta u + (u \cdot \nabla) u + \nabla p \;=\; f \quad \text{in } Q \\ \quad -\operatorname{div} u \;=\; 0 \quad \text{in } Q \\ \quad u \;=\; g \quad \text{on } \Gamma_d \times (0,T] \\ \quad \frac{1}{\mathrm{Re}} \partial_\eta u \;=\; p\eta \quad \text{on } \Gamma_{out} \times (0,T] \\ \quad u(0) \;=\; u_0, \end{cases} \tag{20}$$

with the instationary incompressible Navier-Stokes equations as subsidiary conditions. Here, f denotes the control force acting on the whole of the spatial domain Ω, $Q := \Omega \times (0,T)$ denotes the space-time cylinder, g are boundary values at the

Dirichlet boundary Γ_d and u_0 denotes the initial state. On the outflow boundary Γ_{out} natural boundary conditions are prescribed.

Next, we formally bring the Navier-Stokes equations into the form (2). To begin with we first introduce their variational formulation based on the Leray projection of the stationary Stokes problem, see [10, Lemma 2.1]. We set $V := \{v \in H^1(\Omega, \mathbb{R}^2);\ v_{|\Gamma_d} = 0, \operatorname{div} v = 0\}$ the set of two-dimensional divergence-free vector fields and denote by $(\cdot,\cdot)$ the scalar product in $L^2(\Omega, \mathbb{R}^2)$ $(L^2(\Omega, \mathbb{R}^{2\times 2}))$. The vector field u is called variational solution of the Navier-Stokes equations iff $u(t) \in g(t) + V$, $u(0) = u_0$ and

$$(u_t, v) + \frac{1}{\mathrm{Re}}(\nabla u, \nabla v) + ((u\nabla)u, v) \;=\; (f, v) \quad \forall v \in V, \text{ a.e. in } (0,T). \tag{21}$$

Thus, using the Stokes operator S, this variational formulation may be rewritten as Burgers equation in the space V, i.e. find $u(t) \in g(t) + V$ with $u(0) = u_0$ and

$$u_t + Su + (u\nabla)u \;=\; f. \tag{22}$$

If finally we set $b = b(u, \nabla u) := -(u\nabla)u$, problem (P) has been rewritten in the form (1), (2).

In this setting Algorithm 3.1 applied to the control of flow around a cylinder is equivalent to the semi-implicit time discretization (12) of the dynamical system

$$\dot{u} + Su = b - \frac{\rho}{h}BB^*(u - \bar{u}) - \rho BB^*(b - f), \quad u(0) = u_0, \tag{23}$$

where $f := S\bar{u}$. We note that the operator B is the self-adjoint solution operator of the quasi-Stokes problem in V, i.e.

$$w \;=\; Bh \quad \Longleftrightarrow \quad (I + hS)w = h \quad \text{in } V.$$

5.2. Numerical results

We now present numerical results obtained by the application of Algorithm 3.1 in order to reconstruct the Stokes flow in a given observation volume $\Omega_s \subset \Omega$ behind the cylinder, see Fig. 1. As computational domain we chose a configuration that coincides with the domain used for the two-dimensional benchmark computations in [7]. The Reynolds number is given by the product of the diameter d of the cylinder and the bulk velocity U at the inlet, divided by the kinematic viscosity ν of the fluid. The boundary conditions are no-slip at the top and the bottom walls and *do-nothing* at the outflow boundary. At the inlet a parabolic inflow profile is described. For the spatial discretization the Taylor-Hood finite element is used on a grid containing 5374 triangles, 2840 pressure and 11054 velocity nodes with a-priori local refinement in a neighbourhood of the cylinder. For the discretization of the quasi-Stokes problems the Taylor-Hood finite element is used. Numerical solutions of various problems are performed using an extension of the Navier-Stokes solver developed by Bänsch in [1].

For the computations that follow the Reynolds number is chosen as Re $=$ 100. Furthermore, we use the parameter values $\gamma = 1.e-2$ and $h = 0.0075$. The observation volume is taken as $\Omega_s = [4, 6] \times [-1, 1]$. The time histories of control and of the descent parameters ρ used in Algorithm 3.1 are documented in Tab. 1. In this table ρ^* denotes an approximation of the optimal descent parameter, see Remark 3.1. In Fig. 2 the values of costs (a) and control costs (b) are shown. When

Time	$0 \leq t \leq 1$	$1 \leq t \leq 5$	$5 \leq t \leq 10$	$10 \leq t \leq 20$
ρ	no control	0.1	$\rho \approx \rho^*$	no control
Time	$20 \leq t \leq 40$	$40 \leq t \leq 100$	$100 \leq t \leq 120$	$120 \leq t \leq 200$
ρ	$\rho \approx \rho^*$	no control	$\rho \approx \rho^*$	no control

TABLE 1. Time history of control and the gradient descent parameters ρ used in Algorithm 3.1

control is switched on the costs decrease immediately. If for ρ an approximation of the optimal gradient step-size is used the additional decrease of cost is enormous. As control is relaxed the flow evolves to the uncontrolled state, whose periodic behaviour is very good reflected by the periodic difference to the steady Stokes flow shown in Fig. 1(a). As Fig. 2(b) shows switching on control is accompanied by a large rise of the control costs. Once control is invoked their cost numerically converge to a steady state as is predicted by Theorem 3.6. Note, that in contrast to what is shown in Fig. 2(b), there is no control cost in time intervals without control force applied to the flow. Since the scale of the plot is logarithmic the control cost are set to their last value $\neq 0$.

Fig 1 shows the Stokes flow (a) together with the uncontrolled Navier-Stokes flow (b). The same figure (c) shows the controlled flow together with the control force (d) at $t = 2$. As one can see the recirculation in a neighbourhood of the observation volume Ω_s vanishes completely. This effect is further amplified when control action is kept on.

Remark 5.1. Further numerical investigations on the control of flow around a cylinder can be found in the papers of Choi [2] and Tang & al. [9]. A numerical comparison of the method presented here and the open-loop optimal control approach for backward-facing step flows is presented in [5]. As is shown there, the cost reduction of the method worked out here applied to backward-facing step flows nearly coincides with that obtained by the open-loop optimal control approach, whereas the amount of numerical work for the suboptimal method is substantially less than that for the open-loop optimal approach.

Acknowledgments

The authors acknowledge the support of the Sonderforschungsbereich 557 (Sfb 557) sponsored by the Deutsche Forschungsgemeinschaft.

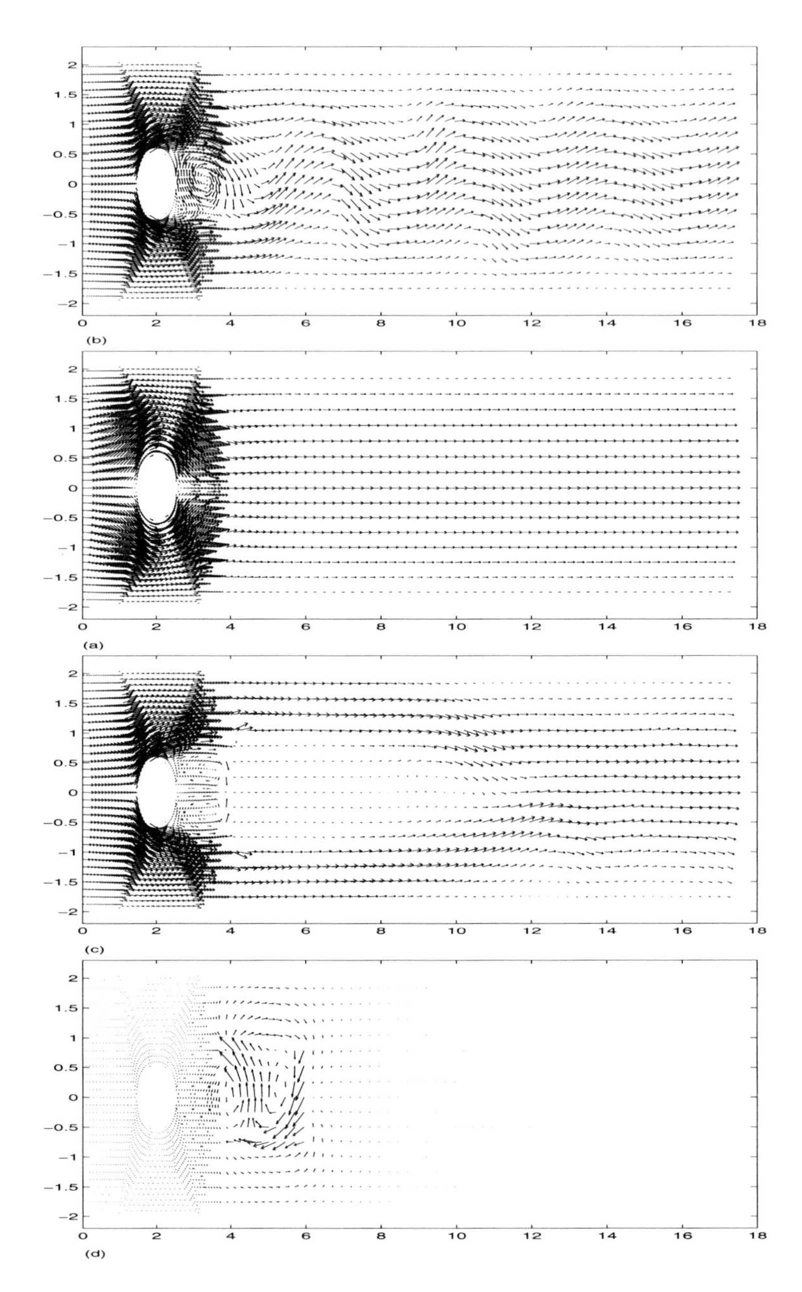

FIGURE 1. Distributed control at Re=100: (a) Stokes flow;(b) Navier-Stokes flow; (c) Controlled flow; (d) Control force

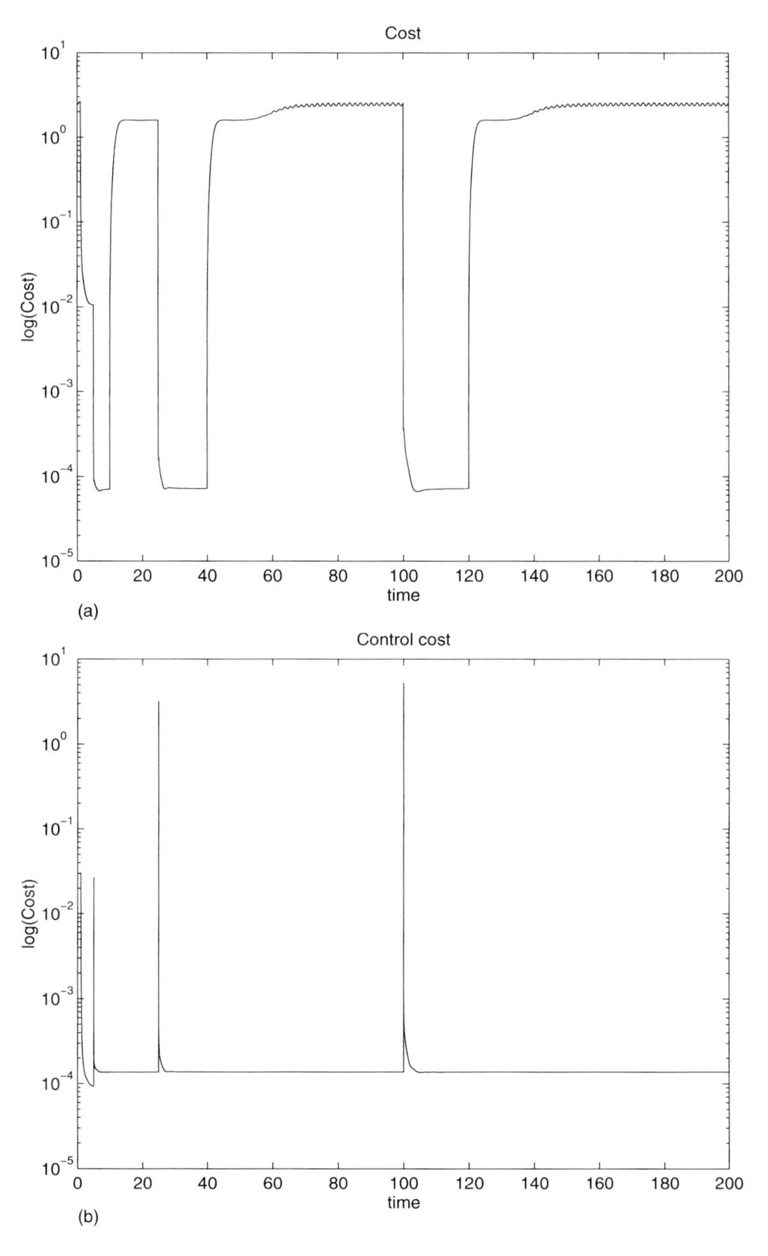

FIGURE 2. Distributed control at Re=100. Observation in $[4,6] \times [-1,1]$: (a) Cost; (b) Control Cost

References

[1] **Bänsch, E.** An adaptive Finite-Element-Strategy for the three-dimensional time-dependent Navier-Stokes-Equations. *J. Comp. Math.*, 36:3–28, 1991.

[2] **Choi, H.** Suboptimal Control of Turbulent Flow Using Control Theory. In *Proceedings of the International Symposium. on Mathematical Modelling of Turbulent Flows*, 1995. Tokyo, Japan.

[3] **Choi, H.; Hinze, M.** & **Kunisch, K.** Instantaneous control of backward-facing-step flows. Preprint No. 571/1997, 1997. Technische Universität Berlin, Deutschland.

[4] **Hinze, M.** & **Kauffmann, A.** A new class of feedback control laws for dynamical sytems. Preprint No. 602/1998, 1998. Fachbereich Mathematik, Technische Universität Berlin.

[5] **Hinze, M.** & **Kunisch, K.** Control strategies for fluid flows - optimal versus suboptimal control. Preprint No. 573/1998, 1998. Technische Universität Berlin, Deutschland, to appear in the Proceedings of the 2nd ENUMATH, Heidelberg.

[6] **Rautert, T.** & **Sachs, E.W.** Computational Design of Optimal Output Feedback Controllers. Forschungsbericht Nr. 95-12, 1995. Institut für Mathematik, Universität Trier.

[7] **Schäfer, M.** & **Turek, S.** Benchmark computations of laminar flow around a cylinder. Report, 1996. IWR, Universität Heidelberg.

[8] **Strehmel, K.** & **Weiner, R.** *Numerik gewöhnlicher Differentialgleichungen.* Teubner Studeinbücher, 1995.

[9] **Tang, K.Y.; Graham, W.R.** & **Peraire, J.** Active Flow Control using a Reduced Order Model and Optimum Control. Preprint, 1996. Computational Aerospace Sciences Laboratory, MIT Department of Aeronautics and Astronautics.

[10] **Temam, R.** *Navier-Stokes Equations.* North-Holland, 1979.

[11] **Wunder, G.** Zur Konstruktion von Reglern für große dynamische Systeme. Diplomarbeit, 1998. Fachbereich Elektrotechnik, Technische Universität Berlin.

Fachbereich Mathematik,
Technische Universität Berlin,
Straße des 17. Juni 136
D-10623 Berlin
E-mail address: `hinze@math.tu-berlin.de, kauffman@math.tu-berlin.de`

International Series of Numerical Mathematics
Vol. 133, © 1999 Birkhäuser Verlag Basel/Switzerland

Homogenization of a Model Describing Vibration of Nonlinear Thin Plates Excited by Piezopatches

Karl-Heinz Hoffmann and Nikolai D. Botkin [1]

Abstract. A model describing vibration of nonlinear von Kármán thin plates excited by actuators made of piezoelectric ceramics is considered. The model contains strong oscillating coefficients due to the piezoelectric actuators. A procedure of homogenization based on the so-called two-scale convergence is applied to the model. This yields a nonlinear system of equations with constant coefficients. The unique solvability of the resulting system is proved. The convergence of all solutions of the original system to the solution of the resulting system as the number of piezoelectric actuators goes to infinity is proved.

1. Introduction

The problem of homogenization of partial differential equations describing vibration of nonlinear thin plates excited by actuators made of piezoelectric ceramics (see [1] and [2]) is considered. It is assumed that the number of the actuators goes to infinity whereas their dimension tends to zero. The procedure of homogenization is based on the theory of two-scale convergence studied in [4]. The specific features of the problem considered are: the evolutional character of the equations, the appearance of the forth spatial derivatives in the first equation describing vertical displacements of the plate, and nonlinearities typical for von Kármán systems. We apply a result of [6] about two-scale convergence of the second derivatives of subsequences of sequences bounded in $L_2(0,T;H_0^2(S))$, which enables us to handle a weak formulation of the problem. Results of [7] and [8] are used. Computer simulations (they are not included into the paper) demonstrate a good approximation of solutions of the original equations by solutions of the homogenized equations if the number of piezoelectric patches is sufficiently large.

[1]This work is supported by DFG, Germany.

2. Notation

$S \subset R^2$	is the domain occupied by the plate.
$S_{P_l} \subset S$	is the domain occupied by the ith piezopatch.
$S_P := \bigcup_{l=1}^m S_{P_l}$	is the domain occupied by all piezopatches.
$S_B := S \setminus S_P$	is the domain occupied by the base material.
$Y := [0,1] \times [0,1]$	is the unit square.
$\langle g \rangle := \int\int_Y g(y)dy$	is the mean value of a function.
$C_\#^\infty(Y) \subset C^\infty(R^2)$	is the subspace of periodic functions that have equal values on opposite faces of the boundary of Y.
$H_\#^m(Y)$	is the completion of $C_\#^\infty(Y)$ for the norm of $H^m(Y)$; it holds: $D^\alpha u\vert_{y_1=0} = D^\alpha u\vert_{y_1=1}$ and $D^\alpha u\vert_{y_2=0} = D^\alpha u\vert_{y_2=1}$ for $\alpha = (\alpha_1, \alpha_2), \alpha_1, \alpha_2 \geq 0, \alpha_1 + \alpha_2 \leq m-1$.
$H_\#^m(Y)/R$	is the quotient space.
$Q := (0,T) \times S$	
$\overset{\circ}{C}{}_T^\infty(Q) \subset C^\infty(Q)$	is the subspace of all functions which vanish on ∂S and at $t = T$ along with all derivatives.
$\overset{\circ}{C}{}_{0,T}^\infty(Q; C_\#^\infty(Y))$	is the space of infinitely differentiable functions from Q into $C_\#^\infty(Y)$ which vanish on ∂S, at $t = 0$, and $t = T$ along with all derivatives.
$H_T^2(0,T; L_2(S))$	is the subspace of all functions from $H^2(0,T; L_2(S))$ which vanish at $t = T$ along with the first time derivative.

$X_\infty^{211} := L_\infty(0,T; H_0^2(S)) \times L_\infty(0,T; H_0^1(S)) \times L_\infty(0,T; H_0^1(S))$

$X_\infty^{100} := L_\infty(0,T; H_0^1(S)) \times L_\infty(0,T; L_2(S)) \times L_\infty(0,T; L_2(S))$

$d_{ij} := 1/2(u_{ix_j} + u_{jx_i} + \xi_{x_i}\xi_{x_j})$ is the in-plane strain tensor.

L^ε	is the set of all functions $(\xi^\varepsilon, u_1^\varepsilon, u_2^\varepsilon)$ that are limits of all possible Galerkin approximations in problem (5) w.r.t. weak-$*$ topology of X_∞^{211}.
$\Vert f \Vert := \Vert f \Vert_{L_2(S)}$	is L_2 norm of a function.
$(f,g) := (f,g)_{L_2(S)}$	is L_2 scalar product.

We assume summation over repeating indices. For example,

$$\|\xi_{x_\alpha}\|^2 = \|\xi_{x_\alpha}\| \cdot \|\xi_{x_\alpha}\| := \sum_{\alpha=1}^{2} \|\xi_{x_\alpha}\| \cdot \|\xi_{x_\alpha}\| = \|\xi\|^2_{H_0^1(S)}.$$

3. Problem setting

Consider a system of nonlinear equations describing oscillations of a thin plate excited by patches made of a piezoelectric ceramic (see [2]). For simplicity, assume that the plate occupies a rectangular domain S and the piezoelectric patches occupy rectangular domains $S_{P_l} \subset S$ (see Figure 1). It is assumed that the patches form a periodic structure of the period ε so that the object is completely defined by ε.

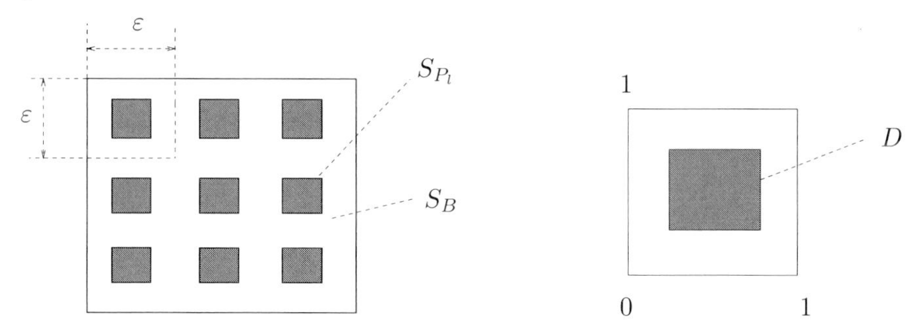

Figure 1. Plate S with patches S_{P_l}. Unit cell scaled from $\varepsilon \times \varepsilon$ to 1×1.

Equations describing the model read:

$$\begin{aligned} \rho\left(\tfrac{x}{\varepsilon}\right)\xi_{tt} - \operatorname{div}\Big(\mu\left(\tfrac{x}{\varepsilon}\right)\nabla\xi_{tt}\Big) + \Delta\Big(\gamma\left(\tfrac{x}{\varepsilon}\right)\Delta\xi\Big) - \tfrac{\partial}{\partial x_\alpha}\Big(\tau_{\alpha\beta}\left(\tfrac{x}{\varepsilon}\right)\xi_{x_\beta}\Big) \\ = F\Delta\Big(K_\varepsilon(t,x)I\left(\tfrac{x}{\varepsilon}\right)\Big) + G\tfrac{\partial}{\partial x_\alpha}\Big(\xi_{x_\alpha}K_\varepsilon(t,x)I\left(\tfrac{x}{\varepsilon}\right)\Big), \end{aligned} \tag{1}$$

$$\rho\left(\tfrac{x}{\varepsilon}\right)u_{\alpha tt} - \tfrac{\partial}{\partial x_\beta}\tau_{\alpha\beta}\left(\tfrac{x}{\varepsilon}\right) = G\tfrac{\partial}{\partial x_\alpha}\Big(K_\varepsilon(t,x)I\left(\tfrac{x}{\varepsilon}\right)\Big),$$

where

$$\tau_{\alpha\beta}\left(\tfrac{x}{\varepsilon}\right) = \ell_{ij\alpha\beta}\left(\tfrac{x}{\varepsilon}\right)\tfrac{1}{2}\left(u_{ix_j} + u_{jx_i} + \xi_{x_i}\xi_{x_j}\right), \quad \alpha,\beta = 1,2.$$

Here ξ and u_α, $\alpha = 1,2$, are transversal and longitudinal displacements of points of the plate. The coefficients ρ, μ, γ, I, and $\ell_{ij\alpha\beta}$ are Y-periodic functions defined on Y as follows: $I(y) = I_D(y)$,

$$\begin{aligned} \rho(y) = I_D(y)\rho_P + (1 - I_D(y))\rho_B, \quad \mu(y) = I_D(y)\mu_P + (1 - I_D(y))\mu_B, \\ \gamma(y) = I_D(y)\gamma_P + (1 - I_D(y))\gamma_B, \quad \ell_{ij\alpha\beta}(y) = I_D(y)\ell^P_{ij\alpha\beta} + (1 - I_D(y))\ell^B_{ij\alpha\beta}, \end{aligned} \tag{2}$$

where I_D is the indicator function of D (see Fig. 1 right). The constants ρ_P, μ_P, γ_P, $\ell^P_{ij\alpha\beta}$, ρ_B, μ_B, γ_B, $\ell^B_{ij\alpha\beta}$ are related to the domains S_P and S_B. It is assumed that $\rho_P > 0$, $\rho_B > 0$, $\mu_P > 0$, $\mu_B > 0$, $\gamma_P > 0$, $\gamma_B > 0$, and

$$\nu\, d_{\alpha\beta}\, d_{\alpha\beta} \le \ell_{ij\alpha\beta}(y)\, d_{ij}\, d_{\alpha\beta} \le N\, d_{\alpha\beta}\, d_{\alpha\beta}$$

with some positive ν and N, for any $y \in Y$ and any symmetric $d_{ij} \in R^{2\times 2}$.

The voltage $v_l(\cdot)$ applied to the lth patch is being chosen as follows. A distribution $K(t,x) \in H^1(0,T; L_2(S))$ of the voltage over the whole plate S is prescribed and we set

$$v_l(t) = \text{meas}(S_{P_l})^{-1} \iint\limits_{S_{P_l}} K(t,x)dx, \quad K_\varepsilon(t,x) = \begin{cases} K(t,x), & x \in S_B, \\ v_l(t), & x \in S_{P_l}. \end{cases} \tag{3}$$

It is obvious that $K_\varepsilon \to K$ in $H^1(0,T;L_2(S))$.

Boundary and initial conditions are:

$$\begin{aligned} &\xi|_{\partial S} = 0, \quad \partial\xi/\partial\vec{n}\big|_{\partial S} = 0, \quad \xi|_{t=0} = \xi_0, \quad \xi_t\big|_{t=0} = \xi_0'. \\ &u_\alpha\big|_\Gamma = 0, \quad u_\alpha\big|_{t=0} = u_{\alpha 0}, \quad u_{\alpha t}\big|_{t=0} = u_{\alpha 0}', \quad \alpha = 1,2. \end{aligned} \tag{4}$$

Definition 3.1. *We say that functions*

$$\xi^\varepsilon \in L_2(0,T;H_0^2(S)), \qquad u_\alpha^\varepsilon \in L_2(0,T;H_0^1(S)), \quad \alpha = 1,2,$$

form a weak solution to system (1), (4) if the following equality holds:

$$\begin{aligned} \int_0^T \iint\limits_S \Big[&\rho\left(\tfrac{x}{\varepsilon}\right)\xi^\varepsilon\varphi_{tt} + \mu\left(\tfrac{x}{\varepsilon}\right)\nabla\xi^\varepsilon\nabla\varphi_{tt} + \gamma\left(\tfrac{x}{\varepsilon}\right)\Delta\xi^\varepsilon\Delta\varphi + \tau_{\alpha\beta}\left(\tfrac{x}{\varepsilon}\right)\xi^\varepsilon_{x_\beta}\varphi_{x_\alpha} \\ &-FK_\varepsilon(t,x)I\left(\tfrac{x}{\varepsilon}\right)\Delta\varphi + GK_\varepsilon(t,x)I\left(\tfrac{x}{\varepsilon}\right)\xi^\varepsilon_{x_\alpha}\varphi_{x_\alpha} \\ &+\rho\left(\tfrac{x}{\varepsilon}\right)u_\alpha^\varepsilon\psi_{\alpha tt} + \tau_{\alpha\beta}\left(\tfrac{x}{\varepsilon}\right)\psi_{\alpha x_\beta} + GK_\varepsilon(t,x)I\left(\tfrac{x}{\varepsilon}\right)\psi_{\alpha x_\alpha}\Big] dxdt \\ + \iint\limits_S \Big[&\rho\left(\tfrac{x}{\varepsilon}\right)\Big(\xi_0\varphi_t(0,x) - \xi_0'\varphi(0,x) + u_{\alpha 0}\psi_{\alpha t}(0,x) - u_{\alpha 0}'\psi_\alpha(0,x)\Big) \\ &+\mu\left(\tfrac{x}{\varepsilon}\right)\Big(\nabla\xi_0\nabla\varphi_t(0,x) - \nabla\xi_0'\nabla\varphi(0,x)\Big)\Big] dx = 0 \end{aligned} \tag{5}$$

for all

$\varphi \in H_T^2(0,T;H_0^1(S)) \cap L_2(0,T;H_0^2(S)), \quad \psi_\alpha \in H_T^2(0,T;L_2(S)) \cap L_2(0,T;H_0^1(S)).$

Proposition 3.2. *Let $\xi^{\varepsilon\, m}$ and $u_\alpha^{\varepsilon\, m}$ be Galerkin approximations computed using definition 3.1. Then:*

a) If $\xi_0 \in H_0^2(S)$, $\xi_0' \in H_0^1(S)$, $u_{\alpha 0} \in H_0^1(S)$, $u_{\alpha 0}' \in L_2(S)$, $K \in H^1(0,T;L_2(S))$, then there exists a constant C independing from ε, m such that

$$\|\xi^{\varepsilon\, m}\|^2_{H_0^2(S)} + \|\xi_t^{\varepsilon\, m}\|^2_{H_0^1(S)} + \|u_\alpha^{\varepsilon\, m}\|^2_{H_0^1(S)} + \|u_{\alpha t}^{\varepsilon\, m}\|^2_{L_2(S)} \le C \tag{6}$$

for any $t \in [0,T]$.

b) If $\xi_0 = 0$, $\xi_0' = 0$, $u_{\alpha 0} = 0$, $u_{\alpha 0}' = 0$, $K \in H^2(0,T;L_2(S))$, and $K(0,\cdot) = 0$, then there exists a constant C independing from ε, m such that

$$\|\xi_t^{\varepsilon\, m}\|^2_{H_0^2(S)} + \|\xi_{tt}^{\varepsilon\, m}\|^2_{H_0^1(S)} + \|u_{\alpha t}^{\varepsilon\, m}\|^2_{H_0^1(S)} + \|u_{\alpha tt}^{\varepsilon\, m}\|^2_{L_2(S)} \le C \tag{7}$$

for any $t \in [0,T)$.

Proposition 3.3. *Let L^ε be the set of all limit points (in weak-$*$ topology of X_∞^{211}) of all possible Galerkin approximations in (5). Then:*

a) $L^\varepsilon \neq \emptyset$ and each $(\xi^\varepsilon, u_1^\varepsilon, u_2^\varepsilon) \in L^\varepsilon$ is a solution of (5).

b) If the first assumption of proposition 3.2 holds, then there exists a constant C independing from ε such that

$$\|\xi^\varepsilon\|^2_{H_0^2(S)} + \|\xi_t^\varepsilon\|^2_{H_0^1(S)} + \|u_\alpha^\varepsilon\|^2_{H_0^1(S)} + \|u_{\alpha t}^\varepsilon\|^2_{L_2(S)} \leq C \tag{8}$$

for any $(\xi^\varepsilon, u_1^\varepsilon, u_2^\varepsilon) \in L^\varepsilon$ and a.e. $t \in [0, T)$.

c) If the second assumption of proposition 3.2 holds, then there exists a constant C independing from ε such that

$$\|\xi_t^\varepsilon\|^2_{H_0^2(S)} + \|\xi_{tt}^\varepsilon\|^2_{H_0^1(S)} + \|u_{\alpha t}^\varepsilon\|^2_{H_0^1(S)} + \|u_{\alpha tt}^\varepsilon\|^2_{L_2(S)} \leq C \tag{9}$$

for any $(\xi^\varepsilon, u_1^\varepsilon, u_2^\varepsilon) \in L^\varepsilon$ and a.e. $t \in [0, T)$.

Proof. Non-emptiness of L^ε follows from (6). The proof that each $(\xi^\varepsilon, u_1^\varepsilon, u_2^\varepsilon) \in L^\varepsilon$ is a solution of (5) can be found in[2]. It follows from (6) and (7) that all necessary time derivatives in (8) and (9) exist, and that (8) and (9) hold. □

4. Homogenization

The following holds for any $(\xi^\varepsilon, u_1^\varepsilon, u_2^\varepsilon) \in L^\varepsilon$ due to proposition 3.3:

$$\|\xi^\varepsilon\|^2_{L_\infty(0,T;H_0^2(S))} + \|u_1^\varepsilon\|^2_{L_\infty(0,T;H_0^1(S))} + \|u_2^\varepsilon\|^2_{L_\infty(0,T;H_0^1(S))} \leq C,$$

with C independent from ε. Therefore, the sequence $(\xi^\varepsilon, u_1^\varepsilon, u_2^\varepsilon)$ contains a weak-$*$ converging subsequence in X_∞^{211}. Now we derive equations that define limit functions of such subsequences (effective equations). We will show that these equations have a unique solution under assumptions of the second part of proposition 3.2, and that the coefficients of the effective equations are independent from the choice of subsequences. This yields that the sequence $(\xi^\varepsilon, u_1^\varepsilon, u_2^\varepsilon) \in L^\varepsilon$ converges weak-$*$ in X_∞^{211} to the solution of the effective system. Similar arguments show that $(\xi_t^\varepsilon, u_{1\,t}^\varepsilon, u_{2\,t}^\varepsilon)$ converges weak-$*$ in X_∞^{211}, and $(\xi_{tt}^\varepsilon, u_{1\,tt}^\varepsilon, u_{2\,tt}^\varepsilon)$ converges weak-$*$ in X_∞^{100} to time derivatives of the solution of the effective system. Then, using corollary 4 of Simon [9] and theorem 16.1 of Lions-Magenes [10], we conclude that $(\xi^\varepsilon, u_1^\varepsilon, u_2^\varepsilon)$ and $(\xi_t^\varepsilon, u_{1t}^\varepsilon, u_{2t}^\varepsilon)$ converges strongly in $C([0,T]; H_0^{2-\alpha}(S)) \times C([0,T]; H_0^{1-\alpha}(S)) \times C([0,T]; H_0^{1-\alpha}(S))$ for any positive real α. In particular, ξ^ε and ξ_t^ε converge uniformly on $[0,T] \times \bar{S}$.

Now we apply two-scale convergence to derivation of the effective equations. For the pioneering works on two-scale convergence for time-independent problems we refer to [3] and [4]. Two-scale convergence for time-dependent problems was considered in [5]. These results were generalized in [6]. Let us reproduce the definition of two-scale convergence for functions depending on additional parameters.

Definition 4.1. (Definition 6.8 of Haller [6]) *Let* $v_\varepsilon \in L_2(Q)$, $v_0 \in L_2(Q\times Y)$. *It is said that* $v_\varepsilon \overset{2-scale}{\longrightarrow} v_0$, *if*

$$\lim_{\varepsilon\to 0}\int_0^T\int_S\!\!\int v_\varepsilon(t,x)\psi(t,x,x/\varepsilon)dxdt = \int_0^T\int_S\!\!\int\int_Y\!\!\int v_0(t,x,y)\psi(t,x,y)dydxdt$$

for all $\psi \in \overset{\circ}{C}{}^\infty_{0,T}(Q; C^\infty_\#(Y))$.

It is proved (theorem 6.15 of Haller [6]) that all properties of two-scale convergence hold, if the test functions in the above definition are replaced by more general test functions of the form: $\psi(t,x,y) = \alpha(t,x)\beta(y)\sigma(t,x,y)$, where $\alpha \in L_\infty(Q)$, $\beta(y) \in L_{\infty\#}(Y)$, and $\sigma \in C^\infty(Q; C^\infty_\#(Y))$ (not necessary vanishes).

Due to theorem 6.12 of Haller [6], there exist: ε_j, $\xi(t,x) \in L_2(0,T;H^2_0(S))$, $u_\alpha(t,x) \in L_2(0,T;H^1_0(S))$, $\bar\xi(t,x,y) \in L_2(Q;H^2_\#(Y))$, and $\bar u_\alpha(t,x,y) \in L_2(Q;H^1_\#(Y))$ such that

$$\xi^{\varepsilon_j} \overset{2-scale}{\longrightarrow} \xi, \quad \nabla\xi^{\varepsilon_j} \overset{2-scale}{\longrightarrow} \nabla\xi, \quad \Delta\xi^{\varepsilon_j} \overset{2-scale}{\longrightarrow} \Delta\xi + \Delta_y\bar\xi.$$

$$u^{\varepsilon_j}_\alpha \overset{2-scale}{\longrightarrow} u_\alpha, \quad u^{\varepsilon_j}_{\alpha x_\beta} \overset{2-scale}{\longrightarrow} u_{\alpha x_\beta} + \bar u_{\alpha y_\beta}.$$

From (8) and from the compactness of the embedding $H^2_0(S) \subset W^1_q(S)$ for any $q>1$, we conclude that $\{\xi^{\varepsilon_j}\}$ is relative compact in $C\big((0,T);W^1_q(S)\big)$ for any $q>1$ (see [9]). So, we can assume that $\xi^{\varepsilon_j} \to \xi$ in $C\big([0,T];W^1_q(S)\big)$ for any $q>1$.

To obtain the effective equation defining ξ and u_α, set

$$\varphi(t,x) = \eta(t,x) + \varepsilon^2\phi(t,x,\tfrac{x}{\varepsilon}), \qquad \psi_\alpha(t,x) = \chi_\alpha(t,x) + \varepsilon\theta_\alpha(t,x,\tfrac{x}{\varepsilon}),$$

where $\eta(t,x),\chi_\alpha(t,x) \in \overset{\circ}{C}{}^\infty_T(Q)$ and $\phi(t,x,y),\theta_\alpha(t,x,y) \in \overset{\circ}{C}{}^\infty_{0,T}(Q;C^\infty_\#(Y))$. Substituting these functions into (5) yields (we omit the index j for brevity):

$$\begin{aligned}
&\int_0^T\int_S\!\!\int \Big\{\rho\left(\tfrac{x}{\varepsilon}\right)\xi^\varepsilon\left[\eta_{tt}+\varepsilon^2\phi_{tt}\right] + \mu\left(\tfrac{x}{\varepsilon}\right)\nabla\xi^\varepsilon\left[\nabla\eta_{tt}+\varepsilon(\dots)\right]\\
&+\gamma\left(\tfrac{x}{\varepsilon}\right)\Delta\xi^\varepsilon\left[\Delta\eta+\Delta_y\phi+\varepsilon(\dots)\right] + \tau_{\alpha\beta}\left(\tfrac{x}{\varepsilon}\right)\xi^\varepsilon_{x_\beta}\left[\eta_{x_\alpha}+\varepsilon(\dots)\right]\\
&-FK_\varepsilon(t,x)I\left(\tfrac{x}{\varepsilon}\right)\left[\Delta\eta+\Delta_y\phi+\varepsilon(\dots)\right] + GK_\varepsilon(t,x)I\left(\tfrac{x}{\varepsilon}\right)\xi^\varepsilon_{x_\alpha}\left[\eta_{x_\alpha}+\varepsilon(\dots)\right]\\
&+\rho\left(\tfrac{x}{\varepsilon}\right)u^\varepsilon_\alpha\left[\chi_{\alpha tt}+\varepsilon\theta_{\alpha tt}\right] + \tau_{\alpha\beta}\left(\tfrac{x}{\varepsilon}\right)\Big[\chi_{\alpha x_\beta}+\theta_{\alpha y_\beta}+\varepsilon(\cdot\cdot)\Big] \qquad (10)\\
&+GK_\varepsilon(t,x)I\left(\tfrac{x}{\varepsilon}\right)\Big[\chi_{\alpha x_\alpha}+\theta_{\alpha y_\alpha}+\varepsilon(\cdot\cdot)\Big]\Big\}dxdt\\
&+\int_S\!\!\int\Big\{\rho\left(\tfrac{x}{\varepsilon}\right)\Big[\xi_0\eta_t(0,x)-\xi_0'\eta(0,x)+u_{\alpha 0}\chi_{\alpha t}(0,x)-u_{\alpha 0}'\chi_\alpha(0,x)\Big]\\
&+\mu\left(\tfrac{x}{\varepsilon}\right)\Big[\nabla\xi_0\nabla\eta_t(0,x)-\nabla\xi_0'\nabla\eta(0,x)\Big]\Big\}dx = 0.
\end{aligned}$$

Now we can pass to the two–scale limit in (10) under the observation that $\varepsilon(\dots)$ and $\varepsilon(\dot{\cdot}\,\cdot)$ converge uniformly to 0, and $K_\varepsilon \to K$ in $H^1(0,T;L_2(S))$. Since the last system contains two test functions, the limit system will be splitted into the two ones. The first, effective, system defining the limit functions ξ and u_α reads:

$$\int_0^T\int_S\int_Y\Big\{\rho(y)\xi\eta_{tt}+\mu(y)\nabla\xi\nabla\eta_{tt}+\gamma(y)[\Delta\xi+\Delta_y\bar{\xi}]\Delta\eta$$
$$+\ell_{ij\alpha\beta}(y)\tfrac{1}{2}\left(u_{ix_j}+u_{jx_i}+\xi_{x_i}\xi_{x_j}+\bar{u}_{iy_j}+\bar{u}_{jy_i}\right)\xi_{x_\beta}\eta_{x_\alpha}$$
$$-FK(t,x)I(y)\Delta\eta+GK(t,x)I(y)\xi_{x_\alpha}\eta_{x_\alpha}+\rho(y)u_\alpha\chi_{\alpha tt}$$
$$+\ell_{ij\alpha\beta}(y)\tfrac{1}{2}\left(u_{ix_j}+u_{ix_j}+\xi_{x_i}\xi_{x_j}+\bar{u}_{iy_j}+\bar{u}_{jy_i}\right)\chi_{\alpha x_\beta}+GK(t,x)I(y)\chi_{\alpha x_\alpha}\Big\}dydxdt$$
$$+\int_S\int_Y\Big\{\rho(y)\Big[\xi_0\eta_t(0,x)-\xi_0'\eta(0,x)+u_{\alpha 0}\chi_{\alpha t}(0,x)-u_{\alpha 0}'\chi_\alpha(0,x)\Big]$$
$$+\mu(y)\Big[\nabla\xi_0\nabla\eta_t(0,x)-\nabla\xi_0'\nabla\eta(0,x)\Big]\Big\}dydx=0.$$

The so-called cell equations defining auxiliary functions $\bar{\xi}$ and $\bar{u}_\alpha$ read:

$$\int_0^T\int_S\int_Y\Bigg\{\gamma(y)[\Delta\xi+\Delta_y\bar{\xi}]\Delta_y\phi-FK(t,x)I(y)\Delta_y\phi+ \tag{11}$$
$$\ell_{ij\alpha\beta}(y)\tfrac{1}{2}\left(u_{ix_j}+u_{jx_i}+\xi_{x_i}\xi_{x_j}+\bar{u}_{iy_j}+\bar{u}_{jy_i}\right)\theta_{\alpha y_\beta}+GK(t,x)I(y)\theta_{\alpha x_\alpha}\Bigg\}dydxdt=0.$$

Here y is the independent variable, whereas x is treated as a parameter. Since the system considered is linear, the superposition principle yields:

$$\bar{\xi}(t,x,y)=N(y)\Delta\xi+M(y)FK(t,x),$$
$$\bar{u}_i(t,x,y)=N_{imn}(y)\tfrac{1}{2}\left(u_{mx_n}+u_{nx_m}+\xi_{x_n}\xi_{x_m}\right)+M_i(y)GK(t,x).$$

where $N,M\in H^2_\#(Y)$ and $N_{imn},M_i\in H^1_\#(Y)$ are unknown functions. Substituting in (11) and some computation yield:

$$\int_0^T\int_S\int_Y\Bigg\{\Delta\xi\gamma(y)[1+\Delta_yN]+FK(t,x)\big[\gamma(y)\Delta_yM-I(y)\big]\Bigg\}\Delta_y\phi dydxdt=0,$$

$$\int_0^T\int_S\int_Y\Bigg\{d_{mn}\ell_{ij\alpha\beta}(y)\left[\delta_{im}\delta_{jn}+\frac{1}{2}\left(\frac{\partial N_{imn}}{\partial y_j}+\frac{\partial N_{jmn}}{\partial y_i}\right)\right]$$
$$+GK(t,x)\left[\ell_{ij\alpha\beta}(y)\frac{1}{2}\left(\frac{\partial M_i}{\partial y_j}+\frac{\partial M_j}{\partial y_i}\right)+\delta_{\alpha\beta}I(y)\right]\Bigg\}\theta_{\alpha y_\beta}dydxdt=0.$$

Because of the superposition principle one can seek N, M, N_{imn}, and M_i separately. Taking the test functions of the form: $\phi(t,x,y) = \phi^1(t,x)\phi^2(y)$, $\theta_\alpha(t,x,y) = \theta^1_\alpha(t,x)\theta^2_\alpha(y)$, we obtain

$$\int\!\!\int_Y \gamma(y)(1+\Delta_y N)\Delta_y\phi^2 dy = 0, \tag{12}$$

$$\int\!\!\int_Y \big(\gamma(y)\Delta_y M - I(y)\big)\Delta_y\phi^2 dy = 0, \tag{13}$$

$$\begin{aligned}
&\int\!\!\int_Y \ell_{ij1\beta}(y)\left[2\delta_{im}\delta_{jn} + \frac{\partial N_{imn}}{\partial y_j} + \frac{\partial N_{jmn}}{\partial y_i}\right]\theta^2_{1y_\beta} dy = 0,\\
&\int\!\!\int_Y \ell_{ij2\beta}(y)\left[2\delta_{im}\delta_{jn} + \frac{\partial N_{imn}}{\partial y_j} + \frac{\partial N_{jmn}}{\partial y_i}\right]\theta^2_{2y_\beta} dy = 0,
\end{aligned} \tag{14}$$

$$\begin{aligned}
&\int\!\!\int_Y \left[\ell_{ij1\beta}(y)\left(\frac{\partial M_i}{\partial y_j} + \frac{\partial M_j}{\partial y_i}\right) + 2\delta_{1\beta}I(y)\right]\theta^2_{1y_\beta} dy = 0\\
&\int\!\!\int_Y \left[\ell_{ij2\beta}(y)\left(\frac{\partial M_i}{\partial y_j} + \frac{\partial M_j}{\partial y_i}\right) + 2\delta_{2\beta}I(y)\right]\theta^2_{2y_\beta} dy = 0
\end{aligned} \tag{15}$$

for all $\phi^2 \in H^2_\#(Y)$, $\theta^2_\alpha \in H^1_\#(Y)$, $\alpha = 1,2$, $(m,n) = (1,1),(2,2),(1,2)$ (the cases $(2,1)$ and $(1,2)$ are equivalent). Equations (12),(13) and systems (14), (15) are uniquely solvable: $N, M \in H^2_\#(Y)/R$, $N_{imn}, M_i \in H^1_\#(Y)/R$.

Substituting into the effective equation yields:

$$\begin{aligned}
&\hat\rho\xi_{tt} - \hat\mu\Delta\xi_{tt} + \hat\gamma\Delta^2\xi - \frac{\partial}{\partial x_\alpha}\Big(\hat\tau_{\alpha\beta}\xi_{x_\beta}\Big) = F\hat I\Delta K(t,x) + G\hat J_{\alpha\beta}\,\frac{\partial}{\partial x_\alpha}\Big(\xi_{x_\beta}K(t,x)\Big),\\
&\hat\rho u_{\alpha tt} - \frac{\partial}{\partial x_\beta}\hat\tau_{\alpha\beta} = G\hat J_{\alpha\beta}\,\frac{\partial}{\partial x_\beta}K(t,x), \text{ where } \hat\tau_{\alpha\beta} = \hat\ell_{ij\alpha\beta}\left(u_{ix_j} + u_{jx_i} + \xi_{x_i}\xi_{x_j}\right).
\end{aligned} \tag{16}$$

The boundary and initial conditions are:

$$\xi|_{\partial S} = 0, \quad \partial\xi/\partial\vec n|_{\partial S} = 0, \quad \xi|_{t=0} = \xi_0, \quad \xi_t|_{t=0} = \xi'_0$$

$$u_\alpha|_{\partial S} = 0, \quad u_\alpha|_{t=0} = u_{\alpha 0}, \quad u_{\alpha t}|_{t=0} = u'_{\alpha 0}$$

The coefficients are:

$$\hat\rho = \langle\rho(y)\rangle, \quad \hat\mu = \langle\mu(y)\rangle, \quad \hat\gamma = \langle 1/\gamma(y)\rangle^{-1}, \quad \hat I = -\langle 1/\gamma(y)\rangle^{-1}\langle I(y)/\gamma(y)\rangle,$$

$$\hat\ell_{mn\,\alpha\beta} = \left\langle \ell_{ij\,\alpha\beta}(y)\left[\delta_{im}\delta_{jn} + \frac{1}{2}\left(\frac{\partial N_{imn}(y)}{\partial y_j} + \frac{\partial N_{jmn}(y)}{\partial y_i}\right)\right]\right\rangle,$$

$$\hat J_{\alpha\beta} = \left\langle \ell_{ij\,\alpha\beta}(y)\frac{1}{2}\left(\frac{\partial M_i(y)}{\partial y_j} + \frac{\partial M_j(y)}{\partial y_i}\right) + \delta_{\alpha\beta}I(y)\right\rangle.$$

The next proposition follows immediately from the homogenization procedure.

Proposition 4.2. *Let $(\xi^{\varepsilon_k}, u_1^{\varepsilon_k}, u_2^{\varepsilon_k})$ be a sequence of weak solutions of (1) converging to some (ξ, u_1, u_2) in the weak$*$ topology of X_∞^{211} (note that such a sequence alweas exists). Then (ξ, u_1, u_2) is a weak solution to problem (16).*

The next theorem states more useful relation between (1) and (16).

Theorem 4.3. *Let $\xi_0 = 0$, $\xi_0' = 0$, $u_{\alpha 0} = 0$, $u_{\alpha 0}' = 0$, $K \in H^2(0,T;L_2(S)) \cap L_\infty(0,T;H^1(S))$, and $K(0,\cdot) = 0$, then any limit point (ξ, u_1, u_2) of L^ε in weak-$*$ topology of X_∞^{211} is a unique strong global solution of (16), that is:*

$$\begin{aligned}\hat{\rho}(\xi_{tt},\varphi) + \hat{\mu}(\nabla\xi_{tt},\nabla\varphi) + \hat{\gamma}(\Delta\xi,\Delta\varphi)\\ +(\hat{\tau}_{\alpha\beta}\xi_{x_\beta},\varphi_{x_\alpha}) - F\hat{I}(K,\Delta\varphi) + \hat{J}_{\alpha\beta}(K\xi_{x_\beta},\varphi_{x_\alpha}) = 0,\end{aligned} \tag{17}$$

$$\hat{\rho}(u_{\alpha tt},\psi_\alpha) + (\hat{\tau}_{\alpha\beta},\psi_{\alpha x_\beta}) + G\hat{J}_{\alpha\beta}(K,\psi_{\alpha x_\beta}) = 0 \tag{18}$$

for any $\varphi \in H_0^2(S)$,$\psi_\alpha \in H_0^1(S)$, and a.e. $t \in [0,T)$. It possesses the properties

$$\begin{aligned}(\xi, u_1, u_2) \in C([0,T];H_0^2(S)) \times C([0,T];H_0^1(S)) \times C([0,T];H_0^1(S)),\\ (\xi_t, u_{1t}, u_{2t}) \in C([0,T];H_0^1(S)) \times C([0,T];L_2(S)) \times C([0,T];L_2(S)),\end{aligned} \tag{19}$$

$$\xi \in L_\infty(0,T;H^3(S)), \quad u_1, u_2 \in L_\infty(0,T;H^2(S))). \tag{20}$$

Proof. Remember that (ξ, u_1, u_2) is the limit point of a sequence $(\xi^{\varepsilon_k}, u_1^{\varepsilon_k}, u_2^{\varepsilon_k}) \in L^{\varepsilon_k}$ in weak-$*$ topology of X_∞^{211}. Using proposition 3.3, we conclude that

$$(\xi, u_1, u_2) \in X_\infty^{211}, \quad (\xi_t, u_{1t}, u_{2t}) \in X_\infty^{211}, \quad (\xi_{tt}, u_{1tt}, u_{2tt}) \in X_\infty^{100} \tag{21}$$

Using theorem 16.1 of [10], we obtain (19). Therefore, (ξ, u_1, u_2) is a strong solution of (17), (18).

Reproducing the proof of lemma 5.2 of [7], one can show that any strong solution of (17), (18) possesses the property (20). This implies uniqueness of strong solutions of (17), (18). The proof of the uniqueness is rather trivial. Assume that there exist two solutions ξ^1, u_α^1 and ξ^2, u_α^2. Using the Gronwall–Lemma and the property (20), show that $\xi^1 - \xi^2 = 0$, $u_\alpha^1 - u_\alpha^2 = 0$. □

Corollary 4.4. *Let the assumptions of theorem 4.3 hold. Then the sets L^ε shrink to a unique solution (ξ, u_1, u_2) of (17), (18) as $\varepsilon \to 0$. That is, any sequence $(\xi^\varepsilon, u_1^\varepsilon, u_2^\varepsilon)$ such that $(\xi^\varepsilon, u_1^\varepsilon, u_2^\varepsilon) \in L^\varepsilon$ converges to (ξ, u_1, u_2) in weak-$*$ topology of $L_\infty(0,T;H_0^2(S)) \times L_\infty(0,T;H_0^1(S)) \times L_\infty(0,T;H_0^1(S))$ as $\varepsilon \to 0$.*

Proof. Assume that this claim is not true. Then one can find a subsequence $(\xi^{\varepsilon_k}, u_1^{\varepsilon_k}, u_2^{\varepsilon_k}) \in L^{\varepsilon_k}$ separated from (ξ, u_1, u_2).Then there exists a sub-subsequence $(\xi^{\varepsilon_{k_l}}, u_1^{\varepsilon_{k_l}}, u_2^{\varepsilon_{k_l}}) \in L^{\varepsilon_{k_l}}$ that converges to some (ξ^*, u_1^*, u_2^*). Theorem 4.3 provides that $(\xi^*, u_1^*, u_2^*) = (\xi, u_1, u_2)$, which is a contradiction. □

Corollary 4.5. *Let* $\xi_0 = 0$, $\xi_0' = 0$, $u_{\alpha 0} = 0$, $u_{\alpha 0}' = 0$, $K \in C^2([0,T]; L_2(S)) \cap C^1([0,T]; H^1(S))$, *and* $K(0,\cdot) = 0$, *then any limit point* (ξ, u_1, u_2) *of* L^ε *is a unique strong global solution of (16) possessing the properties (21) and*

$$\begin{aligned} \xi \in C([0,T]; H^3(S) \cap H_0^2(S)) \cap C^1([0,T]; H_0^2(S)) \cap C^2([0,T]; H_0^1(S)), \\ u_1, u_2 \in C([0,T]; H^2(S) \cap H_0^1(S)) \cap C^1([0,T]; H_0^1(S). \end{aligned} \tag{22}$$

This follows from theorem 5.1 of [7] and from the uniqueness of a strong global solution to (16).

References

[1] H. T. Banks, R. C. Smith and Y. Wang, *Smart material structures: modeling, estimation and control,* J. Wiley & Sons (Chichester-New York) **1996**.

[2] K. -H. Hoffmann and N. D. Botkin, *Oscillations of nonlinear thin plates excited by piezoelectric patches,* ZAMM: Z. angew. Math. Mech., **78** (1998), to appear.

[3] G. Nguetseng, *A general convergence result for a functional related to the theory of homogenization,* SIAM J. Math. Anal., **20 (3)** (1989), 608–623.

[4] G. Allaire, *Homogenization and two-scale Convergence,* SIAM J. Math. Anal., **23 (6)** (1992), 1482–1518.

[5] G. Allaire, *Homogenization of the unsteady Stokes equations in porous media,* in "Progress in partial differential equations: calculus of variations, applications" Pitman Research Notes in Mathematics Series. C. Bandle et al. eds, Longman Higher Education, New York. **267**, (1992), 109–123.

[6] H. Haller, *Verbundwerkstoffe mit Formgedächtnislegierung – Mikromechanische Modellierung und Homogenisierung,* Dissertation, TU-München **1997**.

[7] J-P. Puel and M. Tucsnak, *Global Existence for the Full Kármán System,* Appl. Math. Optimiz., **34** (1996), 139–160.

[8] M.A. Horn and I. Lasiecka, *Global Stabilization of a Dynamic von Kármán Plate with Nonlinear Boundary Feedback,* Appl. Math. Optimiz. **31** (1995), 57–84.

[9] J. Simon, *Compact sets in the space* $L^p(0,T;b)$, Ann. Mat. Pura Appl., **IV (146)** (1987), 65–96.

[10] J. L. Lions and E. Magenes, *Non-homogeneous Boundary Value Problems and applications I,* Springer-Verlag (Berlin-Heidelberg-New York) **1972**.

CAESAR
Center of Advanced European Studies and Research,
Friedensplatz 16,
D-53111 Bonn
E-mail address: hoffmann@caesar.de
E-mail address: botkin@caesar.de

International Series of Numerical Mathematics
Vol. 133,

Stabilization of the Dynamic System of Elasticity by Nonlinear Boundary Feedback

Mary Ann Horn [1]

Abstract. Uniform stabilization for the system of isotropic linear elasticity is established using nonlinear velocity feedback via traction forces on the boundary. Through the use of sharp trace regularity results and a nonlinear compactness/uniqueness argument, a proof is derived without imposition of strong geometric restrictions on the controlled portion of the boundary, thus extending the original work of Lagnese [8].

1. Introduction

Stabilization of the linear system of isotropic elasticity using nonlinear boundary feedback was initially addressed by Lagnese [8]. However, to facilitate the proof of the estimates required, the feedback was modified by including the tangential derivative of the displacement. This modification addressed a mathematical technicality in the proof, but was not motivated by the physical problem modelled by the system. With this feedback, the system is shown to be uniformly stable provided the body satisfies a number of assumptions including the following. First, the body is in a state of plane strain, thus allowing the author to reduce the problem to a two-dimensional system. Second, with control acting on the entire boundary, the domain is assumed to be star shaped.

Following Lagnese's work, considerable attention has been focused on both the questions of controllability and stabilization of the three dimensional system of elasticity within the last decade. Komornik proved uniform decay of the energy assuming modified boundary conditions and modified energy [6] and obtains precise decay rates which are optimal if the domain is a sphere. However, the question of uniform stabilization when feedback control is acting through natural and physically implementable boundary conditions remained open.

To address this question, Alabau and Komornik [1] used a constructive proof based on the method developed in [7] for the wave equation. This technique allowed the authors to obtain precise decay estimates, however, they were forced to assume that the domain was a sphere. But these results are not only valid for isotropic systems, they are applicable to nonisotropic elasticity as well. An extension of

[1] Partially supported by National Science Foundation Grant NSF DMS-9803547.

this work to include nonlinear boundary feedback was later proven by Guesmia [3] with similar geometric assumptions as well as growth assumptions on the nonlinear control.

Although new results were becoming available, all required stringent geometric assumptions on the domain, even in the case when control is acting over the entire boundary. Thus, whether uniform stabilization could be achieved on more general domains remained an open question. An answer was found by Horn [4, 5], where sharp trace regularity estimates are the key to removing the geometric restrictions imposed on the controlled portion of the boundary in earlier work. Additionally, the feedback is a linear function of the velocity and does not involve the tangential derivative of the displacement.

In this paper, these results are extended to nonlinear velocity feedback, thus giving a fuller answer to the question of uniform stabilization initially addressed by Lagnese [8]. Additionally, no growth conditions are required to be imposed at the origin, unlike the restrictions assumed by Guesmia [3].

2. The System of Elasticity and Uniform Stability

Assuming that the elastic body is homogeneous and isotropic, the system of elasticity can be formulated as follows. Let $u = (u_i)$, $1 \leq i \leq n$ be the displacement vector. With the strain tensor (ϵ_{ij}) given by

$$\epsilon_{ij}(u) \equiv \frac{1}{2}\left(\frac{\partial u_i}{\partial x_i} + \frac{\partial u_j}{\partial x_i}\right), \qquad 1 \leq i, j \leq n, \tag{1}$$

the stress-strain relation is

$$\sigma_{ij}(u) = \lambda \sum_{k=1}^{n} \epsilon_{kk}\delta_{ij} + 2\mu\epsilon_{ij} = \lambda(\text{div } u)\delta_{ij} + \mu\left(\frac{\partial u_i}{\partial x_i} + \frac{\partial u_j}{\partial x_i}\right), \tag{2}$$

where λ, $\mu > 0$ are Lamé's coefficients and are assumed to be constant. In the above equation, δ_{ij} is the Kronecker delta, i.e., $\delta_{ij} = 1$ if $i = j$ and $\delta_{ij} = 0$ if $i \neq j$. Let $\Omega \subset \mathbb{R}^n$ be an open domain with smooth boundary $\partial\Omega = \partial\Omega_0 \cup \partial\Omega_1$. Assuming that $\partial\Omega_0 \cap \partial\Omega_1 = \emptyset$, the governing equations are given by

$$u_{tt} - \nabla \cdot \sigma(u) = 0 \quad \text{in } \Omega \times (0,T), \tag{3.a}$$

$$u = 0 \quad \text{on } \partial\Omega_0 \times (0,T), \tag{3.b}$$

$$\sigma(u)\nu = -g(u_t) \quad \text{on } \partial\Omega_1 \times (0,T), \tag{3.c}$$

$$u(x,0) = \psi_0(x);\ u_t(x,0) = \psi_1(x) \quad \text{in } \Omega, \tag{3.d}$$

where ν represents the unit outward normal vector to $\partial\Omega$.

Within the context of this problem, the control, g, is a continuous vector-valued function, with each component assumed to be monotone increasing and zero at the origin and is subject to the following constraints:

$$\begin{array}{ll} g(s) \cdot s > 0 & \text{for } s \neq 0 \\ m|s| \leq |g(s)| \leq M|s| & \text{for } |s| > 1 \end{array} \tag{4}$$

As the questions of existence and uniqueness have been addressed by Lagnese and others (see, e.g., [3, 8]), they are stated without proof. To state these results, the function space $\mathcal{H}^1_{\partial\Omega_0}(\Omega)$ is defined by

$$\mathcal{H}^1_{\partial\Omega_0}(\Omega) \equiv \{v \in (H^1(\Omega))^n | \ v = 0 \text{ on } \partial\Omega_0\}. \tag{5}$$

Theorem 2.1. (Well-posedness.) *Assume that g is continuously differentiable. For any $(\psi_0(x), \psi_1(x)) \in (\mathcal{H}^1_{\partial\Omega_0}(\Omega) \times (L^2(\Omega))^n)$ and $T > 0$, there exists a unique solution (in the sense of distributions),*

$$(u(x,t), u_t(x,t)) \in C([0,T]; \mathcal{H}^1_{\partial\Omega_0}(\Omega) \times (L^2(\Omega))^n) \tag{6}$$

satisfying system (3).

If, addition, $(\psi_0(x), \psi_1(x)) \in ((H^2(\Omega) \cap H^1_{\partial\Omega_1}(\Omega))^n \times \mathcal{H}^1_{\partial\Omega_0}(\Omega))$, and satisfy the compatibility condition,

$$\sigma(\psi_0)\nu = -g(\psi_1) \quad \text{on } \partial\Omega_1, \tag{7}$$

then the solution has the additional regularity,

$$\begin{aligned} &u \in C([0,T]; (H^2(\Omega) \cap H^1_{\partial\Omega_1}(\Omega))^n), \\ &u_t \in C([0,T]; \mathcal{H}^1_{\partial\Omega_0}(\Omega), \\ &u_{tt} \in C([0,T]; (L^2(\Omega))^n), \end{aligned} \tag{8}$$

and

$$g(u_t) \in C([0,T]; (L^2(\Omega))^n). \tag{9}$$

Returning to the goal of establishing uniform decay of the energy to zero, the energy corresponding to the system of elasticity is defined by

$$E(t) \equiv \frac{1}{2}\left\{\|u_t\|^2_{(L^2(\Omega))^n} + (\sigma(u), \epsilon(u))_\Omega\right\}, \tag{10}$$

which is topologically equivalent to the usual topology on $(H^1(\Omega))^n \times (L^2(\Omega))^n$.

To state the stability result, the following definitions are needed. Let the function $\mathcal{G}(s)$ be a concave, strictly increasing, zero at the origin, such that

$$\mathcal{G}(sg(s)) \geq |s|^2 + |g(s)|^2 \quad \text{for } |s| \leq 1. \tag{11}$$

(Such a function be easily constructed, see [9].) Define

$$\tilde{\mathcal{G}}(x) \equiv \mathcal{G}\left(\frac{x}{\text{meas }(\partial\Omega \times (0,T))}\right). \tag{12}$$

Since $\tilde{\mathcal{G}}$ is monotone increasing, for every $c \geq 0$, the operator $cI + \tilde{\mathcal{G}}$ is invertible. Setting

$$p(x) \equiv (cI + \tilde{\mathcal{G}})^{-1}(Kx), \tag{13}$$

where K is a positive constant, p is a positive, continuous, strictly increasing function with $p(0) = 0$. With these definitions, the following theorem holds.

Theorem 2.2. (Uniform stability.) *Let $u(x,t)$ be a solution of system (3) and let $\vec{h}(x) \equiv x - x_0$, where $x_0 \in \mathbb{R}^n$. Assume that*

$$\vec{h}(x) \cdot \nu \leq 0 \text{ on } \partial\Omega_0. \tag{14}$$

Then for some $T_0 > 0$,

$$E(t) \leq C\mathcal{S}\left(\frac{t}{T_0} - 1\right) \quad \text{for } t > T_0, \tag{15}$$

where $\mathcal{S}(t) \to 0$ as $t \to \infty$ and is the solution (contraction semigroup) of the differential equation,

$$\begin{aligned} \tfrac{d}{dt}\mathcal{S}(t) + q(\mathcal{S}(t)) &= 0, \\ \mathcal{S}(0) &= E(0), \end{aligned} \tag{16}$$

and $q(x)$ is given by

$$q(x) \equiv x - (I-p)^{-1}(x) \quad \text{for } x > 0. \tag{17}$$

3. Proof of Theorem 2.2

As the multiplier identities and inequalities proceed analogously to those in Horn [4], only the adjustments needed to account for nonlinear control will be detailed below.

For the system of elasticity given in (3), the following variational formulation is valid:

$$(u_{tt}, \phi)_\Omega + (\sigma(u), \epsilon(\phi))_\Omega + (g(u_t), \phi)_{\partial\Omega_1} - (\sigma(u)\nu, \phi)_{\partial\Omega_0} = 0 \tag{18}$$

for $\phi \in (H^1(\Omega))^n$.

Lemma 3.1. (Dissipativity Identity.) *Let u be a weak solution of system (3). Then for any $s < t$, the following identity holds:*

$$E(t) + \int_s^t (g(u_t), u_t)_{\partial\Omega_1} dt = E(s). \tag{19}$$

Thus, the energy of the system is nonincreasing.

Proof. Let $\phi \equiv u_t$ in the variational formulation (18) and integrate with respect to time. □

3.1. Stability Estimate

To prove uniform stability, a critical step lies in proving the following stability estimate. Once this technical estimate has been established, the proof of uniform stabilization follows using the techniques of Lasiecka and Tataru [9] for nonlinear problems.

Lemma 3.2. (Stability Estimate.) *Let u be a strong solution to system (3). Assume there exists a vector $\mathbf{h}(x)$ such that*

$$\mathbf{h}(x) \cdot \nu \leq 0 \qquad \text{on } \partial\Omega_0. \tag{20}$$

Then there exists a sufficiently large time T and a constant $C_T(E(0))$, dependent upon T and possibly upon the initial energy, $E(0)$, such that the following estimate is satisfied:

$$E(T) \leq C_T(E(0)) \int_0^T \int_{\partial\Omega_1} \left(|u_t|^2 + |g(u_t)|^2\right) dx dt. \tag{21}$$

To prove this stability estimate, the proof is divided into a number of steps. To begin, a preliminary identity is established, giving a bound on the energy in terms of the control, the velocity, traces of the displacement on the boundary and lower order terms.

3.1.1. Multiplier Estimates To establish the estimate in (21), two primary multipliers are used. First, let $\phi = \nabla u \mathbf{h}$ be the test function in the variational formulation (18). Then, integrating with respect to t yields

$$\begin{aligned} \int_0^T (u_{tt}, \nabla u\mathbf{h})_\Omega dt \quad &+ \int_0^T (\sigma(u), \epsilon(\nabla u\mathbf{h}))_\Omega \\ &+ \int_0^T (g(u_t), \nabla u\mathbf{h})_{\partial\Omega_1} dt - \int_0^T (\sigma(u)\nu, \nabla u\mathbf{h})_{\partial\Omega_0} dt = 0. \end{aligned} \tag{22}$$

By integrating the first integral by parts, using the fact that $\mathbf{h}(x)$ is a radial vector, simplifying the term involving $\epsilon(\nabla u\mathbf{h})$ using the identity (see [4]),

$$\epsilon(\nabla u\mathbf{h}) = \epsilon(u) + \mathcal{R}, \tag{23}$$

where the components of the first order tensor $\mathcal{R}$ are given by

$$\mathcal{R}_{ij} \equiv \frac{1}{2}\left(h_k u_{i,kj} + h_k u_{j,ki}\right), \tag{24}$$

and taking advantage of the boundary condition in (3.b), the result is

$$\begin{aligned} (u_t, \nabla u\mathbf{h})_\Omega \big|_0^T \quad &-\tfrac{1}{2}\int_0^T \int_{\partial\Omega_1} |u_t|^2 \mathbf{h}\cdot\nu dx dt + \tfrac{n}{2}\int_0^T \int_\Omega |u_t|^2 dx dt \\ &+ \int_0^T (\sigma(u), \epsilon(u))_\Omega dt + \int_0^T (\sigma_{ij}, u_{i,kj} h_k)_\Omega dt \\ &+ \int_0^T (g(u_t), \nabla u\mathbf{h})_{\partial\Omega_1} dt - \int_0^T (\sigma(u), \epsilon(u)\mathbf{h}\cdot\nu)_{\partial\Omega_0} dt = 0 \end{aligned} \tag{25}$$

In both of the previous equations, the summation convention is adopted to indicate summation with respect to k.

Noting that

$$2\sigma_{ij} u_{i,kj} h_k + n\sigma(u)\cdot\epsilon(u) = (\nabla\cdot\sigma(u))\cdot\epsilon(u)\mathbf{h}, \tag{26}$$

and recalling that $\mathbf{h}\cdot\nu \leq 0$ on $\partial\Omega_0$, the above identity can be further simplified through application of the divergence theorem. Hence, after estimating terms arising from evaluation at $t = 0$ and $t = T$, the resulting inequality is

$$\begin{aligned}\frac{n}{2}\int_0^T\int_\Omega |u_t|^2 dxdt + \int_0^T (\sigma(u),\epsilon(u))_\Omega dt - \frac{n}{2}\int_0^T (\sigma(u),\epsilon(u))_\Omega dt \\ \leq C\left(E(0)+E(T)\right) + \frac{1}{2}\int_0^T\int_{\partial\Omega_1} |u_t|^2 \mathbf{h}\cdot\nu dxdt \\ -\int_0^T (g(u_t),\nabla u\mathbf{h})_{\partial\Omega_1} dt - \int_0^T (\sigma(u),\epsilon(u)\mathbf{h}\cdot\nu)_{\partial\Omega_1} dt.\end{aligned} \tag{27}$$

For the second multiplier, let $\phi = u$ be the test function in the variational formulation. Then by integrating with respect to time and taking into account the boundary conditions, the following identity is obtained:

$$\int_0^T (\sigma(u),\epsilon(u))_\Omega dt = \int_0^T \|u_t\|^2_{(L^2(\Omega))^n} dt - (u_t,u)_\Omega\,\Big|_0^T - \int_0^T (g(u_t),u)_{\partial\Omega_1} dt. \tag{28}$$

Combining the identities in (27) and (28) using trace theory to estimate terms on the boundary and recalling the definition of the energy of the system results in the inequality,

$$\begin{aligned}\int_0^T E(t)dt &\leq C\left(E(0)+E(T)\right) + \int_0^T\int_{\partial\Omega_1} |u_t|^2\mathbf{h}\cdot\nu dxdt \\ &\quad -\int_0^T (g(u_t),\nabla u\mathbf{h})_{\partial\Omega_1} dt - \int_0^T (\sigma(u),\epsilon(u)\mathbf{h}\cdot\nu)_{\partial\Omega_1} dt \\ &\quad +C\left[(u_t,u)_\Omega\,\Big|_0^T + \int_0^T (g(u_t),u)_{\partial\Omega_1} dt\right] \\ &\leq C\left\{E(0)+E(T) + \int_0^T\int_{\partial\Omega_1}\left(|u_t|^2+|g(u_t)|^2\right)dxdt \right. \\ &\quad \left. + \int_0^T\int_{\partial\Omega_1} |\nabla u|^2 dxdt + l.o.t.(u)\right\},\end{aligned} \tag{29}$$

where

$$l.o.t.(u) \equiv \int_0^T \|u\|^2_{(H^{1-\epsilon}(\Omega))^n} dt, \qquad 0<\epsilon<\frac{1}{4}. \tag{30}$$

3.1.2. Absorption of Boundary Traces To avoid the imposition of strong geometric constraints on the controlled portion of the boundary, the boundary traces arising in the preliminary estimate (29) must be bounded in terms of the control. To do so, sharp trace regularity estimates derived in Horn [5] must be used. These estimates permit the tangential derivative of the displacement to be bounded in terms of the control, the velocity and lower order terms and are stated in the following theorem.

Theorem 3.3. (Trace Regularity [5].) *Let u be the solution to (3) and let $0<\alpha<\frac{T}{2}$. Then u satisfies the following inequality:*

$$\begin{aligned}\|\nabla u\tau\|^2_{(L^2(\alpha,T-\alpha,\partial\Omega_1))^n} &\leq C\left\{\|u_t\|^2_{(L^2(0,T;\partial\Omega_1))^n}\right. \\ &\quad \left. + \|\sigma(u)\nu\|^2_{(L^2(0,T;\partial\Omega_1))^n} + l.o.t.(u)\right\},\end{aligned} \tag{31}$$

where ν and τ are, respectively, the unit normal and a unit tangent to the boundary.

While this estimate on the tangential derivative is on a shorter time interval than $[0,T]$, this bound avoids the use of trace theory. If trace theory were directly applied, the bounds on the tangential derivative would involve norms higher than the energy norm, which are not useful in proving (21).

To apply Theorem 3.3, it should be noted that the result of (29) is valid on the interval $(\alpha, T-\alpha)$. Therefore,

$$\begin{aligned}\int_\alpha^{T-\alpha} E(t)dt \quad &\leq C_T \{E(\alpha) + E(T-\alpha) \\ &+ \int_\alpha^{T-\alpha} \int_{\partial\Omega_1} \left(|u_t|^2 + |g(u_t)|^2\right) dxdt \\ &+ \int_\alpha^{T-\alpha} \int_{\partial\Omega_1} |\nabla u|^2 dxdt + l.o.t.(u)\Big\}.\end{aligned} \tag{32}$$

Noting that $\nabla u \equiv \nabla u\nu + \sum_{k=1}^{n-1} \nabla u\tau_k$, where $\{\nu, \tau_k\}_{k=1}^{n-1}$ form an orthonormal set of normal and tangent vectors,

$$\int_\alpha^{T-\alpha} \int_{\partial\Omega_1} |\nabla u|^2 dxdt \leq \int_\alpha^{T-\alpha} \int_{\partial\Omega_1} |\nabla u\nu|^2 dxdt + \sum_{k=1}^{n-1} \int_\alpha^{T-\alpha} \int_{\partial\Omega_1} |\nabla u\tau_k|^2 dxdt \tag{33}$$

To obtain a bound for the normal derivative in terms of the boundary condition, we denote $\mathbf{d}_k \equiv \nabla u\tau_k$ and write

$$\sigma(u)\nu = -g(u_t) \tag{34}$$

$$\nabla u\tau_k = \mathbf{d}_k, \tag{35}$$

resulting in the algebraic linear system,

$$A\mathbf{u} = (-g(u_t), \mathbf{d}_1, \ldots, \mathbf{d}_{n-1})^T, \tag{36}$$

where $\mathbf{u} \equiv (u_{1,1}, \ldots, u_{1,n}, u_{2,1}, \ldots, u_{2,n}, \ldots, u_{n,1}, \ldots, u_{n,n})$ and the determinant of the matrix A is nonzero. Solving the above system and integrating the result over $\partial\Omega_1 \times (\alpha, T-\alpha)$ yields the inequality,

$$\sum_{k=1}^{n} \int_\alpha^{T-\alpha} \int_{\partial\Omega_1} |D_k u|^2 dxdt \leq C \int_\alpha^{T-\alpha} \int_{\partial\Omega_1} \{|g(u_t)|^2 + \sum_{k=1}^{n-1} |\mathbf{d}|^2\} dxdt. \tag{37}$$

This estimate, together with the application of Theorem 3.3 and recollection that the energy of the system is dissipative, yields

$$(T-2\alpha)E(T) \leq C \left\{ E(0) + \int_0^T \int_{\partial\Omega_1} \left(|u_t|^2 + |g(u_t)|^2\right) dxdt + l.o.t.(u) \right\} \tag{38}$$

Recalling the dissipativity identity (19), $E(0)$ may be written in terms of $E(T)$ and a boundary integral. Thus, taking T to be sufficiently large gives

$$E(T) \leq C_T \left\{ \int_0^T \int_{\partial\Omega_1} \left(|u_t|^2 + |g(u_t)|^2\right) dxdt + l.o.t.(u) \right\} \tag{39}$$

3.1.3. Elimination of Lower Order Terms To complete the proof of Lemma 3.2, the lower order terms appearing on the right-hand side of (39) must be bounded in terms of the control and the velocity on the boundary. A nonlinear compactness/uniqueness argument is used which relies on unique continuation results of Dehman and Robbiano [2] for the corresponding static system of elasticity.

Lemma 3.4. *Let u be a solution to system (3). Then there exists a constant, $C_T(E(0))$, dependent upon T and possibly upon the initial energy, $E(0)$, such that the following estimate is satisfied:*

$$l.o.t.(u) \leq C_T(E(0)) \int_0^T \int_{\partial\Omega_1} \left(|u_t|^2 + |g(u_t)|^2\right) dx dt. \tag{40}$$

Proof. Assume that inequality (40) is not valid. Then there exists a sequence of solutions, $\{u_m\}_{m=1}^{\infty}$ satisfying system (3) with initial data $(\psi_{m,0}, \psi_{m,1})$ such that

$$\lim_{n\to\infty} \frac{l.o.t.(u)}{\int_0^T \int_{\partial\Omega_1} (|u_t|^2 + |g(u_t)|^2) \, dx dt} = \infty. \tag{41}$$

From the boundedness of the initial energy, $E(0) \leq M$, the sequence u_n satisfies $E_n(t) \leq M$, where $E_n(t)$ denotes the energy corresponding to the solution $u_n(x,t)$. Therefore,

$$\begin{array}{lll} u_n & \longrightarrow u & \text{in } L^\infty([0,T];(H^1(\Omega))^n) \text{ weakly star,} \\ u_{n,t} & \longrightarrow u_t & \text{in } L^\infty([0,T];(H^1(\Omega))^n) \text{ weakly star,} \end{array} \tag{42}$$

which, due to the compactness of the lower order terms with respect to the topology induced by the energy, implies $l.o.t.(u_n) \longrightarrow l.o.t.(u)$.
Case 1. Assume $l.o.t.(u) \neq 0$. Then

$$\int_0^T \int_{\partial\Omega_1} \left(|u_t|^2 + |g(u_t)|^2\right) dx dt \longrightarrow 0, \tag{43}$$

which, in turn, implies that

$$\begin{array}{ll} u_{n,t} \longrightarrow 0 & \text{in } (L^2(\partial\Omega \times (0,T)))^n, \\ g(u_{n,t}) \longrightarrow 0 & \text{in } (L^2(\partial\Omega \times (0,T)))^n. \end{array} \tag{44}$$

Passing with the limit as $n \to \infty$ on the system for u_n, the limit function, u, must satisfy system (3) with homogeneous boundary data, together with the overdetermined boundary condition,

$$u_t = 0 \qquad \text{on } \partial\Omega_1 \times (0,T). \tag{45}$$

From Horn [4], the only solution of this overdetermined system is $u \equiv 0$, contradicting the assumption, $l.o.t.(u) \neq 0$.
Case 2. Assume $l.o.t.(u) = 0$. Define $c_n \equiv l.o.t.(u_n)$ and $\tilde{u}_n \equiv u_n/c_n$. Then

$$\begin{array}{l} l.o.t.(\tilde{u}_n) = 1, \\ \frac{1}{c_n^2} \int_0^T \int_{\partial\Omega_1} \left(|u_t|^2 + |g(u_t)|^2\right) dx dt \longrightarrow 0, \end{array} \tag{46}$$

which implies

$$\begin{array}{ll} \tilde{u}_{n,t} \longrightarrow 0 & \text{in } (L^2(\partial\Omega \times (0,T)))^n, \\ \frac{1}{c_n} g(u_{n,t}) \longrightarrow 0 & \text{in } (L^2(\partial\Omega \times (0,T)))^n. \end{array} \tag{47}$$

Noting that $\tilde{u}_n$ satisfies the system,

$$\begin{array}{rl} \tilde{u}_{n,tt} - \nabla \cdot \sigma(\tilde{u}_n) = 0 & \text{in } \Omega \times (0,T), \\ \tilde{u}_n = 0 & \text{on } \partial\Omega_0 \times (0,T), \\ \sigma(\tilde{u}_n)\nu = -\frac{1}{c_n} g(u_{n,t}) & \text{on } \partial\Omega_1 \times (0,T), \\ u(x,0) = \psi_0(x);\ u_t(x,0) = \psi_1(x) & \text{in } \Omega, \end{array} \tag{48}$$

the estimate in (39) and convergence in (46) yield

$$\frac{1}{c_n^2} E_n(T) \le C_T(E_n(0)) \left\{ \frac{1}{c_n^2} \int_0^T \int_{\partial\Omega_1} \left(|u_t|^2 + |g(u_t)|^2\right) dxdt + l.o.t.(\tilde{u}_n) \right\} \le M. \tag{49}$$

Therefore,

$$\begin{array}{l} \|\tilde{u}_n(t)\|_{(H^1(\Omega))^n} \le C, \\ \|\tilde{u}_n(t)\|_{(L^2(\Omega))^n} \le C, \end{array} \tag{50}$$

resulting in the convergence,

$$\begin{array}{lll} \tilde{u}_n \longrightarrow \tilde{u} & \text{in } L^\infty([0,T];(H^1(\Omega))^n) \text{ weakly star}, \\ \tilde{u}_{n,t} \longrightarrow \tilde{u}_t & \text{in } L^\infty([0,T];(L^2(\Omega))^n) \text{ weakly star}, \end{array} \tag{51}$$

which implies

$$l.o.t.(\tilde{u}_n) \longrightarrow l.o.t.(\tilde{u}) = 1. \tag{52}$$

Additionally, the above convergences imply that $\tilde{u}$ satisfies system (3) with homogeneous boundary conditions, as well as the overdetermined boundary condition (45). Hence, as in Case 1, the solution must be $\tilde{u} \equiv 0$, contradicting (52). □

3.2. Completion of the Proof of Theorem 2.2

Now that the stability estimate has been established, the remainder of the proof of Theorem 2.2 proceeds as follows. Denote $\Sigma_A \equiv \{(x,t) \in \partial\Omega \times (0,T) | \ |u_t| \le 1\}$ and $\Sigma_B \equiv \{(x,t) \in \partial\Omega \times (0,T) | \ |u_t| > 1\}$. Then, if $\mathcal{G}$ is a concave, strictly increasing function, zero at the origin, such that

$$\mathcal{G}(s \cdot g(s)) \ge |s|^2 + |g(s)|^2 \qquad |s| \le 1, \tag{53}$$

the following inequality holds:

$$\begin{aligned} \int_0^T \int_{\partial\Omega_1} \left(|u_t|^2 + |g(u_t)|^2\right) dxdt \quad &\le \int_{\Sigma_A} \mathcal{G}(u_t \cdot g(u_t)) dxdt + C \int_{\Sigma_B} g(u_t) \cdot u_t dxdt \\ &\le \int_0^T \int_{\partial\Omega} \left(\mathcal{G}(u_t \cdot g(u_t)) + C g(u_t) \cdot u_t\right) dxdt. \end{aligned} \tag{54}$$

Using Jensen's inequality,

$$\begin{aligned} \int_0^T \int_{\partial\Omega_1} \left(|u_t|^2 + |g(u_t)|^2\right) dxdt \quad &\le \mathcal{G}\left(\int_0^T \int_{\partial\Omega} g(u_t) \cdot u_t dxdt\right) \\ &\quad + C \int_0^T \int_{\partial\Omega} g(u_t) \cdot u_t dxdt. \end{aligned} \tag{55}$$

Defining $\mathcal{F} \equiv \int_0^T \int_{\partial\Omega} g(u_t) \cdot u_t dx dt$, the stability inequality (21) becomes

$$E(T) \leq C_T(E(0)) \left(\mathcal{F} + \mathcal{G}(\mathcal{F})\right). \tag{56}$$

Therefore, since $\mathcal{G}$ is monotone,

$$\frac{1}{C_T(E(0))}(I + \mathcal{G})^{-1} E(T) \leq \mathcal{F} = E(0) - E(T), \tag{57}$$

implying that for an appropriately defined monotone function p,

$$p(E(T)) + E(T) \leq E(0). \tag{58}$$

With the monotonicity of p, the results of Lasiecka and Tataru [9] can now be applied to obtain uniform stability of system (3).

References

[1] F. Alabau and V. Komornik, *Boundary observability, controllability and stabilization of linear elastodynamic systems,* Institut de Recherche Mathématique Avancée Preprint Series, **15** (1996).

[2] B. Dehman and L. Robbiano, *La propriété du prolongement unique pour un système elliptique. Le système de Lamé,* J. Math. Pures Appl., **72** (1993), 475–492.

[3] A. Guesmia, *Stabilisation frontière d'un système d'élasticité,* C. R. Acad. Sci. Paris, Série I, **324** (1997), 1355–1360.

[4] M. A. Horn, *Implications of sharp trace regularity results on boundary stabilization of the system of linear elasticity,* J. Math. Anal. Appl., **223** (1998), 126–150.

[5] M. A. Horn, *Sharp trace regularity for the solutions of the equations of dynamic elasticity,* J. Math. Systems, Estim. Control, **8** (1998), 217–219.

[6] V. Komornik, *Boundary stabilization of isotropic elasticity systems,* Control of Partial Differential Equations and Applications, Lecture Notes in Pure and Applied Mathematics, **174** (1996) Marcel Dekker.

[7] V. Komornik and E. Zuazua, *A direct method for the boundary stabilization of the wave equation,* J. Math. Pures Appl., **69** (1990), 33–54.

[8] J. Lagnese, *Uniform asymptotic energy estimates for solutions of the equations of dynamic plane elasticity with nonlinear dissipation at the boundary,* Nonlinear Analysis, Theory, Methods & Applications, **16** (1991), 35–54.

[9] I. Lasiecka and D. Tataru, *Uniform boundary stabilization of semilinear wave equations with nonlinear boundary damping,* Differential and Integral Equations, **6** (1993), 507–533.

Department of Mathematics
Vanderbilt University
1326 Stevenson Center
Nashville, Tennessee 37240
U.S.A.
E-mail address: `horn@math.vanderbilt.edu`

International Series of Numerical Mathematics
Vol. 133,

Griffith Formula and Rice–Cherepanov's Integral for Elliptic Equations with Unilateral Conditions in Nonsmooth Domains

A.M. Khludnev and J. Sokolowski

Abstract. The Poisson equation and linear elasticity equations in two-dimensional case for a nonsmooth domain are considered. The geometrical domain has a cut (a crack) of variable length. At the crack faces, inequality type boundary conditions are prescribed. The behaviour of the energy functional is analyzed with respect to the crack length changes. In particular, the derivative of the energy functional with respect to the crack length is obtained. The associated Griffith formula is derived and properties of the solution are investigated. It is shown that the Rice-Cherepanov's integral defined for the solutions of the unilateral problem defined in the nonsmooth domain is path independent.

1. Poisson equation

Let $D \subset R^2$ be a bounded domain with smooth boundary Γ, and $\Xi_{l+\delta}$ be the set $\{(x_1, x_2)|\ 0 < x_1 < l+\delta,\ x_2 = 0\}$. We assume that this set belongs to the domain D for all sufficiently small δ, and $l > 0$. The domains with cracks $\Xi_{l+\delta}$, Ξ_l are denoted by $\Omega_\delta = D \setminus \overline{\Xi}_{l+\delta}$, $\Omega = D \setminus \overline{\Xi}_l$, respectively.

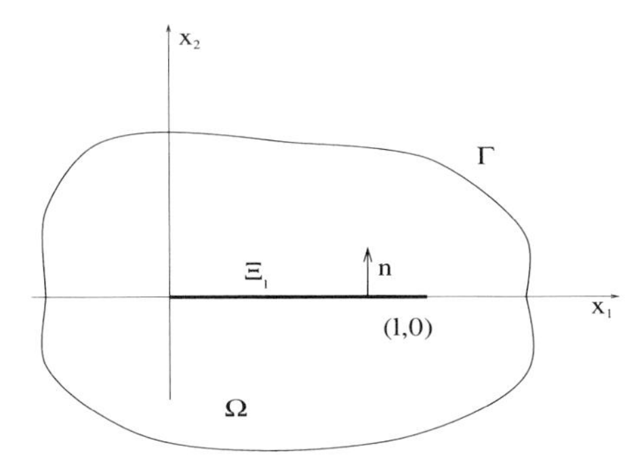

Fig. 1. Domain Ω.

In the domain Ω, we consider the following boundary value problem for a function u which satisfies

$$-\Delta u = f, \tag{1}$$

$$u = 0 \quad \text{on} \quad \Gamma\ , \tag{2}$$

$$[u] \geq 0,\ [u_{x_2}] = 0,\ u_{x_2} \leq 0,\ u_{x_2}[u] = 0 \quad \text{on} \quad \Xi_l. \tag{3}$$

Here $f \in C^1(\bar{D})$ is a given function, $[u] = u^+ - u^-$ is the jump of u across Ξ_l, $u_{x_i} = \frac{\partial u}{\partial x_i}, i = 1, 2$. The vector $n = (0, 1)$ is orthogonal to Ξ_l, and $u^{\pm}$ denote the traces of u on the crack faces $\Xi_l^{\pm}$, corresponding to the positive and negative directions of n. The weak solution to (1)–(3) minimizes the functional

$$I(\phi) = \frac{1}{2}\int\limits_{\Omega} |\nabla\phi|^2 - \int\limits_{\Omega} f\phi$$

over the set

$$K_0 = \{w \in H^1(\Omega)|\ [w] \geq 0 \quad \text{on} \quad \Xi_l;\ w = 0 \quad \text{on} \quad \Gamma\}.$$

The solution of the minimization problem satisfies the variational inequality

$$u \in K_0 : \quad \int\limits_{\Omega} \langle \nabla u, \nabla v - \nabla u\rangle \geq \int\limits_{\Omega} f(v - u) \quad \forall v \in K_0\ .$$

The weak solution of the problem (1)–(3) has an additional regularity up to the crack faces. For any $x \in \Xi_l$ there exists a neighbourhood V of the point x such that $u \in H^2(V \setminus \Xi_l)$. Hence, conditions (3) hold almost everywhere on Ξ_l.

The perturbed problem corresponding to (1)–(3) is formulated as follows. In the domain Ω_δ we want to find a function u^δ such that

$$u^\delta \in K_\delta : \quad \int\limits_{\Omega_\delta} \langle \nabla u^\delta, \nabla v - \nabla u^\delta\rangle \geq \int\limits_{\Omega_\delta} f(v - u^\delta) \quad \forall v \in K_\delta, \tag{4}$$

where

$$K_\delta = \{w \in H^1(\Omega_\delta)|\ [w] \geq 0 \quad \text{on} \quad \Xi_{l+\delta};\ w = 0 \quad \text{on} \quad \Gamma\}.$$

The energy functionals for the problems (1)–(3) and (4) are defined by the formulae

$$J(\Omega) = \frac{1}{2}\int\limits_{\Omega} |\nabla u|^2 - \int\limits_{\Omega} fu,\ J(\Omega_\delta) = \frac{1}{2}\int\limits_{\Omega_\delta} |\nabla u^\delta|^2 - \int\limits_{\Omega_\delta} fu^\delta, \tag{5}$$

where u, u^δ are solutions to (1)–(3) and (4), respectively.

Our aim is to find the derivative

$$\frac{dJ(\Omega_\delta)}{d\delta}\Big|_{\delta=0} = \lim_{\delta\to 0} \frac{J(\Omega_\delta) - J(\Omega)}{\delta} \tag{6}$$

which describes the behaviour of the energy functional $J(\Omega)$ with respect to the variation of the crack length.

In order to find the derivative (6) the transformation of the domain Ω_δ onto the domain Ω is introduced. The transformation is constructed in the following way. Let $\theta \in C_0^\infty(D)$ be any function such that $\theta = 1$ in a neighbourhood of the point $x_l = (l, 0)$. To simplify the arguments the function θ is assumed to be equal to zero in a neighbourhood of the point $(0, 0)$. Consider the transformation of the independent variables

$$y_1 = x_1 - \delta\theta(x_1, x_2), \quad y_2 = x_2, \tag{7}$$

where $(x_1, x_2) \in \Omega_\delta, (y_1, y_2) \in \Omega$. The Jacobian q_δ of this transformation is equal to

$$\left| \frac{\partial(y_1, y_2)}{\partial(x_1, x_2)} \right| = 1 - \delta\theta_{x_1}.$$

For small δ, the Jacobian q_δ is positive, hence the transformation (7) is one-to-one. Therefore, in view of (7) we have $y = y(x, \delta)$, and the inverse mapping is denoted by $x = x(y, \delta)$.

Let $u^\delta(x)$ be the solution of (4), and $u^\delta(x) = u_\delta(y), x = x(y, \delta)$. Then

$$\int_{\Omega_\delta} |\nabla u^\delta|^2 dx = \int_{\Omega} \langle A_\delta \nabla u_\delta, \nabla u_\delta \rangle dy,$$

where $A_\delta = A_\delta(y)$ is the matrix such that

$$A_\delta(y) = \frac{1}{1 - \delta\theta_{x_1}} \begin{pmatrix} (1 - \delta\theta_{x_1})^2 + \delta^2\theta_{x_2}^2 & -\delta\theta_{x_2} \\ -\delta\theta_{x_2} & 1 \end{pmatrix}, \quad \theta = \theta(x(y, \delta)).$$

Note that $A_0(y) = E$ is the identity matrix.

It is easy to find the derivative of $A_\delta(y)$ with respect to δ, namely,

$$A'(y) = \lim_{\delta \to 0} \frac{A_\delta(y) - A_0(y)}{\delta} = \begin{pmatrix} -\theta_{y_1}(y) & -\theta_{y_2}(y) \\ -\theta_{y_2}(y) & \theta_{y_1}(y) \end{pmatrix}. \tag{8}$$

By the change of variables it follows that

$$\int_{\Omega_\delta} f u^\delta dx = \int_{\Omega} f^\delta(y) u_\delta(y) dy, \quad f^\delta(y) = \frac{f(x(y, \delta))}{1 - \delta\theta_{x_1}}.$$

It can be shown

$$f'(y) = \lim_{\delta \to 0} \frac{f^\delta(y) - f^0(y)}{\delta} = (\theta f)_{y_1}(y). \tag{9}$$

Let us denote $w^\delta(x) = w_\delta(y)$. The inclusion $w^\delta \in K_\delta$ implies $w_\delta \in K_0$, and, conversely, $w_\delta \in K_0$ implies $w^\delta \in K_\delta$. This means that the transformation (7) maps K_δ onto K_0, and it is one-to-one.

Lemma. Let u^δ be the solution of (4), $u^\delta(x) = u_\delta(y)$, and u be the solution of (1)–(3). Then there exists a constant $c > 0$ such that

$$\|u_\delta - u\|_{H^1(\Omega)} \leq c\delta.$$

Theorem 1. The derivative of the energy functional is given by the formula

$$\frac{dJ(\Omega_\delta)}{d\delta}|_{\delta=0} = -\frac{1}{2} \int_{\Omega} (\theta_{y_1}(u_{y_1}^2 - u_{y_2}^2) + 2\theta_{y_2} u_{y_1} u_{y_2}) - \int_{\Omega} (\theta f)_{y_1} u. \tag{10}$$

The proof of Theorem 1 [10] is based on the properties of the energy functional, in particular,

$$\limsup_{\delta\to 0} \frac{J(\Omega_\delta)-J(\Omega)}{\delta} \le \frac{1}{2}\int_\Omega \langle A'\nabla u, \nabla u\rangle - \int_\Omega f'u \tag{11}$$

and

$$\liminf_{\delta\to 0} \frac{J(\Omega_\delta)-J(\Omega)}{\delta} \ge \frac{1}{2}\int_\Omega \langle A'\nabla u, \nabla u\rangle - \int_\Omega f'u. \tag{12}$$

therefore

$$\frac{dJ(\Omega_\delta)}{d\delta}|_{\delta=0} = \frac{1}{2}\int_\Omega \langle A'\nabla u, \nabla u\rangle - \int_\Omega f'u. \tag{13}$$

Using (8), (9) in (13) implies (10).

Theorem 1 means that the right-hand side of (10) actually depends on the point x_l and the right-hand side f of (1). This allows us to write (10) as the Griffith formula

$$\frac{dJ(\Omega_\delta)}{d\delta}|_{\delta=0} = k(x_l, f), \tag{14}$$

where k is a functional depending on x_l, f. In particular, we have

$$J(\Omega_\delta) = J(\Omega) + k(x_l, f)\delta + \alpha(\delta)\delta, \tag{15}$$

where $\alpha(\delta)\to 0$ as $\delta\to 0$.

Note that $k(x_l, f) = 0$ provided that the solution u is sufficiently smooth. In particular, the equality $k(x_l, f) = 0$ holds provided that $u \in H^2(B_{x_l})$, where B_{x_l} is a ball centered at x_l.

An additional regularity of u in a vicinity of the point x_l can be shown in some particular cases. For example, assume that the solution u satisfies the condition

$$[u] = 0 \quad \text{on} \quad B_{x_l} \cap \Xi_l,$$

where B_{x_l} is a ball centered at x_l. In this case we can prove that the equation

$$-\Delta u = f$$

holds in B_{x_l} in the sense of distributions, consequently, $k(x_l, f) = 0$.

In addition, the solution u of the problem (1)–(3) satisfies the following boundary condition

$$[u_{x_1} u_{x_2}] = 0 \quad \text{on} \quad \Xi_l. \tag{16}$$

We can write (10) in the form which does not contain θ. To this end, consider a ball $B_{x_l}(r)$ of radius r with a boundary $\Gamma(r)$ such that $\theta = 1$ in $B_{x_l}(r)$. Integration by parts in (10) implies

$$\frac{dJ(\Omega_\delta)}{d\delta}|_{\delta=0} = \int_{\Omega\setminus B_{x_l}(r)} \theta u_{y_1}(\Delta u + f) + \int_{\Xi_l\setminus B_{x_l}(r)} \theta u_{y_2}[u_{y_1}]$$

$$+ \int\limits_{B_{x_l}(r)\setminus\Xi_l} \theta f u_{y_1} + \frac{1}{2} \int\limits_{\Gamma(r)} \theta(\nu_1(u_{y_1}^2 - u_{y_2}^2) + 2\nu_2 u_{y_1} u_{y_2}),$$

where (ν_1, ν_2) is the unit external normal vector to $\Gamma(r)$. Hence, by (1), (16),

$$\frac{dJ(\Omega_\delta)}{d\delta}|_{\delta=0} = \int\limits_{B_{x_l}(r)\setminus\Xi_l} f u_{y_1} + \frac{1}{2} \int\limits_{\Gamma(r)} (\nu_1(u_{y_1}^2 - u_{y_2}^2) + 2\nu_2 u_{y_1} u_{y_2}). \tag{17}$$

Now assume that $f = 0$ in some neighbourhood V of the point x_l. For small r, we have $B_{x_l}(r) \subset V$, and (17) becomes

$$\frac{dJ(\Omega_\delta)}{d\delta}|_{\delta=0} = \frac{1}{2} \int\limits_{\Gamma(r)} (\nu_1(u_{y_1}^2 - u_{y_2}^2) + 2\nu_2 u_{y_1} u_{y_2}). \tag{18}$$

The right-hand side of (18) is independent of r, consequently, we arrive at the following conclusion.

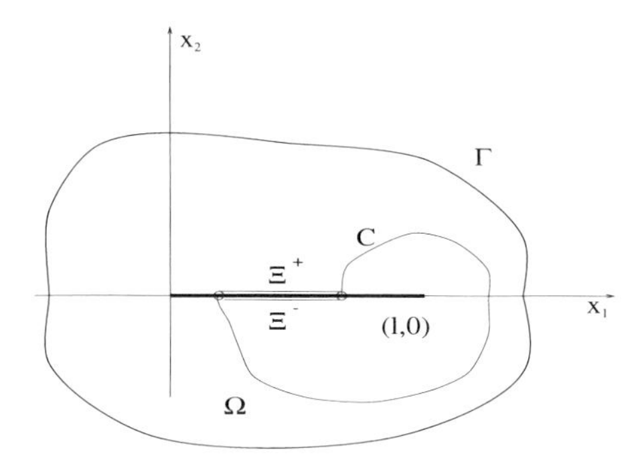

Fig. 2. Curve C.

Let u be the solution of the problem (1)–(3), and f be equal to zero in some neighbourhood of the point x_l. Then the integral

$$I = \int\limits_{\Gamma(r)} (\nu_1(u_{y_1}^2 - u_{y_2}^2) + 2\nu_2 u_{y_1} u_{y_2})$$

is independent of r for all sufficiently small r. Moreover, the above arguments show that the integral

$$I = \int\limits_{C} (\nu_1(u_{y_1}^2 - u_{y_2}^2) + 2\nu_2 u_{y_1} u_{y_2}) \tag{19}$$

does not depend on C for any closed curve C surrounding the point x_l. In this case $\nu = (\nu_1, \nu_2)$ is the normal unit vector to the curve C. A part of this curve may belong to Ξ_l. In such a case we can integrate over Ξ^+ or Ξ^- in (19), since, in view of (3), (16), the jump $[u_{y_1} u_{y_2}]$ is equal to zero on Ξ_l. Here $\Xi^\pm = \Xi_l^\pm \cap C$.

Of course, the path independence of integral (19) takes place provided that f is equal to zero in the domain with the boundary C. The integral of the form (19) is called the Rice–Cherepanov's integral. We have to note that the statement obtained is proved for nonlinear boundary conditions (3).

2. Linear elasticity equations

We use the notations of the previous section. The elasticity system is formulated as follows. We want to find a function $W = (u, v)$ such that

$$-\sigma_{ij,j} = f_i, \quad i = 1, 2, \quad \text{in } \Omega, \tag{20}$$

$$W = 0 \quad \text{on } \Gamma, \tag{21}$$

$$[v] \geq 0, \quad \sigma_{22} \leq 0, \quad [\sigma_{22}] = 0, \quad \sigma_{12} = 0, \quad [v]\sigma_{22} = 0. \tag{22}$$

Here $\sigma_{ij} = \sigma_{ij}(W)$ are the stress tensor components. Hooke's law is assumed to be fulfilled,

$$\sigma_{ij} = 2\mu\varepsilon_{ij} + \lambda \operatorname{div} W \delta_j^i, \quad i, j = 1, 2. \tag{23}$$

By $\lambda \geq 0$, $\mu > 0$ we denote the Lame parameters, $\varepsilon_{ij} = \varepsilon_{ij}(W)$, δ_j^i is the Kronecker symbol,

$$\varepsilon_{11} = u_{x_1}, \quad \varepsilon_{22} = v_{x_2}, \quad \varepsilon_{12} = 1/2\,(u_{x_2} + v_{x_1}).$$

We assume that $f = (f_1, f_2) \in C^1(\bar{D})$. In order to define the weak solution to the problem (20)–(22), we consider the minimization of the functional

$$I(\Omega; U) = \frac{1}{2} \int\limits_\Omega \sigma_{ij}(U)\varepsilon_{ij}(U) - \int\limits_\Omega fU, \;\; U = (u, v),$$

over the set

$$K_0 = \{(u, v) \in H^1(\Omega) \mid \quad u = v = 0 \quad \text{on } \Gamma, \quad [v] \geq 0 \quad \text{on } \Xi_l\}.$$

The solution W of the minimization problem satisfies (20)–(22), in particular the conditions (22) are satisfied a.e. on Ξ_l. The perturbed problem corresponding to (20)–(22) is defined in the same way. We want to find a function $W^\delta = (u^\delta, v^\delta)$ such that

$$-\sigma_{ij,j} = f_i, \quad i = 1, 2, \quad \text{in } \Omega_\delta, \tag{24}$$

$$W^\delta = 0 \quad \text{on } \Gamma, \tag{25}$$

$$[v^\delta] \geq 0, \quad \sigma_{22} \leq 0, \quad [\sigma_{22}] = 0, \quad \sigma_{12} = 0, \quad [v^\delta]\sigma_{22} = 0. \tag{26}$$

Here $\sigma_{ij} = \sigma_{ij}(W^\delta)$, $\varepsilon_{ij} = \varepsilon_{ij}(W^\delta)$, and σ_{ij}, ε_{ij} satisfy Hooke's law (23).

Similar to (20)–(22), the complete formulation of the problem (24)–(26) is variational ie., we minimize the functional

$$I(\Omega_\delta; U) = \frac{1}{2} \int\limits_{\Omega_\delta} \sigma_{ij}(U)\varepsilon_{ij}(U) - \int\limits_{\Omega_\delta} fU$$

over the set

$$K_\delta = \{(u, v) \in H^1(\Omega_\delta) \mid \quad u = v = 0 \quad \text{on } \Gamma, \quad [v] \geq 0 \quad \text{on } \Xi_{l+\delta}\}.$$

Our aim is to establish the existence of derivative of the energy functional

$$\frac{dJ(\Omega_\delta)}{d\delta}\Big|_{\delta=0} = \lim_{\delta\to 0}\frac{I(\Omega_\delta;W^\delta)-I(\Omega;W)}{\delta},$$

where W^δ, W are the solutions of (24)–(26) and (20)–(22), respectively, $J(\Omega_\delta) = I(\Omega_\delta;W^\delta)$.

Theorem 2. The derivative of the energy functional is given by the formula

$$\frac{dJ(\Omega_\delta)}{d\delta}\Big|_{\delta=0} = \frac{1}{2}\int_\Omega\Big((2\mu+\lambda)u_{y_1}^2(-\theta_{y_1}) + \mu u_{y_2}^2\theta_{y_1} + 2\mu u_{y_1}u_{y_2}(-\theta_{y_2})$$

$$+\mu v_{y_1}^2(-\theta_{y_1}) + (2\mu+\lambda)v_{y_2}^2\theta_{y_1} + 2(\lambda+\mu)u_{y_1}v_{y_1}(-\theta_{y_2}) \tag{27}$$

$$+2(2\mu+\lambda)v_{y_1}v_{y_2}(-\theta_{y_2})\Big) - \int_\Omega(\theta f_1)_{y_1}u - \int_\Omega(\theta f_2)_{y_1}v.$$

The Griffith formula (27) gives the derivative of the energy functional with respect to the crack length for two-dimensional elasticity with nonlinear boundary conditions (22).

By [9], the solution of the problem (20)–(22) has an additional regularity up to the crack faces. Namely, for any $x \in \Xi_l$ there exists a neighbourhood V of the point x such that $W \in H^2(V\backslash\Xi_l)$. Consequently, the solution W is continuous up to the crack faces. Note that

$$\sigma_{22} = (2\mu+\lambda)v_{y_2} + \lambda u_{y_1}, \quad \sigma_{12} = \mu(u_{y_2}+v_{y_1}).$$

In addition to (22) we can prove that

$$[\sigma_{22}v_{y_1}] = \sigma_{22}[v_{y_1}] = 0 \quad \text{a.e. on } \Xi_l. \tag{28}$$

Consequently, on Ξ_l we have

$$\mu[u_{y_1}u_{y_2}] + (\lambda+\mu)[u_{y_1}v_{y_1}] + (2\mu+\lambda)[v_{y_1}v_{y_2}] = [\sigma_{12}u_{y_1}] + [\sigma_{22}v_{y_1}]. \tag{29}$$

The right-hand side of (27) is independent of θ, and formula (27) can be written in the form which does not contain the function θ. To show this we choose a ball $B_{x_l}(r)$ of radius r with the boundary $\Gamma(r)$ such that $\theta = 1$ on $B_{x_l}(r)$. In this case the integration by parts in (27) yields

$$\frac{dJ(\Omega_\delta)}{d\delta}|_{\delta=0} = +\frac{1}{2}\int_{\Gamma(r)}\nu_1\left((2\mu+\lambda)(u_{y_1}^2 - v_{y_2}^2) + \mu(v_{y_1}^2 - u_{y_2}^2)\right) \tag{30}$$

$$+ \int_{B_{x_l}(r)\backslash\Xi_l}(f_1u_{y_1} + f_2v_{y_1}) + \int_{\Gamma(r)}\nu_2\Big((2\mu+\lambda)v_{y_1}v_{y_2} + (\lambda+\mu)u_{y_1}v_{y_1} + \mu u_{y_1}u_{y_2}\Big),$$

where (ν_1,ν_2) is the unit normal exterior vector to $\Gamma(r)$.

Let $f = 0$ in some neighbourhood V of the point x_l. For sufficiently small r, $0 < r < r_0$, we have $B_{x_l}(r) \subset V$, and the formula (30) implies

$$\frac{dJ(\Omega_\delta)}{d\delta}|_{\delta=0} = \frac{1}{2} \int\limits_{\Gamma(r)} \nu_1 \left((2\mu+\lambda)(u_{y_1}^2 - v_{y_2}^2) + \mu(v_{y_1}^2 - u_{y_2}^2)\right) \tag{31}$$

$$+ \int\limits_{\Gamma(r)} \nu_2 \Big((2\mu+\lambda) v_{y_1} v_{y_2} + (\lambda+\mu) u_{y_1} v_{y_1} + \mu u_{y_1} u_{y_2}\Big).$$

The right-hand side of (31) does not depend on r, consequently, we have the following property. Let W be the solution of the problem (20)–(22), and f be equal to zero in some neighbourhood of the point x_l. Then the integral

$$I = \frac{1}{2} \int\limits_{C} \nu_1 \left((2\mu+\lambda)(u_{y_1}^2 - v_{y_2}^2) + \mu(v_{y_1}^2 - u_{y_2}^2)\right) \tag{32}$$

$$+ \int\limits_{C} \nu_2 \Big((2\mu+\lambda) v_{y_1} v_{y_2} + (\lambda+\mu) u_{y_1} v_{y_1} + \mu u_{y_1} u_{y_2}\Big)$$

is path independent for any closed curve C surrounding the point x_l, where $\nu = (\nu_1, \nu_2)$ is the normal unit vector to the curve C (see Fig.2). This independence holds provided that f is equal to zero in a domain with the boundary C. Denote $\Xi^\pm = \Xi_l^\pm \cap C$. In this case, by (28), (29), (22), we can integrate over Ξ^+ or Ξ^- in (32).

The well-known path independence of the Rice–Cherepanov's integral was previously proved in elasticity theory for linear boundary conditions $\sigma_{22} = 0, \sigma_{12} = 0$ holding on $\Xi_l^\pm$ (see [2,3]).

In three-dimensional case the Griffith formula is also obtained [11]. In this case the two-dimensional crack is presented as a part of a plane.

Acknowledgment. A.M. Khludnev was supported by the Institut Elie Cartan, Université Henri Poincaré Nancy I during his stay in January 1998 and by the Russian Fund for Basic Research under the grant 97-01-00896.

References

[1] A.M. Khludnev and J. Sokolowski, *Modelling and Control in Solid Mechanics*, Birkhauser, Bazel, Boston, Berlin, **1997**.

[2] V.Z. Parton and E.M. Morozov, *Mechanics of elastoplastic fracture*, Moscow, Nauka, **1985** (in Russian).

[3] G.P. Cherepanov, *Mechanics of Brittle Fracture*, McGraw-Hill, **1979**.

[4] V.G. Mazja V.G. and S.A. Nazarov, *Asymptotics of energy integrals for small perturbation of a boundary near corner and conical points*, Proceedings of Moscow math. society, Moscow State Univ., **1987**, 79–129, (in Russian).

[5] J. Sokolowski and J.P. Zolésio, *Introduction to shape optimization. Shape sensitivity analysis*, Springer-Verlag, **1992**.

[6] A.M. Khludnev, *Contact problem for a plate having a crack of minimal opening*, Control and Cybernetics, **25**, (1996), 605–620.

[7] P. Grisvard, *Elliptic problems in nonsmooth domains*, Pitman, Boston, London, Melbourne, **1985**.

[8] A.M. Khludnev, *The contact problem for a shallow shell with a crack*, J. Appl. Maths. Mechs., **59**, (1995), 299–306.

[9] A.M. Khludnev, *On a Signorini problem for inclusions in shells*, Europ. J. of Appl. Math., **7**, (1996), 499–510.

[10] A.M. Khludnev and J. Sokolowski, *Griffith formula and Rice–Cherepanov's integral for elliptic equations with unilateral conditions in nonsmooth domains*, Les prépublications de l'Institut Élie Cartan, **7**, (1998).

[11] A.M. Khludnev and J. Sokolowski, *Griffith formula for elasticity system with unilateral conditions in domains with cracks*, Rapport de recherche, INRIA, **3447**, (1998).

A.M. Khludnev
Lavrentyev Institute of Hydrodynamics of the Russian Academy of Sciences,
Novosibirsk 630090, Russia
E-mail address: `khlud@hydro.nsc.ru`

J. Sokolowski
Institut Elie Cartan,
Laboratoire de Mathématiques,
Université Henri Poincaré Nancy I,
B.P. 239, 54506 Vandoeuvre lès Nancy Cedex, France
and
Systems Research Institute of the Polish Academy of Sciences,
ul. Newelska 6, 01-447 Warszawa, Poland
E-mail address: `sokolows@iecn.u-nancy.fr`

International Series of Numerical Mathematics
Vol. 133, © 1999 Birkhäuser Verlag Basel/Switzerland

A Domain Optimization Problem for a Nonlinear Thermoelastic System

A. Myśliński and F. Tröltzsch

Abstract. A shape optimization problem for a thermoelastic system with a nonlinear boundary condition is considered. Using the material derivative method as well as the results of regularity of solutions to the state system, the sensitivity analysis of the solution to this system with respect to the variation of the domain is performed and necessary optimality conditions are derived.

1. Introduction

This paper is concerned with a shape optimization problem for an isotropic, homogeneous, elastic body occupying a two-dimensional domain and subjected to a prescribed thermal treatment. The equilibrium state of the body is described by a system of two linear weakly coupled parabolic and elliptic equations where the temperature and the displacement of the body are selected as the state variables. To model the heat exchange of the body with its surrounding medium, a nonlinear boundary condition is assumed in the parabolic equation. Existence and regularity of solutions to this state system were studied in [5, 6, 7, 10].

The shape optimization problem for the considered state system consists in finding an *initial* shape of the domain occupied by the body, such that the *final* shape of this domain after the thermal treatment will resemble a desired prescribed form as closely as possible. Note, that the solid body is subjected to a prescribed thermal treatment. Due to the temperature change, the body undergoes a thermoelastic deformation, that is, the induced thermal stresses force the body to change its shape in time. This problem was introduced by Sprekels and Tröltzsch [9].

Optimal shape design problems for systems governed by instationary equations have attracted much interest in the past years (see [2, 4, 8, 9, 11]). Among others in [9] the temperature change of the thermoelastic system was modelled by a parabolic equation with linear boundary condition. Moreover, the function describing the domain boundary was the variable subject to optimization.

In contrast to this, in this paper, we shall investigate a similar but slightly different model than introduced in [9]. We consider a nonlinear boundary condition on a part of the boundary of the domain occupied by the body. Moreover the objective functional introduced in this paper is slightly modified compared with that in [9]. Furthermore, here, the material derivative framework [8] is employed.

In the paper we shall formulate the optimal shape design problem and the underlying assumptions. Moreover the well-posedness of the state equations is investigated. Using results concerning the regularity of solutions to the thermoelastic systems [6, 7] as well as the material derivative method [8] the directional derivative of the cost functional with respect to variations of the domain is calculated. Moreover, first order necessary optimality conditions are formulated.

Throughout the paper we shall use the following notation: $H^m(\Omega)$, $m = 0, 1, 2$, will denote the Sobolev spaces of order m with norm $\| \cdot \|_{H^m(\Omega)}$ [1], $\mathbf{n}=(n_1, n_2)$ is the unit outward vector to the boundary Γ, $\partial v/\partial n$ is the outward normal derivative of a function v on the boundary Γ of the domain Ω, $\bigtriangledown v$ is the gradient of a function v with respect to a variable x, and $\triangle_x v$ denotes the Laplacian of the function v with respect to x, and $\theta_t = \dfrac{\partial \theta}{\partial t}$.

2. The thermoelastic model

Consider an isotropic, homogeneous, elastic body occupying a domain $\Omega \subset R^2$ in the plane Ox_1x_2 . The domain Ω is bounded and simply connected. Its boundary Γ is supposed to be the type $C^{2,\alpha}$, with some $0 < \alpha < 1$. We assume that the body is loaded by a distributed force $f = f(x)$, $x = (x_1, x_2) \in D$. Hereby, $D \subset R^2$ denotes a hold-all domain containing all admissible domains Ω regarded in the shape optimization problem. We assume without loss of generality that the boundary of D is sufficiently smooth. The boundary Γ is divided into two disjoint parts Γ_1 and Γ_2. We assume $mes\Gamma_1 \geq 0$. Along the boundary Γ_2 the body is subjected to a prescribed thermal treatment, while it is thermally isolated on Γ_1. Γ_2 stands for the part of Γ which is to be shaped.

Let $\theta = \theta(x,t)$, $x \in \Omega$, $t \in [0, \mathcal{T}]$, $\mathcal{T} > 0$ given, denote the temperature of the body. Due to the temperature change the body undergoes a thermoelastic deformation. By $u = u(x,t)$, $u = (u_1, u_2)$, we denote the two-dimensional displacement vector of the body. In an equilibrium state the functions θ and u satisfy the system of state equations:

$$\begin{aligned} \theta_t(x,t) &= \Delta\theta(x,t)\,, \quad \text{in } \Omega\,, \quad 0 < t \leq \mathcal{T} \\ \theta(x,0) &= \theta_0(x) \quad \text{in } \bar{\Omega} \\ \frac{\partial \theta}{\partial n}(x,t) &= g(\theta(x,t),x,t) \quad \text{on } \Gamma \quad 0 < t \leq \mathcal{T} \end{aligned} \tag{1}$$

$$\begin{aligned} \mu\,\Delta u + (\lambda+\mu)\,\nabla\,(\operatorname{div} u) &= -\rho\, f + \beta\,\nabla\theta(.,\mathcal{T}) \quad \text{in } \Omega\,, \\ u &= 0 \quad \text{on } \Gamma_1\,, \\ n_j\,\sigma_{ij} &= \beta\theta(.,\mathcal{T})n_i \quad \text{on } \Gamma_2, \quad i = 1,2,\,. \end{aligned} \tag{2}$$

Note that in (2) and throughout the paper summation convention is used. $\mu > 0$, $\lambda > 0$, $\rho > 0$, and $\beta > 0$ are fixed real constants. Moreover, real-valued functions $f \in L^2(D, R^2)$, and $g : R \times D \times [0, \mathcal{T}]$ are given. In (1)–(2), $\sigma = (\sigma_{ij}(x))$ and $\epsilon = \epsilon_{ij}(x)$ stand for the *stress* and *strain tensor*, respectively.

We assume that g satisfies the following assumptions:

(*A*1) g is continuously differentiable w.r. to (θ, t) on $R \times D \times [0, \mathcal{T}]$
g is continuous w.r to x for all $(\theta, t) \in R \times [0, \mathcal{T}]$.
$| g(0, x, t) | \leq \psi_0 \; \forall (x, t) \in D \times [0, \mathcal{T}]$
$| g_\theta(\theta, x, t) | + | g_t(\theta, x, t) | \leq \psi_M \;\; \forall (\theta, x, t) \in [-M, M] \times D \times [0, \mathcal{T}]$
Here ψ_0 and ψ_M are certain real constants.

(*A*2) $g_\theta(\theta, x, t) \leq 0 \; \forall (\theta, x, t) \in R \times D \times [0, \mathcal{T}]$

(*A*3) Compatibility conditions: $\theta_0(x) = \theta_0$
is constant on D and $g(\theta_0(x), x, 0)) = 0$.

Assumption (A1) ensures in particular that $g(\theta, t, x)$ is well defined on sufficiently smooth boundaries of admissible domains Ω. The requirement that θ_0 is constant was imposed to satisfy the compatibility condition for all admissible boundaries Γ. It should be underlined that this condition is only needed to get higher regularity of the function θ.

2.1. Existence and uniqueness of solutions to the state equations

We shall work with weak solutions θ, u of (1)–(2), $\theta \in W(0, \mathcal{T})$, where the function space $W(0, \mathcal{T}) = \{v \in L^2(0, \mathcal{T}, H^1(\Omega)) \;\; | \;\; v_t \in L^2(0, \mathcal{T}, H^1(\Omega)')\}$ (see [5]). In the sequel, Q denotes the set $\Omega \times (0, \mathcal{T})$. Let H denote the space $L^2(\Omega)$ and $H^k(\Omega) = W_2^k(\Omega)$.

Definition 2.1. *$\theta \in W(0, \mathcal{T}) \cap C(\bar{Q})$ is said to be a* weak solution *of (1), if*

$$\begin{aligned} &\int_0^{\mathcal{T}} \int_\Omega [-\theta(x,t)\, \phi_t(x,t) + \nabla\theta(x,t) \nabla\phi(x,t)]\, dx\, dt \\ &= \int_0^{\mathcal{T}} \int_\Gamma g(\theta(x,t), x, t)\, \phi(x,t)\, dS\, dt + \int_\Omega \theta_o(x) \phi(x, 0)\, dx \end{aligned} \tag{3}$$

for all $\phi \in W_2^{1,1}$ such that $\phi(., \mathcal{T}) = 0$.

The regularity of functions in $W(0, \mathcal{T})$ implies $\theta \in C([0, \mathcal{T}], H)$. In order to have $\nabla\theta(x, \mathcal{T})$ well defined, we need even $\theta \in C([0, \mathcal{T}], H^1(\Omega))$. This is the reason for introducing the quite strong assumptions (A1), (A3). The following result follows from [6, 7]:

Theorem 2.2. *Let the assumptions (A1)–(A3) be satisfied. Then a unique solution $\theta \in W(0, \mathcal{T}) \cap C(\bar{Q})$ of (2) exists. This solution can be repesented in the form*

$$\theta(x, t) = \theta_o(x) + \int_0^t w(x, s)\, ds, \tag{4}$$

where $w \in W(0, \mathcal{T})$.

Existence and uniqueness of $\theta \in W(0, \mathcal{T}) \cap C(\bar{Q})$ follows from Proposition 3.3 and Remark 3.4 in [6]. The representation (4) is given by Theorem 5.2 in [7].

Next we consider the weak form of the displacement equations. For convenience we introduce the space V defined by

$$V = \{u \in H^1(\Omega) \times H^1(\Omega) : u = 0 \text{ on } \Gamma_1\}. \tag{5}$$

Definition 2.3. *A function $u \in V$ is said to be a* weak solution *of (2), if*

$$\int_\Omega \sigma_{ij}(u)\epsilon_{ij}(v)dx = \int_\Omega \rho\, f\, v\, dx + \int_\Omega \beta\theta(.,\mathcal{T})\, div\, v\, dx \;\; \forall v \in V \tag{6}$$

We recall that the strain tensor ϵ is defined by $\epsilon_{ij} = (\partial u^i/\partial x_j + \partial u^j/\partial x_i)/2$, $i,j = 1,2$. The stress tensor $\sigma = \sigma_{ij}(x)$ is related to ϵ by

$$\begin{pmatrix} \sigma_{11} \\ \sigma_{22} \\ \sigma_{12} \end{pmatrix} = \begin{pmatrix} 2\mu+\lambda & \lambda & 0 \\ \lambda & 2\mu+\lambda & 0 \\ 0 & 0 & 2\mu \end{pmatrix} \begin{pmatrix} \epsilon_{11} \\ \epsilon_{22} \\ \epsilon_{12} \end{pmatrix}. \tag{7}$$

Theorem 2.4. *The assumptions (A1)–(A3) imply the existence of a unique weak solution u of (6).*

Proof. In view of Theorem 2.2 we have $\theta(.,\mathcal{T}) \in L^2(\Omega)$. Moreover, $mes\,\Gamma_1 > 0$ permits to apply Korn's first inequality. Existence and uniqueness is a known conclusion [3]. □

To achieve higher regularity of θ we impose the following stronger assumption on g:

$$(A1') \qquad g \in C^1(R \times D \times [0,\mathcal{T}])$$

Lemma 2.5. *If the stronger assumption $(A1')$ is satisfied instead of (A1), then $g(tr\,\theta,.,.)$ belongs to $C([0,\mathcal{T}], H^{1/2}(\Gamma))$ for the trace $tr\,\theta = \theta_{|\Gamma}$ of the solution θ of (3).*

Proof. We know $\theta \in C(\bar{Q})$. Therefore the relation $g(tr\,\theta,.,.) = tr\,g(\theta,.,.)$ is trivially true owing to the continuity assumptions on g. From Theorem 2.2 it can be deduced that $\theta \in C([0,\mathcal{T}], H^1(\Omega))$. Showing that $g(\theta,.,t)$ belongs to $C([0,\mathcal{T}], H^1(\Omega))$ we obtain the desired result. □

Corollary 2.6. *If Γ is sufficiently smooth, then θ belongs to $L^2(0,\mathcal{T}; H^2(\Omega))$.*

Proof. By (3) we know $\theta \in H^1(0,\mathcal{T}; H^1(\Omega))$, hence $\theta_t \in L^2(0,\mathcal{T}; H^1(\Omega))$ as well as $tr\,\theta \in H^1(0,\mathcal{T}; H^{1/2}(\Gamma))$. Lemma 2.5 yields $g(tr\,\theta,.,.) \in C([0,\mathcal{T}], H^{1/2}(\Gamma))$. Writing (1) in the form $\Delta\theta(x,t) = \theta_t(x,t)$ and $\partial\theta/\partial n\,(x,t) + \theta(x,t) = g(\theta(x,t),x,t)$ we obtain from the elliptic regularity theory that $\theta \in L^2(0,\mathcal{T}; H^2(\Omega))$. This follows from the fact that the mapping $(f,g) \to w$, which assigns data (f,g) the solution of $\Delta w = f$, $\partial w/\partial n + w = g$ is continuous from $L^2(\Omega) \times H^{1/2}(\Gamma)$ to $H^2(\Omega)$. □

3. The shape optimization problem

Before we formulate the shape optimization problem we shall introduce a family of admissible domains $\{\Omega_\tau\}$ depending on a parameter τ. The domain Ω_τ will be considered as the image of a mapping T_τ of the reference domain Ω. We shall employ the speed method [8] to describe the mapping T_τ. The shape optimization problem is formulated for the variational system (1)–(2) in the perturbed domain Ω_τ.

Let τ be a real parameter, such that $\tau \in [0,\sigma), \sigma > 0$. We denote by $\mathbf{V}(.,.) : [0,\sigma) \times R^2 \to R^2$ a sufficiently regular vector field, i.e.,

$$\mathbf{V}(\tau,.) \in C^2(R^2,R^2)\ \forall \tau \in [0,\sigma), \quad \mathbf{V}(.,x) \in C([0,\sigma),R^2)\ \forall x \in R^2. \tag{8}$$

By $T_\tau(\mathbf{V}) : R^2 \ni X \to x(\tau, X) \in R^2$ we denote a family of mappings depending on a parameter $\tau \in [0,\sigma)$, where the vector function $x(.,X) = x(.)$ satisfies the ordinary differential equation:

$$\frac{d}{ds}x(s,X) = \mathbf{V}(s,x(s,X)) \quad s \in [0,\sigma), \quad x(0,X) = X,\ X \in R^2. \tag{9}$$

The family Ω_τ of domains depending on the parameter $\tau \in [0,\sigma)$ is defined as follows: $\Omega_0 = \Omega$ and

$$\begin{aligned}\Omega_\tau = T_\tau(\mathbf{V})(\Omega) = \{x \in R^2\ :\ \exists X \in \Omega \subset R^2 \text{ such that } x = x(s), \text{ where}\\ \text{the function } x(.) \text{ satisfies equation (9) for } 0 \le s \le \tau\}.\end{aligned} \tag{10}$$

We shall assume that for any given value of the parameter τ the domain $\Omega_\tau \subset D$ is bounded, simply connected, and has $C^{2,\alpha}$, $0 < \alpha < 1$, continuous boundary Γ_τ. Let us denote

$$Q_\tau = \Omega_\tau \times (0,\mathcal{T}), \quad W_\tau(0,\mathcal{T}) = \{v \in L^2(0,\mathcal{T};H^1(\Omega_\tau)); v_t \in L^2(0,\mathcal{T};H^1(\Omega_\tau)')\}$$

The variational problem (3)–(6) in the domain Ω_τ takes the form: *for given element $g \in C^1(R \times D \times [0,\mathcal{T}])$ and constant $\theta_0 \in L^2(D)$ find functions $\theta_\tau \in W_\tau(0,\mathcal{T}) \cap C(\bar{Q}_\tau)$ and $u_\tau \in V_\tau = \{v \in H^1(\Omega_\tau)\ :\ v = 0 \text{ on } \Gamma_{1\tau}\}$ satisfying*

$$\begin{aligned}&\textstyle\int_0^{\mathcal{T}} \int_{\Omega_\tau} [-\theta_\tau(x,t)\,\phi_t(x,t) + \nabla\theta_\tau(x,t)\nabla\phi(x,t)]\,dx\,dt\\ &= \textstyle\int_0^{\mathcal{T}} \int_{\Gamma_\tau} g(\theta_\tau(x,t),x,t)\,\phi(x,t)\,dS\,dt + \int_{\Omega_\tau} \theta_{0\tau}(x)\phi(x,0)\,dx\end{aligned} \tag{11}$$

for all $\phi \in W_2^{1,1}(Q_\tau)$ such that $\phi(.,\mathcal{T}) = 0$.

$$\int_{\Omega_\tau} \sigma_{\tau ij}(u)\epsilon_{ij}(v)dx = \int_{\Omega_\tau} \rho\, f\, v\, dx + \int_{\Omega_\tau} \beta\theta_\tau(.,\mathcal{T})\, div\, v\, dx \ \ \forall v \in V_\tau \tag{12}$$

For each $\tau \in [0,\sigma), \sigma > 0$ the system (11)–(12) has a unique solution $(\theta_\tau, u_\tau) \in W_\tau(0,\mathcal{T}) \cap C(\bar{Q}_\tau) \cap L^2(0,\mathcal{T};H^2(\Omega_\tau)) \cap L^\infty(Q_\tau) \times V_\tau$.

By U_{ad} we denote the set of admissible domains. We shall consider the following shape optimization problem for the system (11)–(12): *find a domain $\Omega_\tau \in U_{ad}$*

minimizing the cost functional

$$J(\Omega_\tau) = \lambda_1 I_1(\Omega_\tau) + \lambda_2 I_2(\Omega_\tau), \tag{13}$$

$$I_1(\Omega_\tau) = \int_{\Omega_\tau} | u_d - u_\tau |^2 \, dx, \;\; I_2(\Omega_\tau) = \int_{\Gamma_\tau} | z_d - u_\tau |^2 \, ds \tag{14}$$

and $\lambda_1 \geq 0$, $\lambda_2 \geq 0$ $(\lambda_1^2 + \lambda_2^2 \neq 0)$, *are fixed real constants. Moreover, real-valued functions* $u_d \in L^2(D, R^2)$, $z_d \in C(\partial D, R^2)$ *are given. The pair* (θ_τ, u_τ) *is a solution to the system (11)–(12).*

The aim of the optimization problem (13) is to find the domain Ω occupied by the body at the time $t = 0$ such that the final displacement u of the body occupying the domain Ω at the time $t = \mathcal{T}$ comes as close as possible to prescribed functions u_d in Ω and to z_d on Γ. A body occupying the domain Ω is heated according to (1). The result is a thermal elastic deformation described by (2). For the selection of the admissible domains set U_{ad} in numerical applications see [3].

4. Necessary optimality condition

We shall calculate, assuming (A1)–(A3) are satisfied, the derivative of the cost functional (13). Let us recall the definition [8]:

Definition 4.1. *The material derivative* $\dot{u} \in V$ *of the function* $u_\tau \in V_\tau$ *is defined by:*

$$\lim_{\tau \to 0} \| (u_\tau \circ T_\tau - u_0)/\tau - \dot{u} \|_{H^1(\Omega)} = 0 \tag{15}$$

where $u = u_0 \in V$ *and the function* $u^\tau = u_\tau \circ T_\tau \in V$ *is the image of the function* $u_\tau \in V_\tau$ *in the space* V.

Let us recall [8, pp. 111–114] that if the shape derivative $U \in V$ of the function $u_\tau \in V_\tau$ exists, then the following condition holds:

$$U = \dot{u} - \nabla u \mathbf{V}(0) \tag{16}$$

4.1. Sensitivity analysis of solutions to the state system

Using Definition 4.1 we can prove:

Lemma 4.2. *The material derivatives* $\dot{\theta} \in W(0, \mathcal{T})$ *and* $\dot{u} \in V$ *of the functions* $\theta_\tau \in W_\tau(0, \mathcal{T})$ *and* $u_\tau \in V_\tau$ *satisfying the system (11)–(12) are given by:*

$$\begin{aligned}
&\int_0^{\mathcal{T}} \int_\Omega \{[-\dot{\theta}\frac{\partial \varphi}{\partial t} - \theta \frac{\partial \varphi}{\partial t} div \mathbf{V}(0)] + [\nabla \dot{\theta} \nabla \varphi \\
&+ \nabla \theta \nabla (\nabla \varphi \mathbf{V}(0))] + [div \mathbf{V}(0) I - ({}^T D\mathbf{V}(0) + D\mathbf{V}(0)) \nabla \theta \nabla \varphi]\} dx dt \\
&= \int_0^{\mathcal{T}} \int_\Gamma \{\frac{\partial g}{\partial \theta} \dot{\theta} \varphi + g \nabla \varphi \mathbf{V}(0) + \nabla g \varphi \mathbf{V}(0) \\
&+ g\varphi (div \mathbf{V}(0) - (D\mathbf{V}\mathbf{n}, \mathbf{n}))\} ds dt + \int_\Omega \theta_0 \, \varphi(x, 0) \, div \mathbf{V}(0) dx
\end{aligned} \tag{17}$$

$$
\begin{aligned}
&\int_\Omega \{a_{ijkl}[\epsilon_{ij}(\dot{u})\epsilon_{kl}(\varphi) + \bar{\epsilon}_{ij}(u)\epsilon_{kl}(\varphi) + \epsilon_{ij}(u)\bar{\epsilon}_{kl}(\varphi)] \\
&+\bar{a}_{ijkl}\bar{\epsilon}_{ij}(u)\epsilon_{kl}(\varphi)\}dx = -\beta \int_\Omega [-^T D\mathbf{V}(0)\nabla\theta\varphi + \nabla\theta\,\varphi\, divV(0) \\
&+\nabla\dot{\theta}\varphi + \nabla\theta\nabla\varphi\mathbf{V}(0) - (\rho\, f\,\varphi\, div\mathbf{V}(0) + \rho\varphi\,\nabla f\,\mathbf{V}(0)]dx \\
&+\beta \int_\Gamma \{\varphi\dot{\theta}\mathbf{n} + \theta\mathbf{n}\nabla\varphi\mathbf{V}(0) + \theta\varphi(div\mathbf{V}(0)\mathbf{n} - {}^T D\mathbf{V}(0)\mathbf{n})\}ds
\end{aligned} \tag{18}
$$

where $\bar{\epsilon}_{ij} = -\frac{1}{2}\{D\varphi D\mathbf{V}(0) + {}^T D\mathbf{V}(0)^T D\varphi\}$ *and* $\bar{a}_{ijkl} = a_{ijkl}div\mathbf{V}(0) + \nabla a_{ijkl}\mathbf{V}(0)$. *Moreover,* $\mathbf{V}(0) = \mathbf{V}(0, X)$*,* $D\mathbf{V}(0)$ *denotes the Jacobian matrix of the vector* $\mathbf{V}(0)$*,* ${}^T D\mathbf{V}(0)$ *denotes the transposed matrix of* $D\mathbf{V}(0)$*,* I *is the identity matrix.*

Proof. Using formulae for the transport of the function gradient and normal vector into a fixed domain [8, pp. 70–80] we transform system (11)–(12) to the fixed cylinder $\Omega \times [0, \mathcal{T}]$:

$$
\begin{aligned}
&\int_0^{\mathcal{T}} \int_\Omega [-\theta^\tau(x,t)\frac{\partial\varphi}{\partial t}(x,t) + A(\tau)\nabla\theta^\tau\nabla\varphi]\gamma(\tau)dxdt \\
&= \int_0^{\mathcal{T}} \int_{\Gamma_2} g(\theta^\tau, x, t)\varphi(x,t)\omega(\tau)dsdt + \int_\Omega \theta_0^\tau(x)\varphi(x,0)\gamma(\tau)dx \\
&\qquad \forall\varphi \in W_2^{1,1}(Q),\ \varphi(0,\mathcal{T}) = 0
\end{aligned} \tag{19}
$$

$$
\begin{aligned}
&\int_\Omega a_{ijkl}\epsilon_{ij}(u^\tau)\epsilon_{kl}(\eta)A(\tau)dx \\
&= \int_\Omega [\rho f\eta - \beta^T DT_\tau^{-1}\nabla\theta^\tau(.,\mathcal{T})]\eta\gamma(\tau)dx + \int_\Gamma \beta\theta^\tau\eta\omega(\tau)ds
\end{aligned} \tag{20}
$$

for all $\eta \in V$ and a.e. $t \in (0, \mathcal{T})$. DT_τ is the Jacobian matrix of the mapping T_τ, DT_τ^{-1} is the inverse of DT_τ, and ${}^T DT_\tau$ is a transpose of DT_τ. Moreover $\gamma(\tau) = \det DT_\tau$, $A(\tau) = {}^T DT_\tau^{-1} DT_\tau^{-1}\gamma(\tau)$, $\omega(\tau) = \| DT_\tau n \| \gamma(\tau)$.

From Proposition 2.44 and Lemma 2.49 in [8] follows the differentiability of $\gamma(\tau)$, $A(\tau)$, $\omega(\tau)$ with respect to τ in $C^1(D)$, $C^1(D, R^2)$, $C^1(D)$ respectively. Using similar arguments as in [6, 8], from (19) and (20) we obtain the convergence of θ^τ to θ in $L^\infty(Q)$ and of u^τ to u in $H^1(\Omega)$ for $\tau \to 0$. Substracting equation (3) from (19) and (7) from (20) and passing to the limit with $\tau \to 0$ we obtain formulae (17) and (18). □

From Corollary 2.6 it follows that the solution θ to the system (3)–(6) has regularity $\theta \in L^2(0, \mathcal{T}; H^2(\Omega))$, i.e. we have,

$$
\nabla\theta\mathbf{V}(0) \in L^2(0, \mathcal{T}; H^1(\Omega)) \tag{21}
$$

From [7] and Corollary 2.6 we obtain the regularity $u \in L^2(0, \mathcal{T}; H^2(\Omega))$ for the solution u to the system (3)–(6) provided either $\bar{\Gamma} = \bar{\Gamma}_1$ or $\bar{\Gamma} = \bar{\Gamma}_2$. In the case of mixed boundary conditions u may not have regularity $H^2(\Omega)$ on $\bar{\Gamma}_1 \cap \bar{\Gamma}_2$. We assume that the following regularity condition is satisfied,

$$\nabla u \mathbf{V}(0) \in H^1(\Omega). \tag{22}$$

Remark that in our case (22) may be satisfied by choosing a suitable velocity field $\mathbf{V} = 0$ on $\bar{\Gamma}_1 \cap \bar{\Gamma}_2$ (see [8, p. 140]).

Integrating by parts the system (17)–(18) two times, eliminating the terms containing the derivatives of $\mathbf{V}(0)$ and taking into account (16) as well as (21)–(22) we obtain the system of equations determining the shape derivative $(\Theta, U) \in W(0,\mathcal{T}) \times V$ of the solution $(\theta_\tau, u_\tau) \in W_\tau(0,\mathcal{T}) \times V_\tau$ of the system (11)–(12):

$$\begin{aligned}\int_0^{\mathcal{T}}\int_\Omega \{-\Theta\frac{\partial\varphi}{\partial t} + \nabla\Theta\nabla\varphi\}dxdt + \int_0^{\mathcal{T}}\int_\Gamma \{\theta\frac{\partial\varphi}{\partial t}\mathbf{V}(0)n + \nabla\theta\nabla\varphi\mathbf{V}(0)\mathbf{n}\}dsdt \\ = \int_0^{\mathcal{T}}\int_{\Gamma_2}\{\frac{\partial g}{\partial\theta}\Theta\varphi + \frac{\partial g}{\partial\theta}\nabla\theta\mathbf{V}(0)\varphi\}dsdt + \int_0^{\mathcal{T}}\int_{\Gamma_2}[(\frac{\partial g}{\partial n} + g)\varphi \\ + \frac{\partial\varphi}{\partial n}g + Hg\varphi]\mathbf{V}(0)\mathbf{n}dsdt + \int_\Gamma \theta_0\varphi\mathbf{V}(0)\mathbf{n}ds \quad \forall\varphi \in W(0,\mathcal{T}),\ \varphi(.,\mathcal{T}) = 0,\end{aligned} \tag{23}$$

where $H = div\,\mathbf{n}$ denotes the mean curvature of the boundary Γ.

$$\begin{aligned}\int_\Omega a_{ijkl}\epsilon_{ij}(U)\epsilon_{kl}(\eta)dx + \int_{\Gamma_1} a_{ijkl}\epsilon_{ij}(u)\epsilon_{kl}(\varphi)\mathbf{V}(0)\mathbf{n}ds = -\int_\Omega \beta\nabla\Theta\eta dx \\ + \int_\Gamma \beta\{\Theta\mathbf{n}\eta + \rho f\eta\mathbf{V}(0)\mathbf{n} - \nabla\theta\eta\mathbf{V}(0)\mathbf{n} - \theta\eta\mathbf{V}(0)\mathbf{n}\}ds \quad \forall\eta \in V.\end{aligned} \tag{24}$$

4.2. The form of the directional derivative of the cost functional

Let us recall the definition of the Euler derivative [8]:

Definition 4.3. *The Euler derivative $dJ(\Omega,\mathbf{V})$ of the cost functional $J(\Omega)$ at a point Ω in the direction of the vector field $\mathbf{V}$ is determined by*

$$dJ(\Omega,\mathbf{V}) = \limsup_{\tau\to 0}[J(\Omega_\tau) - J(\Omega)]/\tau. \tag{25}$$

Lemma 4.4. *The derivative $dJ(\Omega,\mathbf{V})$ of the cost functional (13) at a point Ω in a direction $\mathbf{V}$, defined by (25) is given by:*

$$\begin{aligned}dJ(\Omega,\mathbf{V}) = -2\lambda_1\int_\Omega (u_d - u, U)d\Omega + \lambda_1\int_\Gamma \mid u_d - u\mid^2 \mathbf{V}(0)\mathbf{n}d\Gamma \\ -2\lambda_2\int_\Gamma (z_d - u, U)d\Gamma - 2\lambda_2\int_\Gamma (z_d - u, \nabla u\mathbf{V}(0)\mathbf{n})d\Gamma \\ + \lambda_2\int_\Gamma \mid z_d - u\mid^2 H\mathbf{V}(0)\mathbf{n}d\Gamma,\end{aligned} \tag{26}$$

where $U = \dfrac{\partial u}{\partial\tau}$ is the shape derivative of the function u_τ defined by (16) and $H = div\,\mathbf{n}$ is the mean curvature of the boundary Γ.

Proof. Follows from (13), (25) as well as Lemma 4.2. □

In order to eliminate U from (26) we introduce an adjoint state $(p,q) \in W(0,\mathcal{T}) \times V$ satisfying the following system of equations:

$$\int_0^{\mathcal{T}} \int_\Omega -\frac{\partial p}{\partial t} \varphi dx dt + \int_0^{\mathcal{T}} \int_\Omega \nabla p \nabla \varphi dx dt - \int_0^{\mathcal{T}} \int_{\Gamma_2} \frac{\partial g}{\partial \theta} p \varphi ds dt$$

$$- \int_\Omega \beta \nabla \varphi q dx + \int_\Gamma \beta q \varphi \mathbf{n} ds = 0 \quad \forall \varphi \in W(0,\mathcal{T}), p(\mathcal{T}) = 0 \tag{27}$$

$$\int_\Omega a_{ijkl} \epsilon_{ij}(q) \epsilon_{kl}(\eta) dx - 2\lambda_1 \int_\Omega (u_d - u, \eta) dx$$

$$-2\lambda_2 \int_\Gamma (z_d - u, \eta) ds = 0 \quad \forall \eta \in V. \tag{28}$$

Lemma 4.5. *The directional derivative $dJ(\Omega, \mathbf{V})$ of the cost functional (13) at a point Ω in the direction $\mathbf{V}$ is given by:*

$$dJ(\Omega, \mathbf{V}) = \int_\Gamma [\lambda_1 (u_d - u, u_d - u) + \lambda_2 (z_d - u, z_d - u) H$$

$$-2\lambda_2 (z_d - u, \nabla u)] \mathbf{V}(0) \mathbf{n} ds - \int_\Gamma \beta \{\rho f q \mathbf{V}(0) \mathbf{n}$$

$$-\nabla \theta q \mathbf{V}(0) \mathbf{n} - \theta q \mathbf{V}(0) \mathbf{n}\} ds + \int_{\Gamma_1} a_{ijkl} \epsilon_{ij}(u) \epsilon_{kl}(q) \mathbf{V}(0) \mathbf{n} ds$$

$$- \int_0^{\mathcal{T}} \int_\Gamma \{\theta \frac{\partial p}{\partial t} \mathbf{V}(0) \mathbf{n} + \nabla \theta \nabla p \mathbf{V}(0) \mathbf{n}\} ds dt + \int_0^{\mathcal{T}} \int_{\Gamma_2} \frac{\partial g}{\partial \theta} \nabla \theta \mathbf{V}(0) \mathbf{n} p ds dt \tag{29}$$

$$+ \int_0^{\mathcal{T}} \int_{\Gamma_2} [(\frac{\partial g}{\partial n} + g) p + \frac{\partial p}{\partial n} g + H g p] \mathbf{V}(0) \mathbf{n} ds dt + \int_\Gamma \theta_0 p \mathbf{V}(0) \mathbf{n} ds$$

where $(\theta, u) \in W(0,\mathcal{T}) \times V$ and $(p,q) \in W(0,\mathcal{T}) \times V$ satisfy, respectively, the systems (3)–(6) and (27)–(28).

Proof. Follows from (23), (24), (26), (27), (28). □

The necessary optimality condition for the problem (13) has the following standard form:

Theorem 4.6. *Assume conditions (A1)–(A3) are satisfied. For all vector fields $\mathbf{V}$ defined by (8)–(9) an optimal solution $\hat{\Omega} \in U_{ad}$ to the problem (13) satisfies the condition*

$$dJ(\hat{\Omega}, \mathbf{V}) \geq 0, \tag{30}$$

where $dJ(\hat{\Omega}, \mathbf{V})$ is given by (29).

Proof. Is standard [3, 8]. □

References

[1] R.A. Adams, *Sobolev spaces*, 1975, Academic Press, New York.

[2] R. Dziri and J. P. Zolesio, *Shape Sensitivity Analysis for Nonlinear Heat Convection*, Applied Mathematics and Optimization, **35** (1997), 1–20.

[3] J. Haslinger and P. Neittaanmäki, *Finite Element Approximation for Optimal Shape Design. Theory and Applications*, 1988, J. Wiley and Sons.

[4] K.H. Hoffmann and J. Sokołowski, *Interface Optimization Problems for parabolic equations*, Control and Cybernetics, **23** (1994), 445–452.

[5] O.A. Ladyshenskaya, W.A. Solonnikov and N.N. Uralcewa, *Linear and quasilinear equations of parabolic type*, 1967, Nauka, Moscow.

[6] J. P. Raymond and H. Zidani, *Hamiltonian Pontriagin's Principles for Control Problems Governed by Semilinear Parabolic Equations*, Preprint, (1996).

[7] E.J.P.G. Schmidt, *Boundary Control for the heat equation with nonlinear boundary condition*, J. Diff. Equations **78** (1989), 89–121.

[8] J. Sokołowski and J.P. Zolesio, *Introduction to Shape Optimization. Shape Sensitivity Analysis*, 1992, Springer, Berlin.

[9] J. Sprekels and F. Tröltzsch, *On an Optimal Shape Design Problem Connected with the Heating of Elastic Bodies*, Report No 359, Fachbereich 10, Bauwesen, GHS Essen, (1992).

[10] F. Tröltzsch, *On convergence of semidiscrete Ritz–Galerkin schemes applied to the boundary control of parabolic equations with non–linear boundary control*, ZAMM **72** (1992), 291–301.

[11] S.EL. Yacoubi and J. Sokołowski, *Domain Optimization Problems for Parabolic Control Systems*, Applied Mathematics and Computer Science, **6** (1996), 277–289.

A. Myśliński
System Research Institute
ul. Newelska 6
01 - 447 Warsaw, Poland
E-mail address: `myslinsk@ibspan.waw.pl`

F. Tröltzsch
Technical University of Chemnitz
Faculty of Mathematics
PSF 964
D-09007 Chemnitz, Germany
E-mail address: `f.troeltzsch@mathematik.tu-chemnitz.de`

International Series of Numerical Mathematics
Vol. 133, © 1999 Birkhäuser Verlag Basel/Switzerland

Approximate Controllability for a Hydro-Elastic Model in a Rectangular Domain

Axel Osses and Jean-Pierre Puel

1. Introduction

Studying the possibility of controlling the movement of a viscous incompressible fluid by an action on the surrounding elastic structure is an important question. At the moment, the general problem involving Navier Stokes equations in domains varying with time, coupled with a nonlinear elastodynamic equation on the boundary is out of possibility.

We present here a much simpler linear model of fluid-structure interaction where the fluid motion is governed by the Stokes system in a fixed domain, coupled with a linear elastodynamic equation for the elastic structure surrounding the fluid.

In [7], Lions and Zuazua consider such a model assuming that the fluid domain Ω is analytic and they study the question of approximate controllability when the control acts as a force on the whole elastic structure. In this context, they give a result of approximate controllability when the eigenvalues of the Laplace operator with Dirichlet boundary conditions in Ω are simple and they exhibit a counterexample when Ω is a ball.

Here, we want to get rid of the analyticity condition and we give a result of approximate controllability in a $2d$-rectangle when the elastic structure corresponds to two opposite sides of the rectangle. We have been able to obtain a positive result in this very special geometry but we conjecture that the result remains true when the boundary of Ω is not analytic and our result has to be considered only as a first step in this direction.

In section 2 we give the precise model that we consider in dimension 2 only for simplicity, and we state the preliminary mathematical results concerning the direct problem.

In section 3, we give the approximate controllability result in the special geometry and its proof.

2. Linear model of fluid-structure interaction

In order to simplify the presentation, we restrict ourselves to the $2d$-case. Let Ω be a bounded connected open set of $\mathbb{R}^2$ with boundary $\Gamma = \overline{\Gamma}_R \cup \overline{\Gamma}_E$ where Γ_R and

Γ_E are two non-empty relatively open subsets of Γ. We assume that each of Γ_R and Γ_E is regular, the joint between them being either smooth or convex corners. We denote by $\mathbf{n}$ the outward unit normal vector at a point of Γ and by τ the unit tangent vector. In Ω we take an incompressible viscous fluid with viscosity $\nu > 0$ satisfying the Stokes system. On Γ_R, called the rigid part of the boundary the fluid satisfies the no-slip boundary condition. The elastic part of the boundary Γ_E is an elastic arch (or beam) clamped on its edges. The deflection w of the structure (displacement normal to Γ_E) is then a function of time t with values in $H_0^2(\Gamma_E)$ satisfying a linear elastodynamic equation. The stored elastic energy is represented by a quadratic form $b(w,w)$ associated with a symmetric bilinear form $b(w,\widetilde{w})$ and a fourth order differential operator $\mathbf{B}w$ such that

$$\forall w, \widetilde{w} \in \mathbf{D}(\Gamma_E), \qquad b(w,\widetilde{w}) = \int_{\Gamma_E} \mathbf{B}w \cdot \widetilde{w}\, d\gamma. \tag{1}$$

We assume a coercivity hypothesis on b, i.e.

$$\exists \alpha > 0,\ \forall w \in H_0^2(\Gamma_E), \qquad b(w,w) \geq \alpha \|w\|^2_{H_0^2(\Gamma_E)}. \tag{2}$$

Coupling between the fluid motion and the elastic structure displacement occurs via continuity of normal velocities and the normal force $\sigma(\mathbf{y},p)\mathbf{n}\cdot\mathbf{n}$ induced by the fluid on Γ_E, where $\mathbf{y}$ and p are the velocity and pressure of the fluid and

$$\sigma_{ij}(\mathbf{y},p) = -p\delta_{ij} + 2\nu e_{ij}(\mathbf{y}) \tag{3}$$

where

$$e_{ij} = \frac{1}{2}\left(\frac{\partial y_i}{\partial x_j} + \frac{\partial y_j}{\partial x_i}\right). \tag{4}$$

The control variable h is a density of force, acting on Γ_E and normal to Γ_E whose support could be, in principle, a subset of Γ_E. We then obtain the following system of equations describing the fluid-structure interaction complemented by initial conditions.

$$\begin{aligned}
&\text{(5a)} & \frac{\partial \mathbf{y}}{\partial t} - \nu\Delta\mathbf{y} + \nabla p &= 0 \quad \text{in} \quad \Omega\times(0,T)\\
&\text{(5b)} & \operatorname{div}\mathbf{y} &= 0 \quad \text{in} \quad \Omega\times(0,T)\\
&\text{(5c)} & \mathbf{y} &= 0 \quad \text{on} \quad \Gamma_R\times(0,T)\\
&\text{(5d)} & \mathbf{y}\cdot\tau &= 0 \text{ on } \Gamma_E\times(0,T)\\
&\text{(5e)} & \mathbf{y}\cdot\mathbf{n} &= \frac{\partial w}{\partial t} \quad \text{on} \quad \Gamma_E\times(0,T)\\
&\text{(5f)} & \mathbf{y}(0) &= \mathbf{y}_0 \quad \text{in} \quad \Omega
\end{aligned}$$

$$\begin{aligned}
&\text{(6a)} & \frac{\partial^2 w}{\partial t^2} + \mathbf{B}w &= -\sigma(\mathbf{y},p)\mathbf{n}\cdot\mathbf{n} + h \quad \text{on} \quad \Gamma_E\times(0,T)\\
&\text{(6b)} & w(t) &\in H_0^2(\Gamma_E) \quad \text{a.e. in} \quad (0,T)\\
&\text{(6c)} & w(0) &= w_0,\ \frac{\partial w}{\partial t}(0) = w_1 \quad \text{in} \quad \Gamma_E.
\end{aligned}$$

Let us notice that because of the incompressibility condition, we have

$$\int_{\Gamma_E} \frac{\partial w}{\partial t}(t) d\gamma = 0 \quad \text{a.e. in} \quad (0,T). \tag{7}$$

We define the spaces

$$\begin{aligned} V &= \{\mathbf{z} \in H^1(\Omega)^2,\ \operatorname{div} \mathbf{z} = 0,\ \mathbf{z} = \mathbf{0} \text{ on } \Gamma_R,\ \mathbf{z} \cdot \tau = 0 \text{ on } \Gamma_E\ \} \\ H &= \{\mathbf{z} \in L^2(\Omega)^2,\ \operatorname{div} \mathbf{z} = 0,\ \mathbf{z} \cdot \mathbf{n} = 0 \text{ on } \Gamma_R\} \\ L_0^2(\Gamma_E) &= \{\psi \in L^2(\Gamma_E),\ \int_{\Gamma_E} \psi \, d\gamma = 0\}. \end{aligned}$$

We easily obtain the following (formal in a first step) energy equality

$$\frac{1}{2}\frac{d}{dt}|\mathbf{y}|_H^2 + \frac{1}{2}\frac{d}{dt}\left|\frac{\partial w}{\partial t}\right|_{L^2(\Gamma_E)}^2 + \frac{1}{2}\frac{d}{dt}b(w,w) + \nu \int_\Omega |\nabla \mathbf{y}|^2 \, dx = \int_{\Gamma_E} h \frac{\partial w}{\partial t} \, d\gamma. \tag{8}$$

This implies the a priori estimate: there exists $C > 0$ such that

$$\begin{aligned} \|\mathbf{y}\|_{C([0,T];H)}^2 &+ \left\|\frac{\partial w}{\partial t}\right\|_{C([0,T];L^2(\Gamma_E))}^2 + \|w\|_{C([0,T];H_0^2(\Gamma_E))}^2 + \|\mathbf{y}\|_{L^2(0,T;V)}^2 \\ &\leq C \ \left(|\mathbf{y}_0|_H^2 + \|w_0\|_{H_0^2(\Gamma_E)}^2 + |w_1|_{L_0^2(\Gamma_E)}^2 + \|h\|_{L^1(0,T;L^2(\Gamma_E))}^2\right). \end{aligned} \tag{9}$$

Using a Galerkin method (for example) together with (9) we can prove an existence and uniqueness result for (5)–(6).

Theorem 2.1. *If $\mathbf{y}_0 \in H$, $w_0 \in H_0^2(\Gamma_E)$, $w_1 \in L_0^2(\Gamma_E)$ and if h belongs to $L^1(0,T;$ $L^2(\Gamma_E))$, there exists a unique solution $(\mathbf{y}, w)$ of (5)–(6) with*

$$\begin{aligned} \mathbf{y} &\in C([0,T];H) \cap L^2(0,T;V), \\ w &\in C([0,T];H_0^2(\Gamma_E)) \cap C^1([0,T];L_0^2(\Gamma_E)). \end{aligned}$$

Remark 2.2. *Because of (7), we have*

$$\frac{d}{dt}\int_{\Gamma_E} w(t) d\gamma = 0 \quad a.e. \ in \quad (0,T), \tag{10}$$

and therefore,

$$\forall t \in [0,T], \qquad \int_{\Gamma_E} w(t) \, d\gamma = \int_{\Gamma_E} w_0 \, d\gamma. \tag{11}$$

3. Approximate controllability result

From the existence result given in Theorem 2.1, for each control $h \in L^1(0,T;$ $L^2(\Gamma_E))$, we can define the state of the system at final time T

$$(\mathbf{y}(T), w(T), \frac{\partial w}{\partial t}(T)) \in H \times H_0^2(\Gamma_E) \times L_0^2(\Gamma_E).$$

Because of Remark 2.2, it is natural to define the linear and affine subspaces

$$W_0 = \{w \in H_0^2(\Gamma_E), \int_{\Gamma_E} w \, d\gamma = 0\} \tag{12}$$

$$W = w_0 + W_0. \tag{13}$$

We now want to study if the set of reachable states of the system at time T, when the control varies, is dense in $H \times W \times L_0^2(\Gamma_E)$. When satisfied, this property will be called approximate controllability. In [7], Lions and Zuazua consider the case of domains Ω with analytic boundaries and they give conditions for having approximate controllability and a counterexample when Ω is a ball. Here, we wish to study the case of non analytic boundaries. At the moment, we only have a result in the $2d$-case for a very specific geometry.

Theorem 3.1. *We suppose that Ω is a rectangle $]0,a[\times]0,b[$ in $\mathbb{R}^2$ and that Γ_E is the union of two opposite sides of the rectangle (e.g. $\Gamma_E =]0,a[\times\{0\}\cup]0,a[\times\{b\}$ and $\Gamma_R = \{0\}\times]0,b[\cup\{a\}\times]0,b[$). Moreover we assume that the control acts on the whole elastic structure, i.e. $Supp(h) = \Gamma_E \times (0,T)$. Then problem* (5)-(6) *is approximately controllable in $H \times W \times L_0^2(\Gamma_E)$, i.e. for every $(\mathbf{y}_0, w_0, w_1) \in H \times H_0^2(\Gamma_E) \times L_0^2(\Gamma_E)$, the set of reachable states*

$$R(T) = \{(\mathbf{y}(T), w(T), \frac{\partial w}{\partial t}(T)), \ h \in L^1(0,T; L^2(\Gamma_E))\} \tag{14}$$

is dense in $H \times W \times L_0^2(\Gamma_E)$.

Proof. As problem (5)–(6) is linear, using a simple translation, we can assume that

$$\mathbf{y}_0 = \mathbf{0}; \ w_0 = 0; \ w_1 = 0,$$

and we then have to prove that $R(T)$ is dense in $H \times W_0 \times L_0^2(\Gamma_E)$. From Remark 2.2 we know that $R(T)$ is a linear subspace of $H \times W_0 \times L_0^2(\Gamma_E)$. Then we can proceed like in [6] and by Hahn Banach Theorem we have to prove that its orthogonal $R(T)^\perp$ in $H \times W_0 \times L_0^2(\Gamma_E)$ is reduced to$\{(\mathbf{0},0,0)\}$. Let $(\mathbf{z}_T, \varphi_T, \psi_T) \in R(T)^\perp$ and let us solve the (backward) adjoint problem

$$-\frac{\partial \mathbf{z}}{\partial t} - \nu\Delta\mathbf{z} + \nabla q = 0 \quad \text{in} \quad \Omega \times (0,T) \tag{15a}$$

$$\operatorname{div} \mathbf{z} = 0 \quad \text{in} \quad \Omega \times (0,T) \tag{15b}$$

$$\mathbf{z} = 0 \quad \text{on} \quad \Gamma_R \times (0,T) \tag{15c}$$

$$\mathbf{z} \cdot \tau = 0 \quad \text{on} \ \Gamma_E \times (0,T) \tag{15d}$$

$$\mathbf{z} \cdot \mathbf{n} = \frac{\partial \varphi}{\partial t} \quad \text{on} \quad \Gamma_E \times (0,T) \tag{15e}$$

$$\mathbf{z}(T) = \mathbf{z}_T \quad \text{in} \quad \Omega \tag{15f}$$

$$\text{(16a)} \qquad \frac{\partial^2 \varphi}{\partial t^2} + \mathbf{B}\varphi - \sigma(\mathbf{z}, q)\mathbf{n} \cdot \mathbf{n} = 0 \quad \text{on} \quad \Gamma_E \times (0,T)$$

$$\text{(16b)} \qquad \varphi(t) \in H_0^2(\Gamma_E) \quad \text{a.e. in} \quad (0,T)$$

$$\text{(16c)} \qquad \varphi(T) = \varphi_T, \; \frac{\partial \varphi}{\partial t}(T) = \psi_T \quad \text{on} \quad \Gamma_E.$$

As was done in the previous section, it can easily be shown that there exists a unique solution $(\mathbf{z}, \varphi)$ to problem (15)–(16) with

$$\mathbf{z} \in C([0,T]; H) \cap L^2(0,T;V),$$
$$\varphi \in C([0,T]; W_0) \cap C^1([0,T]; L_0^2(\Gamma_E)).$$

Moreover, for every $h \in L^1(0,T; L^2(\Gamma_E))$ and $(\mathbf{y}, w)$ solution of (5)–(6) we have

(17)

$$(\mathbf{y}(T), \mathbf{z}_T) + b(w(T), \varphi_T) + \left(\tfrac{\partial w}{\partial t}(T), \psi_T\right)_{L^2(\Gamma_E)} = \int_0^T \int_{\Gamma_E} h \tfrac{\partial \varphi}{\partial t}\, d\gamma dt.$$

Therefore, $(\mathbf{z}_T, \varphi_T, \psi_T) \in R(T)^{\perp}$ is equivalent to

$$\text{(18)} \qquad \int_0^T \int_{\Gamma_E} h \frac{\partial \varphi}{\partial t}\, d\gamma dt = 0, \qquad \forall h \in L^1(0,T; L^2(\Gamma_E)),$$

and to

$$\text{(19)} \qquad \frac{\partial \varphi}{\partial t} = 0 \quad \text{on} \quad \Gamma_E \times (0,T).$$

Our problem is now to study if a solution $(\mathbf{z}, \varphi)$ of (15)–(16) such that (19) holds is necessary $\{\mathbf{0}, 0\}$.

Remark 3.2. *If, instead of taking $h \in L^1(0,T; L^2(\Gamma_E))$ in (18) (which means that the control may act on the whole elastic structure), we only take $h \in L^1(0,T; L^2(\gamma))$ where γ is an open subset of Γ_E, then we obtain instead of (19) : $\frac{\partial \varphi}{\partial t} = 0$ on $\gamma \times (0,T)$. This situation, which is very interesting for practical purposes, leads to a completely open problem.*

Now, as (19) implies $\mathbf{z} = 0$ on $\Gamma \times (0,T)$ (on the whole boundary), system (15), (16) decouples and $\mathbf{z}$ is solution of

$$\text{(20a)} \qquad -\frac{\partial \mathbf{z}}{\partial t} - \nu \Delta \mathbf{z} + \nabla q = 0 \quad \text{in} \quad \Omega \times (0,T)$$

$$\text{(20b)} \qquad \operatorname{div} \mathbf{z} = 0 \quad \text{in} \quad \Omega \times (0,T)$$

$$\text{(20c)} \qquad \mathbf{z} = 0 \quad \text{on} \quad \Gamma \times (0,T)$$

$$\text{(20d)} \qquad \mathbf{z}(T) = \mathbf{z}_T \quad \text{in} \quad \Omega.$$

As $\mathbf{z} \in C([0,T]; H)$, $\mathbf{z}_T$ must satisfy

$$\mathbf{z}_T \cdot \mathbf{n} = 0 \quad \text{on} \quad \Gamma.$$

Moreover, taking the time derivative in (16) and using again (19) we obtain

$$\text{(21)} \qquad \frac{\partial}{\partial t} \sigma(\mathbf{z}, q)\mathbf{n} \cdot \mathbf{n} = 0 \quad \text{on} \quad \Gamma_E \times (0,T).$$

In order to solve (20) we use the Fourier decomposition method. Let

$$H_0 = \{\mathbf{z} \in L^2(\Omega)^N,\ \operatorname{div} \mathbf{z} = 0,\ \mathbf{z} \cdot \mathbf{n} = 0 \text{ on } \Gamma\}.$$

If we write $(\lambda_j)_{j\geq 1}$ the increasing sequence of eigenvalues of the Stokes operator and if $(\mathbf{z}_j)_{j\geq 1}$ are the corresponding eigenfunctions orthonormalized in H_0 (see [9]) so that

$$\begin{aligned}
&\text{(22a)} & -\nu\Delta \mathbf{z}_j + \nabla q_j &= \lambda_j \mathbf{z}_j \quad \text{in} \quad \Omega \\
&\text{(22b)} & \operatorname{div} \mathbf{z}_j &= 0 \quad \text{in} \quad \Omega \\
&\text{(22c)} & \mathbf{z}_j &= \mathbf{0} \quad \text{on} \quad \Gamma \\
&\text{(22d)} & (\mathbf{z}_j, \mathbf{z}_k)_{H_0} &= \delta_{jk},
\end{aligned}$$

then $(\mathbf{z}_j)_{j\geq 1}$ is an orthonormal basis in H_0. If

$$\mathbf{z}_T = \sum_{j\geq 1} a_j \mathbf{z}_j, \tag{23}$$

Then

$$\mathbf{z}(t) = \sum_{j\geq 1} a_j e^{-\lambda_j (T-t)} \mathbf{z}_j, \tag{24}$$

$$q(t) = \sum_{j\geq 1} a_j e^{-\lambda_j (T-t)} q_j + c(t), \tag{25}$$

where $c(\cdot)$ is a function from $(0,T)$ to $\mathbb{R}$, and

$$\sigma(\mathbf{z}, q)\mathbf{n} \cdot \mathbf{n} = \sum_{j\geq 1} a_j e^{-\lambda_j (T-t)} [\sigma(\mathbf{z}_j, q_j)\mathbf{n} \cdot \mathbf{n}] - c(t). \tag{26}$$

Then $[\sigma(\mathbf{z}, q)\mathbf{n} \cdot \mathbf{n} + c(t)]$ is a function which is real analytic in t (with values in $H^{\frac{1}{2}}(\Gamma)$). Let us rewrite $(\mu_k)_{k\geq 1}$ the strictly increasing sequence of eigenvalues of (22) and

$$\mathbf{z}_{\mu_k} = \sum_{j, \lambda_j = \mu_k} a_j \mathbf{z}_j \tag{27}$$

$$q_{\mu_k} = \sum_{j, \lambda_j = \mu_k} a_j q_j. \tag{28}$$

Then $\mathbf{z}_{\mu_k} = 0$ if and only if $a_j = 0$, $\forall j$, $\lambda_j = \mu_k$, and if $\mathbf{z}_{\mu_k} \neq 0$, it is an eigenfunction of the Stokes operator associated to μ_k. We have

$$\sigma(\mathbf{z}, q)\mathbf{n} \cdot \mathbf{n} = \sum_{k\geq 1} e^{-\mu_k (T-t)} [\sigma(\mathbf{z}_{\mu_k}, q_{\mu_k})\mathbf{n} \cdot \mathbf{n}] - c(t). \tag{29}$$

As the μ_k are all distinct and different from 0, and as c(t) is afunction of t only, condition (21) implies

$$\sigma(\mathbf{z}_{\mu_k}, q_{\mu_k})\mathbf{n} \cdot \mathbf{n} = c_k \quad \text{on} \quad \Gamma_E \qquad \forall k \geq 1, \tag{30}$$

where $c_k \in \mathbb{R}$. We will prove the following unique continuation property

Theorem 3.3. *We suppose* Ω *is the rectangle* $\Omega =]0,a[\times]0,b[$ *and that* Γ_E *is the union of two opposite sides of the rectangle (for example* $\Gamma_E =]0,a[\times\{0\}\cup]0,a[\times\{b\}$*). Then the unique solution* $(\mathbf{z}, q)$ *of*

$$-\nu\Delta\mathbf{z} + \nabla q = \mu\mathbf{z} \quad in \quad \Omega \tag{31a}$$

$$\operatorname{div}\mathbf{z} = 0 \quad in \quad \Omega \tag{31b}$$

$$\mathbf{z} = \mathbf{0} \quad on \quad \Gamma \tag{31c}$$

and

$$\sigma(\mathbf{z}, q)\mathbf{n}\cdot\mathbf{n} = -q = c \in \mathbb{R} \quad on \quad \Gamma_E. \tag{32}$$

is

$$\mathbf{z} = \mathbf{0}, \quad q = Cst \quad in \quad \Omega.$$

Let us assume for the moment that this Theorem is proved. Then any solution $\mathbf{z}_{\mu_k}$ of (22) satisfying (30) must be 0.

Then $\mathbf{z}_T = \mathbf{0}$, $\mathbf{z} = \mathbf{0}$ in $\Omega\times(0,T)$ and $q(t) = c(t)$ in $\Omega\times(0,T)$, so that

$$\sigma(\mathbf{z}, q)\mathbf{n}\cdot\mathbf{n} = -c(t).$$

From (19) we have $\psi_T = 0$ and $\varphi(t) = \varphi_T$, $\forall t \le T$. Now (16) gives

$$\begin{aligned} \mathbf{B}\varphi_T &= -c(t) = -c \in \mathbb{R} \quad \text{on} \quad \Gamma_E \\ \varphi_T &\in W_0, \end{aligned}$$

which implies $\varphi_T = 0$.

Therefore the unique continuation property for the eigenvalue problem implies approximate controllability for our original system (5), (6).

Proof of Theorem 3.3. As Ω is a $2d$-domain, we can make the following standard transformation. Let ω be defined as

$$-\Delta\omega = \frac{\partial z_1}{\partial x_2} - \frac{\partial z_2}{\partial z_1} \quad \text{in} \quad \Omega \tag{33}$$

$$\omega = 0 \quad \text{on} \quad \Gamma. \tag{34}$$

Then ω satisfies

$$\Delta^2\omega = -\mu\Delta\omega \quad \text{in} \quad \Omega \tag{35a}$$

$$\omega = 0 \quad \text{on} \quad \Gamma \tag{35b}$$

$$\frac{\partial\omega}{\partial\mathbf{n}} = 0 \quad \text{on} \quad \Gamma \tag{35c}$$

and

$$\frac{\partial}{\partial\mathbf{n}}\Delta\omega = 0 \quad \text{on} \quad \Gamma_E. \tag{36}$$

Now, because of the special geometry of Ω, Γ_E and Γ_R, it is convenient to rewrite (35), (36) as follows

$$\Delta^2\omega = -\mu\Delta\omega \quad \text{in} \quad \Omega \tag{37a}$$

$$\omega = 0 \quad \text{on} \quad \Gamma_R \tag{37b}$$

$$\frac{\partial\omega}{\partial\mathbf{n}} = 0 \quad \text{on} \quad \Gamma_R \tag{37c}$$

$$\frac{\partial\omega}{\partial\mathbf{n}} = 0 \quad \text{on} \quad \Gamma_E \tag{37d}$$

$$\frac{\partial}{\partial\mathbf{n}}\Delta\omega = 0 \quad \text{on} \quad \Gamma_E, \tag{37e}$$

and

$$\omega = 0 \quad \text{on} \quad \Gamma_E. \tag{38}$$

We will first consider all solutions of (37).

Lemma 3.4. *There exists an infinite sequence $(g_n)_n$, $g_n \in C^\infty([0,b])$ (in fact real analytic), and for every n, there exists an infinite sequence $(f_n)^p$, $(f_n)^p \in C^\infty([0,a])$ (in fact real analytic) such that the functions*

$$(\omega_n)^p(x_1,x_2) = g_n(x_2)(f_n)^p(x_1) \tag{39}$$

form an orthonormal basis in $L^2(]0,a[\times]0,b[)$ of eigenfunctions of (37) associated to eigenvalues $(\lambda_n)^p$.

Proof. Let $(g_n)_n$ be the orthonormal basis in $L^2(]0,b[)$ of eigenfunctions of the Neumann problem

$$-\frac{d^2}{dx_2^2}g_n = \alpha_n g_n \quad \text{in} \quad]0,b[\tag{40a}$$

$$\frac{d}{dx_2}g_n(0) = \frac{d}{dx_2}g_n(b) = 0 \tag{40b}$$

$$\int_0^b g_n(x_2)g_m(x_2)dx_2 = \delta_{mn}. \tag{40c}$$

We will look for ω solution of (37) of the form

$$\omega(x_1,x_2) = g_n(x_2)f(x_1). \tag{41}$$

Then f will have to satisfy

$$\frac{d^4}{dx_1^4}f - 2\alpha_n\frac{d^2}{dx_1^2}f + \alpha_n^2 f = \mu\left(-\frac{d^2}{dx_1^2}f + \alpha_n f\right) \tag{42a}$$

$$f(0) = f(a) = 0 \tag{42b}$$

$$\frac{d}{dx_1}f(0) = \frac{d}{dx_1}f(a) = 0 \tag{42c}$$

As $\alpha_n \geq 0$, for each fixed n, (42) admits an infinite sequence of eigenfunctions $(f_n)^p$ associated with eigenvalues $(\lambda_n)^p$ and the $(f_n)^p$ form an orthonormal basis in $L^2(]0,a[)$. If we write

$$(\omega_n)^p(x_1,x_2) = g_n(x_2)(f_n)^p(x_1) \tag{43}$$

then $(\omega_n)^p$ is solution of (37) and they form an orthonormal family in $L^2(]0, a[\times]0,b[)$. Let us show that this family is a basis. Let $v \in L^2(]0,a[\times]0,b[)$ such that

$$\int_0^b \int_0^a v(x_1,x_2)(\omega_n)^p(x_1,x_2)dx_1dx_2 = 0, \qquad \forall n,p \tag{44}$$

If

$$w_n(x_1) = \int_0^b v(x_1,x_2)g_n(x_2)dx_2 \tag{45}$$

Then we have

$$\int_0^a w_n(x_1)(f_n)^p(x_1)dx_1 = 0, \qquad \forall p \tag{46}$$

which implies

$$w_n = 0, \qquad \forall n,$$

and then

$$v = 0.$$

Let us now take ω solution of (37). Then μ is one of the $(\lambda_n)^p$ and may correspond to different values (a finite number) of n and p. Then there exist k_0 and β_i, $i = 1,\ldots,k_0$, such that

$$\omega = \sum_{i=1}^{k_0} \beta_i g_{n_i}(f_{n_i})^{p_i} \tag{47}$$

$$\mu = (\lambda_{n_i})^{p_i}, \qquad \forall i = 1,\ldots,k_0, \tag{48}$$

and ω must satisfy (38). We can expand $(f_{n_i})^{p_i}$ in power series of x_1

$$(f_{n_i})^{p_i}(x_1) = \sum_{j\geq 0} a_i^j x_1^j. \tag{49}$$

Writing this in (42), we obtain $a_i^0 = O$, $a_i^1 = 0$ and

$$(j+4)!a_i^{j+4} + (j+2)!(\mu - 2\alpha_{n_i})a_i^{j+2} + p!(\alpha_{n_i}^2 - \mu\alpha_{n_i})a_i^j = 0. \tag{50}$$

We write

$$\gamma_i = \beta_i g_{n_i}(0), \tag{51}$$

so that from (38) we have

$$\sum_{i=1}^{k_0} \gamma_i a_i^j = 0, \qquad \forall j \geq 0. \tag{52}$$

Taking $j=0$, then $j=1$ in (50) gives

$$\sum_{i=1}^{k_0} \alpha_{n_i}\gamma_i a_i^2 = 0, \tag{53}$$

and

$$\sum_{i=1}^{k_0} \alpha_{n_i}\gamma_i a_i^3 = 0, \tag{54}$$

Let us argue by recurrence.
First case : $k=2p$.

Let us assume that for every $l=1,\dots,p$

$$\sum_{i=1}^{k_0} \alpha_{n_i}^{1+j}\gamma_i a_i^{2l-2j} = 0, \qquad \forall j=0,\dots,l-1. \tag{55}$$

Let us prove the same relation for $l=p+1$.

We obtain from (50) taken with $j=2p,2p-2,\dots,2,0$ a linear $(p+1,p+1)$ system

$$M_{p+1}A_{p+1}=0,$$

with

$$A_{p+1} = \begin{pmatrix} u_0 \\ \vdots \\ u_p \end{pmatrix},$$

where

$$u_j = \sum_{i=1}^{k_0} \alpha_{n_i}^{1+j}\gamma_i a_i^{2p+2-2j},$$

and

$$M_{p+1} = \begin{bmatrix} -2(2p+2)! & (2p)! & & & \\ (2p+2)! & -2(2p)! & (2p-2)! & & \\ & (2p)! & -2(2p-2)! & (2p-4)! & \\ & \ddots & \ddots & \ddots & \\ & & \ddots & \ddots & \ddots \\ & & 6! & -2\cdot 4! & 2! \\ & & & 4! & -2\cdot 2! \end{bmatrix}.$$

We have

$$\det M_{p+1} = (2p+2)!(2p)!\dots 2!\Delta_{p+1},$$

where

$$\Delta_{k+1} = \begin{vmatrix} -2 & 1 & & & & \\ 1 & -2 & 1 & & & \\ & \ddots & \ddots & \ddots & & \\ & & \ddots & \ddots & \ddots & \\ & & & 1 & -2 & 1 \\ & & & & 1 & -2 \end{vmatrix} \neq 0.$$

Therefore

$$A_{p+1} = \begin{pmatrix} 0 \\ \vdots \\ 0 \\ \vdots \\ 0 \end{pmatrix}.$$

Second case : $k = 2p+1$.

We assume that for every $l = 1, \dots, p$

$$\sum_{i=1}^{k_0} \alpha_{n_i}^{1+j} \gamma_i a_i^{2l+1-2j} = 0, \qquad \forall j = 0, \dots, l-1. \tag{56}$$

and we want to show that

$$\sum_{i=1}^{k_0} \alpha_{n_i}^{1+j} \gamma_i a_i^{2p+3-2j} = 0, \qquad \forall j = 0, \dots, l-1. \tag{57}$$

Here we take (50) with $j = 2p+1, 2p-1, \dots, 3, 1$ and we obtain a linear $(p+1, p+1)$ system

$$N_{p+1} B_{p+1} = 0,$$

$$B_{p+1} = \begin{pmatrix} v_0 \\ \vdots \\ v_p \end{pmatrix},$$

$$v_j = \sum_{i=1}^{k_0} \alpha_{n_i}^{1+j} \gamma_i a_i^{2p+3-2j},$$

and

$$N_{p+1} = \begin{bmatrix} -2(2p+3)! & (2p+1)! & & & \\ (2p+3)! & -2(2p+1)! & (2p-1)! & & \\ & (2p+1)! & -2(2p-1)! & (2p-3)! & \\ & \ddots & \ddots & \ddots & \\ & & \ddots & \ddots & \ddots \\ & & 7! & -2\cdot 5! & 3! \\ & & & 5! & -2\cdot 3! \end{bmatrix}.$$

Then

$$\det N_{p+1} = (2p+3)!(2p+1)!\dots 3!\Delta_{p+1} \neq 0$$

and the linear system implies

$$B_{p+1} = \begin{pmatrix} 0 \\ \vdots \\ 0 \\ \vdots \\ 0 \end{pmatrix}.$$

Therefore we have proved, in particular that

$$\sum_{i=1}^{k_0} \alpha_{n_i} \gamma_i a_i^k = 0, \qquad \forall k \geq 0 \tag{58}$$

which means that

$$\Delta\omega(x_1, 0) = 0, \tag{59}$$

so that ω satisfies

$$\begin{aligned}
\Delta^2\omega &= -\mu\Delta\omega \quad \text{in} \quad \Omega && \text{(60a)}\\
\omega &= 0 \quad \text{on} \quad \Gamma && \text{(60b)}\\
\frac{\partial\omega}{\partial\mathbf{n}} &= 0 \quad \text{on} \quad \Gamma && \text{(60c)}\\
\Delta\omega(x_1,0) &= 0 \quad \text{on} \quad (0,a), && \text{(60d)}\\
\frac{\partial}{\partial\mathbf{n}}\Delta\omega(x_1,0) &= 0 \quad \text{on} \quad (0,a), && \text{(60e)}
\end{aligned}$$

and this implies

$$\omega = O$$

from usual continuation results. The proofs of Theorem 3.3 and Theorem 3.1 are now complete.

In fact the hypothesis on the geometry of Ω is used to prove real analyticity in a neighborhood of a rectangular corner and we hope to be able to obtain our result assuming only that Γ_E and Γ_R meet at a rectangular corner. We also conjecture that approximate controllability is still valid whenever Γ is not analytic and a first step in this direction could be to prove the result when Γ has a convex corner. We hope to be able to use results on singularities like the ones of [5] to get a precise behaviour of the eigenfunctions near a convex corner. A second conjecture is, when Γ is analytic, that the only counterexample is the case of a ball. A counterexample corresponds to the existence of an eigenfunction for the Stokes operator which induces a constant pressure on a part of the boundary. There is of course a strong relation between our problem and the ones treated by Berenstein [1],[2], Beretta and Vogelius [3], Vogelius [10], Williams [11] and others [4], [8] related to the Pompeiu problem.

References

[1] C. Berenstein: An inverse spectral theorem and its relation to the Pompeiu problem, *J. Anal. Math.*, **37**, 1980, 128–144.

[2] C. Berenstein: *The Pompeiu problem, what's new?* Deville, R. (ed.) et al., Complex analysis, harmonic analysis and applications. Proceedings of a conference in honour of the retirement of Roger Gay, June 7–9, 1995, Bordeaux, France. Harlow: Longman. Pitman Res. Notes Math. Ser. 347, 1996, 1–11.

[3] E. Beretta, M. Vogelius: An inverse problem originating from magnetohydrodynamics. III: Domains with corners of arbitrary angles, *Asymptotic Anal.*, **11**(3), 1995, 289–315.

[4] L. Brown, B.M. Schreiber, B.A. Taylor: Spectral synthesis and the Pompeiu problem, *Ann. Inst. Fourier*, **23**(3), 1973, 125–154.

[5] P. Grisvard: *Elliptic problems in nonsmooth domains*, Monographs and Studies in Mathematics, 24. Pitman, Boston-London-Melbourne, 1985.

[6] J.-L. Lions:Remarques sur la contrôlabilité approchée, Control of distributed systems, Span.-Fr. Days, Malaga/Spain 1990, Grupo Anal. Mat. Apl. Univ. Malaga 3, 1990, 77–87.

[7] J.-L. Lions, E. Zuazua: Approximate controllability of a hydro-elastic coupled system, *ESAIM: Control, Optimization and Calculus of Variations*, **1**, 1995, 1–15.

[8] J. Serrin: A symmetry problem in potential theory, *Arch. Rational. Mech. Anal.*, **43**, 1971, 304–318.

[9] R. Temam: *Navier–Stokes Equations*, North–Holland, Amsterdam, 1977.

[10] M. Vogelius: An inverse problem for the equation $\Delta u = -cu - d$, *Ann. Inst. Fourier*, **44**(4), 1994, 1181–1209.

[11] S.A. Williams: Analyticity of the boundary for Lipschitz domains without the Pompeiu property, *Indiana Univ. Math. J.*, **30**, 1981, 357–369.

A. Osses
Centre de Mathématiques Appliquées
Ecole Polytechnique
91128 Palaiseau Cedex, France
and Universidad de Chile
Departamento de Ingeniería Matemática
casilla 170/3, correo 3, Santiago, Chile
E-mail address: axosses@cmapx.polytechnique.fr

Jean-Pierre Puel
Université de Versailles Saint-Quentin
and
Centre de Mathématiques Appliquées
Ecole Polytechnique
91128 Palaiseau Cedex, France
E-mail address: jppuel@cmapx.polytechnique.fr

International Series of Numerical Mathematics
Vol. 133,

Noncooperative Games with Elliptic Systems

Tomáš Roubíček

Abstract. Noncooperative games with "slightly" nonlinear systems and "sufficiently" uniformly convex individual cost functionals may admit a relaxation having a globally convex structure, which guarantees existence of its Nash equilibria as well as existence of approximate Nash equilibria (in a suitable sense) for the original game. The relaxation is made by a continuous extension on a suitable convex (local) compactification. This will be illustrated by semilinear elliptic systems. The regularity will be employed, too.

1. Introduction and problem formulation

Noncooperative games with linear systems and convex individual cost functionals (or the concave "payoff" functionals) possess, under mild data qualification, the Nash equilibria [8]. If the system (resp. the costs) is nonlinear (resp. nonconvex) in controls, a relaxation can convexify the problem but, if nonlinear/nonconvex phenomena are exhibited with respect to state, even approximate Nash equilibria need not exist in general. Yet, sometimes the nonlinearity of the controlled system can be compensated by a uniform convexity of the costs. We want to illustrate this on the following game with a semilinear elliptic system of m 2nd-order equations:

$$
(\mathsf{G})\left\{\begin{array}{lll}
\begin{array}{l}\text{Find Nash}\\ \text{equilibrium}\end{array} & \left\{\begin{array}{l}\displaystyle\int_\Omega g_1(x,y)+h_1(x,u_1)-\varphi(x,u_1,u_2)\,\mathrm{d}x\,, \\ \displaystyle\int_\Omega g_2(x,y)+h_2(x,u_2)+\varphi(x,u_1,u_2)\,\mathrm{d}x\,,\end{array}\right. & \begin{array}{r}(1^{\text{st}}\text{ player cost})\\ (2^{\text{nd}}\text{ player cost})\end{array}\\
\text{subject to} & \displaystyle\sum_{j,l=1}^{n}\sum_{i=1}^{m}\frac{\partial}{\partial x_l}\left(a_{ijkl}(x)\frac{\partial y_i}{\partial x_j}\right)=b_k(x,y) & (\text{state equations})\\
 & \qquad +f_{1,k}(x,u_1)+f_{2,k}(x,u_2)\quad \text{on }\Omega, & k=1,\dots,m,\\
 & u_1(x)\in S_1(x),\ \ u_2(x)\in S_2(x)\ \ \text{for }x\in\Omega, & (\text{control constraints})\\
 & y\in W_0^{1,2}(\Omega;\mathbb{R}^m),\quad u_1\in L^{p_1}(\Omega;\mathbb{R}^{m_1}),\quad u_2\in L^{p_2}(\Omega;\mathbb{R}^{m_2}), &
\end{array}\right.
$$

where $\Omega\subset\mathbb{R}^n$ is a bounded domain with a smooth boundary $\partial\Omega$, $a_{ijkl}:\Omega\to\mathbb{R}$, $g_\alpha:\Omega\times\mathbb{R}^m\to\mathbb{R}$, $b:\Omega\times\mathbb{R}^m\to\mathbb{R}^m$, $h_\alpha:\Omega\times\mathbb{R}^{m_\alpha}\to\mathbb{R}$, $f_\alpha:\Omega\times\mathbb{R}^{m_\alpha}\to\mathbb{R}^m$ and $\varphi:\Omega\times\mathbb{R}^{m_1}\times\mathbb{R}^{m_2}\to\mathbb{R}$, and moreover $S_\alpha:\Omega\rightrightarrows\mathbb{R}^{m_\alpha}$ are multivalued mappings, where $\alpha=1,2$ distinguishes the particular players. The notation for the Sobolev space $W_0^{1,2}(\Omega;\mathbb{R}^m) := \{y \in L^2(\Omega;\mathbb{R}^m);\ \nabla y \in L^2(\Omega;\mathbb{R}^{m\times n}),\ y|_{\partial\Omega} = 0\}$ and for the Lebesgue space $L^p(\Omega;\cdot)$ is standard. The controlled system may arise, e.g., from Lamé's system in classical elasticity with small displacements – then $m = n$ and y can be interpreted as a (small) deformation of the reference body Ω.

Besides the mentioned nonlinearity of the controlled system, the following special phenomena should be highlighted (in contrast with [14] as far as 1. and 2. concerns):

1. The multidimensional character of distributed parameter systems requires regularity investigations both of the state and of the adjoint state to quantify an "admissible nonlinearity" of the controlled system.
2. Strategies bounded only in L^p- (but not L^∞-) norm require special relaxation technique, more general than classical Young measures used, e.g., in Warga [18].
3. General nonlinearities with respect to the controls cause failure of joint weak* continuity of the extended costs. This can be admitted only in the zero-sum term (cf. $\pm\varphi$ in (G)), otherwise additive coupling is necessary.

The controls u_1 and u_2 in (G) represent the strategies of the particular players. The game (G) has got the global structure of finding Nash equilibria of the costs $\Phi_1, \Phi_2 : U_1 \times U_2 \to \mathbb{R}$ defined by

$$\Phi_\alpha(u_1,u_2) := \int_\Omega g_\alpha(x, y(u_1,u_2)) + h_\alpha(x,u_\alpha) + (-1)^\alpha \varphi(x,u_1,u_2)\,\mathrm{d}x \tag{1.1}$$

with $y = y(u_1,u_2) \in W_0^{1,2}(\Omega;\mathbb{R}^m)$ the unique weak solution to the semilinear system

$$\mathrm{div}(A(x)\nabla y) = b(x,y) + f_1(x,u_1) + f_2(x,u_2)\ , \tag{1.2}$$

where $A(x) : \mathbb{R}^{m\times n} \to \mathbb{R}^{m\times n} : [\xi_{kl}] \mapsto [\sum_{j=1}^n \sum_{i=1}^m a_{ijkl}(x)\xi_{ij}]$, while the sets of admissible strategies U_1 and U_2 are defined by

$$U_\alpha := \{u_\alpha \in L^\infty(\Omega;\mathbb{R}^{m_\alpha});\ u_\alpha(x) \in S_\alpha(x) \text{ for a.a. } x \in \Omega\}. \tag{1.3}$$

Let us recall that the pair of strategies $(u_1,u_2) \in U_1 \times U_2$ is called a Nash equilibrium for the game (G) if

$$\Phi_1(u_1,u_2) = \min_{v_1\in U_1} \Phi_1(v_1,u_2) \quad \& \quad \Phi_2(u_1,u_2) = \min_{v_2\in U_2} \Phi_2(u_1,v_2). \tag{1.4}$$

Such equilibria, however, do not exist unless quite strong data qualification are imposed. Instead, it is practically satisfactory to find at least an approximate equilibrium. In analogy with minimizing sequences used standardly in cooperative situations, here it is natural to speak about the so-called equilibrium sequences $\{(u_1^k,u_2^k)\}_{k\in\mathbb{N}}$ introduced in [13] as

$$\exists \Psi_1 : U_1 \to \mathbb{R} : \quad \lim_{k\to\infty} \Phi_1(\cdot, u_2^k) = \Psi_1 \ \text{ point-wise,} \tag{1.5a}$$

$$\exists \Psi_2 : U_2 \to \mathbb{R} : \quad \lim_{k\to\infty} \Phi_2(u_1^k, \cdot) = \Psi_2 \ \text{ point-wise,} \tag{1.5b}$$

$$\lim_{k\to\infty} \Psi_1(u_1^k) = \inf_{u\in U_1} \Psi_1(u) \quad \& \quad \lim_{k\to\infty} \Psi_2(u_2^k) = \inf_{u\in U_2} \Psi_2(u)\ , \tag{1.5c}$$

cf. also [14] for a comparison with other concepts. It should be emphasized that even the approximate equilibria need not exist (cf. Patrone [12; Example 3]), which makes the presented results nontrivial.

There is not much literature about games with distributed-parameter systems, we refer to Benssousan [2], Díaz and Lions [4], Roxin [16], or Warga [18]. Special interplay of the data admitting nonlinear phenomena was treated by Chawla [3], Lenhart at al. [6], Mordukhovich and Zhang [7], Szpiro and Protopopescu [17] for zero-sum games with partial differential equations.

Considering $n = 2$ or $n = 3$, $q > n/2$, $r < 2n/(n-2)$, $p_\alpha > 1$, $a_p \in L^p(\Omega)$, $c \in \mathbb{R}$, $\gamma : \mathbb{R} \to \mathbb{R}^+$ arbitrary continuous increasing, and $\varepsilon_1, \varepsilon_2, \varepsilon_3, \delta > 0$, for $\alpha = 1, 2$, the basic data qualifications we will need are the following:

$$\Omega \subset \mathbb{R}^n \text{ is a bounded } C^{1,1}\text{-domain,} \tag{1.6a}$$

$$g_\alpha,\ b,\ h_\alpha,\ f_\alpha,\ \varphi \text{ are Carathéodory functions,}$$
$$\text{i.e. measurable in } x\in\Omega \text{ and continuous in the resting variables,} \tag{1.6b}$$

$$|b(x,y)| \le a_q(x) + c|y|^{r/q}, \qquad |f_\alpha(x,u)| \le a_q(x) + c|u|^{p_\alpha/q}, \tag{1.6c}$$

$$|g_\alpha(x,y)| \le a_1(x) + \gamma(|y|), \qquad |h_\alpha(x,u)| \le a_1(x) + c|u|^{p_\alpha}, \tag{1.6d}$$

$$a_{ijkl} \in W^{1,\infty}(\Omega)\ , \qquad \forall \xi\in\mathbb{R}^{m\times n} : \sum_{j,l=1}^{n}\sum_{i,k=1}^{m} a_{ijkl}(x)\xi_{ij}\xi_{kl} \ge \delta|\xi|^2\ , \tag{1.6e}$$

$$\forall y_1, y_2 \in \mathbb{R}^m : \quad (b(x,y_1) - b(x,y_2)) \cdot (y_1 - y_2) \ge 0, \tag{1.6f}$$

$$S_\alpha \text{ admits a measurable } p_\alpha\text{-integrable selection, i.e. } U_\alpha \ne \emptyset, \tag{1.6g}$$

$$h_\alpha(x,u) \ge \varepsilon_\alpha |u|^{p_\alpha}, \quad g_\alpha \ge 0, \quad |\varphi(x,u_1,u_2)| \le a_{1+\varepsilon_3}(x). \tag{1.6h}$$

2. A relaxed game and its structure

We will extend continuously the game (G) on a suitable convex locally (sequentially) compact hull of U_α created by means of suitable linear spaces of Carathéodory integrands containing all possible nonlinearities, e.g.

$$H_\alpha := \Big\{ a(x)h_\alpha(x,s) + f(x,s);\ a\in C(\bar{\Omega}),\ f\in L^1(\Omega; C(\mathbb{R}^{m_\alpha})) \Big\}, \tag{2.1}$$

where $C(\cdot)$ stands for a space of bounded continuous functions. It is natural to equip H_α with

$$\|h\|_\alpha := \inf_{|h(x,s)|\le a(x)+c|s|^{p_\alpha}} \|a\|_{L^1(\Omega)} + c \tag{2.2}$$

which is a norm (see [13]) making H_α separable. Then we embed $L^{p_\alpha}(\Omega;\mathbb{R}^{m_\alpha})$ continuously into H_α^* by

$$i_\alpha : u_\alpha \mapsto \left(h \mapsto \int_\Omega h(x, u_\alpha(x))\mathrm{d}x \right) \tag{2.3}$$

and define the set of the so-called generalized Young functionals by $Y_{H_\alpha}^{p_\alpha}(\Omega;\mathbb{R}^{m_\alpha}) := \mathrm{w}*\text{-}\mathrm{cl}\, i_\alpha(L^{p_\alpha}(\Omega;\mathbb{R}^{m_\alpha}))$. It is known [13] that, as a consequence of (2.1) with (1.6h), $Y^{p_\alpha}_{H_\alpha}(\Omega;\mathbb{R}^{m_\alpha})$ is a convex locally (sequentially) compact envelope of $L^{p_\alpha}(\Omega;\mathbb{R}^{m_\alpha})$. The sets of admissible relaxed controls are then defined by

$$\bar{U}_\alpha := \text{w*-cl}\, i_\alpha(U_\alpha). \tag{2.4}$$

Thanks to the special form (1.3), each $\bar{U}_\alpha$ is convex and locally compact in H^*_α too; of course, we consider the weak* topology on H^*_α.

We will need a continuous extension on $\bar{U}_\alpha$ of the Nemytskiĭ mapping $u \mapsto f_\alpha(x,u(x)) : L^{p_\alpha}(\Omega;\mathbb{R}^{m_\alpha}) \to L^q(\Omega;\mathbb{R}^m)$ defined, for $\eta \in Y^{p_\alpha}_{H_\alpha}(\Omega;\mathbb{R}^{m_\alpha})$, by

$$f_\alpha \bullet \eta \in L^q(\Omega;\mathbb{R}^m): \qquad \int_\Omega a(x)\cdot[f_\alpha \bullet \eta](x)\mathrm{d}x \equiv \langle f_\alpha \bullet \eta, a\rangle = \langle \eta, a\cdot f_\alpha\rangle \tag{2.5}$$

for any $a \in C(\bar{\Omega};\mathbb{R}^m)$, where $[a\cdot f_\alpha](x,u) := \sum_{k=1}^m a_k(x) f_{\alpha,k}(x,u)$. Note that $a \mapsto a\cdot f_\alpha : C(\bar{\Omega};\mathbb{R}^m) \to H_\alpha$ is continuous because $\|a\cdot f_\alpha\|_\alpha \le \|a\|_{C(\bar{\Omega};\mathbb{R}^m)}\|f_\alpha\|_\alpha$ and $\eta \mapsto f_\alpha \bullet \eta$ is linear; cf. [13; Example 3.6.3]. Obviously, $f_\alpha \bullet i_\alpha(u) = f_\alpha(u)$.

Analogously, we define a continuous extension of the Nemytskiĭ mapping $u \mapsto h_\alpha(x,u(x)) : L^{p_\alpha}(\Omega;\mathbb{R}^{m_\alpha}) \to L^1(\Omega)$, but now

$$h_\alpha \bullet \eta \in \mathrm{rca}(\bar{\Omega}): \qquad \int_{\bar{\Omega}} a(x)[h_\alpha \bullet \eta](\mathrm{d}x) \equiv \langle h_\alpha \bullet \eta, a\rangle = \langle \eta, a\cdot h_\alpha\rangle \tag{2.6}$$

for any $a \in C(\bar{\Omega})$, where $\mathrm{rca}(\bar{\Omega}) \cong C(\bar{\Omega})^*$ denotes standardly "regular countably additive" set functions (i.e. Radon measures) on $\bar{\Omega}$ (i.e. the closure of Ω).

Eventually, we need also an extension on $\bar{U}_1 \times \bar{U}_2$ of the two-argument Nemytskiĭ mapping $(u_1,u_2) \mapsto \varphi(x,u_1(x),u_2(x)) : L^{p_1}(\Omega;\mathbb{R}^{m_1}) \times L^{p_2}(\Omega;\mathbb{R}^{m_2}) \to L^{1+\varepsilon_3}(\Omega)$ defined by

$$\varphi\bullet\eta_1\bullet\eta_2 := (\varphi\bullet_1\eta_1)\bullet_2\eta_2 \quad\text{where}\quad [\varphi\bullet_1\eta_1](x,u_2) := \varphi(\cdot,\cdot,u_2)\bullet_1\eta_1 \tag{2.7}$$

with $\bullet_1$ and $\bullet_2$ defined by (2.5) with $f_1 \in H_1$ and $f_2 \in H_2$ given by $f_1 \equiv f_{1,u_2} := \varphi(\cdot,\cdot,u_2)$ and $f_2(x,u_2) := [f_{1,u_2}\bullet_1\eta_1](x)$, respectively; cf. [13; Lemma 3.6.14].

The relaxed game is then created by the continuous extension of the original game (G) from $U_1 \times U_2$ to $\bar{U}_1 \times \bar{U}_2$, which gives:

$$(\mathsf{RG})\qquad \begin{cases} \text{Find Nash equilibrium} & \begin{cases} \bar{J}_1(\eta_1,\eta_2,y) := \displaystyle\int_{\bar{\Omega}} g_1(y) + h_1\bullet\eta_1 - \varphi\bullet\eta_1\bullet\eta_2 \,\mathrm{d}x\,, \\ \bar{J}_2(\eta_1,\eta_2,y) := \displaystyle\int_{\bar{\Omega}} g_2(y) + h_2\bullet\eta_2 + \varphi\bullet\eta_1\bullet\eta_2 \,\mathrm{d}x\,, \end{cases} \\ \text{subject to} & \mathrm{div}(A(x)\nabla y) = b(y) + f_1\bullet\eta_1 + f_2\bullet\eta_2\,, \\ & y \in W^{1,2}_0(\Omega;\mathbb{R}^m), \quad \eta_1 \in \bar{U}_1, \quad \eta_2 \in \bar{U}_2. \end{cases}$$

To investigate the structure of the relaxed game (RG) in more detail, let us define the extended costs $\bar{\Phi}_1, \bar{\Phi}_2 : \bar{U}_1 \times \bar{U}_2 \to \mathbb{R}$ by

$$\bar{\Phi}_\alpha(\eta_1,\eta_2) := \bar{J}_\alpha(\eta_1,\eta_2,y(\eta_1,\eta_2))\,, \qquad \alpha = 1,2, \tag{2.8}$$

where $y = y(\eta_1, \eta_2) \in W_0^{1,2}(\Omega; \mathbb{R}^m)$ denotes the unique weak solution to the boundary-value problem

$$\operatorname{div}(A(x)\nabla y) = b(y) + f_1 \bullet \eta_1 + f_2 \bullet \eta_2 \ . \tag{2.9}$$

Obviously, (RG) just represents a Nash equilibrium search over $\bar{U}_1 \times \bar{U}_2$ of the extended costs $\bar{\Phi}_1$ and $\bar{\Phi}_2$; this means: we are to find $(\eta_1, \eta_2) \in \bar{U}_1 \times \bar{U}_2$ such that

$$\bar{\Phi}_1(\eta_1, \eta_2) = \min_{\tilde{\eta}_1 \in \bar{U}_1} \bar{\Phi}_1(\tilde{\eta}_1, \eta_2) \quad \text{and} \quad \bar{\Phi}_2(\eta_1, \eta_2) = \min_{\tilde{\eta}_2 \in \bar{U}_2} \bar{\Phi}_2(\eta_1, \tilde{\eta}_2). \tag{2.10}$$

The main results about (RG) and relations between (RG) and (G) are supported by the properties stated in the following six assertions.

Lemma 2.1. *Let (1.6) hold and $\varepsilon > 0$. Then there are ϱ_1 and ϱ_2 such that*

$$\forall u_2 \in U_2: \quad \Phi_1(u_1, u_2) \leq \inf \Phi_1(U_1, u_2) + \varepsilon \ \Rightarrow \ \|u_1\|_{L^{p_1}(\Omega;\mathbb{R}^{m_1})} \leq \varrho_1 \ , \tag{2.11a}$$

$$\forall u_1 \in U_1: \quad \Phi_2(u_1, u_2) \leq \inf \Phi_2(u_1, U_2) + \varepsilon \ \Rightarrow \ \|u_2\|_{L^{p_2}(\Omega;\mathbb{R}^{m_2})} \leq \varrho_2 \ . \tag{2.11b}$$

Proof. By the uniform coercivity (1.6h), we have $\Phi_1(u_1, u_2) \geq \varepsilon_1 \|u_1\|^{p_1}_{L^{p_1}(\Omega;\mathbb{R}^{m_1})} - \|a_{1+\varepsilon_3}\|_{L^1(\Omega)}$ while $\sup \Phi_1(v_1, U_2) < +\infty$ for $v_1 \in U_1$ taken arbitrarily but fixed. Altogether,

$$\begin{aligned} \varepsilon_1 \|u_1\|^{p_1}_{L^{p_1}(\Omega;\mathbb{R}^{m_1})} - \|a_{1+\varepsilon_3}\|_{L^1(\Omega)} &\leq \Phi_1(u_1, u_2) \\ &\leq \inf \Phi_1(U_1, u_2) + \varepsilon \leq \Phi_1(v_1, u_2) + \varepsilon \leq \sup \Phi_1(v_1, U_2) + \varepsilon. \end{aligned}$$

From this, (2.11a) follows with $\varrho_1 = ((\sup \Phi_1(v_1, U_2) + \|a_{1+\varepsilon_3}\|_{L^1(\Omega)} + \varepsilon)/\varepsilon_1)^{1/p_1}$. The implication (2.11b) is just by a symmetry. □

Corollary 2.2. *All relaxed Nash optimal pairs, if exist, must be contained in $\bar{U}_{1,\varrho_1} \times \bar{U}_{2,\varrho_2}$ where*

$$\bar{U}_{\alpha,\varrho_\alpha} := \text{w*-cl}\, i_\alpha(U_{\alpha,\varrho_\alpha}) \ , \qquad \textit{with} \qquad U_{\alpha,\varrho_\alpha} := \{u \in U_\alpha;\ \|u\|_{L^{p_\alpha}(\Omega;\mathbb{R}^{m_\alpha})} \leq \varrho_\alpha\},$$

is a convex weakly compact set in H^*_α, $\alpha = 1, 2$.*

Lemma 2.3. *Let (1.6a-c,e-f) be valid. Then the mapping $(\eta_1, \eta_2) \mapsto y(\eta_1, \eta_2) : Y^{p_1}_{H_1}(\Omega; \mathbb{R}^{m_1}) \times Y^{p_2}_{H_2}(\Omega; \mathbb{R}^{m_2}) \to W_0^{1,2}(\Omega; \mathbb{R}^m)$, where $y(\eta_1, \eta_2)$ denotes the unique weak solution to (2.9), is (weak*×weak*,weak)-continuous. Moreover,*

$$\sup_{(\eta_1,\eta_2) \in \bar{U}_{1,\varrho_1} \times \bar{U}_{2,\varrho_2}} \|y(\eta_1, \eta_2)\|_{L^\infty(\Omega;\mathbb{R}^m)} \leq C < +\infty \ . \tag{2.12}$$

Proof. The claimed continuity follows from (weak*×weak*,weak)-continuity of the mapping $(\eta_1, \eta_2) \mapsto f_1 \bullet \eta_1 + f_2 \bullet \eta_2 : Y^{p_1}_{H_1}(\Omega; \mathbb{R}^{m_1}) \times Y^{p_2}_{H_2}(\Omega; \mathbb{R}^{m_2}) \to L^q(\Omega; \mathbb{R}^m)$, from compactness of the embedding $L^q(\Omega; \mathbb{R}^m) \subset W^{-1,2}(\Omega; \mathbb{R}^m)$, and from (norm,-weak)-continuity of the mapping $f \mapsto y : W^{-1,2}(\Omega; \mathbb{R}^m) \to W_0^{1,2}(\Omega; \mathbb{R}^m)$ with y being the weak solution to $\operatorname{div}(A\nabla y) - b(y) = f := f_1 \bullet \eta_1 + f_2 \bullet \eta_2$. Moreover, the standard energy estimate bounds $\|y\|_{W^{1,2}(\Omega;\mathbb{R}^m)}$ uniformly for $\eta_1 \in \bar{U}_{1,\varrho_1}$ and $\eta_2 \in \bar{U}_{2,\varrho_2}$. Then, by (1.6c) we also have $\|b(y) + f_1 \bullet \eta_1 + f_2 \bullet \eta_2\|_{L^q(\Omega;\mathbb{R}^m)}$ bounded. By $W^{2,2}$-regularity for the linear system $\operatorname{div}(A\nabla y) = f := b(y) + f_1 \bullet \eta_1 + f_2 \bullet \eta_2$

with smooth A and homogeneous Dirichlet boundary condition on the $C^{1,1}$-domain (see Nečas [9]) and interpolation between Hilbertian Sobolev spaces, we have $\|y\|_{W^{s,2}(\Omega;\mathbb{R}^m)}$ bounded if $L^q(\Omega) \subset W^{s-2,2}(\Omega)$. By the standard embedding, we eventually get (2.12) provided $s > n/2$. $\square$

Remark. The situation in the proof of Lemma 2.3 can schematically be displayed on the following diagram, where the vertical arrows indicate how the mapping $f \mapsto y$ operates, while $\supset$ (possibly rotated vertically to $\cap$) stands for inclusion, and the horizontal arrows only indicates needed conditions:

$$
\begin{array}{lccccccccl}
 & & & L^{r/(r-1)} & \supset & L^q & & & & \text{<- - - - - - - - - - if } r > \frac{q}{q-1} \\
\text{if } r < \frac{2n}{n-2} \text{ - - - - ->} & & & \cap & & \cap & & & & \text{<- - - - - - - - - if } q > \frac{2n}{n+4-2s} \\
 & f & \in & W^{-1,2} & \supset & W^{s-2,2} & \supset & L^2 & & \\
 & \downarrow & & \downarrow & & \downarrow & & \downarrow & & \text{<- - - if } 2 \geq s \geq 1 \\
 & y & \in & W_0^{1,2} & \supset & W_0^{s,2} & \supset & W_0^{2,2} & & \\
 & & & & & \cap & & & & \text{<- - - - - - - - if } s > \frac{n}{2} \\
 & & & & & L^\infty & & & &
\end{array}
$$

The existence of $s \in [1,2]$ with $s > n/2$ and $2n/(n+4-2s) < q$ makes the restriction $n < 4$ and $q > n/2$ imposed in Section 1. Moreover, having $r < 2n/(n-2)$ choosen large enough (which makes (1.6c) weak enough), the requirement $r > q/(q-1)$ needed here can be satisfied, too.

To investigate the geometrical properties of $\bar{\Phi}_\alpha$ with $\alpha = 1,2$, we will have to calculate its Gâteaux differential with respect to the geometry coming from the linear space H_α^* containing $\bar{U}_\alpha$. This is, in fact, a standard task undertaken within the derivation of the Pontryagin maximum principle for the relaxed controls, for which we need to assume growth and Lipschitz continuity of the partial derivative of g_α and g with respect to the variable y, denoted by g'_α and b', respectively:

$$
|g'_\alpha(x,y)| \leq a_q(x) + \gamma(|y|), \qquad |b'(x,y)| \leq a_{rq/(q-r)}(x) + \gamma(|y|), \tag{2.13a}
$$
$$
|b'(x,y_1) - b'(x,y_2)| \leq (a_1(x) + \gamma(|y_1|) + \gamma(|y_2|))|y_1 - y_2| \,, \tag{2.13b}
$$
$$
|g'_\alpha(x,y_1) - g'_\alpha(x,y_2)| \leq (a_1(x) + \gamma(|y_1|) + \gamma(|y_2|))|y_1 - y_2| \,, \tag{2.13c}
$$

where $\alpha = 1,2$ and also $r \geq q$ is assumed with possibly only a further nonrestrictive requirement for the choice of r for (1.6c). The so-called adjoint boundary-value problem is defined as:

$$
\operatorname{div}(A^{\mathrm{T}}(x)\nabla\lambda_\alpha) = b'(x,y)\lambda_\alpha + g'_\alpha(x,y) \,, \qquad \lambda_\alpha|_{\partial\Omega} = 0 \,, \tag{2.14}
$$

where $A^{\mathrm{T}}(x) : \mathbb{R}^{m\times n} \to \mathbb{R}^{m\times n} : [\xi_{kl}] \mapsto \sum_{j=1}^n \sum_{i=1}^m a_{klij}(x)\xi_{ij}$.

Lemma 2.4. *Let (1.6) and (2.13a) be valid. Then (2.14) with $y = y(\eta_1,\eta_2)$ possesses precisely one solution $\lambda_\alpha(\eta_1,\eta_2)$ and*

$$
\sup_{(\eta_1,\eta_2)\in\bar{U}_{1,\varrho_1}\times\bar{U}_{2,\varrho_2}} \|\lambda_\alpha(\eta_1,\eta_2)\|_{L^\infty(\Omega;\mathbb{R}^m)} \leq C^* < +\infty. \tag{2.15}
$$

Proof. First, $\lambda_\alpha(\eta_1,\eta_2)$ is bounded in $W_0^{1,2}(\Omega;\mathbb{R}^m)$ and hence also in $L^r(\Omega;\mathbb{R}^m)$ uniformly for $\eta_1 \in \bar{U}_1$ and $\eta_2 \in \bar{U}_2$. By (2.13a) with (2.12), $b'(y)\lambda_\alpha + g'_\alpha$ is bounded in $L^q(\Omega;\mathbb{R}^m)$. Then (2.15) follows by the $W^{2,2}$-regularity for $\operatorname{div}(A^{\mathrm{T}}\nabla\lambda) = f$ together with interpolation like in the proof of Lemma 2.3. $\square$

Adapting a procedure by Gabasov and Kirillova [5; Section VII.2] developed to prove sufficiency of the maximum principle for optimal control problems, we can establish the following incrementation formula in Lemma 2.5 and to use it in Lemma 2.6 for specifying conditions guaranteeing convexity of the extended costs in the geometry of H^*_α. Let us handle $\bar{\Phi}_1(\cdot,\eta_2)$, the other case $\bar{\Phi}_2(\eta_1,\cdot)$ being entirely analogous.

Lemma 2.5. *Let (1.6a–f) and (2.13) be valid, let $\eta_1, \tilde{\eta}_1 \in \bar{U}_1$ and $\eta_2 \in \bar{U}_2$, let $y = y(\eta_1,\eta_2) \in W_0^{1,2}(\Omega;\mathbb{R}^m)$ be defined by (2.4), and let $\lambda_1 \in W_0^{1,2}(\Omega;\mathbb{R}^m)$ solve (2.14) with $\alpha = 1$. Then*

$$\bar{\Phi}_1(\tilde{\eta}_1,\eta_2) - \bar{\Phi}_1(\eta_1,\eta_2) = \int_{\bar{\Omega}} \mathsf{H}^{\eta_2}_{1,\lambda_1} \bullet (\tilde{\eta}_1 - \eta_1)\mathrm{d}x + \int_\Omega \left[D_{g_1}(x) + \lambda_1(x) D_b(x)\right] \mathrm{d}x \tag{2.16}$$

where the “Hamiltonian” $\mathsf{H}^{\eta_2}_{1,\lambda}$ is defined by

$$\mathsf{H}^{\eta_2}_{1,\lambda}(x,v) := \lambda(x) f_1(x,v) + h_1(x,v) - [\varphi(\cdot,v,\cdot) \bullet \eta_2](x)\ , \tag{2.17}$$

and the second-order correcting terms D_{g_1} and D_b are defined by

$$D_{g_1}(x) := g_1(x,\tilde{y}(x)) - g_1(x,y(x)) - g_1'(x,y(x))(\tilde{y}(x) - y(x))\ , \tag{2.18a}$$

$$D_b(x) := b(x,\tilde{y}(x)) - b(x,y(x)) - b'(x,y(x))(\tilde{y}(x) - y(x))\ , \tag{2.18b}$$

with $\tilde{y} = y(\tilde{\eta}_1,\eta_2) \in W_0^{1,2}(\Omega;\mathbb{R}^m)$ being the solution to the initial-value problem (2.4) with $\tilde{\eta}_1$ in place of η_1.

Proof. Using successively the formula (2.17), the equation (2.9) both for η_1 and for $\tilde{\eta}_1$, the Green formula twice, and the adjoint equation (2.14), we can calculate:

$$\begin{aligned}
\bar{\Phi}_1(\tilde{\eta}_1,\eta_2) - \bar{\Phi}_1(\eta_1,\eta_2) - \int_{\bar{\Omega}} \mathsf{H}^{\eta_2}_{1,\lambda_1} \bullet [\tilde{\eta}_1 - \eta_1]\mathrm{d}x &= \int_\Omega g_1(\tilde{y}) - g_1(y) - \lambda_1 f_1 \bullet [\tilde{\eta}_1 - \eta_1]\mathrm{d}x \\
&= \int_\Omega \Big[g_1(\tilde{y}) - g_1(y) + \lambda_1 \Big(b(\tilde{y}) - b(y) - \operatorname{div}(A\nabla(\tilde{y} - y))\Big)\Big]\mathrm{d}x \\
&= \int_\Omega \Big[g_1(\tilde{y}) - g_\alpha(y) + \lambda_1(b(\tilde{y}) - b(y)) + (\nabla\lambda_1)^{\mathrm{T}} A\nabla(\tilde{y} - y)\Big]\mathrm{d}x \\
&= \int_\Omega \Big[g_1(\tilde{y}) - g_1(y) + \lambda_1(b(\tilde{y}) - b(y)) - \operatorname{div}(A^{\mathrm{T}}\nabla\lambda_1)(\tilde{y} - y)\Big]\mathrm{d}x \\
&= \int_\Omega \Big[g_1(\tilde{y}) - g_1(y) - g_1'(y)(\tilde{y} - y) + \lambda_1(b(\tilde{y}) - b(y) - b'(y)(\tilde{y} - y))\Big]\,\mathrm{d}x.
\end{aligned}$$

$\square$

Lemma 2.6. *Let (1.6) and (2.13) be valid, and let $g_\alpha(x,\cdot)$ be uniformly convex on the ball of radius C from (2.12) in the sense*

$$\max(|y|,|\tilde{y}|)\le C \quad\Rightarrow\quad g_\alpha(x,\tilde{y})-g_\alpha(x,y)-g'_\alpha(x,y)(\tilde{y}-y)\ge c_\alpha(x)|\tilde{y}-y|^2 \quad (2.19)$$

with a modulus c_α satisfying

$$c_\alpha(x)\ge\Big(\frac{1}{2}a_1(x)+\gamma(C)\Big)C^* \qquad (2.20)$$

for $\alpha=1,2$ with a_1 and γ from (2.13b) and C^* from (2.15). Then, for any $\eta_1\in\bar{U}_{1,\varrho_1}$ and $\eta_2\in\bar{U}_{2,\varrho_2}$, the extended costs $\bar{\Phi}_1(\cdot,\eta_2):\bar{U}_{1,\varrho_1}\to\mathbb{R}$ and $\bar{\Phi}_2(\eta_1,\cdot):\bar{U}_{2,\varrho_2}\to\mathbb{R}$ defined in (2.3)–(2.4) are convex.

Proof. Let us show the case $\bar{\Phi}_1(\cdot,\eta_2)$; the case $\bar{\Phi}_2(\eta_1,\cdot)$ being analogous. By (2.12), (2.13b) and the Taylor expansion, we can estimate

$$|b(x,\tilde{y}(x))-b(x,y(x))-b'(x,y(x))(\tilde{y}(x)-y(x))|\le(\frac{1}{2}a_1(x)+\gamma(C))|\tilde{y}(x)-y(x)|^2.$$

Then (2.19) with (2.20) and (2.15) ensure

$$\int_\Omega D_{g_1}+\lambda_1 D_b\,\mathrm{d}x\ge\int_\Omega\Big(c_1-(\frac{a_1}{2}+\gamma(C))|\lambda_1|\Big)\,|\tilde{y}-y|^2\,\mathrm{d}x\ge 0$$

so that the second right-hand side term in (2.16) is non-negative. By (2.16) we can calculate the Gâteaux differential of $\bar{\Phi}_1(\cdot,\eta_2):\bar{U}_1\to\mathbb{R}$, i.e.

$$[\nabla_{\eta_1}\bar{\Phi}_1(\eta_1,\eta_2)](\tilde{\eta}_1-\eta_1):=\lim_{\varepsilon\searrow 0}\frac{\bar{\Phi}_1(\eta_1+\varepsilon(\tilde{\eta}_1-\eta_1),\eta_2)-\bar{\Phi}_1(\eta_1,\eta_2)}{\varepsilon} \qquad (2.21)$$

$$=\int_\Omega \mathsf{H}^{\eta_2}_{1,\lambda_1}\bullet[\tilde{\eta}_1-\eta_1]\mathrm{d}x+\lim_{\varepsilon\searrow 0}(I_\varepsilon+J_\varepsilon)=\int_\Omega\mathsf{H}^{\eta_2}_{1,\lambda_1}\bullet[\tilde{\eta}_1-\eta_1]\mathrm{d}x$$

with $\eta_1,\tilde{\eta}_1\in\bar{U}_1$ arbitrary, and with the adjoint state λ_1 and the Hamiltonian $\mathsf{H}^{\eta_2}_{1,\lambda}$ defined by (2.14) and (2.17), respectively. We used, for $y_\varepsilon=y(\eta_1+\varepsilon(\tilde{\eta}_1-\eta_1),\eta_2)$,

$$I_\varepsilon:=\left|\int_\Omega g_1(y_\varepsilon)-g_1(y)-g'_1(y)(y_\varepsilon-y)\,\mathrm{d}x\right|\le\int_\Omega\ell\,|y_\varepsilon-y|^2\,\mathrm{d}x\to 0$$

and

$$J_\varepsilon:=\left|\int_\Omega\lambda_1\left(b(y_\varepsilon)-b(y)-b'(y)(y_\varepsilon-y)\right)\,\mathrm{d}x\right|\le\int_\Omega\ell|\lambda_1|\,|y_\varepsilon-y|^2\,\mathrm{d}x\to 0$$

by Lebesgue's dominated convergence theorem; we took $\ell:=(\frac{1}{2}a_1+\gamma(C))\in L^1(\Omega)$ from (2.13b,c) and used the fact $\lim_{\varepsilon\searrow 0}\|y_\varepsilon-y\|_{C(\bar{\Omega};\mathbb{R}^m)}=0$. Therefore we obtained

$$\bar{\Phi}_1(\tilde{\eta}_1,\eta_2)-\bar{\Phi}_1(\eta_1,\eta_2)-[\nabla_{\eta_1}\bar{\Phi}_1(\eta_1,\eta_2)](\tilde{\eta}_1-\eta_1)\ge 0 \qquad (2.22)$$

for all $\eta_1,\tilde{\eta}_1\in\bar{U}_{1,\varrho_1}$. Replacing the role of $\tilde{\eta}_1$ and η_1 and adding the corresponding inequalities (2.22), we get $\big[\nabla_{\eta_1}\bar{\Phi}_1(\tilde{\eta}_1,\eta_2)-\nabla_{\eta_1}\bar{\Phi}_1(\eta_1,\eta_2)\big]\,(\tilde{\eta}_1-\eta_1)\ge 0$, which just says that $\bar{\Phi}_1(\cdot,\eta_2)$ has a monotone differential and, thus, is convex on $\bar{U}_{1,\varrho_1}$. □

3. Main results

We are now ready to formulate the main achievements: existence of the Nash equilibria for the relaxed game (RG), relations between (RG) and the original game (G), as well as existence of equilibrium sequences for (G). We will use the Nikaidô and Isoda generalization [10] of the classical Nash theorem [8].

Theorem 3.1. *Let (1.6), (2.13) and (2.19)–(2.20) for $\alpha = 1, 2$ be valid. Then:*

1. *The relaxed game* (RG) *possesses at least one Nash equilibrium point.*
2. *Every Nash equilibrium point of the relaxed game* (RG) *can be attained by an equilibrium sequence for the original game* (G) *embedded via $i_1 \times i_2$.*
3. *Conversely, every equilibrium sequence for the original game* (G) *(embedded via $i_1 \times i_2$) has a weakly* convergent subsequence and the limit of every such a subsequence is a Nash equilibrium for the relaxed game* (RG).

Proof. By Lemmas 2.1, 2.3 and 2.6, we can see that the relaxed problem (RG) represents a game over the convex compact sets $\bar{U}_{1,\varrho_1}$ and $\bar{U}_{2,\varrho_2}$ for the separately continuous costs $\bar{\Phi}_1$ and $\bar{\Phi}_2$ whose sum is jointly continuous and such that $\bar{\Phi}_1(\cdot, \eta_2)$ and $\bar{\Phi}_2(\eta_1, \cdot)$ are convex. This just guarantees, by Nikaidô and Isoda [10], that (RG) has at least one Nash equilibrium point, as claimed in (i).

The points (ii) and (iii) follow by Lemmas 2.1 and 2.3. In fact, $\Psi_1 : U_1 \to \mathbb{R}$ and $\Psi_2 : U_2 \to \mathbb{R}$ in (1.5) are defined by

$$\Psi_1 = \bar{\Phi}_1(\cdot, \eta_2) \circ i_1 \qquad \& \qquad \Psi_2 = \bar{\Phi}_2(\eta_1, \cdot) \circ i_2 \;, \tag{3.1}$$

for details see [13; Proposition 7.1.1] and realize the metrizability of the weak* topology on $\bar{U}_{1,\varrho_1}$ and $\bar{U}_{2,\varrho_2}$, which allows us to work in terms of sequences. □

Corollary 3.2. *Under the assumptions of Theorem 3.1, the original game* (G) *possesses an equilibrium sequence, i.e. a sequence $\{(u_1^k, u_2^k)\}_{k\in\mathbb{N}}$ satisfying (1.5).*

Proof. It follows straightforwardly by the points (i) and (ii) of Theorem 3.1. □

Corollary 3.3. *Let (1.6), (2.13) and (2.19)–(2.20) for $\alpha = 1, 2$ be valid. Then $(\eta_1, \eta_2) \in \bar{U}_1 \times \bar{U}_2$ is a Nash equilibrium for* (RG) *if and only if*

$$\eta_1 \in \bar{U}_{1,\varrho_1} \text{ and } \eta_2 \in \bar{U}_{2,\varrho_2} \text{ with } \varrho_1 \text{ and } \varrho_2 \text{ from (2.11)}, \tag{3.2a}$$

$$\int_{\bar{\Omega}} \mathsf{H}^{\eta_2}_{1,\lambda_1} \bullet \eta_1 \, \mathrm{d}x = \inf_{u\in U_1} \int_{\Omega} \mathsf{H}^{\eta_2}_{1,\lambda_1}(x, u(x)) \, \mathrm{d}x \;, \text{ and} \tag{3.2b}$$

$$\int_{\bar{\Omega}} \mathsf{H}^{\eta_1}_{2,\lambda_2} \bullet \eta_2 \, \mathrm{d}x = \inf_{u\in U_2} \int_{\Omega} \mathsf{H}^{\eta_1}_{2,\lambda_2}(x, u(x)) \, \mathrm{d}x \;. \tag{3.2c}$$

Proof. Take $(\eta_1, \eta_2) \in \bar{U}_1 \times \bar{U}_2$ a Nash equilibrium for (RG). By Lemma 2.1, we have (3.2a). As η_1 minimizes $\bar{\Phi}_1(\cdot, \eta_2)$ over $\bar{U}_1$, it holds $[\nabla_{\eta_1} \bar{\Phi}_1(\eta_1, \eta_2)](\tilde{\eta}_1 - \eta_1) \geq 0$ for any $\tilde{\eta}_1 \in \bar{U}_1$. By $i_1(U_1) \subset \bar{U}_1$, we also have $[\nabla_{\eta_1} \bar{\Phi}_1(\eta_1, \eta_2)](i(u_1) - \eta_1) \geq 0$ for any $u \in U_1$, which just gives (3.2b) if also Lemma 2.5 and the density of $i_1(U_1)$ in $\bar{U}_1$ are taken into account. By symmetry, (3.2c) follows analogously.

Conversely, suppose that (3.2) holds. By Lemma 2.5, (3.2a,b) implies that $[\nabla_{\eta_1}\bar{\Phi}_1(\eta_1,\eta_2)](i(u_1)-\eta_1)\geq 0$ for any $u\in U_{1,\varrho_1}$. By the density of $i_1(U_1)$ in $\bar{U}_1$, we also have $[\nabla_{\eta_1}\bar{\Phi}_1(\eta_1,\eta_2)](\tilde{\eta}_1-\eta_1)\geq 0$ for any $\tilde{\eta}_1\in\bar{U}_{1,\varrho_1}$. By Lemma 2.6, this is a sufficient condition for η_1 to minimize $\bar{\Phi}_1(\cdot,\eta_2)$ on $\bar{U}_{1,\varrho_1}$. Yet, by the uniform coercivity of $\Phi_1(\cdot,u_2)$ (cf. also Lemma 2.1), η_1 minimizes $\bar{\Phi}_1(\cdot,\eta_2)$ even on the whole $\bar{U}_1$. By symmetry, (3.2a,c) makes η_2 a minimizer $\bar{\Phi}_2(\eta_1,\cdot)$ on $\bar{U}_2$. Thus (η_1,η_2) is a Nash equilibrium for (RG). □

Remark 3.4. By [13], $\{|u_\alpha^k|^{p_\alpha}\}_{k\in\mathbb{N}}$ with $\{(u_1^k,u_2^k)\}_{k\in\mathbb{N}}$ an equilibrium sequence is always relatively weakly compact in $L^1(\Omega)$; cf. [13; Propositions 4.4.2(iv) and 3.4.16]. This enables us to localize (3.2b,c) point-wise if S_α qualified appropriately.

Remark 3.5. If the controlled system is linear with respect to the state, i.e. $b(t,\cdot)$ is affine, then obviously $a_1\equiv 0$ and $\gamma\equiv 0$ in (2.13b) and one can take $c_\alpha\equiv 0$ in (2.19) which then just requires $g_1(t,\cdot)$ and $g_2(t,\cdot)$ to be merely convex; cf. [13; Section 7.4] for $m=1$ or Bensoussan [2], Díaz and Lions [4], and also Balder [1] or Nowak [11] for games with ordinary differential equations.

Remark 3.6. In fact, (2.19)–(2.20) represent a restriction on the nonlinearity of $b(x,\cdot)$, cf. [15; Section 5] for the case of integral equations.

Acknowledgements. The author appreciates discussion about regularity of elliptic systems with prof. J. Nečas, dr. J. Jarušek and dr. J. Málek. Besides, the author is thankful to anonymous referees for their valuable comments. This research was partly supported by the grants No. 201/96/0228 (GA ČR) and No. A 107 5705 (GA AV ČR).

References

[1] Balder, E.J.: On a Useful Compactification for Optimal Control Problems. *J. Math. Anal. Appl.* **72** (1979), 391–398.

[2] Bensoussan, A.: Points de Nash dans le cas de fonctionnelles quadratiques et jeux différentiels linéaires a n personnes. *SIAM J. Control* **12** (1974), 460–499.

[3] Chawla, S.: A minmax problem for parabolic systems with competitive interactions. IMA Preprint Series No.1448, Uni. Minnesota, Minneapolis, 1996.

[4] Díaz, J.I., Lions, J.L.: On the approximate controllability of Stackelberg-Nash strategies. In: Proc. *2nd Diderot forum: Math. & Environment.* Springer, 1999 (to appear).

[5] Gabasov, R., Kirillova, F.: *Qualitative Theory of Optimal Processes.* Nauka, Moscow, 1971.

[6] Lenhart, S., Protopopescu, V., Stojanovic, S.: A minimax problem for semilinear nonlocal competitive systems. *Appl. Math. Optim.* **28** (1993), 113–132.

[7] Mordukhovich, B.S., Zhang, K.: Minimax control of parabolic systems with Dirichlet boundary conditions and state constraits. *Appl. Math. Optim.* **36** (1997), 323–360.

[8] Nash, J.: Non-cooperative games. *Annals of Math.* **54** (1951), 286–295.

[9] Nečas, J.: *Les méthodes directes en théorie des équations elliptiques.* Academia, Prague, 1967.

[10] Nikaidô, H., Isoda, K.: Note on non-cooperative equilibria. *Pacific J. Math.* **5** (1955), 807–815.

[11] Nowak, A.S.: Correlated relaxed equilibria in nonzero-sum linear differential games. *J. Math. Anal. Appl.* **163** (1992), 104–112.

[12] Patrone, F.: Well-posedness for Nash equilibria and related topics. In: *Recent Developments in Well-Posed Variational Problems.* (Eds.R.Lucchetti et al.) Kluwer, 1995, pp.211–227.

[13] Roubíček, T.: *Relaxation in Optimization Theory and Variational Calculus.* W. de Gruyter, Berlin, 1997.

[14] Roubíček, T.: On noncooperative nonlinear differential games. *Kybernetika* (submit.)

[15] Roubíček, T., Schmidt, W.H.: Existence of solutions to certain nonconvex optimal control problems governed by nonlinear integral equations. *Optimization* **42** (1997), 91–109.

[16] Roxin, E.O.: Differential games with partial differential equations. In: *Diff. Games and Appl.* (Eds. P. Hagedorn et al.) Lecture Notes on Control Inf. Sci. **3**, Springer, Berlin, 1997, pp.186–204.

[17] Szpiro, A.A., Protopescu, V.: Parabolic systems with competitive interactions and control on initial conditions. *Appl. Math. Comp.* **59** (1993), 215–245.

[18] Warga, J.: *Optimal Control of Differential and Functional Equations.* Academic Press, New York, 1972.

Mathematical Institute
Charles University
Sokolovská 83
CZ-186 75 Praha 8
and
Institute of Information Theory and Automation
Academy of Sciences
Pod vodárenskou věží 4
CZ-182 08 Praha 8, Czech Republic.
E-mail address: roubicek@karlin.mff.cuni.cz

International Series of Numerical Mathematics
Vol. 133, © 1999 Birkhäuser Verlag Basel/Switzerland

Incomplete Indefinite Decompositions as Multigrid Smoothers for KKT Systems

Volker H. Schulz

Abstract. The key point for the efficient simultaneous multigrid solution of linear systems resulting from linearizations of Karush-Kuhn-Tucker (KKT) systems is the construction of an appropriate smoother. In this paper new incomplete decompositions of Bunch-Parlett type for linear KKT systems are developed as natural generalizations of incomplete LU decompositions, in order to be applied as smoothers within a multigrid algorithm. Numerical results are presented for a model problem.

1. Introduction

Multigrid methods are iterative solution methods for systems of equations with the potential of optimal (i.e. linear) complexity with respect to (w.r.t.) the number of variables involved, which is in contrast to general direct solution methods and most other iterative solution methods, e.g., of Krylov type. They are understood best for systems of equations which result from discretizations of scalar elliptic differential equations. For a reference book on multigrid methods see, e.g., [Hac2] (Due to page limitations a basic introduction into multigrid methods cannot be provided here. For this we refer to the first chapters of the book mentioned above.) The present paper is part of a current research effort to make the computational potential of multigrid methods also available for systems of equations arising from linearizations of Karush-Kuhn-Tucker (KKT) conditions.

We consider direct discretizations [Sch3] of optimal control problems of the form

$$\min f_h(x_h, u_h) \tag{1}$$

$$\text{s.t.} \quad c_h(x_h, u_h) = 0, \tag{2}$$

where $x_h \in \mathbb{R}^{n_x}$ denotes the discretized state variables, $u_h \in \mathbb{R}^{n_u}$ the discretized control variables, $c_h \in \mathfrak{C}^2(\mathbb{R}^{n_x} \times \mathbb{R}^{n_u}, \mathbb{R}^{n_x})$ the discretized state (differential) equation and $f_h \in \mathfrak{C}^2(\mathbb{R}^{n_x} \times \mathbb{R}^{n_u}, \mathbb{R})$ the discretized objective functional. We assume that $\partial c_h / \partial x_h$ is nonsingular. The index h indicates that all these magnitudes are derived from discretizations. In order to solve the discretized optimization problem, we apply an SQP method, which essentially means applying a Newton-like method to the necessary Karush-Kuhn-Tucker (KKT) conditions

$$\nabla_{x_h} \mathfrak{L}(x_h, u_h, \lambda) = 0, \quad \nabla_{u_h} \mathfrak{L}(x_h, u_h, \lambda) = 0, \quad \nabla_{\lambda} \mathfrak{L}(x_h, u_h, \lambda) = 0,$$

where $\mathfrak{L}(x_h, u_h, \lambda) := f_h(x_h, u_h) + c_h(x_h, u_h)^\top \lambda$ denotes the Lagrangian for $\lambda \in \mathbb{R}^{n_x}$. This necessitates the solution of linear systems of the form

$$\begin{bmatrix} H_{xx} & H_{xu} & C_x^\top \\ H_{ux} & H_{uu} & C_u^\top \\ C_x & C_u & 0 \end{bmatrix} \begin{pmatrix} \Delta x \\ \Delta u \\ \Delta \lambda \end{pmatrix} = - \begin{pmatrix} \nabla_{x_h} \mathfrak{L}(x_h, u_h, \lambda) \\ \nabla_{u_h} \mathfrak{L}(x_h, u_h, \lambda) \\ \nabla_{\lambda} \mathfrak{L}(x_h, u_h, \lambda) \end{pmatrix}, \tag{3}$$

where H with indices denotes (approximations to) second derivatives of the Lagrangian $\mathfrak{L}(x_h, u_h, \lambda)$ and C with indices (approximations to) first derivatives of c_h.

The focus of this paper is on the construction of efficient multigrid methods for KKT systems of the type (3). Only few literature is devoted to this topic. For a general overview see [Sch4]. The numerical solution of optimization problems following the direct discretization approach and using multigrid methods in various ways has been performed, e.g., in [Sch1, Sch2, SBH, SW].

Most other multigrid approaches (e.g., [Hac1]) to the solution of system (3) assume that an exact solver of low computational complexity is available for linear systems with system matrix C_x and construct essentially multigrid methods only w.r.t. the Δu-variables, which for general problems either may result in a loss of optimal complexity or in nested multigrid methods (i.e. a multigrid method for C_x-systems within a multigrid method for the Δu-variables). In contrast to that, we aim at the construction of non-nested simultaneous multigrid methods for all variables ($\Delta x, \Delta u$ and $\Delta \lambda$ together), which preserve linear computational complexity. The main challenge within this class of algorithms lies in the choice of appropriate smoothing iterations, since most standard smoothing techniques (Jacobi, Gauss-Seidel, ILU, etc.) are not directly applicable to systems of type (3).

In the present paper we propose a new incomplete decomposition of system (3) to be used in a defect correcting smoothing iteration for the whole system. The existence of the incomplete decomposition is shown for problems where C_x possesses typical properties resulting from discretizations of PDE models. The linear complexity of the resulting multigrid algorithm is investigated numerically for a model problem in section 3.

2. An ILΔL$^\top$ decomposition

Since their development [MV], incomplete LU-decompositions (ILU) have been modified in various ways and used as smoothers in multigrid methods (c.f., e.g., [Wit]). Typically, these decompositions are applied without pivoting and avoid fill-in – at least in their basic version. In this section we introduce a novel incomplete decomposition for matrices of optimization saddle point problems, which omits in a similar way pivoting and fill-in. The existence of the decomposition is proved for the case that the constraints of the saddle point problem possess typical properties resulting from discretizations of PDE models. These indefinite decompositions are also introduced in order to be used as smoothers in a multigrid method.

In [Saa] incomplete decompositions for unsymmetric and indefinite matrices are also considered. However, in contrast to the methods proposed here, these decompositions do not reflect the special structure of matrices from optimization saddle point problems, nor do they completely avoid fill-in.

ILU decompositions are derived from the usual Gaussian decomposition by fixing the pivoting and omitting fill-in. In a similar way we develop incomplete indefinite decompositions from the Bunch/Parlett resp. Bunch/Kaufmann factorization of indefinite symmetric matrices [BP, BK]. In [BP] is shown that for each nonsingular symmetric matrix K there is a permutation matrix P and a lower triangular matrix L which allow a factorization of the form

$$PKP^\top = L\Delta L^\top.$$

There, Δ is a block-diagonal matrix with 1×1 or 2×2 diagonal blocks and

$$K := \left[\begin{array}{cc|c} H_{xx} & C_x^\top & H_{xu} \\ C_x & 0 & C_u \\ \hline H_{ux} & C_u^\top & H_{uu} \end{array}\right].$$

Here we have arranged the KKT-matrix K in this special way in order to indicate that we aim at null-space related methods rather than at range space methods [Sch4]. In the sequel we call the indefinite factorization derived from this direct factorization ILΔL$^\top$factorization. The construction of the factorization relies on the recursive construction of Schur complements w.r.t. appropriately chosen 1×1 blocks k_{ii} or 2×2 blocks

$$\begin{bmatrix} k_{ii} & k_{ij} \\ k_{ji} & k_{jj} \end{bmatrix}$$

of the respective Schur complement rest matrix. In general, the choice of these blocks strongly influences the stability of the decomposition. Here we fix this pivoting a priori in the following way. The construction of the ILΔL$^\top$decomposition is performed in two steps. First, Schur complements w.r.t. 2×2 blocks are formed. For this step we choose

$$\begin{bmatrix} h_{11}^{xx} & c_{11}^{x} \\ c_{11}^{x} & 0 \end{bmatrix}$$

as the first pivot block of the optimization matrix

$$K = \left[\begin{array}{ccccc|c} \boldsymbol{h^{xx}_{11}} & \dots & h^{xx}_{1n} & \boldsymbol{c^{x}_{11}} & \dots & c^x_{n1} & \\ \vdots & & \vdots & \vdots & & \vdots & H_{xu} \\ h^{xx}_{n1} & \dots & h^{xx}_{nn} & c^x_{1n} & \dots & c^x_{nn} & \\ \boldsymbol{c^x_{11}} & \dots & c^x_{1n} & \boldsymbol{0} & \dots & 0 & \\ \vdots & & \vdots & \vdots & & \vdots & C_u \\ c^x_{n1} & \dots & c^x_{nn} & 0 & \dots & 0 & \\ \hline & H^\top_{xu} & & & C^\top_u & & H_{uu} \end{array}\right]. \tag{4}$$

In the Schur complement w.r.t. 2×2 blocks mentioned above, we obtain for the corresponding entries in the $\begin{bmatrix} H_{xx} & C_x^\top \\ C_x & 0 \end{bmatrix}$ block of K the following update for $i = 2, \dots, n$ and $j = 2, \dots, i$:

$$\begin{bmatrix} \tilde{h}^{xx}_{ij} & \tilde{c}^x_{ji} \\ \tilde{c}^x_{ij} & 0 \end{bmatrix} = \begin{bmatrix} h^{xx}_{ij} & c^x_{ji} \\ c^x_{ij} & 0 \end{bmatrix} - \begin{bmatrix} h^{xx}_{i1} & c^x_{1i} \\ c^x_{i1} & 0 \end{bmatrix} \begin{bmatrix} h^{xx}_{11} & c^x_{11} \\ c^x_{11} & 0 \end{bmatrix}^{-1} \begin{bmatrix} h^{xx}_{1j} & c^x_{j1} \\ c^x_{1j} & 0 \end{bmatrix} \tag{5}$$

$$= \begin{bmatrix} h^{xx}_{ij} & c^x_{ji} \\ c^x_{ij} & 0 \end{bmatrix} - \begin{bmatrix} \frac{c^x_{1i} h^{xx}_{1j}}{c^x_{11}} + \frac{c^x_{1j} h^{xx}_{1i}}{c^x_{11}} - \frac{c^x_{1i} c^x_{1j} h^{xx}_{11}}{(c^x_{11})^2} & \frac{c^x_{1i} c^x_{j1}}{c^x_{11}} \\ \frac{c^x_{1j} c^x_{i1}}{c^x_{11}} & 0 \end{bmatrix}. \tag{6}$$

That means that in the Schur complement a $\begin{bmatrix} \tilde{H}_{xx} & \tilde{C}_x^\top \\ \tilde{C}_x & 0 \end{bmatrix}$ structure is preserved as well. Thus we obtain the factorization

$$PKP^\top = L \left[\begin{array}{c|cccccc|c} \begin{matrix} h^{xx}_{11} & c^x_{11} \\ c^x_{11} & 0 \end{matrix} & & & & 0 & & & \\ \hline & \boldsymbol{\tilde{h}^{xx}_{11}} & \dots & \tilde{h}^{xx}_{1\tilde{n}} & \boldsymbol{\tilde{c}^x_{11}} & \dots & \tilde{c}^x_{\tilde{n}1} & \\ & \vdots & & \vdots & \vdots & & \vdots & \tilde{H}_{xu} \\ & \tilde{h}^{xx}_{\tilde{n}1} & \dots & \tilde{h}^{xx}_{\tilde{n}\tilde{n}} & \tilde{c}^x_{1\tilde{n}} & \dots & \tilde{c}^x_{\tilde{n}\tilde{n}} & \\ 0 & \boldsymbol{\tilde{c}^x_{11}} & \dots & \tilde{c}^x_{1\tilde{n}} & \boldsymbol{0} & \dots & 0 & \\ & \vdots & & \vdots & \vdots & & \vdots & \tilde{C}_u \\ & \tilde{c}^x_{\tilde{n}1} & \dots & \tilde{c}^x_{\tilde{n}\tilde{n}} & 0 & \dots & 0 & \\ \cline{2-8} & & \tilde{H}^\top_{xu} & & & \tilde{C}^\top_u & & \tilde{H}_{uu} \end{array}\right] L^\top,$$

where P is the permutation matrix which swaps the 2nd with the $(n+1)$st row, $\tilde{n} = n - 1$ and

$$L^\top = \left[\begin{array}{c|c} I_2 & \begin{bmatrix} h^{xx}_{11} & c^x_{11} \\ c^x_{11} & 0 \end{bmatrix}^{-1} \begin{bmatrix} h^{xx}_{12} & \dots & h^{xx}_{1n} & c^x_{21} & \dots & c^x_{n1} & h^{xu}_{12} & \dots & h^{xu}_{1n_u} \\ c^x_{12} & \dots & c^x_{1n} & 0 & \dots & 0 & c^u_{12} & \dots & c^u_{1n_u} \end{bmatrix} \\ \hline 0 & I_{2\tilde{n}+n_u} \end{array}\right].$$

Here, because of (6) the Schur complement possesses the same structure as the original matrix in (4) with n substituted by $\tilde{n} = n - 1$. Therefore, formally we can apply the same factorization step again until after n such steps the first part of the factorization is finished yielding a decomposition of the form

$$\bar{P}K\bar{P}^\top = \bar{L} \left[\begin{array}{ccccc|c} \bar{h}^{xx}_{11} & \bar{c}^x_{11} & & & & \\ \bar{c}^x_{11} & 0 & & & & \\ & & \ddots & & & 0 \\ & & & \bar{h}^{xx}_{nn} & \bar{c}^x_{nn} & \\ & & & \bar{c}^x_{nn} & 0 & \\ \hline & & 0 & & & \bar{H}_{uu} \end{array}\right] \bar{L}^\top \tag{7}$$

with correspondingly defined new entries indicated by a $(\bar{\ })$.

The decomposition considered up to now still is a complete decomposition. We show an existence result for this, which will later be generalized to the incomplete case analogously to [MV].

Lemma 2.1. *We assume that C_x is an M-matrix, and we perform the recursion defined above leading to the factorization (7). Then the matrices $\tilde{C}_x$ appearing in the $\begin{bmatrix} \tilde{H}_{xx} & \tilde{C}_x^\top \\ \tilde{C}_x & 0 \end{bmatrix}$ block of the respective Schur complement matrices satisfy the sign condition*

$$\tilde{c}^x_{ii} > 0, \forall i, \quad \tilde{c}^x_{ij} \leq 0, \forall i \neq j$$

and the recursion is well defined.

Proof. We consider the recursion steps only w.r.t. the $\begin{bmatrix} H_{xx} & C_x^\top \\ C_x & 0 \end{bmatrix}$-blocks. The Schur complement w.r.t. the $\begin{bmatrix} h^{xx}_{11} & c^x_{11} \\ c^x_{11} & 0 \end{bmatrix}$ pivot block is formed by the update (6). Since C_x is an M-matrix, it satifies the sign condition mentioned in the assertion. Therefore all nonzero entries satisfy:

$$\tilde{c}^x_{ij} = c^x_{ij} - \frac{c^x_{1j}c^x_{i1}}{c^x_{11}} \leq c^x_{ij} \leq 0, \quad i \neq j,$$

which proofs inductively the first part of the assertion. The new diagonal entries $\tilde{c}^x_{ii}$ are identical with one Gaussian elimination step for C_x w.r.t. c^x_{11} as the pivot. Let $\det(\{c^x_{kl}\}^{i,i}_{1,1})$ denote the determinant of the i-th main submatrix of C_x. Since C_x is an M-matrix, all these main subdeterminants are positive, which results in the estimation

$$\tilde{c}^x_{ii} \geq \det(\{c^x_{kl}\}^{i,i}_{1,1}) / \det(\{c^x_{kl}\}^{i-1,i-1}_{1,1}) > 0. \tag{8}$$

Note that the last argument does not hold recursively. Instead, equation (8) is valid for all Gaussian elimination steps, where on the right-hand side are always main submatrices of C_x and not the ones of the respective previously constructed Schur complement. $\square$

If C_x itself is already symmetric, the following more general statement holds.

Lemma 2.2. *If C_x is symmetric, then the recursion defined above leads to a decomposition of C_x of the form*

$$C_x = LDL^\top, \quad \textit{with} \quad L = \begin{bmatrix} 1 & & 0 \\ & \ddots & \\ * & & 1 \end{bmatrix}, \; D = \begin{bmatrix} 1 & & 0 \\ & \ddots & \\ 0 & & 1 \end{bmatrix}.$$

Proof. The representation (6) shows that the matrix $\tilde{C}_x$ of the Schur complement is symmetric again. Therefore, we only need to consider its lower triangle. This is symmetric with the lower triangle of an LU-decomposition, which proofs the assertion. □

In the sequel we consider incomplete decompositions. We define index sets $I_x := \{1, \ldots, n_x\}$ and $I_u := \{1, \ldots, n_u\}$ and a subset

$$E := (E_{xx}, E_{xu}, E_{uu}) \subset (I_x \times I_x) \times (I_x \times I_u) \times (I_u \times I_u).$$

For ease of presentation we assume that E_{xx} is symmetric, i.e. for $(i,j) \in E_{xx}$, $(j,i) \in E_{xx}$ holds, too.

Now, we formulate in an algorithmic way the first step of an ILΔL$^\top$-decomposition analogously to the complete decomposition above. We use a syntax related to the programming language C with minor modifications to increase the readability.

for $(i = 1;\ i \le n_x;\ i{+}{+})$
{
  for $(j = i+1;\ j \le n_x;\ j{+}{+})$
    if $(j,i) \in E_{xx}$ then
    {

(*) $$\Pi = \begin{bmatrix} h^{xx}_{ji} & c^x_{ij} \\ c^x_{ji} & 0 \end{bmatrix} \begin{bmatrix} h^{xx}_{ii} & c^x_{ii} \\ c^x_{ii} & 0 \end{bmatrix}^{-1};$$

      for $(k = i+1;\ k \le j;\ k{+}{+})$
        if $(j,k) \in E_{xx}$ then $\begin{bmatrix} h^{xx}_{jk} & c^x_{kj} \\ c^x_{jk} & 0 \end{bmatrix} -= \Pi \begin{bmatrix} h^{xx}_{ik} & c^x_{ki} \\ c^x_{ik} & 0 \end{bmatrix};$

(**) store Π on $\begin{bmatrix} h^{xx}_{ji} & c^x_{ij} \\ c^x_{ji} & 0 \end{bmatrix};$

    }
  for $(j = 1;\ j \le n_u;\ j{+}{+})$
    if $(i,j) \in E_{xu}$ then
    {

$$\pi = \begin{bmatrix} h^{xx}_{ii} & c^x_{ii} \\ c^x_{ii} & 0 \end{bmatrix}^{-1} \begin{pmatrix} h^{xu}_{ij} \\ c^u_{ij} \end{pmatrix};$$

      for $(k = i+1;\ k \le n_x;\ k{+}{+})$
        if $(k,j) \in E_{xu}$ then $\begin{pmatrix} h^{xu}_{kj} \\ c^u_{kj} \end{pmatrix} -= \begin{bmatrix} h^{xx}_{ki} & c^x_{ik} \\ c^x_{ki} & 0 \end{bmatrix} \pi;$

```
            for (k = 1; k ≤ j; k++)
(***)           if (j,k) ∈ E_uu then h^uu_jk −= π^T (h^xu_ik ; c^u_ik);
            store π on (h^xu_ij ; c^u_ij);
        }
}
```

Remarks:

- Note as a technical detail that in step (**) of the algorithm the pivot matrix Π of step (*) indeed can be stored on $\begin{bmatrix} h^{xx}_{ji} & c^x_{ij} \\ c^x_{ji} & 0 \end{bmatrix}$, since

$$\Pi = \begin{bmatrix} h^{xx}_{ji} & c^x_{ij} \\ c^x_{ji} & 0 \end{bmatrix} \begin{bmatrix} h^{xx}_{ii} & c^x_{ii} \\ c^x_{ii} & 0 \end{bmatrix}^{-1} = \begin{bmatrix} \frac{c^x_{ij}}{c^x_{ii}} & \frac{h^{xx}_{ji}}{c^x_{ii}} - \frac{c^x_{ij}h^{xx}_{ii}}{(c^x_{ii})^2} \\ 0 & \frac{c^x_{ji}}{c^x_{ii}} \end{bmatrix}.$$

- The pivot matrices Π and vectors π define the lower triangle $\bar{L}$ in (7).
- The algorithm changes only the lower tringles of H_{xx} and H_{uu}. Only these parts of the matrices must be stored, since the Schur complements are symmetric as well.

The following theorem proofs that the incomplete algorithm defined above is well defined.

Theorem 2.3. *Let E_{xx} contain the diagonal. Assume furthermore that the matrix C_x is an M-matrix or that it is symmetric and an ILU decomposition w.r.t. the pattern E_{xx} exists . Then all diagonal entries needed in the algorithm above are positive, i.e. $c_{ii} \neq 0$, and the algorithm is well defined.*

Proof. First we assume that C_x is an M-matrix. We compare inductively each incomplete recursion step with a corresponding complete one and denote the matrix entries of the incompletely recurred matrix by $\mathfrak{c}^x$. Then, analogously to lemma 2.1, for $i \neq j$

$$\tilde{\mathfrak{c}}^x_{ij} = \begin{cases} c^x_{ij} - \frac{c^x_{1j}c^x_{i1}}{c^x_{11}} \leq c^x_{ij} \leq 0, & \text{if } (i,j) \in E_{xx} \\ c^x_{ij} \leq 0, & \text{if } (i,j) \notin E_{xx} \end{cases}$$

holds. Therefore, the diagonal entries satisfy

$$\tilde{\mathfrak{c}}^x_{ii} \geq \tilde{c}^x_{ii} \geq \det(\{c^x_{kl}\}^{i,i}_{1,1}) / \det(\{c^x_{kl}\}^{i-1,i-1}_{1,1}) > 0.$$

If C_x is symmetric, we conclude from lemma 2.2 that an ILU-decomposition is performed during the algorithm, which exists by the assumptions. □

The algorithm described above produces a matrix $\bar{S}$ with the pattern E_{uu} that approximates the Schur complement of the optimization matrix. Besides symmetry we do not know any further properties of this matrix. Since it approximates the exact Schur complement S, we may expect that $\bar{S}$ is positive definite, which can be observed in numerical investigations.

In the second step of the incomplete decomposition we might perform an ILU-decomposition for $\bar{S}$, which is modified analogously to [GM] in order account for the fact that an unmodified ILU decomposition need not necessarily exist for it.

```
for (i = 1; i ≤ n_u; i++)
{
    θ = max{|s_ji| | j = i+1, ..., n_u};
    d = max{δ, |s_ii|, θ²/α²};
    s_ii = d;
    for (j = i+1; j ≤ n_u; j++)
        if (i,j) ∈ E_uu then
        {
            p = s_ji/d
            for (k = i+1; k ≤ j; k++)
                if (j,k) ∈ E_uu then s_jk −= p · s_ki;
        }
}
```

By this modified decomposition we guarantee that after the decomposition

$$s_{ii} > \delta \quad \text{and} \quad s_{ji}/\sqrt{s_{ii}} < \alpha, \; j > i$$

holds, where $s_{ji}/\sqrt{s_{ii}}$ corresponds exactly to the off-diagonal entries of a Cholesky decomposition. We choose $\delta > 0$ (small) and α before the decomposition in such a way that it holds that

$$\alpha^2 = \max \left\{ \begin{array}{l} \max\{|s_{ii}| \mid i = 1, \ldots, n_u\} \\ \max\{|s_{ji}| \mid j > i\}/\sqrt{n_r^2 - 1} \end{array} \right\},$$

where n_r is the number of non-zero entries per row in $\bar{S}$. For further details on the topic of (complete) modified Cholesky decompositions we refer to [GM, GMW]. Thus we can compute α during the last step in the recursion above.

Of course, besides the modification within the ILΔL$^\top$-decomposition mentioned above, further modifications are possible analogously to the ILU$_\beta$-variant in [Wit], to be applied in step (***) and denoted by ILΔL$_\beta^\top$. For further details see [Sch4].

3. Numerical results

As a simple test problem we consider a linear-quadratic distributed control problem, which is similar to the problem investigated in [Hac1]. We solve in $\Omega :=$

$[0,1]^2 \subset \mathbb{R}^2$

$$\min_{y,p} \frac{1}{2} \int_\Omega (y(x) - z(x))^2 dx + \frac{\delta}{2} \int_\Omega u(x)^2 dx \tag{9}$$

$$\text{s.t.} \qquad \triangle y = u \tag{10}$$

$$y(x) = 0 \; \forall x \in \Gamma := \partial\Omega, \tag{11}$$

where $\triangle := \partial^2/\partial^2_{x_1} + \partial^2/\partial^2_{x_2}$ means the Laplace operator in $\mathbb{R}^2$ and the function $z(x_1, x_2) \equiv 1, \quad \forall (x_1, x_2) \in \Omega$ is to be approximated. The second term in the objective functional serves for regularization purposes for $\delta > 0$.

In a finite element approach we use linear triangular elements on a regular grid for the states y as well as for the controls u, where the number of nodes N_ℓ on each grid level ℓ of the multigrid method satisfies the relationship $N_\ell = (2^{\ell+1}+1)^2$. The coarsest grid is always grid $\ell = 0$. Only there the QP is solved exactly. On all other grids only smoothing iterations using the ILΔL$^\top$decomposition described above are performed. The sparsity patterns are chosen according to the finite element discretization of the Laplacian operator. Figure 1 shows the average convergence rate κ for a W-cycle and a V-cycle with one pre- and two post-smoothing steps for different discretization levels ($\delta = 10^{-4}$). Essentially we can observe a level independent convergence rate of the multigrid method using the ILΔL$^\top$-smoother.

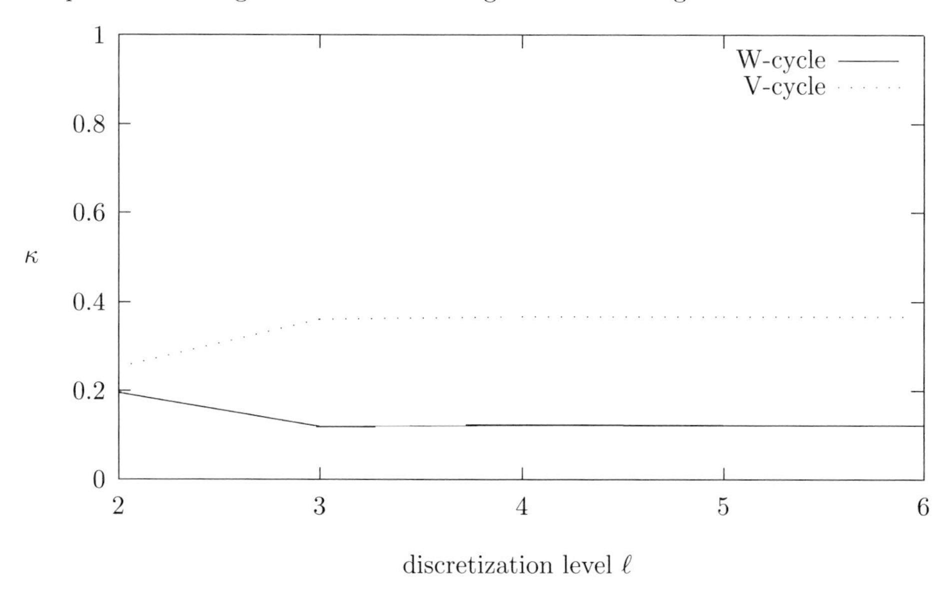

FIGURE 1. Average convergence rates κ for different multigrid cycle types

From figure 1 we conclude that the proposed incomplete decomposition is a viable approach to efficient smoothing techniques for the simultaneous multigrid solution of optimal control problems. For the convergence theory correlated with the multigrid method sketched above see [Sch4].

References

[BK] J.R. Bunch and L. Kaufmann. Some stable methods for calculating inertia and solving symmetric linear systems. *Math. Comp.*, 31:163–179, 1977.

[BP] J.R. Bunch and B.N. Parlett. Direct methods for solving indefinite systems of linear equations. *SIAM Journal on Numerical Analysis*, 8(3):639–655, 1971.

[GM] P.E. Gill and W. Murray. Newton-type methods for unconstrained and linearly constrained optimization. *Mathematical Programming*, 28:311–350, 1974.

[GMW] P.E. Gill, W. Murray, and M.H. Wright. *Practical optimization.* Academic Press, London, 1981.

[Hac1] W. Hackbusch. Fast solution of elliptic control problems. *Journal of Optimization Theory and Applications*, 31:565–581, 1980.

[Hac2] W. Hackbusch. *Multigrid methods and applications.* Springer, 1985.

[MV] J.A. Meijerink and H.A. van der Vorst. An iterative solution method for linear systems of which the coefficient matrix is a symmetric m-matrix. *Mathematics of Computation*, 31:148–162, 1977.

[Saa] Y. Saad. Preconditioning techniques for nonsymmetric and indefinite linear systems. *Journal of Computational and Applied Mathematics*, 24:89–105, 1988.

[SBH] V. Schulz, A. Bardossy, and R. Helmig. Conditional statistical inverse modeling in groundwater flow by multigrid methods. *Computational Geosciences*, 1999. (to appear).

[Sch1] V.H. Schulz. Numerical optimization of the cross-sectional shape of turbine blades. *Zeitschrift für angewandte Mathematik und Mechanik*, 76(S1):207–210, 1996.

[Sch2] V.H. Schulz. Solving discretized optimization problems by partially reduced SQP methods. *Computing and Visualization in Science*, 1(2):83–96, 1998.

[Sch3] V.H. Schulz (guest-editor). Special issue on SQP-based direct discretization methods for practical optimal control problems. *Journal of Computational and Applied Mathematics*, 1999. (to appear).

[Sch4] V.H. Schulz. Mehrgittermethoden für Optimierungsprobleme bei partiellen Differentialgleichungen. Habilitation thesis, University of Heidelberg, 1999, (in preparation).

[SW] V.H. Schulz and G. Wittum. Multigrid optimization methods for stationary parameter identification problems in groundwater flow. In W. Hackbusch and G. Wittum, editors, *Multigrid Methods V, Lecture Notes in Computational Science and Engineering 3*, pages 276–288. Springer, 1998.

[Wit] G. Wittum. On the robustness of ILU smoothing. *SIAM Journal on Scientific and Statistical Computing*, 10:699–717, 1989.

Interdisciplinary Center for Scientific Computing (IWR),
University of Heidelberg,
Im Neuenheimer Feld 368,
D-69120 Heidelberg, Germany
E-mail address: vschulz@na-net.ornl.gov

International Series of Numerical Mathematics
Vol. 133, © 1999 Birkhäuser Verlag Basel/Switzerland

Domain Optimization for the Navier-Stokes Equations by an Embedding Domain Method

Thomas Slawig

Abstract. Domain optimization problems for the two-dimensional stationary flow of incompressible linear-viscous fluids, i.e. the Navier-Stokes equations, are studied. An embedding domain technique which provides an equivalent formulation of the problem on a fixed domain is introduced. Existence of a solution to the domain optimization problem and Fréchet differentiability with respect to the variation of the domain of tracking type cost functionals for the velocity field are proved. A simply computable formula for the derivative of the cost functional is presented. Numerical examples show the advantages of the embedding domain method and the reliability of the derivative formula.

1. Introduction

In this work we consider domain optimization problems for the stationary Navier-Stokes equations. The aim of domain optimization is to find the shape of a domain which is optimal in the sense that a given cost functional is minimized subject to the constraint that some (partial) differential equation is satisfied. Typical features of such problems are the highly non-linear dependence and the lack of sensitivity of the cost functionals with respect to the variation of the domain.

Here we consider these problems in the field of fluid mechanics. Therefore typical constraints are the Navier-Stokes equations. We restrict our study to incompressible fluids and stationary problems in two space dimensions.

There are different types of cost functionals which are of interest in fluid mechanics, and in many types of geometrical settings domain optimization might be applied. Two examples are channel flow with a sudden expansion in machines where one tries to avoid back-flow and optimal design of airfoils for drag reduction.

The methods presented in this work allow significant reduction of computational effort necessary to perform an iterative optimization process for state equations whose solution itself is rather time-consuming, as it is the case for the Navier-Stokes equations.

The embedding domain method we used reduces the effort of discretization and assembling of the discrete systems for the changing domains during the optimization process. Moreover it provides us with a formula for the derivative of the cost functional with respect to the domain which is efficient and numerically stable.

This paper summarizes results of the author's PhD thesis [7]. It is based on a work by Kunisch and Peichl [6] who applied the same technique to the Laplace equation with mixed boundary conditions.

The outline of this paper is the following: In the second section we formulate the domain optimization problem with its geometrical setting and summarize the used basic results on the theory of the Navier-Stokes equations. In the next section we introduce the embedding domain method and derive an equivalent formulation of the Navier-Stokes equations on a fixed domain. In the fourth section we prove continuous dependence of the solution of the state equations with respect to the variation of the domain and the existence of a solution to the domain optimization problem. The central part is the fifth section which is concerned with Fréchet differentiability and the explicit formula for the derivative of the cost functional. In the last section we describe the applied numerical methods and present some results.

2. The Domain Optimization Problem

Before we formulate the domain optimization problem we describe the geometrical setting:

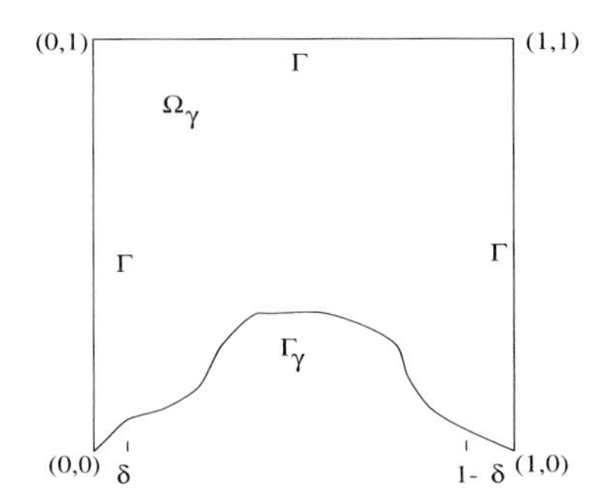

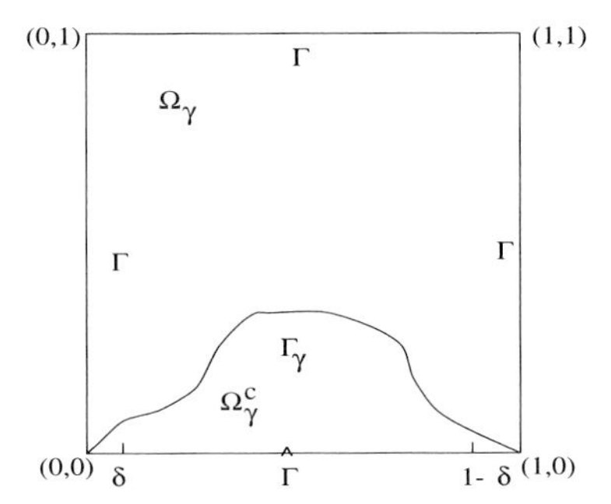

FIGURE 1.

We consider domains $\Omega_\gamma := \Omega(\gamma) \subset \mathbb{R}^2$ where the control parameter γ is a function defined on $I = (0,1)$ whose graph is one part (denoted by Γ_γ) of the boundary of Ω_γ, compare the picture on the left in Fig. 1.

The remaining part $\Gamma := \partial\Omega_\gamma \setminus \overline{\Gamma}_\gamma$ is fixed and consists of the three segments $[(0,0),(0,1)]$, $[(0,1),(1,1)]$, $[(1,1),(1,0)]$. The variable part Γ_γ therefore connects the two end points of Γ, namely $(0,0)$ and $(1,0)$.

To apply the embedding domain method and to prove the explicit formula of the derivative of the cost functional we need a combination of smooth and convex polygonal boundary. Clearly Γ is a convex polygon, for Γ_γ we assume C^2 regularity.

To preserve the convexity of Ω_γ near the two transition points $(0,0),(1,0)$ we assume that γ is linear in neighbourhoods of these points. Furthermore Γ_γ shall always be in the unit square $(0,1)\times(0,1)$. Working in Sobolev spaces we assure the regularity by choosing $\gamma \in H^3(I)$, and to get existence of a solution of (2) we need boundedness in this space. Summarizing we define the set of admissible functions γ defining the variable boundary parts Γ_γ and thus the admissible domains Ω_γ by

$$\begin{aligned}\mathcal{S} \ :=\ & \Big\{\gamma \in H^3(I) : \|\gamma\|_{H^3(I)} \le c_{\mathcal{S}}, \gamma(0)=\gamma(1)=0, c_0 \le \gamma(x) \le c_1, \\ & \qquad x \in (\delta, 1-\delta), \gamma'|_{(0,\delta)} = c^0, \gamma'|_{(1-\delta,1)} = c^1\Big\}. \end{aligned} \tag{1}$$

Here $c_0, c_1 \in (0,1), \delta \in (0,\frac{1}{2}), c_{\mathcal{S}}, c^0 \in \mathbb{R}^+, c^1 \in \mathbb{R}^-$ are fixed.

Given an observation region Ω_C which is a subset of Ω_γ for all $\gamma \in \mathcal{S}$ and a velocity field $\mathbf{u}_d \in L^2(\Omega_C)^2$ we study the following domain optimization problem:

$$\min_{\gamma\in\mathcal{S}} \mathcal{J}(\gamma) \ :=\ \frac{1}{2}\|\mathbf{u}_\gamma - \mathbf{u}_d\|^2_{L^2(\Omega_C)^2} \tag{2}$$

where $\mathbf{u}_\gamma$ is the velocity component of a variational solution $(\mathbf{u}_\gamma, p_\gamma) \in H^1(\Omega_\gamma)^2 \times L^2_0(\Omega_\gamma)$ of the Navier-Stokes equations

$$\begin{aligned} -\nu\triangle\mathbf{u}_\gamma + \mathbf{u}_\gamma\cdot\nabla\mathbf{u}_\gamma + \nabla p_\gamma &= \mathbf{f}_\gamma \quad \text{in} \quad \Omega_\gamma \\ \nabla\cdot\mathbf{u}_\gamma &= 0 \quad \text{in} \quad \Omega_\gamma \\ \mathbf{u}_\gamma &= \Phi \quad \text{on} \quad \Gamma \\ \mathbf{u}_\gamma &= \mathbf{0} \quad \text{on} \quad \Gamma_\gamma. \end{aligned} \tag{3}$$

In (2) an additional regularization term may be included. The space for the pressure p_γ is defined as $L^2_0(\Omega_\gamma) := \{q \in L^2(\Omega_\gamma) : \int_{\Omega_\gamma} q\,dx = 0\}$. For the inhomogeneity we assume $\mathbf{f}_\gamma \in L^\infty(\Omega_\gamma)^2$. The function Φ describing the boundary values of the velocity on Γ shall have a divergence-free extension onto Ω_γ which is in $H^2(\Omega_\gamma)^2$. We define

$$\begin{aligned} H(\Gamma) \ :=\ & \big\{\Phi \in L^2(\Gamma)^2 : \text{ there exists } \mathbf{u}^0_\gamma \in H^2(\Omega_\gamma)^2 : \nabla\cdot\mathbf{u}^0_\gamma = 0 \text{ in } \Omega_\gamma, \\ & \mathbf{u}^0_\gamma|_{\Gamma_\gamma} = \mathbf{0}, \mathbf{u}^0_\gamma|_\Gamma = \Phi\big\}. \end{aligned}$$

We summarize some needed theoretical results for the Navier-Stokes equations:

Theorem 2.1. *Let $\gamma \in \mathcal{S}, \mathbf{f}_\gamma \in L^2(\Omega_\gamma)^2$, and $\Phi \in H(\Gamma)$.*

(a) *Then there exists a variational solution $(\mathbf{u}_\gamma, p_\gamma) \in H^2(\Omega_\gamma)^2 \times [H^1(\Omega_\gamma) \cap L^2_0(\Omega_\gamma)]$ to (3) which for some $C > 0$ independent of $\gamma, \mathbf{f}_\gamma$, and Φ satisfies*

$$\|\mathbf{u}_\gamma\|_{H^2(\Omega_\gamma)^2} + \|p_\gamma\|_{H^1(\Omega_\gamma)} \ \le\ C\left(\|\mathbf{f}_\gamma\|_{L^2(\Omega_\gamma)^2} + \|\Phi\|_{L^\infty(\Gamma)^2}\right).$$

(b) *If $\nu > \nu_0 = \nu_0(\gamma, \mathbf{f}_\gamma, \Phi)$ the solution is unique.*

Proof. Regularity and uniqueness results for completely smooth C^2 or convex polygonal boundary are standard, see e.g. [3], [5]. From [5] it can be deduced that the regularity remains valid also in our case where both boundary types are mixed. The uniform regularity for the polygonal part is stated in the same reference, whereas for the smooth part it is shown e.g. in [2, Section IV.5.]. □

3. The Embedding Domain Method

Concerning the effort of discretization and assembling of the system matrices in the numerical solution of the state equations it is much more efficient to solve them on a simple-shaped and fixed domain rather than on the changing Ω_γ during an iterative optimization process. Thus we choose a so-called "fictitious domain" $\hat{\Omega}$ satisfying $\Omega_\gamma \subset \hat{\Omega}$ for all $\gamma \in \mathcal{S}$. In our case we take $\hat{\Omega}$ as the unit square.

Remark 3.1. *A smooth transition between Γ and Γ_γ would result in a highly irregular complementary domain $\Omega_\gamma^c = \hat{\Omega} \setminus \overline{\Omega}_\gamma$ with not even Lipschitz boundary. Thus there would be no existence result for the state equations on Ω_γ^c.*

We now derive an equivalent formulation of the Navier-Stokes equations on $\hat{\Omega}$. For this purpose we introduce the trace operator τ_γ onto the boundary Γ_γ and extend the inhomogeneity $\mathbf{f}_\gamma$ by zero to $\hat{\Omega}$. The former boundary condition $\tau_\gamma \mathbf{u}_\gamma = \mathbf{u}_\gamma|_{\Gamma_\gamma} = \mathbf{0}$ is treated as a constraint using a Lagrange multiplier g_γ.

The *fictitious domain formulation* of the Navier-Stokes equations then is to find $(\hat{\mathbf{u}}_\gamma, \hat{p}_\gamma, g_\gamma) \in H^1(\hat{\Omega})^2 \times L_0^2(\hat{\Omega}) \times H_\gamma$ such that

$$\begin{array}{rcll} -\nu\triangle\hat{\mathbf{u}}_\gamma + \hat{\mathbf{u}}_\gamma \cdot \nabla\hat{\mathbf{u}}_\gamma + \nabla\hat{p}_\gamma - \tau_\gamma^* g_\gamma &=& \tilde{\mathbf{f}}_\gamma & \text{in } H^{-1}(\hat{\Omega})^2 \\ \nabla \cdot \hat{\mathbf{u}}_\gamma &=& 0 & \text{in } L_0^2(\hat{\Omega}) \\ \tau_\gamma \hat{\mathbf{u}}_\gamma &=& \mathbf{0} & \text{in } H_\gamma. \end{array} \tag{4}$$

where τ_γ^* denotes the adjoint of the trace operator, and $H_\gamma := H_{00}^{1/2}(\Gamma_\gamma)^2$ is an abbreviation for the space

$$H_{00}^{1/2}(\Gamma_\gamma)^2 = \left\{ \mathbf{h} \in H^{1/2}(\Gamma_\gamma)^2 : \text{there exists } \tilde{\mathbf{h}} \in H^{1/2}(\partial\Omega_\gamma)^2 : \tilde{\mathbf{h}}|_{\Gamma_\gamma} = \mathbf{h}, \tilde{\mathbf{h}}|_\Gamma = \mathbf{0} \right\}.$$

We can now prove the equivalence of problems (3) and (4):

Theorem 3.2. *Let $\gamma \in \mathcal{S}, \mathbf{f}_\gamma \in L^2(\Omega_\gamma)^2$ and $\Phi \in H(\Gamma)$. Then $(\hat{\mathbf{u}}_\gamma, \hat{p}_\gamma, g_\gamma) \in H^1(\hat{\Omega})^2 \times L_0^2(\hat{\Omega}) \times H_\gamma^*$ is a solution of (4) if and only if*

- $(\mathbf{u}_\gamma, p_\gamma) := (\hat{\mathbf{u}}_\gamma, \hat{p}_\gamma)|_{\Omega_\gamma} \in H^2(\Omega_\gamma)^2 \times [H_1(\Omega_\gamma) \cap L_0^2(\Omega_\gamma)]$ *solves (3),*
- $(\hat{\mathbf{u}}_\gamma, \hat{p}_\gamma)|_{\Omega_\gamma^c} = (\mathbf{0}, 0)$,
- $g_\gamma = \left(\nu \dfrac{\partial \mathbf{u}_\gamma}{\partial \mathbf{n}_\gamma} - p_\gamma \mathbf{n}_\gamma \right)\Bigg|_{\Gamma_\gamma}$ *in $H^{1/2}(\Gamma_\gamma)^2$ where $\mathbf{n}_\gamma$ denotes the outer (with respect to Ω_γ) normal vector on Γ_γ.*

Proof. The result is proved by testing the weak formulation of (3) with appropriate functions that vanish on Ω_γ^c, applying a uniqueness result for the homogeneous Navier-Stokes equations and Green's formula. The regularity of g_γ follows from the regularity of $\mathbf{u}_\gamma, p_\gamma$, compare Theorem 2.1. For more details see [7, Th. 3.5]. □

Remark 3.3. *Here we see the first advantage of the embedding domain method: If γ changes only Γ_γ but not the whole domain has to be re-discretized. To get the discrete form of (4) only the discretized trace operator (which in principal is a one dimensional mass matrix) has to be re-assembled, and the discrete right-hand side has to be set to zero on Ω_γ^c. The rest of the system remains unchanged.*

4. Continuous Dependence of the Solution on the Shape of the Domain

To study convergence with respect to γ of the Lagrange multipliers $g_\gamma \in H_\gamma^*$ we introduce on H_γ the mapping

$$\mathcal{I}_\gamma \mathbf{h}(x) \;:=\; \mathbf{h}(x,\gamma(x)) \quad \mathbf{h} \in H_\gamma, x \in I,$$

which can be shown (see [7, Th. 2.4]) to be an isomorphism between H_γ and

$$H_I \;:=\; \left\{ \mathbf{g} \in H^{1/2}(I)^2 : \int_I \frac{\|\mathbf{g}(t)\|_2^2}{t(1-t)}\, dt < \infty \right\}.$$

We define the adjoint of $\mathcal{I}_\gamma^{-1}$ by

$$\begin{aligned} \left(\mathcal{I}_\gamma^{-1}\right)^* \;&:\; H_\gamma^* \to H_I^* \\ \langle \left(\mathcal{I}_\gamma^{-1}\right)^* g, \mathbf{g}\rangle_{H_I^*,H_I} \;&:=\; \langle g, \mathcal{I}_\gamma^{-1}\mathbf{g}\rangle_{H_\gamma^*,H_\gamma} \qquad g \in H_\gamma^*,\ \mathbf{g} \in H_I \end{aligned}$$

where $\langle\cdot,\cdot\rangle$ denotes dual pairings. Now we can formulate the following result:

Theorem 4.1. *Let $\gamma, \gamma_n \in \mathcal{S}$ with*

$$\gamma_n \to \gamma \;\; in\;\; W^{1,\infty}(I), \qquad \tilde{\mathbf{f}}_{\gamma_n} \to \tilde{\mathbf{f}}_\gamma \;\; in\;\; L^2(\hat{\Omega})^2,$$

and let the condition $\nu > \nu_0$ for the uniqueness of the solution to the Navier-Stokes equations be fulfilled. Then the solutions of problem (4) satisfy

$$\begin{aligned} \hat{\mathbf{u}}_{\gamma_n} \;&\to\; \hat{\mathbf{u}}_\gamma && in\; H^1(\hat{\Omega})^2, \\ \hat{p}_{\gamma_n} \;&\to\; \hat{p}_\gamma && in\; L_0^2(\hat{\Omega}), \\ \left(\mathcal{I}_{\gamma_n}^{-1}\right)^* g_{\gamma_n} \;&\stackrel{*}{\rightharpoonup}\; \left(\mathcal{I}_\gamma^{-1}\right)^* g_\gamma && in\; H_I^*. \end{aligned}$$

For $\nu > \nu_1 = \nu_1(\mathbf{f},\Phi)$ the mapping $\gamma \mapsto \hat{\mathbf{u}}_\gamma$ is Lipschitz continuous, i.e. there exists L independent of $\gamma, \bar{\gamma}$ such that

$$\|\hat{\mathbf{u}}_{\bar{\gamma}} - \hat{\mathbf{u}}_\gamma\|_{H^1(\hat{\Omega})^2} \;\leq\; L\|\bar{\gamma}-\gamma\|_{L^\infty(I)} \qquad for\ all\ \bar{\gamma},\gamma \in \mathcal{S}.$$

Proof. First step is to show uniform boundedness of the family of solutions for $\gamma \in \mathbb{S}$ which implies weak convergence for a subsequence and weak-$*$ convergence of the Lagrange multipliers. To show strong convergence of velocity and pressure we exploit the weak form of (4) with appropriate test functions. The different terms can be estimated exploiting their regularity. See [7, Th.3.7]. □

As a consequence of this Theorem and the boundedness of $\mathcal{S}$ in $H^3(I)$ which is compactly embedded in $C^2(\bar{I})$ we now obtain:

Corollary 4.2. *The domain optimization problem has at least one solution $\gamma \in \mathcal{S}$.*

5. Fréchet Differentiability and Derivative Formula

To show differentiability we use the solution of the adjoint system of the domain optimization problem (2). We introduce a Lagrangian with two multipliers $\lambda_\gamma, \mu_\gamma$ for the constraints of the momentum and continuity equation, respectively. Then we compute the necessary optimality conditions for a saddle point of this Lagrangian which form the adjoint equations. Roughly speaking they are linearized Navier-Stokes equations: Find $(\lambda_\gamma, \mu_\gamma) \in H_0^1(\Omega_\gamma)^2 \times L_0^2(\Omega_\gamma)$ such that

$$\begin{array}{rcll} -\nu\triangle\lambda_\gamma + \nabla\mathbf{u}_\gamma \cdot \lambda_\gamma - \mathbf{u}_\gamma \cdot \nabla\lambda_\gamma + \nabla\mu_\gamma &=& -D_u\mathcal{J}(\gamma) & \text{in } \Omega_\gamma \\ \nabla \cdot \lambda_\gamma &=& 0 & \text{in } \Omega_\gamma \end{array} \tag{5}$$

where $\mathbf{u}_\gamma$ is the velocity component of a solution to (3). Again we derive an equivalent fictitious domain formulation by introducing an additional Lagrange multiplier χ_γ corresponding to the constraint $\tau_\gamma\lambda_\gamma = \mathbf{0}$. Then we obtain the problem:
Find $(\hat{\lambda}_\gamma, \hat{\mu}_\gamma, \chi_\gamma) \in H_0^1(\hat{\Omega})^2 \times L_0^2(\hat{\Omega}) \times H_\gamma^*$ such that

$$\begin{array}{rcll} -\nu\triangle\hat{\lambda}_\gamma + \nabla\hat{\mathbf{u}}_\gamma \cdot \hat{\lambda}_\gamma - \hat{\mathbf{u}}_\gamma \cdot \nabla\hat{\lambda}_\gamma + \nabla\hat{\mu}_\gamma - \tau_\gamma^*\chi_\gamma &=& -D_u\mathcal{J}(\gamma) & \text{in } \hat{\Omega} \\ \nabla \cdot \hat{\lambda}_\gamma &=& 0 & \text{in } \hat{\Omega} \\ \tau_\gamma\hat{\lambda}_\gamma &=& \mathbf{0}. & \end{array} \tag{6}$$

For the solution of the adjoint problem we can show:

Theorem 5.1. *Let $\gamma \in \mathcal{S}$ and let $(\mathbf{u}_\gamma, p_\gamma)$ be a solution to (3). Then we have:*

(a) *Problem (5) has a solution $(\lambda_\gamma, \mu_\gamma) \in [H^2(\Omega_\gamma)^2 \cap H_0^1(\Omega_\gamma)^2] \times [H^1(\Omega_\gamma) \cap L_0^2(\Omega_\gamma)]$. The regularity is uniform in γ.*
(b) *The solution is unique for ν sufficiently large.*
(c) *$(\hat{\lambda}_\gamma, \hat{\mu}_\gamma, \chi_\gamma) \in H^1(\hat{\Omega})^2 \times L_0^2(\hat{\Omega}) \times H_\gamma^*$ is a solution of (6) if and only if*
- *$(\lambda_\gamma, \mu_\gamma) := (\hat{\lambda}_\gamma, \hat{\mu}_\gamma)|_{\Omega_\gamma}$ is a solution of (5),*
- *$(\hat{\lambda}_\gamma, \hat{\mu}_\gamma)|_{\Omega_\gamma^c} = (\mathbf{0}, 0)$,*
- *$\chi_\gamma = \left(\nu \dfrac{\partial\lambda_\gamma}{\partial\mathbf{n}_\gamma} - \mu_\gamma\mathbf{n}_\gamma\right)\Big|_{\Gamma_\gamma}$ in $H^{1/2}(\Gamma_\gamma)^2$.*

Proof. Part (a) is shown in [4], (b) follows from a uniqueness result for saddle point problems, and (c) is shown as Theorem 3.2. See [7, Th. 3.9]. □

To show differentiability we characterize the set of admissible directions as

$$\mathcal{S}' := \left\{\bar{\gamma} \in H^3(I) : \text{there exists } \{t_n\}_{n\in\mathbb{N}} : t_n \downarrow 0, \gamma + t_n\bar{\gamma} \in \mathcal{S}, n \in \mathbb{N}\right\}.$$

Now we can state the main result of this paper, the Fréchet differentiability of the cost functional with respect to γ. This can be shown if the parameter ν is large enough to ensure Lipschitz continuity of the velocity vectors $\hat{\mathbf{u}}_\gamma$ with respect to γ, compare Theorem 4.1.

Theorem 5.2. *Let* $\mathbf{f} \in L^\infty(\hat{\Omega})^2, \mathbf{f}_\gamma := \mathbf{f}|_{\Omega_\gamma}$ *for* $\gamma \in \mathcal{S}$, *and* $\nu > \nu_1$ *as is Theorem 4.1. Then* $\mathcal{J}$ *is Fréchet differentiable with respect to* γ, *and the derivative in* γ *satisfies*

$$D_\gamma \mathcal{J}(\gamma)\bar{\gamma} = \frac{1}{\nu}\int_I \left[g_\gamma(x,\gamma(x)) \cdot \chi_\gamma(x,\gamma(x)) - p_\gamma(x,\gamma(x))\mu_\gamma(x,\gamma(x))\right]\bar{\gamma}(x)\,dx \quad (7)$$

for all $\bar{\gamma} \in \mathcal{S}'$.

Proof. Using the variational forms of the Navier-Stokes and the adjoint equations with appropriate test functions it can be shown that the directional derivative satisfies (7). Clearly the right-hand side of (7) is a bounded linear operator on $\mathcal{S}'$ and therefore the Gateaux differential. Finally it can be shown that the mapping $\gamma \mapsto D_\gamma \mathcal{J}(\gamma)$ is continuous which proves that it is the Fréchet derivative. For the details see [7, Th. 3.10]. □

We want to emphasize a second advantage of the embedding domain method:

Remark 5.3. *The Lagrange multipliers* g_γ, χ_γ *introduced by the embedding domain method allow to compute the derivative of* $\mathcal{J}$ *as the one-dimensional integral in (7) without computing normal derivatives of the velocities. In the discrete case (e.g. for finite element basis functions) the integral can be computed exactly by a simple quadrature rule.*

The assumption on ν can be generalized in the following way:

Remark 5.4. *The result of the last Theorem remains valid if the assumption* $\nu > \nu_1$ *is replaced by Hölder continuity of the mapping* $\gamma \mapsto \hat{\mathbf{u}}_\gamma$ *in the* L^4 *norm with an exponent* $p > \frac{1}{2}$, *i.e. there exist* $L > 0, p > \frac{1}{2}$ *independent of* $\bar{\gamma}, \gamma \in \mathcal{S}$ *such that*

$$\|\hat{\mathbf{u}}_{\bar{\gamma}} - \hat{\mathbf{u}}_\gamma\|_{L^4(\hat{\Omega})^2} \leq L\|\bar{\gamma} - \gamma\|^p_{L^\infty(I)} \qquad \text{for all } \bar{\gamma}, \gamma \in \mathcal{S}.$$

6. Numerical Methods and Results

The numerical examples presented below were computed using the formula (7) for the derivative. The state and adjoint equations were discretized using stabilized finite elements (see e.g. [1]). The discretized Navier-Stokes equations were solved using a semi-implicit scheme presented in [3]. The linear systems were solved using sparse direct solvers. As optimization routine we used a SQP method.

As example we studied a flow separation problem in a driven cavity at a Reynolds number of $Re = \frac{1}{\nu} = 500$: The computational domain is the unit square where a part of the right lateral boundary is variable. On the top boundary we have a constant horizontal velocity which is positive in y direction, on the other boundaries we have zero velocity. If the right lateral boundary is a straight line at $x = 1$ the resulting flow shows one big vortex turning clockwise.

Aim of the optimization was to split this vortex into two which are separated by a horizontal line at $y = 0.5$. To achieve this we chose this line as observation

It.	Ev.	$\mathcal{J}$	$\gamma(0.125)$	$\gamma(0.25)$	$\gamma(0.375)$	$\gamma(0.5)$	$\gamma(0.625)$	$\gamma(0.75)$
	1	1.961e-04	0.9999	0.9999	0.9999	0.9999	0.9999	0.9999
1	2	1.4893e-06	0.9999	0.9999	0.9999	0.9999	0.9999	0.7500
2	3	1.211e-06	0.9999	0.9999	0.9999	0.7822	0.7500	0.7500
3	5	1.134e-06	0.9999	0.8750	0.8750	0.7661	0.7500	0.7500
4	7	1.1296e-06	0.9812	0.8758	0.8742	0.7622	0.7512	0.7500
...								
7	52	1.1296e-06	0.9812	0.8758	0.8742	0.7622	0.7512	0.7500
		1.2472e-06	0.7500	0.7500	0.7500	0.7500	0.7500	0.7500

TABLE 1. Convergence behaviour. In the last line the result with Γ_γ being a straight line at 0.75 is given. The cost functional value obtained by the optimization is still ten percent better. Most of the function evaluations (Ev.) were needed in the line search in the last iterations when no significant cost functional reduction was achieved.

region Ω_C and minimized the cost functional

$$\mathcal{J}(\gamma) \quad := \quad \int_{\Omega_C} \|\mathbf{u}_\gamma\|_2^2 \, dx$$

The right wall between $y = 0.125$ and 0.75 was allowed to vary in the range $x \in [0.75, 1)$ with six control parameters. As start curve we used a straight line near the right lateral wall. We used no regularization.

The optimization using the gradient computed with the formula (7) reduced the cost functional to less than one percent in a few iterations. The achieved solution was "optimal" in the sense that a comparison with the solution when Γ_γ was a straight line at $x = 0.75$ showed the superiority of the optimized solution. For detailed results see Table 1 and Figure 2.

Acknowledgements

The author would like to thank Prof. K.-H. Hoffmann, Prof. G. Leugering and Prof. F. Tröltzsch for the invitation to the Chemnitz conference. He also wants to thank Prof. K. Kunisch for the guidance during the author's work on his PhD thesis.

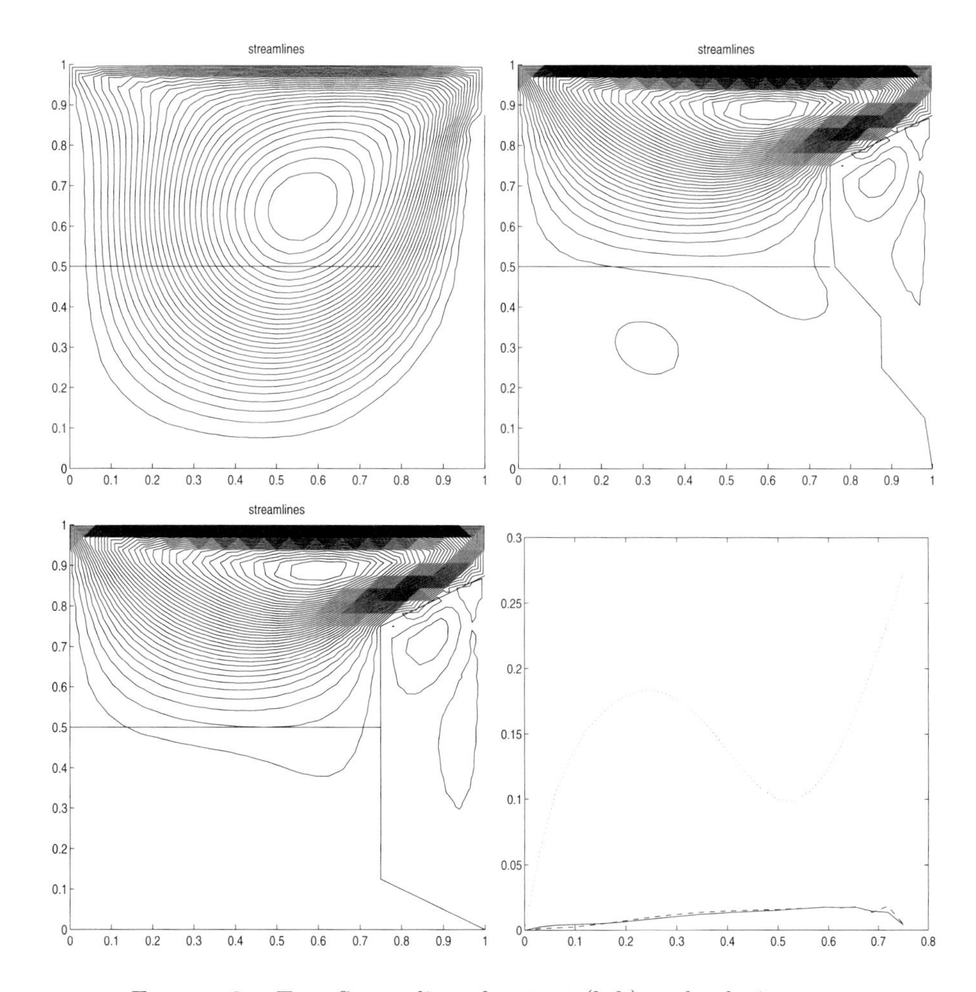

FIGURE 2. Top: Streamlines for start (left) and solution curve (right). The horizontal line marks the observation region Ω_C. Bottom left: Streamlines for straight line at $x = 0.75$, compare Table 1. Bottom right: Euclidean norm of velocity vector on Ω_C. Dotted: start curve, solid: solution, dashed: straight line at 0.75.

References

[1] L.P. Franca, S.L. Frey: Stabilized Finite Element Methods: II. The Incompressible Navier-Stokes Equations, *Comp. Meth. in Appl. Mech. Eng.* **99** 1992, pp. 209–233.

[2] G.P. Galdi: *An Introduction to the Mathematical Theory of the Navier-Stokes Equations, Vol.1* Springer New York 1994.

[3] V. Girault, P.-A. Raviart: *Finite Element Methods for Navier-Stokes Equations*, Springer Series in Comp. Math., Berlin New York 1986.

[4] M.D. Gunzburger, L. Hou, T.P. Svobodny: Analysis and Finite Element Approximation of Optimal Control Problems for the Stationary Navier-Stokes Equations with Distributed and Neumann Control, *Math. of Comp.* **57**, No. 195, 123–151 (1991).

[5] R.B. Kellogg, J.E. Osborn: A Regularity Result for the Stokes Problem in a Convex Polygon, *Journal on Functional Analysis*, **21**, pp. 397–431, 1976.

[6] K. Kunisch, G. Peichl: Shape Optimization for Mixed Boundary Value Problems Based on an Embedding Domain Method, to appear in *Dynamics of Continuous, Discrete and Impulsive Systems.*

[7] T. Slawig: *Domain Optimization for the stationary Stokes and Navier-Stokes Equations by an Embedding Domain Technique*, Shaker Aachen 1998.

Technische Universität Berlin,
Fachbereich Mathematik MA 6-2,
Strasse des 17. Juni 136,
D-10623 Berlin, Germany
E-mail address: slawig@math.tu-berlin.de

International Series of Numerical Mathematics
Vol. 133,

On the Approximation and Optimization of Fourth Order Elliptic Systems

J. Sprekels and D. Tiba

Abstract. We review several recent results devoted to this topic, and we present a new approach to a variational inequality with unilateral conditions on the boundary, associated with a partially clamped plate.

AMS Classification Code: 49J20, 49K20, 49M30, 93C20.

1. Introduction

In the recent papers Arnautu, Langmach, Sprekels and Tiba [1], Sprekels and Tiba [8], Sprekels and Tiba [9], Sprekels and Tiba [10], we have studied fourth order elliptic equations of the form

$$\Delta(u^3 \Delta y) = f \quad \text{in } \Omega, \tag{1.1}$$

where $\Omega \subset \mathbb{R}^N, N \in \mathbb{N}$, is a bounded smooth domain. To (1.1) various boundary conditions may be added:

$$y = \Delta y = 0 \quad \text{on} \quad \partial\Omega\,, \tag{1.2}$$

$$y = \frac{\partial y}{\partial n} = 0 \quad \text{on} \quad \partial\Omega \tag{1.3}$$

$$\begin{aligned} y &= \frac{\partial y}{\partial n} = 0 \quad \text{on} \quad \Gamma_1\,, \\ \Delta y &= \frac{\partial}{\partial n}(u^3 \Delta y) = 0 \quad \text{on} \quad \Gamma_2\,, \end{aligned} \tag{1.4}$$

where $\partial\Omega = \Gamma_1 \cup \Gamma_2, \Gamma_1 \cap \Gamma_2 = \emptyset$. In space dimensions one or two, equation (1.1) represents a simplified model for beams or plates, see Bendsoe [2]. The boundary conditions (1.2)–(1.4) correspond to simply supported, clamped, respectively partially clamped (in space dimension one: cantilevered) plates (beams). In this interpretation, y is the deflection (deformation) of the plate, u is the thickness and f is the load.

The existence of a unique weak solution y in $H^2(\Omega) \cap H_0^1(\Omega)$ or in $H_0^2(\Omega)$ is standard if $f \in L^2(\Omega)$ and if the boundary conditions (1.2), respectively (1.3), are considered. Conditions (1.4) are discussed in Section 3.

One essential ingredient in our approach is to transform (1.1) in various ways such that the solution y remains unchanged. These types of transformations are called "resizing" rules, according to a procedure used in the engineering literature (Haftka, Kamat and Gürdal [5]) for the design of beams and plates. Some examples are given in Section 2, where we shall provide a brief review of our recent work devoted to the existence, approximation and optimization of the system (1.1)–(1.4).

In Section 3, we study the difficult case of a partially clamped plate, and we prove new existence and approximation results.

2. Preliminaries

Let q denote the strong solution to $\Delta q = f$ in $\Omega, q = 0$ on $\partial\Omega$. Then, (1.1) can be equivalently rewritten as

$$\Delta y = (q + h)\ell \quad \text{in} \quad \Omega , \tag{2.1}$$

$$\Delta h = 0 \quad \text{in} \quad \Omega , \tag{2.2}$$

where $\ell = u^{-3} \in L^\infty(\Omega)_+$. Appropriate boundary conditions have to be added. If (1.2) is considered, then (2.1) gives (formally) that $h = 0$ on $\partial\Omega$, since q and Δy satisfy the same boundary condition and ℓ is strictly positive. By (2.2), we see that $h = 0$ in Ω, and the boundary value problem (1.1), (1.2) is equivalent to

$$\Delta y = q\ell \quad \text{in} \quad \Omega , \tag{2.3}$$

$$y = 0 \quad \text{on} \quad \partial\Omega . \tag{2.4}$$

In the other cases, the boundary conditions (1.3) or (1.4) must be added to the system (2.1), (2.2). Further such transformations have been discussed in the papers Sprekels and Tiba [8], Sprekels and Tiba [9]. Here, we indicate several applications.

We associate to the state equation (1.1) two classical optimization problems:

$$\text{Min} \int_\Omega u(x)dx , \tag{2.5}$$

$$\text{Min} \frac{1}{2} \int_\Omega [y(x) - y_d(x)]^2 dx , \tag{2.6}$$

where $y_d \in L^2(\Omega)$ represents some observed or desired deflection. Natural control and state constraints have to be added (m, M, τ are positive constants):

$$0 < m \leq u(x) \leq M \quad \text{a.e. in} \quad \Omega , \tag{2.7}$$

$$y(x) \geq -\tau \quad \text{a.e. in} \quad \Omega . \tag{2.8}$$

A very general state constraint that can be taken into account is

$$y \in A \,, \tag{2.9}$$

with $A \subset L^2(\Omega)$ being a closed, but not necessarily convex set.

Problem (2.5) is a "minimization-of-volume (weight)"-problem, while (2.6) is an "identification"-type problem.

If the above transformation is also applied to (2.5)–(2.9), then (2.6) and (2.8), (2.9) remain unchanged, while (2.5), (2.7) become

$$\text{Min} \int_\Omega \ell^{-\frac{1}{3}}(x)dx \,, \tag{2.5'}$$

$$0 < M^{-3} \leq \ell(x) \leq m^{-3} \quad \text{a.e. in} \quad \Omega \,. \tag{2.7'}$$

For the simply supported plate, although the original minimization problem (2.5) is nonconvex, the transformed problem is even strictly convex (if the general state constraint (2.9) is imposed, then A has to be convex, as well). We have the following result (cf. Sprekels and Tiba [9]).

Theorem 2.1. *The problem (2.3)–(2.5), (2.7), (2.9) has a unique global minimum u^*.*

While existence is obvious in this situation, in the general case compactness conditions on the set of admissible controls u have usually been imposed in the literature, for instance, that $\{\nabla u\}$ is bounded in some space (see Bendsoe [3], Neto and Polak [7]).

Using the decomposition (2.1), (2.2) it is possible to take advantage of the fact that the functions h are harmonic. In particular, when analyzing a minimizing sequence $[u_n, y_n]$, or equivalently, $[\ell_n, h_n, y_n]$, one may pass to the limit in the product $\ell_n h_n$ without such hypotheses on u_n, ℓ_n.

In Sprekels and Tiba [9], the following general existence result has been proved:

Theorem 2.2. *The optimization problems (2.5) or (2.6) associated to the state system (1.1) and (1.2) or (1.3) and to the constraints (2.7), (2.9) have at least one optimal pair $[y^*, u^*]$.*

Moreover, again for simply supported plates, a detailed description of the optimal pairs in the form of bang-bang results, may be obtained by this approach, see Sprekels and Tiba [10]:

Theorem 2.3. *If M is "big", $f \in L^p(\Omega), p > \frac{N}{2}, f \neq 0$ a.e. in Ω and $y_d \in L^s(\Omega), s > \frac{N}{2}$, then for any optimal pair $[y^*, u^*]$ of the problem (2.3), (2.4), (2.6)–(2.8) there is a subset $\Omega^* \subset \Omega$ of null measure such that*

$$\Omega = \Omega^* \cup \{x \in \Omega; u^*(x) = M\} \cup \{x \in \Omega; u^*(x) = m\}$$
$$\cup \{x \in \Omega; y^*(x) = -\tau\} \cup \{x \in \Omega; y^*(x) = y_d(x)\} .$$

This result shows that for the optimal pairs at least one of the constraints or the "objective" function y_d are active, a.e. in Ω. **Theorem 2.3** is called a "generalized bang-bang result", according to Tröltzsch [11]. Classical bang-bang results have also been proved in Sprekels and Tiba [10] for a variant of (2.5).

The decomposition (2.1), (2.2), when the thickness u is fixed, also suggests a new method for the computation of the solution of (1.1). We underline that this is an abstract mathematical approach and it is not related to the notion of energy and the Dirichlet principle. We penalize the Neumann boundary condition from (1.3) and obtain the first term in (2.10). Since (2.1), (2.2) has a similar form as some well-known optimality system, with y being the state and h the adjoint state, and since in unconstrained distributed control problems the adjoint state "coincides" with the control, we obtain the following optimization problem:

$$\text{Min} \left\{ \frac{1}{\epsilon} \int_{\partial\Omega} (\frac{\partial y}{\partial n})^2 d\sigma + \frac{1}{2} \int_{\Omega} \ell h^2 dx \right\} , \quad \epsilon > 0 , \tag{2.10}$$

subject to (2.1), (2.4). Now, $\ell \in L^\infty(\Omega)_+$ is fixed, and $h \in L^2(\Omega)$ is the unconstrained control parameter.

The optimality conditions for the unique solution $[y_\epsilon, h_\epsilon]$ of the problem (2.10) are given by (2.1), (2.4), by the adjoint system,

$$\Delta p_\epsilon = 0 \quad \text{in} \quad \Omega , \tag{2.11}$$

$$p_\epsilon = \frac{1}{\epsilon} \frac{\partial y_\epsilon}{\partial n} \quad \text{on} \quad \Omega , \tag{2.12}$$

and by Pontryagin's maximum principle,

$$p_\epsilon + h_\epsilon = 0 \quad \text{a.e.} \quad \text{in} \quad \Omega . \tag{2.13}$$

By eliminating p_ϵ via (2.13), it is obvious that (2.1), (2.4), (2.11), (2.12) is a penalization of the system (2.1), (2.2), (1.3) or, equivalently, of the equations (1.1), (1.3). We have the following result (cf. Arnautu, Langmach, Sprekels and Tiba [1]):

Theorem 2.4. *For $\epsilon \longrightarrow 0, \{h_\epsilon\}$ is bounded in $L^2(\Omega), \{y_\epsilon\}$ is bounded in $H^2(\Omega) \cap H_0^1(\Omega)$, and $\frac{\partial y_\epsilon}{\partial n} \longrightarrow 0$ strongly in $L^2(\partial\Omega)$. The weak limits $[\tilde{y}, \tilde{h}]$ in $H^2(\Omega) \times L^2(\Omega)$ of $[y_\epsilon, h_\epsilon]$, satisfy (2.1), (2.2) and (1.3).*

Remark. In Arnautu, Langmach, Sprekels and Tiba [1] it has been shown that the following variational inequality, which has been studied in Glowinski [4], may be approximated using the same approach:

$$\int_\Omega \Delta y\, u^3(\Delta y - \Delta w)dx \le \int_\Omega f(y-w)dx\,,$$

$$y \in \mathcal{K}, \forall w \in \mathcal{K} = \{v \in H_0^2(\Omega)\,; \alpha \le \Delta v \le \beta \text{ a.e. in } \Omega\}\,.$$

Here, $u \in L^\infty(\Omega)_+$ is the given thickness of the plate, $f \in L^2(\Omega)$ is the load as in (1.1) and α, β are some real obstacles.

The advantage of **Theorem 2.4** over previous results is that at the computational level piecewise linear, continuous finite elements can be used. Many numerical examples are reported in Arnautu, Langmach, Sprekels and Tiba [1]. In Section 3, we study the same approach in the difficult case of a partially clamped plate.

Finally, we mention that, in the paper by Sprekels and Tiba [8], the case of ordinary differential equations (beams) is investigated by specific methods and a global duality–type algorithm is indicated together with numerical results.

3. Partially clamped plates

Let us now introduce the control problem (compare with (2.10), (2.1), (2.4)):

$$\text{(3.1)} \qquad \text{Min}\Big\{\frac{1}{2\epsilon}\int_{\Gamma_1}(\frac{\partial y}{\partial n})^2 + \frac{1}{2}\int_\Omega \ell h^2\Big\}\,, \quad \epsilon > 0\,,$$

subject to

$$\text{(3.2)} \qquad \Delta y = (g+h)\ell \quad \text{in} \quad \Omega\,,$$

$$\text{(3.3)} \qquad y = 0 \quad \text{on} \quad \Gamma_1\,,$$

$$\text{(3.4)} \qquad y = v \quad \text{on} \quad \Gamma_2\,,$$

$$\text{(3.5)} \qquad -r \le v \le +r \quad \text{a.e. on} \quad \Gamma_2\,, \quad r > 0\,.$$

Here it is assumed that Ω is a bounded domain with smooth boundary $\partial\Omega = \Gamma_1 \cup \Gamma_2, \bar\Gamma_1 \cap \bar\Gamma_2 = \emptyset$, meas $(\Gamma_i) > 0$. The mappings $g \in L^2(\Omega), \ell \in L^\infty(\Omega), \ell(x) \ge m > 0$ a.e. in Ω, are given, and $h \in L^2(\Omega), v \in L^2(\Gamma_2)$ are control parameters satisfying the control constraint (3.5). The solution of the boundary value problem (3.2)–(3.4) should be understood in the transposition sense, and hence we only have $y \in H^{\frac{1}{2}}(\Omega)$. However, we have $y \in H^2(\mathcal{V})$ for some open neighbourhood $\mathcal{V} \subset \bar\Omega$ of Γ_1, so that the cost functional (3.1) makes sense. The existence of optimal triples $[y_\epsilon, h_\epsilon, v_\epsilon]$ is obvious by the coercivity in h of (3.1), and by (3.5).

Proposition 3.1. *The optimal triple* $[y_\epsilon, h_\epsilon, v_\epsilon] \in H^{\frac{1}{2}}(\Omega) \times L^2(\Omega) \times L^\infty(\Gamma_2)$ *is unique.*

Proof. Assuming that this is not true, we denote by $[\hat{y}_\epsilon, \hat{h}_\epsilon, \hat{v}_\epsilon]$ another possible optimal triple for the problem (3.1). By the strict convexity of the norm, we get

$$\frac{\partial y_\epsilon}{\partial n} = \frac{\partial \hat{y}_\epsilon}{\partial n} \quad \text{in} \quad \Gamma_1 , \tag{3.6}$$

$$h_\epsilon = \hat{h}_\epsilon \quad \text{in} \quad \Omega . \tag{3.7}$$

Then, (3.2), (3.3), (3.6), (3.7) show that $y_\epsilon, \hat{y}_\epsilon$ are the solutions of a Cauchy problem and, by uniqueness, we get $y_\epsilon = \hat{y}_\epsilon$ in Ω. Consequently, $v_\epsilon = \hat{v}_\epsilon$ in $L^2(\Gamma_2)$, as well. This follows from the definition of the transposition solution and from the density of the set $\{\frac{\partial \varphi}{\partial n}|_{\Gamma_2}\}$ in $L^2(\Gamma_2)$, where $\Delta\varphi = \psi$ in Γ, $\varphi = 0$ on $\partial\Omega$, and where ψ is arbitrary in $L^2(\Omega)$.

Proposition 3.2. *The triples* $[y_\epsilon, h_\epsilon, v_\epsilon]$ *are bounded in* $H^{\frac{1}{2}}(\Omega) \times L^2(\Omega) \times L^\infty(\Gamma_2)$ *with respect to* $\epsilon > 0$. *Moreover,* $\frac{\partial y_\epsilon}{\partial n} \longrightarrow 0$ *strongly in* $L^2(\Gamma_1)$.

Proof. The triple $[0, -g, 0]$ is admissible for any $\epsilon > 0$, and we have

$$\frac{1}{2\epsilon} \int_{\Gamma_1} \left(\frac{\partial y_\epsilon}{\partial n}\right)^2 + \frac{1}{2} \int_\Omega \ell h_\epsilon^2 \leq \frac{1}{2} \int_\Omega \ell g^2 , \quad \forall \epsilon > 0 .$$

Hence $\{h_\epsilon\}$ is bounded in $L^2(\Omega)$, since $\ell(x) \geq m > 0$ a.e. in Ω, and $\frac{\partial y_\epsilon}{\partial n} \longrightarrow 0$ strongly in $L^2(\Gamma_1)$. By (3.5), $\{v_\epsilon\}$ is bounded in $L^\infty(\Gamma_2)$, and (3.2)–(3.4) yield the boundedness of $\{y_\epsilon\}$ in $H^{\frac{1}{2}}(\Omega)$.

Proposition 3.3. *The optimal triple* $[y_\epsilon, h_\epsilon, v_\epsilon]$ *satisfies the optimality conditions given by (3.2)–(3.4) and:*

$$h_\epsilon = -p_\epsilon \quad \text{a.e. in} \quad \Omega ,$$

$$\int_{\Gamma_2} \frac{\partial p_\epsilon}{\partial n}(v - v_\epsilon) \geq 0 \quad \forall v \in U_{ad} ,$$

where $p_\epsilon \in H^1(\Omega)$ *is the solution of the adjoint equation,*

$$\Delta p_\epsilon = 0 \quad \text{in} \quad \Omega , \qquad p_\epsilon = \frac{1}{\epsilon}\frac{\partial y_\epsilon}{\partial n} \quad \text{on} \quad \Gamma_1 , \qquad p_\epsilon = 0 \quad \text{on} \quad \Gamma_2 .$$

Proof. Let $K \in L^2(\Omega)$ and $w \in U_{ad} - v_\epsilon$ be given, where

$$U_{ad} = \{v \in L^2(\Gamma_2);\ -r \leq v \leq r \text{ a.e. on } \Gamma_2\} .$$

Denote by $z \in H^{\frac{1}{2}}(\Omega)$ the transposition solution of

$$\Delta z = \ell K \quad \text{in} \quad \Omega , \tag{3.8}$$

$$z = 0 \quad \text{on} \quad \Gamma_1 , \tag{3.9}$$

$$z = w \quad \text{on} \quad \Gamma_2 . \tag{3.10}$$

Thus, for any $\lambda \in [0,1]$,

$$\frac{1}{2\epsilon}\int_{\Gamma_1}\left(\frac{\partial y_\epsilon}{\partial n}\right)^2 + \frac{1}{2}\int_{\Omega}\ell h_\epsilon^2 \leq \frac{1}{2\epsilon}\int_{\Gamma_1}\left(\frac{\partial y_\epsilon}{\partial n} + \lambda\frac{\partial z}{\partial n}\right)^2 + \frac{1}{2}\int_{\Omega}\ell(h_\epsilon + \lambda K)^2 ,$$

and it follows that

$$\frac{1}{\epsilon}\int_{\Gamma_1}\frac{\partial y_\epsilon}{\partial n}\frac{\partial z}{\partial n} + \int_{\Omega}\ell h_\epsilon K \geq 0 , \quad \forall\, z, K, w . \tag{3.11}$$

If w is smooth, we can multiply (3.8) by $p_\epsilon \in H^1(\Omega)$ and integrate twice by parts to obtain

$$\frac{1}{\epsilon}\int_{\Gamma_1}\frac{\partial y_\epsilon}{\partial n}\frac{\partial z}{\partial n} = \int_{\Gamma_2}\frac{\partial p_\epsilon}{\partial n}w + \int_{\Omega}\ell p_\epsilon K .$$

This relation remains valid for $w \in L^2(\Gamma_2)$, by passage to the limit. Using it in (3.11), we get

$$\int_{\Gamma_2}\frac{\partial p_\epsilon}{\partial n}w + \int_{\Omega}\ell(p_\epsilon + h_\epsilon)K \geq 0 , \quad \forall K \in L^2(\Omega) , \forall w \in U_{ad} - v_\epsilon ,$$

whence the assertion follows. □

From the optimality system given in **Proposition 3.3** we notice a supplementary property of the optimal triple. To this end, recall that $\ell = u^{-3}$ (u is the thickness of the plate), and put

$$\begin{aligned} V &:= \{w \in H^2(\Omega);\ w = \tfrac{\partial w}{\partial n} = 0 \quad \text{on } \Gamma_1\} \\ \mathcal{K} &:= \{w \in V : w|_{\Gamma_2} \in [-r, r] \quad \text{on } \Gamma_2\} . \end{aligned}$$

The state equation (3.2) may be rewritten as

$$u^3 \Delta y_\epsilon = g + h_\epsilon ,$$

and multiplying it by $\Delta y_\epsilon - \Delta w$, $\forall w \in \mathcal{K}$, yields

$$\begin{aligned} \int_{\Omega} u^3\Delta y_\epsilon(\Delta y_\epsilon - \Delta w) &= \int_{\Omega}(g + h_\epsilon)(\Delta y_\epsilon - \Delta w) \\ = -\int_{\Omega}\nabla(g + h_\epsilon)\nabla(y_\epsilon - w) &+ \int_{\partial\Omega}(g + h_\epsilon)\frac{\partial}{\partial n}(y_\epsilon - w) \\ = \int_{\Omega}\Delta(g + h_\epsilon)(y_\epsilon - w) &+ \int_{\partial\Omega}(g + h_\epsilon)\frac{\partial}{\partial n}(y_\epsilon - w) \\ -\int_{\partial\Omega}\frac{\partial}{\partial n}(g + h_\epsilon)(y_\epsilon - w) . & \end{aligned}$$

Here, we have assumed that g is in $H^2(\Omega)$, $g = \frac{\partial g}{\partial n} = 0$ in Γ_2, and y_ϵ is replaced by its smoothing (again denoted by y_ϵ) obtained by taking a smooth approximation of v_ϵ in (3.4).

By **Proposition 3.3**, we have $h_\epsilon = -p_\epsilon$, and this implies

(3.12)
$$\int_\Omega u^3 \Delta y_\epsilon (\Delta y_\epsilon - \Delta w) = \int_\Omega \Delta g (y_\epsilon - w) + \int_{\Gamma_1} (g + h_\epsilon) \frac{\partial}{\partial n} y_\epsilon - \int_{\Gamma_2} \frac{\partial}{\partial n} h_\epsilon (y_\epsilon - w) .$$

Notice that in this relation y_ϵ is again the optimal state of the problem (3.1), since it is possible to pass to the limit in the above–mentioned regularization of $v_\epsilon = y_\epsilon|_{\Gamma_2}$.

From the maximum principle, and from (3.12), we infer the inequality

$$(3.13)\quad \begin{aligned} \int_\Omega u^3 \Delta y_\epsilon (\Delta y_\epsilon - \Delta w) \quad &\le \int_\Omega \Delta g (y_\epsilon - w) \ + \int_{\Gamma_1} (g + h_\epsilon) \frac{\partial}{\partial n} y_\epsilon \\ &= \int_\Omega \Delta g (y_\epsilon - w) \ + \int_{\Gamma_1} g \frac{\partial}{\partial n} y_\epsilon - \frac{1}{\epsilon} \int_{\Gamma_1} \left(\frac{\partial}{\partial n} y_\epsilon \right)^2 \\ &\le \int_\Omega \Delta g (y_\epsilon - w) \ + \int_{\Gamma_1} g \frac{\partial}{\partial n} y_\epsilon . \end{aligned}$$

We obtain the following result.

Theorem 3.1. *Assume that the load* $f \in L^2(\Omega)$ *has the form* $f = \Delta g$ *with* $g \in H^2(\Omega)$, $g = \frac{\partial g}{\partial n} = 0$ *on* Γ_2. *Then, the variational inequality*

$$(3.14) \qquad \int_\Omega u^3 \Delta y (\Delta y - \Delta w) \le \int_\Omega f (y - w) , \quad \forall w \in \mathcal{K}_1 ,$$

$$\mathcal{K}_1 = \left\{ \mu \in H^{\frac{1}{2}}(\Omega) \; ; \; \Delta \mu \in L^2(\Omega) \ \ with \ \ \mu = \frac{\partial \mu}{\partial n} = 0 \ \ on \ \ \Gamma_1 \, , -r \le \mu|_{\Gamma_2} \le r \right\} ,$$

has a unique solution $y \in \mathcal{K}_1$, *and* $y_\epsilon \longrightarrow y$ *strongly in* $L^2(\Omega)$.

Proof. By **Proposition 3.2**, we can pass to the limit in (3.13) to obtain (3.14) for any $w \in \mathcal{K}$. The inequality remains valid for $w \in \mathcal{K}_1$, by density.

Suppose now that $y_1, y_2 \in \mathcal{K}_1$ are two solutions of (3.14). By choosing in turn $y = y_1, w = y_2$, and $y = y_2, w = y_1$, we obtain that

$$\int_\Omega u^3 (\Delta y_1 - \Delta y_2)^2 = 0 ,$$

i.e., $\Delta(y_1 - y_2) = 0$ in Ω. As $y_1 - y_2$ satisfies Cauchy type boundary conditions by the definition of $\mathcal{K}_1$, we get $y_1 = y_2$ in Ω, and the uniqueness follows. $\square$

Remark. Relation (3.14) models a plate which is partially clamped on Γ_1 and is subject to condition (3.5) on Γ_2 (the deflection y should be between the obstacles $-r$ and r). The approach via control problems governed by second order elliptic equations presented in this section is constructive and can be employed for the computation of the solution of (3.14). Another useful variant is to replace (3.1) by the minimization of the functional

$$\frac{1}{2\varepsilon}\int_{\Gamma_1} y^2 + \frac{1}{2}\int_{\Omega} \ell h^2 , \quad \varepsilon > 0 .$$

Then, (3.3) has to be replaced as well by

$$\frac{\partial y}{\partial n} = 0 \quad \text{on} \quad \Gamma_1 .$$

The convergence results proved in this section hold for this variant in a similar way. Moreover, variational inequalities with unilateral conditions on $\frac{\partial y}{\partial n}$, on Γ_2, may be formulated by changing the definition of $\mathcal{K}_1$ in **Theorem 3.1** and the same method may be applied.

Remark. The intrinsic difficulty in the variational inequality (3.14) comes from the fact that the corresponding bilinear form,

$$\int_{\Omega} u^3 \Delta y \, \Delta w ,$$

is not coercive on the space V. Indeed, otherwise the Cauchy problem (associated with the Laplace operator) with Cauchy conditions on Γ_1 would be well-posed, which is false, in general. This fact demonstrates the flexibility and broad range of applicability of the control approach proposed here.

References

[1] V. Arnautu, H. Langmach, J. Sprekels and D. Tiba: On the approximation and optimization of plates, Preprint No. 357, WIAS, Berlin (1997); submitted to Numerische Mathematik.

[2] M. Bendsoe: Optimization of structural topology, shape and material, Springer-Verlag, New York (1995).

[3] M. Bendsoe: Existence proofs for a class of plate optimization problems, LNCIS 59, Springer-Verlag, Heidelberg, 773–779 (1984).

[4] R. Glowinski, L. Marini and M. Vidraşcu: Finite element approximation and iterative solutions of a fourth order elliptic variational inequality, IMA J. Num. Anal. 4 (1984), 127–167.

[5] R. T. Haftka, P.M. Kamat and Z. Gürdal: Elements of structural optimization, Kluwer Academic Press, Boston (1990).

[6] I. Hlavacek, I. Bock and J. Lovisek: Optimal control of a variational inequality with applications to structural analysis. I. Optimal design of a beam with unilateral supports, Appl. Math. Optimiz. 11 (1984), 111–143.

[7] C. K. Neto and E. Polak: On the use of consistent approximations for the optimal design of beams, SIAM J. Control Optimiz. 34 (1996), 1891–1913.

[8] J. Sprekels and D. Tiba: A duality-type method for the design of beams, Preprint No. 222, WIAS, Berlin (1996); accepted for publication in Adv. Math. Sci. Appl.

[9] J. Sprekels and D. Tiba: A duality approach in the optimization of beams and plates, Preprint No. 335, WIAS, Berlin (1997); accepted for publication in SIAM J. Control Optimiz.

[10] J. Sprekels and D. Tiba: Propriétés de bang-bang généralisées dans l'optimisation des plaques, CRAS Paris, t. 327, Série I, p. 705–710 (1998).

[11] F. Tröltzsch: Optimality conditions for parabolic control problems and applications, Teubner Verlag, Leipzig (1984).

J. Sprekels
Weierstrass Institute for Applied Analysis and Stochastics
Mohrenstrasse 39
D-10117 Berlin, Germany

D. Tiba
Institute of Mathematics
Romanian Academy
P.O.B. 1-764
RO-70700 Bucharest, Romania

International Series of Numerical Mathematics
Vol. 133,

On the Existence and Approximation of Solutions for the Optimal Control of Nonlinear Hyperbolic Conservation Laws

Stefan Ulbrich

Abstract. Optimal control problems for possibly discontinuous entropy solutions of nonlinear multidimensional conservation laws with controls in source term and initial condition are considered. The control-to-state-mapping is analyzed by using monotone difference schemes and existence results for optimal controls are proven. Moreover, a result on the convergence of optimal solutions of finite dimensional approximations to solutions of the original problem is given. In the 1-D case the theory of compensated compactness is used to prove that the control-to-state-mapping is compact from L^∞ to $C([0,T];L^1_{loc})$ which ensures the existence of optimal controls under very weak assumptions.

1. Introduction

We consider optimal control problems

(P) minimize $J(y,u)$ subject to $u \in \mathcal{U}_{ad}$,

where $u=(u_0,u_1)$ is the control and $y=y(u)$ is entropy solution of the nonlinear scalar inhomogeneous conservation law

$$\begin{aligned} &y_t + \operatorname{div} f(y) = g(t,x,y,u_1), \quad (t,x) \in (0,T)\times\mathbb{R}^n =: \Omega_T \\ &y(0,x) = u_0(x), \quad x \in \mathbb{R}^n \end{aligned} \tag{1}$$

We will work either with the control space $\mathcal{U} = (L^1 \cap L^\infty)(\mathbb{R}^n) \times (L^1\cap L^\infty)(\Omega_T)^r$ or $\mathcal{U} = L^\infty(\mathbb{R}^n)\times L^\infty(\Omega_T)^r$ and assume that

(A1) The flux $f:\mathbb{R}\to\mathbb{R}^n$ is locally Lipschitz. $g \in (L^1\cap L^\infty)(\Omega_T; C^{0,1}_{loc}(\mathbb{R}\times\mathbb{R}^r))$ and for all $M_u>0$ there are $C_1, C_2 > 0$ with

$$g(t,x,y,u_1)\operatorname{sgn}_\pm(y) \le C_1 + C_2|y| \quad \forall (t,x,y,u_1)\in\Omega_T\times\mathbb{R}\times[-M_u,M_u]^r,$$

where $\operatorname{sgn}_\pm(0)=\pm 1$, $\operatorname{sgn}_\pm(y) = y/|y|$ for $y\neq 0$.

(A2) The *admissible set* $\mathcal{U}_{ad}$ is bounded in $\mathcal{U}$ and closed in $L^1_{loc}(\mathbb{R}^n)\times L^1_{loc}(\Omega_T)^r$.

It is well known that even for arbitrarily smooth u_0, u_1, g the solution of (1) may develop discontinuities after a finite time. In view of the results in Kružkov [9] entropy solutions have to be considered in order to ensure global existence and uniqueness. A function $y \in L^\infty(\Omega_T)$ is an *entropy solution* of (1) if for all $c \in \mathbb{R}$ and $\eta_c(\lambda) := |\lambda - c|$, $q_c(\lambda) := \operatorname{sgn}(\lambda - c)(f(\lambda) - f(c))$

$$\eta_c(y)_t + \operatorname{div} q_c(y) \le \operatorname{sgn}(y - c) g(t, x, y, u_1) \quad \text{in } \mathcal{D}'(\Omega_T) \tag{2}$$

and if the initial data $u_0 \in L^\infty(\mathbb{R}^n)$ are assumed in the sense

$$\lim_{t \to 0+} \frac{1}{t} \int_0^t \|y(\tau, .) - u_0\|_{1,K} \, d\tau = 0 \quad \forall\, K \subset\subset \mathbb{R}^n. \tag{3}$$

The functions η_c, q_c in (2) form a special family of convex entropy pairs. We recall that (η, q) is a *convex entropy pair* for (1) if $\eta : \mathbb{R} \to \mathbb{R}$ is convex and the *entropy flux* $q : \mathbb{R} \to \mathbb{R}^n$ satisfies $\nabla q = \eta' \nabla f$. Hereby, η' shall be defined everywhere in such a way that it is monotone increasing. Taking the limit of appropriate nonnegative linear combinations of (2) shows that the Lax entropy inequality

$$\eta(y)_t + \operatorname{div} q(y) \le \eta'(y)\, g(t, x, y, u_1) \quad \text{in } \mathcal{D}'(\Omega_T) \tag{4}$$

holds in fact for all convex entropy pairs (η, q).

In [5], [8] the problem of identifying the flux f in homogeneous conservation laws is considered. In [1] a boundary control problem for the 1-D Problem with $g = 0$, $f'' > 0$ is analyzed by using an explicit solution formula of Le Floch.

In this work we consider the inhomogeneous multidimensional problem (P), (1). It covers, e.g., model problems for the optimal design of a duct in 1-D and the identification of friction parameters from flow observations. The purpose of this work is to establish the existence of optimal controls in a general framework and to give a convergence result for finite dimensional approximations of (P). Due to space limitations we will only consider the Cauchy problem (1) although most of our results can also be extended to problems with additional boundary controls. In a forthcoming paper we will continue the investigations by presenting a sensitivity- and adjoint-calculus in the 1-D case.

The paper is organized as follows. In §2 we state an existence and stability result for the state equation. In §3 we establish the existence of optimal controls by a compactness argument. In §4 the convergence of finite dimensional approximations is considered. For the 1-D case we generalize in §5 the existence result of §3 considerably by applying the theory of compensated compactness. In §6 we sketch the proof of the stability result in §2 by using monotone difference schemes.

2. Existence and stability results for the state equation

We recall the following uniqueness result which is a straightforward generalization of the one in [9] concerning the smoothness requirements on f, g:

Theorem 2.1. *Let $\mathcal{U} = L^\infty(\mathbb{R}^n) \times L^\infty(\Omega_T)^r$ and let (A1), (A2) hold. Then for $u \in \mathcal{U}_{ad}$ there is at most one entropy solution $y = y(u) \in L^\infty(\Omega_T)$ satisfying (2),*

(3). *Moreover, let* $y = y(u_0, u_1), \hat{y} = y(\hat{u}_0, u_1) \in L^\infty(\Omega_T)$ *be entropy solutions with* $\|y\|_\infty, \|\hat{y}\|_\infty \le M$. *Define for* $(\bar{t}, \bar{x}) \in \Omega_T$ *and* $R > 0$ *the propagation cone*

$$K(\bar{t}, \bar{x}, R) := \{(\tau, x)\,;\ 0 \le \tau \le \bar{t}, \|x - \bar{x}\|_2 \le R + M_{f'}(\bar{t} - \tau)\} \tag{5}$$

with $M_{f'} = \operatorname{ess\,sup}_{|\lambda| \le M} \|\nabla f(\lambda)\|_2$ *and denote by* S_t *the cross-section of* $K(\bar{t}, \bar{x}, R)$ *at* $\tau = t$. *Then for all* $t \in [0, \bar{t}]$ *the local stability estimate holds*

$$\|y(t) - \hat{y}(t)\|_{1,S_t} \le \|u_0 - \hat{u}_0\|_{1,S_0} e^{Lt},$$

where L *is the Lipschitz constant of* $g(t, x, y, u_1)$ *w.r.t.* y *on* $[-M, M]$.

Proof. A reinspection of the proof in [9] shows that the above result holds if $(t, x) \longmapsto g(t, x, ., u_1(t, x))$ is in $L^\infty(\Omega_T; C^{0,1}_{loc}(\mathbb{R}))$. For $u \in \mathcal{U}_{ad}$ we have $\|u_1\|_\infty \le M_u < \infty$ and hence $\|g(t, x, ., u_1(t, x))\|_{C^{0,1}_{loc}(\mathbb{R})} \le \|g(t, x, ., .)\|_{C^{0,1}_{loc}(\mathbb{R} \times [-M_u, M_u]^r)}$. Thus, $(t, x) \longmapsto g(t, x, ., u_1(t, x)) \in (L^1 \cap L^\infty)(\Omega_T; C^{0,1}_{loc}(\mathbb{R}))$ by (A1). □

Under stronger regularity assumptions on the source term than provided by (A1), (A2) the existence of an entropy solution was proven in [9] by the vanishing viscosity method, but the dependence of the solution on the source term is not analyzed in detail. The following existence and stability result collects important properties of the control-to-state mapping $u \mapsto y(u)$ implicitly defined by (1). The proof will be sketched in §6 by using monotone difference approximations.

Theorem 2.2. *With* $\mathcal{U} = (L^1 \cap L^\infty)(\mathbb{R}^n) \times (L^1 \cap L^\infty)(\Omega_T)^r$ *let* (A1), (A2) *hold. Fix* $M_u > 0$ *such that* $u \in \mathcal{U}_{ad}$ *implies* $\|u\|_\infty \le M_u$ *and set*

$$M = \left(M_u + C_1(1 - e^{-C_2 T})/C_2\right) e^{C_2 T}. \tag{6}$$

Denote by L_f *the Lipschitz constant of* f *on* $[-M, M]$ *and by* L, L_u *the Lipschitz constants of* g *w.r.t.* y *and* u_1 *on* $[-M, M] \times [-M_u, M_u]^r$, *resp. Then for all* $u = (u_0, u_1) \in \mathcal{U}_{ad}$ *there exists a unique entropy solution* $y = y(u) \in (L^1 \cap L^\infty)(\Omega_T)$. *Moreover,* $y \in C^0([0, T]; L^1(\mathbb{R}^n))$ *and with* $\Omega_t := (0, t) \times \mathbb{R}^n$, $0 \le t \le T$, *holds*

1) $\|y(t)\|_\infty \le \left(\|u_0\|_\infty + C_1(1 - e^{-C_2 t})/C_2\right) e^{C_2 t} =: M_\infty(t)$
2) $\|y(t)\|_1 \le (\|u_0\|_1 + L_u\|u_1\|_{1,\Omega_t} + \|g(., 0, 0)\|_{1,\Omega_t}) e^{Lt} =: M_1(t)$

If under the same assumptions $\hat{y}$ *is a solution for* $\hat{u}$, $\hat{g}$ *instead of* u, g *then*

3) $\|y(t) - \hat{y}(t)\|_1 \le \left(\|u_0 - \hat{u}_0\|_1 + L_u\|u_1 - \hat{u}_1\|_{1,\Omega_t} + \|(g - \hat{g})(., y, u_1)\|_{1,\Omega_t}\right) e^{Lt}$

If in addition $u_0 \in BV(\mathbb{R}^n)$[†], $u_1 \in L^1([0, T]; BV(\mathbb{R}^n))$, g *has a Lipschitz constant* L_x *and compact support* $\operatorname{supp}_x(g) \subset K \subset\subset \mathbb{R}^n$ *w.r.t.* x *then for* $0 \le \hat{t} \le t \le T$

4) $\|y(t)\|_{BV} \le (\|u_0\|_{BV} + L_u\|u_1\|_{L^1([0,t];BV)} + 2nL_x\mu(K)(1 - e^{-Lt})/L) e^{Lt} =: M_{BV}(t)$
5) $\|y(t) - y(\hat{t})\|_1 \le (t - \hat{t})\,(CL_f M_{BV}(t) + LM_1(t)) + \|g(., 0, u_1)\|_{1,[\hat{t},t] \times K}$

[†]The Banach-space $BV(\mathbb{R}^n)$ of functions of bounded variation is equipped with the norm $\|\,.\,\|_1 + \|\,.\,\|_{BV}$ where $\|v\|_{BV} := \sum_{i=1}^{n} \sup_{h_i \ne 0} \int_{\mathbb{R}^n} \frac{|v(x_1, \ldots, x_{i-1}, x_i + h_i, x_{i+1}, \ldots, x_n) - v(x)|}{|h_i|}\, dx.$

If $\mathcal{U} = L^\infty(\mathbb{R}^n) \times L^\infty(\Omega_T)^r$ *and* $L^1 \cap L^\infty$ *is replaced by* L^∞ *in* (A1) *then there still exists a unique entropy solution* $y \in L^\infty(\Omega_T)$. *Moreover,* 1) *remains valid,* $y \in C^0([0,T]; L^1_{loc}(\mathbb{R}^n))$, *and local versions of* 1)–5) *hold: if the norms on the left-hand side of* 1)–5) *are restricted to the ball* $B_R(\bar{x}) = \{\|x - \bar{x}\|_2 \le R\}$ *then* u, g *may be modified outside of the propagation cone* $K(t, \bar{x}, R)$ *given in* (5).

In the next section we apply a compactness framework to establish the existence of an optimal solution to (P) for the general case. In §5 a more general result will be proven for the one-dimensional case by compensated compactness.

3. Existence of optimal controls

The aim of this section is to give sufficient conditions for the existence of an optimal solution to the problem (P). Let (A1), (A2) hold with the control space $\mathcal{U} := L^\infty(\mathbb{R}^n) \times L^\infty(\Omega_T)^r$. Then the local versions of Theorem 2.2, 1)–3) enable us to prove the existence of optimal controls under the additional assumptions

- (A3) $J : C^0([0,T]; L^p_{loc}(\mathbb{R}^n)) \times \mathcal{U}_{ad} \subset (L^p_{loc}(\mathbb{R}^n) \times L^p_{loc}(\Omega_T)^r) \longrightarrow \mathbb{R}$ is sequentially lower semicontinuous for some $p \in [1, \infty)$.
- (A4) $\mathcal{U}_{ad}$ is compact in $L^1_{loc}(\mathbb{R}^n) \times L^1_{loc}(\Omega_T)^r$.

In fact, we have the following result:

Theorem 3.1. *Let* (A1), (A2) *hold. Then for all* $R > 0$ *and* $K_R := K(T, 0, R)$ *defined by* (5) *with* S_τ *denoting the cross section of* K_R *at* $t = \tau$ *holds:*

- i) $u \in \mathcal{U}_{ad} \subset (L^1(S_0) \times L^1(K_R)^r) \longmapsto y(u) \in C^0([0,T]; L^p(S_T))$ *is* $1/p$-*Hölder-continuous for* $p \in [1, \infty)$ *and thus Lipschitz for* $p = 1$.
- ii) *If* J *satisfies* (A3) *then* $u \in \mathcal{U}_{ad} \subset L^1_{loc}(\mathbb{R}^n) \times L^1_{loc}(\Omega_T)^r \longmapsto J(y(u), u)$ *is well defined and sequentially lower semicontinuous.*
- iii) *If* (A3), (A4) *hold then* (P) *has a solution* $\bar{u} \in \mathcal{U}_{ad}$.

Proof. i): By Theorem 2.2 there exists a unique entropy solution $y \in L^\infty(\Omega_T) \cap C^0([0,T]; L^1_{loc}(\mathbb{R}^n))$ with $\|y\|_\infty \le M$ for all $u \in \mathcal{U}_{ad}$. The local version of 3) on K_R yields the Lipschitz continuity of $u \in \mathcal{U}_{ad} \subset (L^1(S_0) \times L^1(K_R)^r) \longmapsto y(u) \in C^0([0,T]; L^1(S_T))$. Moreover, $y \in C^0([0,T]; L^p_{loc}(\mathbb{R}^n))$, since $\|\,.\,\|_p \le \|\,.\,\|_1^\theta \|\,.\,\|_\infty^{1-\theta}$, $\theta = 1/p$, and also the assertion on the Hölder-continuity of $u \longmapsto y(u)$ for $p \in [1, \infty)$ follows from this interpolation inequality.
ii): By i) the mapping $u \in \mathcal{U}_{ad} \subset (L^1_{loc}(\mathbb{R}^n) \times L^1_{loc}(\Omega_T)^r) \longmapsto J(y(u), u)$ is well defined. Now let the sequence $(u^k) \subset \mathcal{U}_{ad}$ converge in $L^1_{loc}(\mathbb{R}^n) \times L^1_{loc}(\Omega_T)^r$ to $\bar{u}$. Then by interpolation also $u^k \to \bar{u}$ in L^p_{loc}, $1 \le p < \infty$, and $\bar{u} \in \mathcal{U}_{ad}$ due to (A2). By i) $y(u^k)$ converge in $C^0([0,T]; L^p(S_T))$ to $y(\bar{u})$ for all K_R, $R > 0$. Hence, $y(u^k) \to y(\bar{u})$ in $C^0([0,T]; L^p_{loc}(\mathbb{R}^n))$. From (A3) we thus obtain as asserted $\liminf_{k\to\infty} J(y(u^k), u_k) \ge J(y(\bar{u}), \bar{u})$.
iii): Let (u^k) be a minimizing sequence for (P). By (A4) a subsequence converges in $L^1_{loc}(\mathbb{R}^n) \times L^1_{loc}(\Omega_T)^r$ to some $\bar{u} \in \mathcal{U}_{ad}$. Now $\bar{u}$ solves (P) by ii). □

4. Convergence of approximate solutions

For practical applications it is important to know under which conditions the optimal solutions of finite dimensional approximations to (P) converge to the optimal solution of the original problem. The following result can be applied to a large class of discretizations:

Theorem 4.1. *Let* (A1)–(A4) *hold. For* $l \to \infty$ *let* (P^l) *be the approximations of* (P) *obtained by replacing the control-to-state-mapping* $u \longmapsto y(u)$ *defined in* (1) *by a continuous finite dimensional approximation* $U^l \in \mathcal{U}^l_{ad} \subset L^1_{loc} \longmapsto Y^l \in \mathcal{Y}^l$, $\mathcal{Y}^l$ *denoting a subspace of* $L^\infty([0,T];L^p_{loc})$, *and the objective functional* $J(y(u),u)$ *by* $J(S_l Y^l, U^l)$ *with continuous operators* $S_l : \mathcal{Y}^l \to C^0([0,T];L^p_{loc})$, *such that*

a) $\mathcal{U}^l_{ad}$ *are closed and uniformly bounded in* $\mathcal{U}$,
$$\lim_{l\to\infty} \sup_{v\in\mathcal{U}^l_{ad}} \inf_{u\in\mathcal{U}_{ad}} \|v-u\|_{1,loc} = 0, \quad \lim_{l\to\infty} \operatorname{dist}\{u,\mathcal{U}^l_{ad}\}_{L^1_{loc}} = 0 \quad \forall u \in \mathcal{U}_{ad}.$$

b) $\lim_{l\to\infty} \|U^l - u\|_{L^1_{loc}} = 0 \Longrightarrow \lim_{l\to\infty} \|S_l Y^l - y(u)\|_{C^0([0,T];L^p_{loc})} = 0.$

c) *For all* $u \in \mathcal{U}_{ad}$ *there exist* $U^l \in \mathcal{U}^l_{ad}$ *such that* $J(S_l Y^l, U^l) \to J(y(u),u)$.

Then (P^l) *have optimal solutions* $\bar{U}^l$. *A subsequence* $\bar{U}^{l'}$ *converges in* L^1_{loc} *to an optimal solution* $\bar{u}$ *of* (P). *The associated discrete states* $S_{l'}\bar{Y}^{l'}$ *tend in* $C^0([0,T];L^p_{loc})$ *to* $y(\bar{u})$ *and* $J(S_{l'}\bar{Y}^{l'},\bar{U}^{l'}) \to J(y(\bar{u}),\bar{u})$.

Proof. By a), the finite dimensional subsets $\mathcal{U}^l_{ad}$ of L^1_{loc} are compact in the spaces L^p_{loc}. Moreover, $U^l \in \mathcal{U}^l_{ad} \subset L^p_{loc} \longmapsto J(S_l Y^l, U^l)$ is sequentially lower semicontinuous by (A3) and the assumptions on the discretization. Thus, (P^l) have optimal solutions $\bar{U}^l$. Let $\bar{U}^l$ be any such sequence. By a) there are $u^l \in \mathcal{U}_{ad}$ with $\|\bar{U}^l - u^l\|_{1,loc} \to 0$. By (A2) a subsequence $u^{l'}$ converges in L^1_{loc} to some $\bar{u} \in \mathcal{U}_{ad}$ and thus $\bar{U}^{l'} \to \bar{u}$ in L^1_{loc} and by interpolation also in L^p_{loc}, $1 \le p < \infty$, since $\bar{U}^{l'}$ are by a) uniformly bounded in $\mathcal{U}$. Hence, $S_{l'}\bar{Y}^{l'} \to y(\bar{u})$ in $C^0([0,T];L^p_{loc})$ and $J(y(\bar{u}),\bar{u}) \le \liminf_{l\to\infty} J(S_l\bar{Y}^l,\bar{U}^l)$ by b) and (A3). We conclude the proof by showing that $\bar{u}$ solves (P). Theorem 3.1 ensures the existence of an optimal solution $\tilde{u} \in \mathcal{U}_{ad}$ of (P). By c) there are $\tilde{U}^l \in \mathcal{U}^l_{ad}$ with $J(S_l\tilde{Y}^l,\tilde{U}^l) \to J(y(\tilde{u}),\tilde{u})$. Clearly $J(S_l\tilde{Y}^l,\tilde{U}^l) \ge J(S_l\bar{Y}^l,\bar{U}^l)$ and thus $J(y(\tilde{u}),\tilde{u}) \ge J(y(\bar{u}),\bar{u})$. Since $\tilde{u}$ is optimal, we thus have $J(S_l\bar{Y}^l,\bar{U}^l) \to J(y(\tilde{u}),\tilde{u}) = J(y(\bar{u}),\bar{u})$. □

Remark 4.2. By a), b) condition c) is satisfied if $J : C^0([0,T];L^p_{loc}(\mathbb{R}^n)) \times (\mathcal{U}_{ad} \subset \mathcal{V}) \longrightarrow \mathbb{R}$ is continuous for a Banach-space $\mathcal{V} \hookrightarrow L^p_{loc}(\mathbb{R}^n) \times L^p_{loc}(\Omega_T)^r$ and if $\lim_{l\to\infty} \operatorname{dist}\{u,\mathcal{U}^l_{ad}\}_{\mathcal{V}} = 0$ for all $u \in \mathcal{U}_{ad}$. If $\mathcal{V}$ is not separable, it suffices for the case $J(y,u) = H(y) + \alpha|u|_*$ with a semi-norm $|\,.\,|_*$ on $\mathcal{V}$ that for all $u \in \mathcal{U}_{ad}$ there are $U^l \in \mathcal{U}^l_{ad}$ with $\|U^l - u\|_{1,loc} + ||U^l|_* - |u|_*| \to 0$. In [3] it is shown that for an equivalent seminorm $|\,.\,|_*$ of $\|.\|_{BV}$ this property holds if U^l are piecewise constant.

Condition b) is satisfied by monotone difference approximations, cf. §6 and Remark 6.3. An extension of the framework in [5] to nonhomogeneous equations can be used to establish b) for high resolution schemes by the theory of measure-valued solutions.

5. Existence of optimal controls via compensated compactness

In §3 the existence of optimal controls was obtained by using a compactness argument. In the 1-D case more general results can be proven by the theory of compensated compactness. We consider the problem

(P′) minimize $J(y, g(.,y,u))$ subject to $u \in \mathcal{U}_{ad}$,

where $y = y(u)$ is an entropy solution of the Cauchy problem

$$\begin{aligned} & y_t + f(y)_x = g(t,x,y,u), \quad (t,x) \in]0,T[\times\mathbb{R} =: \Omega_T \\ & y(0,x) = y_0(x), \quad x \in \mathbb{R}, \quad y_0 \in L^\infty(\mathbb{R}). \end{aligned} \tag{7}$$

We take as control space $\mathcal{U} := L^\infty(\Omega_T)^r$ and require instead of (A2), (A4) only

(A2′) $\mathcal{U}_{ad} \subset \mathcal{U}$ is weak*-sequentially compact and $u \in \mathcal{U}_{ad} \longmapsto g(.,y,u)$ is L^∞-weak*-continuous for all $y \in L^\infty(\Omega_T)$.

or

(A2″) $\mathcal{U}_{ad} = \{u \in \mathcal{U};\ u(t,x) \in \mathcal{C}(t,x) \text{ a.e.}\}$ where $\mathcal{C} : (t,x) \in \Omega_T \longmapsto \mathcal{C}(t,x) \subset \mathbb{R}^r$ is a uniformly bounded measurable set-valued map with nonempty closed connected images.

Moreover, we replace (A3) by

(A3′) $J : C^0([0,T]; L^p_{loc}(\mathbb{R})) \times L^\infty(\Omega_T) \longrightarrow \mathbb{R}$ is sequentially lower semicontinuous in the following sense: if $y^k \to \bar{y}$ strongly in $C^0([0,T]; L^p_{loc}(\mathbb{R}))$ and $g^k \rightharpoonup^* \bar{g}$ in $L^\infty(\Omega_T)$-weak* then $J(\bar{y}, \bar{g}) \le \liminf_{k\to\infty} J(y^k, g^k)$.

We will use the following compensated compactness result of Tartar [11]:

Theorem 5.1. *Let f be continuously differentiable and not affine on any interval. If $y^k \rightharpoonup^* \bar{y}$ in $L^\infty(\Omega_T)$ and if for all convex entropy pairs (η, q) holds*

(CC) $\eta(y^k)_t + q(y^k)_x \in$ *compact subset of* $H^{-1}_{loc}(\Omega_T)$

then $y^k \to \bar{y}$ strongly in $L^p_{loc}(\Omega_T)$ for all $p \in [1,\infty)$.

We will apply this result to the sequence $y^k = y(u^k)$ of states associated with a minimizing sequence. First, we use the entropy inequality (4) to show that (CC) actually holds for any sequence $y(u^k)$ associated with admissible controls u^k.

Lemma 5.2. *Let (A1) and (A2′) or (A2″) hold. Then for each sequence $u^k \in \mathcal{U}_{ad}$ the unique entropy solutions $y^k = y(u^k)$ of (7) satisfy: for all convex entropy pairs (η, q) there exists a bounded set $A \subset W^{-1,\infty}(\Omega_T)$, A precompact in $W^{-1,q}_{loc}(\Omega_T)$ for all $q \in [1,\infty)$, such that $\eta(y^k)_t + q(y^k)_x \in A$.*

Proof. (A2′) or (A2″) imply $\|u^k\|_\infty \le M_u$ for M_u large enough. By the local version of Theorem 2.2 there exist unique entropy solutions $y^k = y(u^k)$ with $\|y^k\|_\infty \le M$. Moreover, for convex entropy pairs (η, q) holds

$$\nu^k := \eta(y^k)_t + q(y^k)_x - \eta'(y^k) g(t,x,y^k,u^k) \le 0 \text{ in } \mathcal{D}'(\Omega_T). \tag{8}$$

Now $\|\eta(y^k)_t + q(y^k)_x\|_{W^{-1,\infty}(\Omega_T)} \le \|\eta(y^k)\|_\infty + \|q(y^k)\|_\infty$ and is thus uniformly bounded[‡]. For open bounded domains Q the embedding $W^{-1,\infty}(Q) \hookrightarrow W^{-1,p}(Q)$, $p \ge 1$, is continuous. Thus, $\eta(y^k)_t + q(y^k)_x$ lies in a bounded subset of $W^{-1,p}(Q)$ for all $p \ge 1$. Since η' is monotone increasing and hence locally bounded, we have $\|\eta'(y^k)g(.,y^k,u^k)\|_\infty \le M_1$ and obtain with (8) that ν^k lies in a bounded set of $H^{-1}(Q)$ and is ≤ 0 in $H^{-1}(Q)$. By [10] this implies that ν^k lies in a precompact set of $W^{-1,q}_{loc}(Q)$ for $q \in [1,2)$. Now the embedding $L^\infty(Q) \hookrightarrow W^{-1,q}(Q)$ is compact and hence $\eta'(y^k)g(.,y^k,u^k)$ lies in a precompact set of $W^{-1,q}_{loc}(Q)$ as well. Hence, $\eta(y^k)_t + q(y^k)_x \in$ precompact set of $W^{-1,q}_{loc}(Q)$ for $q \in [1,2)$ and thus $\eta(y^k)_t + q(y^k)_x \in$ precompact set of $W^{-1,q}_{loc}(\Omega_T)$. By using the interpolation result $\|\cdot\|_{-1,q} \le c(\theta)\|\cdot\|^\theta_{-1,p}\|\cdot\|^{1-\theta}_{-1,r}$, $p,r \in (1,\infty)$, $\frac{1}{q} = \frac{\theta}{p} + \frac{1-\theta}{r}$, we conclude, e.g., with $p = 2q$, $r = \frac{3}{2}$, that $\eta(y^k)_t + q(y^k)_x \in$ precompact set of $W^{-1,q}_{loc}(\Omega_T)$, $q \in [1,\infty)$. □

We state now our main result:

Theorem 5.3. *Let* (A1), (A2′) *or* (A2″) *and* (A3′) *hold. If f is continuously differentiable and not affine on any interval then the following holds:*

i) *Any minimizing sequence $u^k \in \mathcal{U}_{ad}$ contains a subsequence (u^k) such that $y^k = y(u^k)$ converges in $L^p_{loc}(\Omega_T)$ to some $\bar{y} \in L^\infty(\Omega_T)$, $1 \le p < \infty$. There is $\bar{u} \in \mathcal{U}_{ad}$ such that $\bar{y} = y(\bar{u})$, i.e., $\bar{y}$ is the unique entropy solution associated with $\bar{u}$. Further, $\bar{y} \in C^0([0,T]; L^p_{loc}(\mathbb{R}))$ for all $p \in [1,\infty)$, $y^k \to \bar{y}$ in $C^0([0,T]; L^p_{loc}(\mathbb{R}))$ and $\bar{u}$ satisfies $g(.,\bar{y},u^k) \rightharpoonup^* g(.,\bar{y},\bar{u})$ in $L^\infty(\Omega_T)$.*
ii) *$\bar{u}$ is optimal for* (P′).

Proof. i): Let $u^k \in \mathcal{U}_{ad}$ be a minimizing sequence. Since $\mathcal{U}_{ad} \subset L^\infty(\Omega_T)$ is bounded by (A2′) or (A2″), we have $\|u^k\|_\infty \le M_u$ for all k. Thus, Theorem 2.2, 1) yields $M > 0$ with $\|y^k\|_\infty \le M$ for all k. Consequently, (y^k) contains a subsequence (again denoted by (y^k)) with $y^k \rightharpoonup^* \bar{y}$ for some $\bar{y} \in L^\infty(\Omega_T)$. By Lemma 5.2 Theorem 5.1 can be applied and yields that $y^k \to \bar{y}$ strongly in $L^p_{loc}(\Omega_T)$ for all $p \in [1,\infty)$. Now let (η,q) be an arbitrary convex C^2-entropy-pair. Since η, q are locally Lipschitz, we get $\eta(y^k)_t + q(y^k)_x \to \eta(\bar{y})_t + q(\bar{y})_x$ in $\mathcal{D}'(\Omega_T)$. Moreover, $y \longmapsto \eta'(y)g(t,x,y,u^k(t,x))$ has a uniform Lipschitz constant on $[-M,M]$ for all k by (A1) and hence $\eta'(y^k)g(.,y^k,u^k) - \eta'(\bar{y})g(.,\bar{y},u^k) \to 0$ in $\mathcal{D}'(\Omega_T)$ and also in L^∞-weak*. Since $\|g(.,\bar{y},u^k)\|_\infty \le M_g$ for all k by (A1), we can again extract a subsequence (u^k) with $g(.,\bar{y},u^k) \rightharpoonup^* \bar{g}$. Since $\eta'(\bar{y}) \in L^1_{loc}(\Omega_T)$ we conclude that

$$\eta(\bar{y})_t + q(\bar{y})_x \le \eta'(\bar{y})\bar{g} \text{ in } \mathcal{D}'(\Omega_T). \tag{9}$$

Approximation with $\eta^\varepsilon_c(y) = \sqrt{(y-c)^2 + \varepsilon^2}$ yields that (9) holds also for the family of entropies $\eta_c(y) = |y-c|$ with $q_c(y) = \operatorname{sgn}(y-c)(f(y)-f(c))$, $c \in \mathbb{R}$.

Next, we show that there is $\bar{u} \in \mathcal{U}_{ad}$ with $\bar{g} = g(.,\bar{y},\bar{u})$. Under assumption (A2′) this is true for $\bar{u} = \text{w}^* - \lim_{k\to\infty} u^k$. It remains to consider the case (A2″). We have $g(.,\bar{y},u^k) \rightharpoonup^* \bar{g}$ in $L^\infty(\Omega_T)$ and hence for any bounded domain

‡ $W^{-1,p}(Q) = (W^{1,p'}_0(Q))'$, $1/p + 1/p' = 1$

$Q \subset \Omega_T$ $g^k := g(.,\bar{y},u^k)|_Q \rightharpoonup \bar{g}|_Q$ in $L^2(Q)$. By Mazur's theorem there exists a sequence of convex combinations $(\bar{g}^l := \sum_{k=1}^l \alpha_{k,l} g^k)$ that converge strongly to $\bar{g}|_Q$ in $L^2(Q)$. For a.a. $(t,x) \in Q$ the map $u \in \mathbb{R}^r \longmapsto g(t,x,\bar{y}(t,x),u)$ is continuous by (A1) and thus $\mathcal{H}(t,x) = g(t,x,\bar{y}(t,x),\mathcal{C}(t,x))$ is convex, because $\mathcal{C}(t,x)$ is connected. Since $g^k(t,x) \in \mathcal{H}(t,x)$, we thus find $v^l(t,x) \in \mathcal{C}(t,x)$ with $\bar{g}^l(t,x) = g(t,x,\bar{y}(t,x),v^l(t,x))$. For a subsequence $\bar{g}^l$ converges a.e. on Q to $\bar{g}$. For any such (t,x) we can pick a further subsequence such that $v^l(t,x)$ converge to some $v(t,x) \in \mathcal{C}(t,x)$, because $\mathcal{C}(t,x)$ is bounded and closed. Since $Q \subset \Omega_T$ was an arbitrary bounded domain, we have thus shown that

$$\bar{g}(t,x) \in g(t,x,\bar{y}(t,x),\mathcal{C}(t,x)) \text{ for a.a. } (t,x) \in \Omega_T. \tag{10}$$

(10) together with the fact that $(t,x,u) \longmapsto g(t,x,\bar{y}(t,x),u)$ is a Carathéodory map by (A1) yield by a theorem of Filippov, see [2, Thm. 8.2.10], the existence of a measurable selection $\bar{u}$ of $\mathcal{C}$ with $\bar{g} = g(.,\bar{y},\bar{u})$. We have thus proven the existence of $\bar{u} \in \mathcal{U}_{ad}$ such that $\eta_c(\bar{y})_t + q_c(\bar{y})_x \le \eta_c'(\bar{y}) g(t,x,\bar{y},\bar{u})$ in $\mathcal{D}'(\Omega_T)$ for all $c \in \mathbb{R}$.

We still have to show that $\bar{y}$ has initial data y_0: from the local version of Theorem 2.2, 3) we get for all $t > 0$ sufficiently small and all $K \subset\subset \mathbb{R}$ $\|y^1(t) - y^k(t)\|_{1,K} \le C_K\, t\, \|u^1 - u^k\|_\infty \le 2C_K M_u t$. Since $y^1 \in C^0([0,T]; L^1_{loc}(\mathbb{R}))$ with $y^1(0) = y_0$ and $y^k \to \bar{y}$ in $L^1_{loc}(\Omega_T)$ it follows $\operatorname{ess\,lim}_{t\to 0+} \|y_0 - \bar{y}(t)\|_{1,K} = 0$. Hence, $\bar{y} = y(\bar{u})$ is the unique entropy solution associated with $\bar{u}$.

We now prove that $y^k \to \bar{y}$ even in $C^0([0,T]; L^p_{loc}(\mathbb{R}))$, $1 \le p < \infty$. Let $a > 0$ be arbitrary. Since $\bar{y} \in C^0([0,T]; L^p_{loc}(\mathbb{R}))$, for any $\varepsilon > 0$ there is $\Delta = T/N > 0$ and $\psi_i \in \mathcal{D}([-a,a])$, $i = 1,\dots,N$, with $\|\psi_i\|_\infty \le M$, $\|\bar{y}(t) - \psi_i\|_{1,[-a,a]} \le \varepsilon$, for all $t \in [(i-1)\Delta, i\Delta]$. Fix any $\hat{t} \in [(i-1)\Delta, i\Delta)$ and $\bar{t} \in [\hat{t}, T]$. Our aim is to estimate $\|y^k(\bar{t}) - y^k(\hat{t})\|^2_{2,[-a,a]}$ using the entropy inequality (8) with $\eta(\lambda) = \lambda^2$ and $\eta(\lambda) = \pm\lambda$. To this end define for $\delta > 0$, $\alpha < \beta$ the function $\phi^\delta_{\alpha,\beta}(t) := \max\{0, \min\{(t-\alpha)/\delta, 1, (\beta - t)/\delta\}\}$. Then $\phi^\delta_{\hat{t},\bar{t}}(t)\phi^\delta_{-a,a}(x)$ is Lipschitz with support $[\hat{t},\bar{t}] \times [-a,a]$ and may be used as test function in the entropy inequality (8). This yields after integration by parts

$$\begin{aligned}
&\frac{1}{\delta}\int_{\bar{t}-\delta}^{\bar{t}} \int_{-a}^{a} \phi^\delta_{-a,a}(x)\eta(y^k)\,dxdt - \frac{1}{\delta}\int_{\hat{t}}^{\hat{t}+\delta}\int_{-a}^{a} \phi^\delta_{-a,a}(x)\eta(y^k)\,dxdt \\
&\le \int_{\hat{t}}^{\bar{t}} \left(\int_{-a}^{a} |\eta'(y^k) g(t,x,y^k,u^k)|\,dx + \frac{1}{\delta}\int_{-a}^{\delta-a} |q(y^k)|\,dx + \frac{1}{\delta}\int_{a-\delta}^{a} |q(y^k)|\,dx \right) dt.
\end{aligned}$$

Since $y^k \in C^0([0,T]; L^p_{loc}(\mathbb{R}))$, this yields for $\eta(\lambda) = \lambda^2$ and $\delta \to 0+$ a constant $C_1 > 0$ only depending on a and the bounds M, M_u for $\|y^k\|_\infty$, $\|u^k\|_\infty$ with

$$\|y^k(\hat{t}+\tau)\|^2_{2,[-a,a]} - \|y^k(\hat{t})\|^2_{2,[-a,a]} \le C_1\tau \tag{11}$$

for all $\tau \in [0, T - \hat{t}]$. Using the test function $2\phi^\delta_{\hat{t},\bar{t}}(t)\psi_i(x)$ in (8) for $\eta(\lambda) = \pm\lambda$ gives analogously a constant $C_2 > 0$ only depending on a, M, M_u, with

$$(-2\psi_i, y^k(\hat{t}+\tau) - y^k(\hat{t}))_{2,[-a,a]} \le C_2\tau\,(1 + \|(\psi_i)_x\|_1) \tag{12}$$

where $(.,.)_{2,[-a,a]}$ is the inner product on $L^2([-a,a])$. For a.a. $\hat{t} \in [0,T]$ we have $y^k(\hat{t}) \to \bar{y}(\hat{t})$ in L^1_{loc}. For such a $\hat{t}$ we find by the choice of ψ_i some $K_{\hat{t}} \in \mathbb{N}$ with

$$\|y^k(\hat{t}) - \psi_i\|_{1,[-a,a]} \leq \|y^k(\hat{t}) - \bar{y}(\hat{t})\|_{1,[-a,a]} + \|\bar{y}(\hat{t}) - \psi_i\|_{1,[-a,a]} \leq 2\varepsilon$$

for all $k \geq K_{\hat{t}}$. Hence, we have with (12)

$$(-2y^k(\hat{t}), y^k(\hat{t}+\tau) - y^k(\hat{t}))^2_{2,[-a,a]} \leq C_2\tau\,(1 + \|(\psi_i)_x\|_1) + 4\varepsilon M. \tag{13}$$

Now choose $\tau_\varepsilon > 0$ so small, that $C_1\tau + C_2\tau\,(1 + \|(\psi_i)_x\|_1) \leq M\varepsilon$ for all $\tau \in [0,\tau_\varepsilon]$ and $i = 1,\ldots,N$. Adding (11) and (13) gives

$$\|y^k(\hat{t}+\tau) - y^k(\hat{t})\|^2_{2,[-a,a]} \leq 5M\varepsilon \quad \forall\, \tau \in [0,\tau_\varepsilon],\ k \geq K_{\hat{t}}. \tag{14}$$

Since τ_ε does not depend on $\hat{t}$, we obtain $\sup_{t\in[0,T]} \|y^k(t) - \bar{y}(t)\|_{2,[-a,a]} \leq \sqrt{5M\varepsilon}$ for all sufficiently large k. In fact, set $\hat{t}_0 = 0$ and choose $\hat{t}_r \in [r\tau_\varepsilon/2, (r+1)\tau_\varepsilon/2]$ such that $y^k(\hat{t}_r) \to \bar{y}(\hat{t}_r)$ in L^1_{loc}, $1 \leq r \leq R$, R minimal with $(R+1)\tau_\varepsilon/2 \geq T$. Then $[\hat{t}_r, \hat{t}_r + \tau_\varepsilon]$, $0 \leq r \leq R$, cover $[0,T]$ and with $K = \max K_{\hat{t}_r}$ holds $\|y^k(\hat{t}_r) - \bar{y}(\hat{t}_r)\|_{1,[-a,a]} \leq \varepsilon$ for all r and all $k \geq K$. Hence, (14) is satisfied for all $\hat{t} = \hat{t}_r$ giving the uniform estimate in t.
ii): Since for a minimizing subsequence u^k by i) $y^k \to y(\bar{u})$ in $C^0([0,T]; L^p_{loc}(\mathbb{R}))$ and $g(.,y^k,u^k) \rightharpoonup^* g(.,\bar{y},\bar{u})$, (A3′) shows the optimality of $\bar{u}$. □

Remark 5.4. If in Theorem 4.1 (A1)–(A4) are replaced by the assumptions of Theorem 5.3, $g(.,y,u)$ is approximated in the fashion of §6, (16) and if for all sequences $U^l \in \mathcal{U}^l_{ad}$ holds $\|Y^l\|_\infty \leq M$ and (CC) is true for Y^l instead of y^k then similar arguments as above yield that any solution sequence $\bar{U}^l$ of the discretizations (P′^{l}) of (P′) contains a subsequence $\bar{U}^{l'}$ with $g(.,\bar{Y}^{l'},\bar{U}^{l'}) \rightharpoonup^* g(.,\bar{y},\bar{u})$ in L^∞ and $\bar{Y}^{l'} \to \bar{y}$ in $L^\infty([0,T]; L^p_{loc}(\mathbb{R}^n))$, $p \in [1,\infty)$, where $\bar{u}$, $\bar{y}$ is an optimal pair for (P′). Many monotone schemes satisfy this framework, cf. Remark 6.4.

6. Proof of the existence and stability result

In this section we sketch the proof of Theorem 2.2 by using monotone difference schemes. We refer to [4] for a comprehensive treatment and extend the results of this paper to get a precise result concerning the dependence on the source term.

6.1. Monotone difference schemes

For $\lambda_i > 0$, $\Delta t = \delta > 0$ set $\Delta x_i = \Delta t/\lambda_i$ and define the grid $\Delta := \{j\Delta x\,;\ j \in \mathbb{Z}^n\}$ on $\mathbb{R}^n$ and $\Delta' := \{t_k\,;\ k \in \mathbb{N}\} \times \Delta$ on Ω_T where $j\Delta x = (j_1\Delta x_1, \ldots, j_n\Delta x_n)$, $t_k = k\Delta t$. A grid-function Y on Δ' is defined by $Y(t,x) = Y^k_j$ on $[t_k, t_{k+1}) \times R_j$ with $R_j := \bigotimes_i [(j_i - \frac{1}{2})\Delta x_i, (j_i + \frac{1}{2})\Delta x_i)$. Moreover, let $Y^k := Y(t_k,.)$. Vice versa, given a function $v \in L^1_{loc}(\mathbb{R}^n)$ we define the grid-function v_Δ by setting $(v_\Delta)_j = \mu(R_j)^{-1}\int_{R_j} v(x)\,dx$ where μ is the Lebesgue measure. $y_{\Delta'}$ for $y \in L^1_{loc}((0,\infty)\times\mathbb{R}^n)$ is defined analogously. We set $j + 1_i := (j_1, \ldots, j_i + 1, \ldots, j_n)$ and introduce difference operators Δ^i_+ by $(\Delta^i_+ Y^k)_j := Y^k_{j+1_i} - Y^k_j$.

We consider explicit difference schemes

$$Y_j^{k+1} = \mathcal{G}(Y_{j-p}^k, \dots, Y_{j+r}^k) + \Delta t\, G_j^k(Y_j^k). \tag{15}$$

where $G(Y)$ and the initial data are given by

$$G(Y) := (g(.,Y(.),U_1(.)))_{\Delta'},\ Y^0 = U_0 := (u_0)_\Delta,\ U_1 := (u_1)_{\Delta'}. \tag{16}$$

It is easy to see that if (A1) holds, $y \in L^\infty(\Omega_T)$ and $Y^{\delta_l} \to y$ boundedly a.e. as $\delta_l \to 0+$ then $G^{\delta_l} = g(.,Y^{\delta_l}, U_1^{\delta_l}) \to g(.,y,u_1)$ in $L^1_{loc}(\Omega_T)$. The scheme in terms of grid functions will be written by $Y^{k+1} = \vec{\mathcal{G}}(Y^k) + \Delta t\, G^k(Y^k)$. The scheme is *conservative* if $\mathcal{G}$ can be put in the form

$$\mathcal{G}(Y_{j-p}^k, \dots, Y_{j+r}^k) = Y_j^k - \sum_{i=1}^n \lambda_i \Delta_+^i F_i(Y_{j-p}^k, \dots, Y_{j+r^i}^k). \tag{17}$$

The numerical fluxes F_i are consistent with f_i if they are locally Lipschitz and satisfy $F_i(u,\dots,u) = f_i(u),\ 1 \le i \le n$. Finally, the difference approximation is called *monotone* on $[a,b]$ if $\mathcal{G} : [a,b]^N \longrightarrow \mathbb{R}$ is a monotone nondecreasing function of each argument and we call it *strongly monotone* on $[a,b]$ if in addititon the right-hand side of (15) is monotone nondecreasing in Y_j^k.

In [4] it is shown for the homogeneous case that monotone schemes satisfy a discrete entropy inequality with the numerical entropy fluxes

$$Q_c^i(Y_{j-p}^k,\dots,Y_{j+r^i}^k) := F_i([Y_{j-p}^k,c]_+,\dots,[Y_{j+r^i}^k,c]_+) - F_i([Y_{j-p}^k,c]_-,\dots,[Y_{j+r^i}^k,c]_-)$$

where $[\lambda,c]_+ := \max(\lambda,c)$, $[\lambda,c]_- := \min(\lambda,c)$. Clearly, Q_c^i are consistent with $(q_c)_i$. Using the monotonicity of $\mathcal{G}$ as in [4], one deduces that in our setting

$$|Y_j^{k+1} - c| \le |Y_j^k - c| - \sum_{i=1}^n \lambda_i \Delta_+^i Q_c^i(Y_{j-p}^k,\dots) + \Delta t\,\mathrm{sgn}(Y_j^{k+1} - c) G_j^k(Y_j^k). \tag{18}$$

(18) yields the following variant of the Lax-Wendroff theorem:

Theorem 6.1. *Let* (A1),(A2) *hold. If* $\mathcal{G}$ *in* (17) *is monotone on* $[a,b]$ *and* Y^{δ_l} *from* (15) *satisfy* $a \le Y^{\delta_l} \le b$ *a.e. and tend to* y *a.e. for* $\delta_l \to 0+$, *then* y *satisfies* (2).

6.2. Proof of Theorem 2.2

We will prove the existence and stability result by applying Helly's theorem to sequences of grid-functions given by (15), (17). We recall from [4] the estimates

$$\|v_\Delta\|_1 \le \|v\|_1,\ \|v_\Delta\|_\infty \le \|v\|_\infty,\ \|v_\Delta\|_{BV} \le \|v\|_{BV} \tag{19}$$

and the following properties of monotone difference operators:

Proposition 6.2. *Let the difference operator* $\mathcal{G}$ *in* (17) *be monotone on* $[a,b]$. *Then for all grid functions* U, V *on* Δ *with* $a \le U, V \le b$ *the following holds true:*

i) $\min_{j-p\le s\le j+r} U_s \le (\vec{\mathcal{G}}(U))_j \le \max_{j-p\le s\le j+r} U_s$ ii) $\|\vec{\mathcal{G}}(U) - \vec{\mathcal{G}}(V)\|_1 \le \|U - V\|_1$

iii) $\|\vec{\mathcal{G}}(U)\|_{BV} \le \|U\|_{BV}$ iv) $\|\vec{\mathcal{G}}(U) - U\|_1 \le \Delta t L_F N \|U\|_{BV}$

where L_F *is a Lipschitz constant for all numerical fluxes* F_i *on* $[a,b]^N$.

Proof of Theorem 2.2. Let the assumptions of Theorem 2.2 hold. For M defined by (6) let (15), (17) be strictly monotone as long as $U_1 \in [-M_u, M_u]$. We show that for any sequence $\delta_l \to 0+$ the approximate solutions Y^{δ_l} defined by (15), (17) tend in $L^\infty([0,T]; L^1_{loc}(\mathbb{R}^n))$ to the unique entropy solution $y = y(u)$ and that y satisfies the assertions of Theorem 2.2.

We extend g by zero for $t > T$. For arbitrary δ_l set $Y = Y^{\delta_l}$, $U_1 = U_1^{\delta_l}$. Then Y is defined by (15) where $\mathcal{G}$ is monotone on $[-M, M]$ with M as in 1). We have $\|Y^0\|_\infty \le \|u_0\|_\infty \le M$ and as long as $\|Y^k\|_\infty \le M$ by the strong monotonicity of (15) and Proposition 6.2, i) with $Y^k_{j,\pm} := [Y^k_j, 0]_\pm$

$$\begin{aligned} \pm Y^{k+1}_{j,\pm} &\le \pm \mathcal{G}(Y^k_{j-p,\pm}, \dots, Y^k_{j+r,\pm}) + \Delta t \,\operatorname{sgn}(Y^{k+1}_{j,\pm}) G^k_j(Y^k_{j,\pm}) \\ &\le \|Y^k\|_\infty + \Delta t \,\operatorname{sgn}(Y^{k+1}_{j,\pm}) G^k_j(Y^k_{j,\pm}). \end{aligned} \tag{20}$$

Now we have by (16), (19) for any grid-function V

$$\|G^k(V)\|_i \le \frac{1}{\Delta t} \int_{t_k}^{t_{k+1}} \|(g(t,.,V,U_1^k))_\Delta\|_i \, dt \le \frac{1}{\Delta t} \int_{t_k}^{t_{k+1}} \|g(t,.,V,U_1^k)\|_i \, dt \tag{21}$$

for $i = 1, \infty, BV$. Thus, $\pm G^k_j(Y^k_{j,\pm}) \le C_1 + C_2 |Y^k_j|$ by (A1) and we get from (20)

$$\|Y^{k+1}\|_\infty \le (1 + \Delta t \, C_2) \|Y^k\|_\infty + \Delta t \, C_1.$$

By (16), (19) we have $\|Y^0\|_\infty \le \|u_0\|_\infty$ and the Gronwall lemma yields

$$\|Y(t)\|_\infty \le (\|u_0\|_\infty + C_1(1 - e^{-C_2 t})/C_2) e^{C_2 t} =: M_\infty(t). \tag{22}$$

Hence, $\|Y^{k+1}\|_\infty \le M_\infty(T) \le M$ as long as $t_{k+1} < \bar{T}$ and the difference approximations are well defined. Moreover, we have by Proposition 6.2

$$\|Y^{k+1}\|_i \le \|\vec{\mathcal{G}}(Y^k)\|_i + \Delta t \|G^k\|_i \le \|Y^k\|_i + \Delta t \|G^k\|_i, \; i = 1, \infty, BV, \tag{23}$$

where $G^k = G^k(Y^k)$. By (A1), (19), (21) we obtain with $\Omega_k := [t_k, t_{k+1}) \times \mathbb{R}^n$

$$\|G^k\|_1 \le \Delta t^{-1} \|g(.,0,0)\|_{1,\Omega_k} + \Delta t^{-1} L_u \|u_1\|_{1,\Omega_k} + L \|Y^k\|_1.$$

Since $\|Y^0\|_1 \le \|u_0\|_1$ by (16), (19), we deduce from (23) and Gronwall's Lemma $\|Y(t)\|_1 \le M_1(t)$ with $M_1(t)$ defined in 2). Let $\hat{Y}$ be the difference approximation for $\hat{u}$, $\hat{g}$ instead of u, g. Using the L^1-contraction iii) in Proposition 6.2 we get

$$\|Y^{k+1} - \hat{Y}^{k+1}\|_1 \le \|Y^k - \hat{Y}^k\|_1 + \Delta t \|G^k - \hat{G}^k\|_1. \tag{24}$$

With $\delta u_0 := u_0 - \hat{u}_0$, $\delta u_1 := u_1 - \hat{u}_1$, $\delta g := g - \hat{g}$ we obtain as above

$$\|G^k - \hat{G}^k\|_1 \le \Delta t^{-1} \|\delta g(., Y^k, U^k)\|_{1,\Omega_k} + \Delta t^{-1} L_u \|\delta u_1\|_{1,\Omega_k} + L \|Y^k - \hat{Y}^k\|_1$$

and thus from (24) by Gronwall

$$\|Y(t) - \hat{Y}(t)\|_1 \le \left(\|\delta u_0\|_1 + L_u \|\delta u_1\|_{1,\Omega_t} + \|\delta g(., Y, U_1)\|_{1,\Omega_t} \right) e^{Lt}. \tag{25}$$

Now let in addition $u_0 \in BV(\mathbb{R}^n)$, $u_1 \in L^1([0,T]; BV)$, and let g have a Lipschitz constant L_x and compact support K w.r.t. x. From (21) we deduce with (19)

$$\|G^k\|_{BV} \le 2n L_x \mu(K) + \Delta t^{-1} L_u \|u_1\|_{L^1([t_k, t_{k+1}); BV)} + L \|Y_k\|_{BV}.$$

Hence, an application of Gronwall's Lemma to (23) yields as above $\|Y(t)\|_{BV} \leq M_{BV}(t)$ with $M_{BV}(t)$ given in 4). Finally, using the L^1-contraction of $\mathcal{G}$ and iv) in Proposition 6.2 we get for $0 \leq \hat{t} < t \leq T$, l, m such that $\hat{t} \in [t_l, t_{l+1})$, $t \in [t_m, t_{m+1})$

$$\begin{aligned}
\|Y(t) - Y(\hat{t})\|_1 &\leq \sum_{k=l}^{m-1} \|Y^{k+1} - Y^k\|_1 \leq \sum_{k=l}^{m-1} \left(\|\vec{\mathcal{G}}(Y^k) - Y^k\|_1 + \Delta t \|G^k\|_1\right) \\
&\leq L_F N \|Y\|_{L^1([t_l,t_m);BV(\mathbb{R}^n))} + \|g(.,0,U_1)\|_{1,[t_l,t_m)\times K} + L\|Y\|_{1,[t_l,t_m)\times K} \\
&\leq (t - \hat{t} + \Delta t)\left(L_F N M_{BV}(T) + L M_1(T)\right) + \|g(.,0,U_1)\|_{1,[\hat{t}-\Delta t,t]\times K} \\
&\leq (|t - \hat{t}| + \Delta t)(C + \mu(K)(\|g(.,0,0)\|_\infty + L_u\|u_1\|_\infty)) =: \omega(|t-\hat{t}| + \Delta t).
\end{aligned}$$

Consider first the case $\operatorname{supp}(u_0) \subset\subset \mathbb{R}^n$. We fix λ_i such that the scheme is strictly monotone on $[-M, M]$. Then for all $\delta > 0$ small enough the solutions Y^δ satisfy $\operatorname{supp}_x Y^\delta(t,.) \subset \bar{K} \subset\subset \mathbb{R}^n$. Independent of δ and $t, \hat{t} \in [0,T]$ we have

$$\|Y^\delta(t)\|_i \leq M_i(T), \ i = 1, BV, \infty \quad \|Y^\delta(t) - Y^\delta(\hat{t})\|_1 \leq \omega(|t-\hat{t}| + \delta). \tag{26}$$

Let (δ_l) tend to zero. Since $BV \hookrightarrow L^1_{loc}$ is compact, there is a diagonal sequence $(\delta_{l'})$ such that $Y^{\delta_{l'}}(\tau_k)$ converges in L^1 on a dense subset $\{\tau_k\,;\ k \in \mathbb{N}\}$ of $[0,T]$. The equicontinuity-like property (26) yields that $Y^{\delta_{l'}}$ is a Cauchy sequence in $L^\infty([0,T]; L^1(\mathbb{R}^n))$. The limit y satisfies $\|y(t) - y(\hat{t})\|_1 \leq \omega(|t - \hat{t}|)$ by (26) and thus $y \in C^0([0,T]; L^1(\mathbb{R}^n))$. By construction $y(0) = \lim_{l\to\infty} Y_0^{\delta_{l'}} = u_0$ in $L^1(\mathbb{R}^n)$. Thus, $y = y(u)$ by Theorem 6.1 and 2.1. Furthermore, the lower semicontinuity of $\|\cdot\|_\infty$ and $\|\cdot\|_{BV}$ under L^1-convergence show that 1)–5) hold.

For $u \in \mathcal{U}_{ad}$, $g \in (L^1 \cap L^\infty)(\Omega_T; C^{0,1}_{loc}(\mathbb{R}\times\mathbb{R}^r))$, there are $g^r \in C_0^1(\Omega_T; C^{0,1}_{loc}(\mathbb{R}\times\mathbb{R}^r))$ having the same local Lipschitz constants L, L_u w.r.t. y, u_1 as g, $u_0^r \in C_0^1(\mathbb{R}^n)$, $u_1^r \in C_0^1(\Omega_T)$ with $\|u_i^r\|_\infty \leq \|u_i\|_\infty$, $i = 0,1$, $\|g^r(.,0,0)\|_\infty \leq \|g(.,0,0)\|_\infty$, and

$$\lim_{r\to\infty} \|u_0^r - u_0\|_1 + \|u_1^r - u_1\|_1 + \left\| \sup_{|y|\leq M, |v|\leq M_u} |(g^r - g)(.,y,v)| \right\|_{1,\Omega_T} = 0. \tag{27}$$

The previous results hold for $Y_r^{\delta_l}$ associated with u^r, g^r. We find a diagonal sequence $\delta_{l'}$ such that $Y_r^{\delta_{l'}}$ converge in $L^\infty([0,T]; L^1(\mathbb{R}^n))$ for all r to the unique entropy solution y^r for u^r, g^r, satisfying 1)–5). From 3) we get for all $t \in [0,T]$

$$\|y^r(t) - y^s(t)\|_1 \leq (\|u_0^r - u_0^s\|_1 + L_u\|u_1^r - u_1^s\|_1 + \|(g^r - g^s)(.,y^r,u^r)\|_{1,\Omega_T})e^{LT}.$$

Hence, (y^r) converges by (27) in $C^0([0,T]; L^1(\mathbb{R}^n))$ to $y \in L^\infty(\Omega_T)$ with $y(0) = u_0$. y inherits the validity of (2), 1)–3) and, if applicable, 4), 5) from y^r and is hence the unique entropy solution for u, g. y is actually the limit of the sequence $Y^{\delta_{l'}}$ generated by (15) for u, g, since (22), (25) hold for $Y^{\delta_{l'}}$ and for $t \in [0,T]$

$$\|Y^{\delta_{l'}}(t) - y(t)\|_1 \leq \|y^r(t) - y(t)\|_1 + \|Y^{\delta_{l'}}(t) - Y_r^{\delta_{l'}}(t)\|_1 + \|Y_r^{\delta_{l'}}(t) - y^r(t)\|_1$$

Uniformly in t, the first and by (25), (27) also the second term can be made arbitrarily small by choosing r large, then the last by choosing l large. The uniqueness of y and a subsequence-argument yield that the whole sequence Y^{δ_l} tends to y.

If $\mathcal{U} = L^\infty(\mathbb{R}^n) \times L^\infty(\Omega_T)^r$ and $L^1 \cap L^\infty$ is replaced by L^∞ in (A1) then again 1) holds. The solution $Y^\delta(\bar{t}, \bar{x})$ only depends on the definition of u, g in the numerical propagation cone depending on λ_i. Hence, we get by localization that Y^{δ_l} converges in $L^\infty([0,T];L^1_{loc}(\mathbb{R}^n))$ to an entropy solution $y \in L^\infty(\Omega_T) \cap C^0([0,T];L^1_{loc}(\mathbb{R}^n))$. Local variants of 1)–5) hold, because y_0 and g can be modified outside the propagation cone given in Theorem 2.1. □

Remark 6.3. It is obvious from the above proof that the solutions Y^{δ_l} associated with $Y_0^{\delta_l} = U_0^{\delta_l}$ and $U_1^{\delta_l}$ converge in $L^\infty([0,T];L^p_{loc})$, $1 \le p < \infty$, to $y(u)$ if $(U_0^{\delta_l}, U_1^{\delta_l})$ tend to $u = (u_0, u_1)$ in L^1_{loc} and $\|U_i^{\delta_l}\|_\infty \le M_u$, $i = 0, 1$. Hence, monotone schemes satisfy the assumptions of Theorem 4.1.

Remark 6.4. In the 1-D-case one can derive from (18) for most strongly monotone 3-point schemes (but also for some high resolution schemes, cf. [5]) that $\|Y^{\delta_l}(.,. + \lambda_1\delta_l) - Y^{\delta_l}\|_{1,[0,T]\times[-a,a]} \le C(a)\delta_l^\alpha$ for some $\alpha > 0$ and all $a > 0$, see, e.g., [6]. From this, one can deduce as in [6] that (CC) is satisfied for Y^{δ_l} instead of y^k if $\|U^{\delta_l}\|_\infty \le M_u$. Hence, Remark 5.4 applies.

References

[1] F. Ancona and A. Marson, *On the attainable set for scalar nonlinear conservation laws with boundary control*, SIAM J. Control Optim., **36** (1998), 290–312.

[2] J.-P. Aubin and H. Frankowska, *Set-valued analysis*, Birkhäuser, Boston, **1990**.

[3] E. Casas, K. Kunisch, and C. Pola, *Some applications of BV functions in optimal control and calculus of variations*, (1997), Preprint.

[4] M. Crandall and A. Majda, *Monotone difference approximations for scalar conservation laws*, Math. Comp., **34** (1980), 1–21.

[5] F. Coquel and Ph. LeFloch, *Convergence of finite difference schemes for conservation laws in several space dimensions: A general theory*, SIAM J. Numer. Anal., **30** (1993), 675–700.

[6] R. J. DiPerna, *Convergence of approximate solutions to conservation laws*, Arch. Rational Mech. Anal., **82** (1983), 27–70.

[7] B. G. Fitzpatrick, *Parameter estimation in conservation laws*, J. Math. Syst. Estim. Control, **3** (1993), 413–425.

[8] F. James amd M. Sepúlveda, *Convergence results for the flux identification in a scalar conservation law*, (1998), to appear in SIAM J. Control Optim.

[9] S. N. Kružkov, *First order quasilinear equations in several independent variables*, Math. USSR Sb., **10** (1970), 217–243.

[10] F. Murat, *L'injection du cône positif de H^{-1} dans $W^{-1,q}$ est compacte pour tout* $q < 2$, J. Math. Pures Appl., **60** (1981), 309–322.

[11] L. C. Tartar, *Compensated compactness and applications to partial differential equations*, in: R. J. Knops, Ed., Research Notes in Mathematics, Nonlinear Analysis and Mechanics, Heriot-Watt Symposium **4**, (Pitman, New York) (1979), 136–212.

Zentrum Mathematik,
Technische Universität München,
D-80290 München, Germany
E-mail address: sulbrich@mathematik.tu-muenchen.de

International Series of Numerical Mathematics
Vol. 133,

Identification of Memory Kernels in Heat Conduction and Viscoelasticity

L. v. Wolfersdorf and J. Janno

Abstract. In the general field equations of linear heat conduction and viscoelasticity time-dependent memory kernels occur which are mostly unknown in practice. Identification problems for such kernels are equivalent to nonlinear Volterra integral equations of the first kind. These first kind equations can be approximated through corresponding equations of the second kind by some type of Lavrentiev regularization or reduced to equations of the second kind by differentiation. Further the least squares method with Tikhonov regularization is applied to the identification problem for obtaining convergence rates by a method of Engl, Kunisch and Neubauer.

1. Statement of problems in heat conduction

Let D be a bounded domain in $\mathbb{R}^p$, $p \geq 1$, with smooth boundary S. In general linear heat conduction we use the *constitutive relations*

$$e(x,t) = \beta(x)(u(x,t) + \int_0^t n(t-\tau)u(x,\tau)d\tau) \tag{1}$$

$$\underline{q}(x,t) = -\gamma(x)(\nabla u(x,t) - \int_0^t m(t-\tau)\nabla u(x,\tau)d\tau) \tag{2}$$

where u is the temperature of a rigid isotropic body in $Q = D \times (0,T)$, which is taken as zero for $t < 0$, e the internal energy, $\underline{q}$ the heat flux, β heat capacity, γ heat conduction coefficient, and m, n are the memory kernels of internal energy and heat flux.

Further, there holds the *heat balance equation*

$$e_t(x,t) + \operatorname{div} \underline{q}(x,t) = f(x,t) \tag{3}$$

with f the heat supply.

Relations (1)–(3) imply the *generalized heat equation*

$$\begin{aligned} &\beta(x)(u_t + \frac{\partial}{\partial t}\int_0^t n(t-\tau)u(x,\tau)d\tau) \\ &= \operatorname{div}\,(\gamma(x)\nabla u) - \int_0^t m(t-\tau)\operatorname{div}\,(\gamma(x)\nabla u(x,\tau))d\tau + f \ \text{ in } Q. \end{aligned} \tag{4}$$

In the *direct problem* for u equation (4) is to be solved with the initial condition

$$u(x,0) = \varphi(x) \quad \text{on } D \tag{5}$$

and one of the boundary conditions

$$u = 0 \quad \text{or} \quad \gamma(x)\partial u/\partial\nu + \lambda(x)u = 0 \qquad \text{on } S\times(0,T) \tag{6}$$

where ν is the outer normal of S and λ the heat exchange coefficient.

In the *inverse problem* for m, n via u we have additional conditions

$$\Psi_j[u](t) = h_j(t) \quad \text{on } [0,T] \ \ (j=1,\dots,q) \tag{7}$$

where $\Psi_j, j=1,\dots,q$ are given linear functionals on $u(\cdot,t), t\in[0,T]$; for instance, q measurements of temperature

$$\Psi_j[u](t) = \int_D \rho_j(x)u(x,t)dx \qquad (j=1,\dots,q) \tag{8}$$

with non-negative weight-functions $\rho_j \in L_2(D)$.

Taking $F(z) = (F_1(z),\dots,F_q(z))$ with $F_j(z) = \Psi_j[u]$, where $u = u[z]$ is the (generalized) solution of (4)–(6) for given $z = (m,n)$, the *operator equation* of the inverse problem writes

$$F(z) = y_0 \quad , \quad y_0 = (h_1,\dots,h_q). \tag{9}$$

2. Reduction to Volterra integral equation of first kind

The (formal) solution of the direct problem is given by the Fourier series

$$u[z](x,t) = \sum_{k=1}^{\infty} A_k[z](t)v_k(x) \tag{10}$$

where v_k are the in $L_2(D)$ orthonormal eigenfunctions of

$$\operatorname{div}\,(\gamma(x)\nabla v) + \mu\beta(x)v = 0 \ \text{ in } \ D$$

with (6) and eigenvalues $\mu_k > 0$, and $A_k = A_k[z]$ are the solutions of the linear Volterra integral equations

$$A_k(t) = L_k[z] * A_k(t) + \Phi_k(t) \tag{11}$$

where $*$ denotes convolution and

$$L_k[z](t) = \mu_k e^{-\mu_k t} * (m+n)(t) - n(t)$$
$$\Phi_k(t) = \varphi_k e^{-\mu_k t} + e^{-\mu_k t} * r_k(t)$$

with φ_k and $r_k(t)$ the Fourier coefficients of φ and $r = f/\beta$.

By (10) the operator of the inverse problem F has the components

$$F_j(z)(t) = \sum_{k=1}^{\infty} \gamma_{j,k} A_k[z](t), \qquad t \in [0,T] \tag{12}$$

with the "observation coefficients" $\gamma_{j,k} = \Psi_j[v_k]$. In view of (11) the F_j can be written as *nonlinear Volterra integral operators of convolution type.* In particular, for the case of one memory kernel $z = m, n \equiv 0$ with $q = 1$ additional condition (7) we have the *equation of the first kind*

$$\int_0^t K[m](t-s)m(s)ds = g(t), \quad t \in [0,T] \tag{13}$$

where

$$K[m](t) = \sum_{k=1}^{\infty} \gamma_k \mu_k \int_0^t e^{\mu_k(\tau - t)} A_k(\tau) d\tau$$
$$g(t) = h(t) - \sum_{k=1}^{\infty} \gamma_k \Phi_k(t)\,.$$

3. Methods of inverse problems

3.1. Regularization of Lavrentiev type

Under suitable assumptions on the data $d = (\varphi_k, r_k)$ of the direct problem and the observation coefficients γ_k the kernel $K[m]$ in (13) is *2-smoothing*, i.e. $K \in \{C[0,T] \to C^2[0,T]\}$ with

$$K[m](0) = 0,\ K'[m](0) = \sum_{k=1}^{\infty} \gamma_k \mu_k \varphi_k \equiv \lambda_1 \neq 0 \quad \text{(by assumption)}.$$

Equation (13) can be regularized by the *equation of the second kind*

$$\begin{aligned} &\varepsilon^2 m_{\varepsilon,\delta}(t) + 2\varepsilon \int_0^t m_{\varepsilon,\delta}(s)ds + \frac{1}{\lambda_1}\int_0^t K[m_{\varepsilon,\delta}](t-s)m_{\varepsilon,\delta}(s)ds \\ &= \frac{1}{\lambda_1} g_\delta(t) + \varepsilon^2 m_0 + 2\varepsilon m_o t \end{aligned} \tag{14}$$

where $m_0 = m(0) = (1/\lambda_1)\ddot{g}(0)$ and g_δ is a perturbation of g with $\|g - g_\delta\|_{C[0,T]} \leq \delta$. Then if the solution m of (13) is Lipschitz continuous there exist $m_{\varepsilon,\delta}$ which converge to m in $C[0,T]$. More precisely, we have

$$\|m_{\varepsilon,\delta} - m\|_{C[0,T]} \leq \text{ Const } \delta^{\frac{1}{3}} \tag{15}$$

for $\varepsilon = \varepsilon(\delta) = \text{Const } \delta^{\frac{1}{3}}$ (cf. [6]).

3.2. Reduction to integral equation of second kind

Assuming $g \in C^2[0,T]$ two times differentiation of equation (13) with the 2-smoothing kernel $K[m]$ yields the equivalent *nonlinear Volterra integral equation of the second kind*

$$m(t) + \frac{1}{\lambda_1} \int_0^t K''[m](t-s)m(s)ds = \frac{1}{\lambda_1} g''(t) \tag{16}$$

together with the necessary conditions $g(0) = g'(0) = 0$.

Applying a contraction principle for a scale of weighted norms with exponential weights, existence, uniqueness and stability of the solution $m \in C[0,T]$ to (16) in any finite time interval $[0,T]$ can be proved (cf. [7], [11]). This also holds for weakly singular kernels $m \in L_p(0,T), 1 < p \leq \infty$ and $m \in C_p[0,T], 0 < p \leq 1$ with $C_p[0,T] = \{t^{1-p}m \in C[0,T]\}$ (cf. [8], [19]).

4. Least squares methods

For applying the least squares method to the operator equation (9) we choose the spaces $X = (H^1(0,T))^2, Y = (L_2(0,T))^q$ and $F : D(F) \subset X \to Y$ with $D(F)$ a closed convex subset of $X, y_0 \in Y$. The solution of the direct problem $u = u[z] \in H^{1,0}(Q)$ for $\varphi \in L_2(D), f \in L_2(Q)$, and Ψ_j are linear bounded functionals on $H^1(D)$. Then F is continuous and weakly closed which are the basic requirements of the theory of Engl, Kunisch and Neubauer (cf. [3], [4] and [5]).

In the *Tikhonov regularization* equation (9) is replaced by the condition

$$\min_{z \in D(F)} \{\|F(z) - y_\delta\|^2 + \alpha\|z - z^*\|^2\} \tag{17}$$

where $\alpha > 0$ is a regularization parameter, $z^* \in X$, and the approximation $y_\delta \in Y$ of y_0 satisfies $\|y_0 - y_\delta\| \leq \delta$. Further we generalize the minimum condition (17) to the problem of finding $z_\alpha^{\delta,\eta} \in D(F)$ such that

$$\begin{aligned} &\|F(z_\alpha^{\delta,\eta}) - y_\delta\|^2 + \alpha\|z_\alpha^{\delta,\eta} - z^*\|^2 \\ &\leq \|F(z) - y_\delta\|^2 + \alpha\|z - z^*\|^2 + \eta \quad \text{for } z \in D(F) \end{aligned} \tag{18}$$

where $\eta \geq 0$ is a small parameter.

We make the main assumption that y_0 *is attainable*, i.e. there exists $z_0 = (m_0, n_0) \in D(F)$ with $F(z_0) = y_0$. Then there exists a z^*-minimum norm solution

(z^*-MNS) of (9), i.e. a solution z_0 of $F(z_0) = y_0$ with minimal distance to z^*. If $\alpha = \alpha(\delta, \eta)$ in (18) fulfills

$$\alpha \to 0, \frac{\delta^2}{\alpha} \to 0, \frac{\eta}{\alpha} \to 0 \quad \text{for} \quad \delta \to 0, \eta \to 0$$

then every sequence of solutions $\{z_{\alpha_k}^{\delta_k,\eta_k}\}, \alpha_k = \alpha(\delta_k, \eta_k)$, of (18) for $\delta_k \to 0, \eta_k \to 0$ has a subsequence which strongly converges to a z^*-MNS.

Operator F has the *Fréchet derivative* F' in X defined by

$$F'(z)\omega = (\Psi_1 U[z]\omega, \ldots, \Psi_q U[z]\omega), \;\; \omega = (a, b) \in X \tag{19}$$

where $U[z]$ is the Fréchet derivative of solution operator $u : X \to H^{1,0}(Q)$ of the direct problem. $U[z]$ is given by the Fourier series

$$U[z]\omega(x,t) = \sum_{k=1}^{\infty} B_k[z]\omega(t) v_k(x) \tag{20}$$

where $B_k = B_k[z]\omega$ are the solutions of the linear Volterra integral equations

$$B_k(t) = L_k[z] * B_k(t) + \Phi_k^1[z]\omega(t)$$

with $L_k[z]$ as in (11) and

$$\Phi_k^1[z]\omega(t) = \mu_k e^{-\mu_k t} * (a + b) * A_k[z](t) - b * A_k[z](t).$$

F' fulfills the *Lipschitz condition*

$$\|F'(z_0) - F'(z)\| \leq L\|z_0 - z\| \quad \text{for} \quad z \in D(F)$$

with $\|z_0 - z\| \leq p$ for any $p > 2\|z_0 - z^*\|$, and $L = \Lambda(\|z_0\| + p)$ or $L = \Lambda(\|z^*\| + \frac{3}{2}p)$ where the function $\Lambda(Z)$ is explicity given by the data (cf. [20] for details).

Further we suppose the *source representation*

$$z_0 - z^* = F'(z_0)^*\chi, \quad \chi \in Y \tag{21}$$

where $L\|\chi\| < 1$. Also the adjoint operator $F'(z_0)^* : Y \to X$ of F' can be explicitly given. In particular for the functionals (8) with first boundary condition in (6) we have

$$F'(z_0)^*\chi = (B^{-1}H_1, B^{-1}H_2) \tag{22}$$

where

$$B : D(B) = \{w \in H^2(0,T) | w'(0) = w'(T) = 0\} \to L_2(0,T)$$

is defined by $Bw = -w'' + w$ and

$$\begin{aligned} H_1(s) &= \int_s^T \int_D \nabla v(x,t) \nabla u_0(x, t-s) dx dt \in L_2(0,T) \\ H_2(s) &= \int_s^T \int_D v_t(x,t) u_0(x, t-s) dx dt \in L_2(0,T) \end{aligned}$$

with $u_0 = u[z_0]$ and the adjoint state function $v = v[z_0, \chi] \in H^1(Q)$ (cf. [15] and [20] again).

The main theorem of Engl, Kunisch and Neubauer [4] now yields the *convergence rate*

$$\|z_\alpha^{\delta,\eta} - z_0\| = O(\sqrt{\delta}) \tag{23}$$

for the solution $z_\alpha^{\delta,\eta}$ of (18) by the choice $\alpha \sim \delta$ and $\eta = O(\delta^2)$.

The convergence rate (23) can be improved to

$$\|z_\alpha^\delta - z_0\| = O(\delta^{\frac{2}{3}}) \tag{24}$$

for $z_\alpha^\delta = z_\alpha^{\delta,0}$ if $z_0 \in \overset{\circ}{D}(F)$, nonempty interior of $D(F)$, instead of (21) the source representation

$$z_0 - z^* = F'(z_0)^* F'(z_0)\omega, \quad \omega = (a,b) \in X \tag{25}$$

with (19) holds and $\alpha = \alpha(\delta)$ is suitably chosen [13].

If y_0 *is not attainable* we assume the domain $D(F)$ as bounded, too. Then a least squares solution to (9) exists, i.e. a solution of (17) for $\alpha = 0$. Then for a sequence $\{y_{\delta_k}\} \in Y$ with $\|y_0 - y_{\delta_k}\| \leq \delta_k$ and $\delta_k, \alpha_k \to 0$ where $\delta_k = O(\alpha_k)$ or $\delta_k = o(\alpha_k)$, respectively, the sequence $\{z_{\alpha_k}^{\delta_k}\}$ of solutions to the minimum problem (17) for $y_\delta = y_{\delta_k}, \alpha = \alpha_k$ has a subsequence which converges strongly to a least squares solution or a minimum least squares solution of (9), respectively. For convergence rates in the case of non-attainability see the papers [1], [14].

5. Problems in viscoelasticity

Let again D be a bounded domain in $\mathbb{R}^p, p \geq 1$, with smooth boundary S. We deal with the *linear hyperbolic integrodifferential equation*

$$\beta(x)u_{tt} = \operatorname{div}(\gamma(x) \nabla u) - \int_0^t m(t-\tau) \operatorname{div}(\gamma(x) \nabla u(x,\tau))d\tau + f \tag{26}$$

in $Q = D \times (0,T)$ with β, γ positive continuous functions on $\overline{D}$, $f \in L_2(Q)$, and the memory kernel m. In the particular case $p = 1$ equation (26) is the general wave equation of viscoelasticity.

In the *direct problem* for u equation (26) is to be solved with the initial conditions

$$u(x,0) = \varphi(x), \quad u_t(x,0) = \psi(x) \quad \text{on} \quad D \tag{27}$$

and one of the boundary conditions (6). In the *inverse problem* for m via u we have additional conditions of the form (7). The operator equation of the inverse problem is given by (9) for $z = m$ where $u = u[m]$ is the (generalized) solution of (26), (27), (6) for given m.

This solution is represented by the Fourier series (10) where $A_k = A_k[m]$ are the solutions of the Volterra equations (11) with the kernel

$$L_k[m](t) = m(0)(1 - \cos\lambda_k t) + m' * (1 - \cos\lambda_k t), \quad \lambda_k = \sqrt{\mu_k},$$

and the right-hand side

$$\Phi_k(t) = \varphi_k \cos\lambda_k t + \frac{\psi_k}{\lambda_k}\sin\lambda_k t + \frac{1}{\lambda_k}\sin\lambda_k t * r_k(t).$$

The components F_j of the operator F in (9) have again the form (12) and can be written as *nonlinear Volterra integral operators of convolution type*. The kernel in the equation of the first kind analogous to (13) is now 3-smoothing and the equation can be reduced to an integral equation of the second kind by three times differentiation in the generic case. For the treatment of this integral equation in the case of smooth and weakly singular memory kernels, respectively, we refer to our papers [9], [10], [19].

The *least squares method* can be applied as in the heat conduction case if we choose the spaces $X = H^2(0,T), Y = (L_2(0,T))^q$, take the solution of the direct problem $u = u[m] \in H^1(Q)$ for $\varphi \in \overset{\circ}{H}{}^1(D)$ or $H^1(D)$, respectively, $\psi \in L_2(D)$ and the Ψ_j again as linear bounded functionals on $H^1(D)$. See [20] for details.

References

[1] A. Binder, H.W. Engl, C.W. Groetsch, A. Neubauer and O. Scherzer, *Weakly closed nonlinear operators and parameter identification in parabolic equations by Tikhonov regularization*, Appl. Anal. 55 (1994), 215–234.

[2] G. Chavent and K. Kunisch, *Convergence of Tikhonov regularization for constrained ill-posed inverse problems*, Inverse Problems 10 (1994), 63–76.

[3] H.W. Engl, M. Hanke and A. Neubauer, *Regulatization of Inverse Problems*, Kluwer, Dordrecht, 1996.

[4] H.W. Engl, K. Kunisch and A. Neubauer, *Convergence rates for Tikhonov regularisation of non-linear ill-posed problems*, Inverse Problems 5 (1989), 523–540.

[5] R. Gorenflo and B. Hofmann, *On autoconvolution and regularization*, Inverse Problems 10 (1994), 353–373.

[6] J. Janno and L. v. Wolfersdorf, *Regularization of a class of nonlinear Volterra equations of a convolution type*, J. Inv. Ill-Posed Problems 3 (1995), 249–257.

[7] J. Janno and L. v. Wolfersdorf, *Inverse problems for identification of memory kernels in heat flow*, J. Inv. Ill-Posed Problems 4 (1996), 39–66.

[8] J. Janno and L. v. Wolfersdorf, *Identification of weakly singular memory kernels in heat conduction*, Z. Angew. Math. Mech. 77 (1997), 243–257.

[9] J. Janno and L. v. Wolfersdorf, *Inverse problems for identification of memory kernels in viscoelasticity*, Math. Meth. Appl. Sci. 20 (1997), 291–314.

[10] J. Janno and L. v. Wolfersdorf, *Identification of weakly singular memory kernels in viscoelasticity*, Z. Angew. Math. Mech. 78 (1998), 391–403.

[11] J. Janno and L. v. Wolfersdorf, *Identification of memory kernels in general linear heat flow*, J. Inv. Ill-Posed Problems (submitted).

[12] J. Janno and L. v. Wolfersdorf, *Inverse problems for identification of memory kernels in thermo- und poro-viscoelasticity*, Math. Meth. Appl. Sci. (to appear).

[13] A. Neubauer, *Tikhonov regularisation for non-linear ill-posed problems: optimal convergence rates and finite-dimensional approximation*, Inverse Problems 5 (1989), 541–557.

[14] U. Tautenhahn, *Tikhonov regulatization for identification problems in differential equations*, Parameter Identification and Inverse Problems in Hydrology, Geology and Ecology, Edited by J. Gottlieb and P. Duchateau, Kluwer Acad. Publ. 1996, 261–270.

[15] F. Unger and L. v. Wolfersdorf, *On a control problem for memory kernels in heat conduction*, Z. Angew. Math. Mech. 75 (1995), 365–370.

[16] L. v. Wolfersdorf, *On identification of memory kernels in linear viscoelasticity*, Math. Nachr. 161 (1993), 203–217.

[17] L. v. Wolfersdorf, *On optimality conditions in some control problems for memory kernels in viscoelasticity*, Z. Anal. Anwend. 12 (1993), 745–750.

[18] L. v. Wolfersdorf, *On identification of memory kernels in linear theory of heat conduction*, Math. Meth. Appl. Sci. 17 (1994), 919–932.

[19] L. v. Wolfersdorf, *Inverse problems for memory kernels in heat flow and viscoelasticity*, J. Inv. Ill-Posed Problems 4 (1996), 267–282.

[20] L. v. Wolfersdorf and J. Janno, *On Tikhonov regularization for identifying memory kernels in heat conduction and viscoelasticity*, TU Bergakademie Freiberg, Fakultät für Mathematik und Informatik, Preprint 98-1, 1–19.

L. v. Wolfersdorf
Fakultät für Mathematik und Informatik
TU Bergakademie Freiberg
D-09596 Freiberg, Germany
E-mail address: `wolfersdorf@mathe.tu-freiberg.de`

J. Janno
Institute of Cybernetics
Estonian Academy of Sciences
21, Akademia tee
EE-0026 Tallin, Estonia
E-mail address: `janno@ioc.ee`

International Series of Numerical Mathematics
Vol. 133, © 1999 Birkhäuser Verlag Basel/Switzerland

Variational Formulation for Incompressible Euler Equation by Weak Shape Evolution

Jean-Paul Zolésio

Abstract. This paper is concerned with a variational formulation for the Euler equation in a non cylindrical domain (free boundary). We make use of the flow associated to a divergence free field V introduced in [3] and the non cylindrical evolution domain Q_V built by that flow (we recall here the main results concerning the shape evolution associated with non smooth vector field). We consider the extremization of an energy functional in the form of the kinetic energy in that tube and we show that under smoothness assumption the incompressible Euler equation is solved. The computation of the derivative uses the tube's derivative, characterized by the field Z which buidt the *transverse* tube $s \to T_t(V+sW)(\Omega)$. Introducing the adjoint equation we compute the gradient of the functional and the Euler problem. That analysis is done in that first work under smoothness assumption on the extremal field V but we introduce some elements in order to develop that sensitivity analysis without smoothness asumption. Nevertheless without any smoothness assumption we give an existence result of minimizing fields V for an augmented functional $J(V) + \sigma\, p(Q_V)$ where p stands for the time-space perimeter. That term is a modelling of a surface tension effect on the free boundary. The optimality condition for that functional is characterized and consists in a new modelling for fluid-tubes coupled structures such as *arteries*.

1. Introduction

We consider a bounded domain D with Lipschitzian boundary and the following Euler equation; given a domain Ω_0 in D and V_0, f defined on D

$$\frac{\partial}{\partial t}V + DV(t).V(t) + \nabla p = f \text{ in } Q_V, \quad V(0) = V_0 \text{ on } \Omega_0 \tag{1}$$

$$divV = 0 \quad \text{in } Q_V, \tag{2}$$

Where Q_V is the non cylindrical evolution domain built by the field V. That is, in weak form,

$$\frac{\partial}{\partial t}\chi_{Q_V} + \nabla\chi_{Q_V}.V = 0, \quad \chi_{Q_V}(0) = \chi_{\Omega_0} \tag{3}$$

Where χ_{Q_V} is the characteristic function of the set Q_V, a time-space function. The Euler incompressible flow is then the solution u, p of (1)–(3). We consider the functional

$$J_\sigma(V) = \int_0^\tau \int_{Q_V} ||V(t,x)||^2 \; - f.V \, dx + \; \sigma p(V) \tag{4}$$

where f is given in $L^2(I \times D)^3$ (with non zero $curl f$ and $I = [0,\tau]$), $p(V)$ is the time-space perimeter of the non cylindrical evolution domain Q_V built by V, initiated at Ω_0 and relative to the cylinder $[0,\tau] \times D$. Consider the extremization problem of $J_\sigma(V)$ over the set of vector fields V verifying the initial condition $V(0) = V_0$ in $\Omega(0) = \Omega_0$. We will give a meaning to the necessary condition associated with such extremality (assuming first the vector field "smooth enough") and show that we obtain solution to the problem (1)–(3). In the last section we derive an existence result to the optimization problem associated to a "two fluids" problem. In the first section we generalize the "speed method" to nn smooth vector fields "V" in order to give a sense to problem (3), we begin by recalling the situation for smooth fields for which the global flow mapping $T_t(V)$ enables us to define the tube Q_V as $\cup_{0<t<T} \; \{t\} \times T_t(V)(\Omega_0)$.

1.1. Continuous Field

The domain D is not necessarely bounded in that step, for simplicity we assume it is bounded later. Consider the Frechet linear space

$$C^{k,\infty}(D, R^N) = \{\phi \,|\phi \in C^k(\bar{D}, \mathbb{R}^N) \,, \; \phi \in L^\infty(D, \mathbb{R}^N) \,, \; < \phi, n_{\partial D} >= 0 \; \} \tag{5}$$

Given an integer k, $k \geq 1$, $E^{k,\infty}(D) = C^0([0,\infty[, C^{k,\infty}(D, \mathbb{R}^N))$. A sequence V_n converges to an element V in $E^{k,\infty}(D)$ if and only if for each compact subset K of D, any $\tau > 0$ and any integer $\alpha, \alpha \leq k$, we have the uniforme convergence of $\nabla^\alpha V_n(.,.)$ on the compact set $[0,\tau] \times K$ and also the uniforme convergence over$[0,\tau] \times D$ of the sequence$V_n(.,.)$.

Denote here $V(t,x)$ the value of the field $V(t)$ at the point X of D, that is to say that we set $V(t,x) = [V(t)](x)$.

Theorem 1.1. *([2]) Let V belongs to $E^{k,\infty}(D)$ then the flow mapping $T(V) : t \longrightarrow T_t(V)$ is defined from $[0,\infty[$ in $C^0(\bar{D}, \bar{D})$ and we have the following regularity.*

$$T(V) \in E^{1,k,\infty}(D) \tag{6}$$

$$E^{1,k,\infty}(D) = \{T \mid (t \to T_t) \; and \; (t \to \frac{d}{dt}(T_t) \in C^0([0,\infty[, C^{k,\infty}(\bar{D}, R^N)) \; \} \tag{7}$$

When D is bounded $C^{k,\infty}(D, R^N)$ is a Banach space.

1.2. Field in L^p

We turn now to the situation when the vector field V has less regularity both in time variable t and in the space variable x. The domain D being bounded in $\mathbb{R}^N$ and given $p > 1$, consider the following linear space:

(8) $$\mathcal{E}^p(D) = \{ \ V \in L^0(0,\infty, Lip(D,\mathbb{R}^N)) \mid \text{ there exist } C_V, \rho_V \ , \forall t$$

$$\int_t^{t+\rho} ||V(s)||^p_{Lip(\bar{D},\bar{D})} ds \leq C \ , \quad < V(t,.), n_{\partial D} >= 0 \ a.e.t \ \}$$

the following result is as follows.

Theorem 1.2. *Given* $p > 1$ *and a vector field* V *in* $\mathcal{E}^p(D)$ *then the flow mapping* $T(V)$ *is defined over* $Q = [0,\infty[\times\bar{D}$ *and we have for all* τ, $\tau > 0$*:*

(9) $$T(V) \in L^\infty(0,\tau,\ Lip(D,\mathbb{R}^N))$$

The mapping $T_t(V)$*is bijective from* D *onto itself and the inverse mapping is a flow mapping associated to the following vector field:*

(10) $$\forall t \in [0,\tau], \forall s \in [0,t], \ V_t(s,y) = -V(t-s,y)$$

(11) $$T_t(V)^{-1} = T_t(V_t) \in Lip(D,D)$$

1.3. Shape Convection of the Characteristic Function

The evolution of a set can be modeled using the flow mapping of a smooth vector field. The perturbed domain is then $\Omega_t = T_t(V)(\Omega)$ whose characteristic function χ_t can be written as $\chi_{\Omega_t} = \chi_\Omega \, o \, T_t(V)^{-1}$. The measure $\nabla\chi(t)$ must be understood as the distribution over D defined by:

$$\forall W \in C^0_{comp}(D,R^N), \quad < \nabla\chi(t) \ , \ W \ >= -\int_{\Omega_t} div\, W \ dx$$

And in the case where the domain is smooth enough, via Stoke's formula we get

$$\forall W \in C^0_{comp}(D,R^N), \quad < \nabla\chi(t) \ , \ W \ >= -\int_{\partial\Omega_t} W.n_t \ d\Gamma_t$$

For each element $V \in E$ the product $\nabla\chi_\Omega.V$ is given by (as $divV = 0$):

(12) $$\forall\varphi \in C^1_{comp}(D), \ < \nabla\chi_\Omega.V \ , \varphi >= -\int_\Omega \nabla\varphi.V \ dx$$

Finally for any element $\chi \in L^\infty(]0,\tau[\times D)$ and any $V \in L^2(0,\tau,L^2(D,R^N))$ with $divV(t,.) = 0 \ a.e.t$, define by the same formula the element $\nabla\chi.V$ as an element of $L^2(0,\tau,H^{-1}(D))$.

(13) $$\forall\phi \in L^2(0,\tau,H^1_0(D)), \ < \nabla\chi.V \ , \phi >= -\int_0^\tau\int_D \chi(t)\nabla\phi(t).V(t) \ dxdt$$

Proposition 1.3. *Let* $V \in E^{k,\infty}(D)$*, then the solution* $\chi(t) = \chi_{\Omega_t}$ *of (1) verifies:*

(14) $$\chi \in C^0([0,\tau],L^2(D)) \cap C^1([0,\tau],H^{-1}(D)), \ \chi(0) = \chi_\Omega, \ \chi_t + \nabla\chi.V = 0$$

2. Weak Convection of Characteristic Functions

2.1. The Galerkin Approximation

We consider the dynamical system

$$u(0) = \phi, \quad \frac{\partial}{\partial t}u(t) + < V(t), \nabla u(t) > = f \tag{15}$$

with initial condition $\phi \in L^2(D)$ and right-hand side $f \in L^1(0, \tau, L^2(D))$.

Proposition 2.1. *Let $V \in L^1_{loc}(0, \infty, L^2(D, R^3))$ with $divV \in L^2_{loc}(0, \infty, L^2(D, R^3))$ verifying the following uniform integrability condition:*

$$\text{There exist } T_0 > 0,\ \rho < 1,\ s.t. \forall a \geq 0, \quad \int_a^{a+T_0} ||V(t)||_{L^2(D,R^3)} dt \ \leq \rho < 1$$

verifying also the following condition on the positive part of the divergence, $divV = (divV)^+ - (divV)^-$, verifies

$$||(divV(t))^+||_{L^\infty(D,R^3)} \in L^1_{loc}(0, \infty),$$

We also assume $< V(t,.), n > = 0$ (as an element of $L^1_{loc}(0, \infty, H^{-\frac{1}{2}}(\partial D))$, $f \in L^1_{loc}(0, \infty, L^2(D)$ and initial condition $\phi \in L^2(D)$. Then there exists a unique solution,

$$u \in L^\infty_{loc}(0, \infty, L^2(D)) \cap W^{1,p^*}_{loc}(0, \infty, W^{-1,3}(D)) \subset C^0([0, \infty[, W^{-\frac{1}{2},\frac{3}{2}}(D))$$

where $\frac{1}{p} + \frac{1}{p^*} = 1$, and there exists a constant M such that:

$$\forall \tau, \ ||u||_{L^\infty(0,\tau,L^2(D,R^N))} \leq M\{ \ ||\phi||_{L^2(D)} + ||f||_{L^1(0,\tau,L^2(D))} \ \} \tag{16}$$

$$(1 + \int_0^\tau (||(divV(s))^+||_{L^\infty(D,R^3)} + ||f(s)||_{L^2(D,R^3)})$$

$$\int_s^\tau (||(divV(\sigma))^+||_{L^\infty(D,R^3)} + ||f(\sigma)||_{L^2(D,R^3)}) d\sigma) ds$$

If the field V is smoother, $V \in L^p(0, \tau, H^1(D, R^N))$, these solutions verify

$$u \in L^\infty(0, \tau, L^2(D)) \cap W^{1,p^*}(0, \tau, W^{-1,3}(D)) \subset C^0([0, \tau], W^{-\frac{1}{2},\frac{3}{2}}(D))$$

Moreover, in both situations, if the initial condition is a characteristic function

$$\phi = \chi_{\Omega_0} \in L^2(D)$$

and $f = 0$, then the unique solution is itself a characteristic function:

$$a.e.(t, x), \quad u(t, x)\ (1 - u(t, x)\ ) = 0 \quad \text{that is } u = \chi_{Q_V}$$

Where Q_V is a non cylindrical measurable set in $]0, \tau[\times D$. For a.e.t, we set

$$\Omega_t(V) = \{x \in D | \quad (t, x) \in Q_V \ \}.$$

If V is a free divergence field, $divV(t, x) = 0$ for a.e. $t, \in]0, \tau[$, then the set $\Omega_t(V)$ verifies a.e.t, $meas(\Omega_t(V)) = meas(\Omega_0)$. Notice that for $p = 1$ the results hold locally in time (i.e. for τ small enough, but not a priori given).

Proof. Let us consider $V \in L^2(0, \tau, H_0^1(D))$ and a dense family $e_1, \ldots, e_m, \ldots$ in $H_0^1(D)$ with each $e_i \in C^\infty_{comp}(D, R^3)$. Consider the approximated solution

$$u^m(t,x) = \Sigma_{i=1,\ldots,m} \quad u_i^m(t)\, e_i(x)$$

with $U^m = (u_1^m, \ldots, u_m^m)$ solution of the following linear ordinary differential system:

$$\forall t, \int_D (\frac{\partial}{\partial t} U^m(t) + < V(t), \nabla U^m(t) >)\, e_j(x)\, dx = \int_D f(t,x) e_j(x)\, dx, \; j = 1, \ldots, m$$

That is

$$\frac{\partial}{\partial t} U^m(t) + M^{-1}.A(t).U^m(t) \; = \; F(t) \tag{17}$$

where

$$M_{i,j} = \int_\Omega \; e_i(x)\, e_j(x)\, dx$$

$$A_{i,j}(t) = \int_D \; < V(t), \nabla e_i(x) > \; e_j(x)\, dx$$

That is an ordinary linear differential systems possessing a global solution when $V \in L^p(0, \tau, L^2(D, R^N))$ for some p, $p > 1$. By classical energy estimate, as $\int_D < V(t), \nabla u^m(t) > u^m(t)\, dx = -\frac{1}{2} \int_D < u^m(t), u^m(t) > divV(t)\, dx$, *a.e.t* we get:

$$\forall \tau, \; \tau \le T, \; ||u^m(\tau)||^2_{L^2(D)} \le ||u^m(0)||^2_{L^2(\Omega)}$$
$$+ \int_0^\tau \int_D < u^m(t,x), u^m(t,x) > (divV(t,x))^+ \; dtdx$$
$$+ 2 \int_0^\tau \int_D f(t,x) u(t,x) dtdx$$

Setting

$$\psi(t) = ||(divV(t,.))^+||_{L^\infty(D,R^3)}$$

When $f = 0$,

$$\frac{1}{2} \int_D u^m(t,x)^2 dx \le \frac{1}{2} \int_D u^m(0,x)^2 dx$$
$$+ \frac{1}{2} \int_0^t \psi(s) \int_D \; ||u^m(t,x)||^2 \, dx$$

by use of the Gronwall's lemma we get:

$$\int_D u^m(t,x)^2 dx \le \int_D u^m(0,x)^2 dx \; (1 \; + \; \int_0^t \psi(s) exp\{ \int_s^t \psi(\sigma) d\sigma \;\} \, ds \;\;)$$

By the choice of the initial conditions in the ordinary differential system we get

$$M > 0, s.t. \forall \tau, \le T, \; ||u^m(\tau)||_{L^2(D)}$$

$$\le \; M \; ||\phi||_{L^2(D)} \; (1 + \int_0^t \psi(s) exp\{ \int_s^t \psi(\sigma) d\sigma \;\} \, ds \;\;)$$

When $\psi = 0$, we get

$$\forall \tau,\ \tau \leq T,\ ||u^m(\tau)||^2_{L^2(D)} \leq ||u^m(0)||^2_{L^2(\Omega)}$$
$$+2\int_0^\tau \int_D f(t,x)u(t,x)dtdx$$

In the general case, we use

$$||u^m|| \leq 1 + ||u^m||^2$$

and we derive the following estimate:

$$\forall \tau,\ \tau \leq T,\ ||u^m(\tau)||^2_{L^2(D)} \leq ||u^m(0)||^2_{L^2(D,R^3)}$$
$$+\int_0^\tau \int_D < u^m(t,x), u^m(t,x) > (divV(t,x))^+\ dtdx$$
$$+2\int_0^\tau \int_D f(t,x)u(t,x)dtdx$$
$$\leq\ ||u^m(0)||^2_{L^2(D,R^3)} + \int_0^\tau ||f(t)||_{L^2(D,R^3)}\,dt$$
$$+\int_0^\tau x < u^m(t,x), u^m(t,x) > (\psi(t) + ||f(t)||_{L^2(D,R^3)}\,)\,dtdx$$
$$\leq\ M\,(||u_0||^2_{L^2(D,R^3)} + ||f||_{L^1(0,\tau,L^2(D,R^3)}\ )$$
$$+\int_0^t\ (\psi(t) + ||f(t)||_{L^2(D,R^3)})||u^m(s)||^2_{L^2(D)}ds$$

From Gronwall's inequality we derive:

$$||u^m(\tau)||^2_{L^2(D)}\ \leq M(||u_0||^2_{L^2(D,R^3)} + ||f||_{L^1(0,\tau,L^2(D,R^3)}\ )$$
$$\{\ 1 + \int_0^t\ [\ (\psi(s) + ||f(s)||_{L^2(D,R^3)}) \int_s^t (\psi(\sigma) + ||f(\sigma)||_{L^2(D,R^3)})d\sigma\ ]ds\ \}$$

In all cases n u^m remains bounded in $L^\infty(0,\tau,L^2(D,R^N))$ and there exists an element u in that space and a subsequence still denoted u^m which weakly-* converges to u. In the limit u itself verifies the previous estimate from which the uniqueness follows. It can be verified that u solves the problem in distribution sensex. That is

$$\forall \phi \in H^1_0(0,\tau,L^2(D,R^3)) \cap L^2(0,\tau,H^1_0(D,R^3)),\ \ \phi(0) = 0,$$
$$-\int_0^\tau \int_D\ u(\frac{\partial}{\partial t}\phi + div(\ \phi V)\ )dxdt = \int_D\ \phi(0)\,u_0\,dx\ + \int_0^\tau \int_D\ < f, \phi > dxdt$$

When $V(t) \in L^2(D,R^3)$ the duality brackets $< \frac{\partial}{\partial t}u, \phi >$ are defined as soon as $\nabla\phi$ belongs to $L^\infty(D,R^3)$, this is verified, for example, when $\phi \in H^3_0(D)$ so that u_t is identified to an element of the dual space $H^{-3}(D)$. When $V(t) \in H^1(D,R^3)$ we get $a.e.t,\ \ u(t,.) \in L^2(D),\ \ V(t,.) \in L^6(D),$ then $\nabla\phi(t)$ should be in $L^3(D,R^3)$,that is $\phi(t) \in W^{1,3}_0(D)$ and then, for $a.e.t$, the element u_t is in the dual space $W^{-1,\frac{3}{2}}(D)$

while $u_t \in L^{p^*}(0,\tau, W^{-1,\frac{3}{2}}(D))$ and $u \in W^{1,p^*}(0,\tau, W^{-1,\frac{3}{2}}(D))$. Then we have $u \in L^\infty(0,\tau, L^2(D)) \cap$

$$\cap W^{1,p^*}(0,\tau\, W^{-1,\frac{3}{2}}(D)) \subset L^2(0,\tau, W^{0,\frac{3}{2}}(D)) \cap W^{1,p^*}(0,\tau, W^{-1,\frac{3}{2}}(D))$$
$$\subset C^0([0,\tau], W^{-\frac{1}{2},\frac{3}{2}}(D))$$

When the initial data is a characteristic function, $u_0 = \chi_{\Omega_0}$ we shall verify that u^2 is also a solution which by uniqueness implies $u^2 = u$ *a.e.*. We introduce $u_0^n \to u_0$ in $L^2(D, R^3)$ with $u_0^n \in C^\infty(D, R^3)$ and $u_0^n(x) \leq 1$. We also consider $V^n \to V$ in $L^2(0,\tau, L^2(D, R^3))$ with $V \in C^\infty$ and $div(V^n)^- \in L^\infty([0,\tau] \times \bar{D})$. The solution u^n associated to these data is obtained through the flow of V^n as follows:

$$u^n(t,x) = (u_0) o T_t(V^n)^{-1}(x)$$

as a consequence $u^n \in C^\infty$, then $(u^n)^2$ is classically defined and, when $f = 0$, it is obviously solution to the equation associated to the initial condition u_0^2 and the field V^n. Now as $(u_0^n)^2 \leq u_0^n \leq 1$ we get $(u^n)^2 \leq u^n \leq 1$, then we can find a subsequence such that u^n is weakly converging to some element v, while $(u^n)^2$ is weakly converging to an element w, weakly in $L^2(0,\tau, L^2(D, R^3))$.

$$\forall \phi \in H_0^1(0,\tau, L^2(D,R^3)) \cap L^2(0,\tau, H_0^1(D,R^3)), \ \ \phi(0) = 0,$$
$$-\int_0^\tau \int_D < u^n, \frac{\partial}{\partial t}\phi + \nabla\phi.V > dxdt = \int_D \phi(0)\, u_0^n\, dx$$

And

$$-\int_0^\tau \int_D < (u^n)^2, \frac{\partial}{\partial t}\phi + \nabla\phi.V > dxdt = \int_D \phi(0)\, (u_0^n)^2\, dx$$

In the limit we get

$$\forall \phi \in H_0^1(0,\tau, L^2(D,R^3)) \cap L^2(0,\tau, H_0^1(D,R^3)), \ \ \phi(0) = 0,$$
$$-\int_0^\tau \int_D < v, \frac{\partial}{\partial t}\phi + \nabla\phi.V > dxdt = \int_D \phi(0)\, u_0\, dx$$

And

$$-\int_0^\tau \int_D < w, \frac{\partial}{\partial t}\phi + \nabla\phi.V > dxdt = \int_D \phi(0)\, (u_0)^2\, dx$$

then we derive that $v = u$ and w is the solution associated with the initial condition $u_0^2 = \chi_{\Omega_0}$ and the field V. As $u = \chi_{\Omega_0}$ is a characteristic function it can be verified that the convergence of u^n to $u = \chi_{\Omega_0}$ is strong: we verify the behavior of the norm

$$\lim_{n\to\infty} \int_0^\tau \int_D (u^n)^2\ dxdt \leq \lim_{n\to\infty}, \int_0^\tau \int_D (u^n)\ dxdt = \int_0^\tau \int_D (u^2)\ dxdt$$

(as $u^2 = u = \chi_{\Omega_0}$). Then $(u^n)^2$ itself is strongly converging to w, as a consequence we get $w = u^2 = \chi_{\Omega_0}$.

Proposition 2.2. *Let $V^n \to V$ in $L^2(0,\tau, L^2(D,R^3))$ with $divV^n(t,.) = divV(t,.) =$ 0,*

$$\chi_{Q_{V^n}} \to \chi_{Q_{V^n}} \ \ in\ L^2(0,\tau, L^2(D))$$

With the choice of smooth fields V^n we get the density result:

Corollary 2.3. *Let Ω be a smooth open set in D, say with smooth boundary $\partial\Omega$ of class C^k, $k \geq 1$ and $V \in L^2(0,\tau,L^2(D,R^3))$ with $divV(t,.) = 0$, $< V(t,.), n >= 0$ on ∂D. Then there exists a sequence of smooth tubes Q_n such that a.e.$t \in [0,\tau]$, $\Omega_t^n =: \{x \in D |\ (t,x) \in Q_n\ \}$ verifies $\chi_{\Omega_t^n} \to \Omega_t(V)$ in $L^2(D)$.*

We consider the related Sobolev space associated with the measurable set $\Omega_t(V)$:

$$\mathcal{H}^1(\Omega_t(V)(\text{resp.}\mathcal{H}_0^1(\Omega_t(V)) = \{\Phi = (\phi,\psi_1,\ldots,\psi_N) \in L^2(D,R^{N+1}) \mid \text{ there exists}$$
$$\Phi_n \in H^1(\Omega_t(V^n))(\text{resp. } H_0^1(\Omega_t(V^n))\ )\ \text{ with}$$
$$\chi_{\Omega_t(V^n)}\phi_n \ \text{ weakly converges in } L^2(D) \text{ to } \chi_{\Omega_t(V)}\phi, \ \text{ and for all } i = 1,\ldots,n,$$
$$\chi_{\Omega_t(V^n)}\frac{\partial}{\partial x_i}\Phi_n \ \text{ weakly converges in } \ L^2(D)^N \ \text{ to } \chi_{\Omega_t(V)}\psi\,\}.$$

Following [1] this defines a Hilbert space and that when the domain is smooth we recover the usual Sobolev space. Let $V \in L^2(0,\tau,L^2(D,R^3))$ with $divV = 0$ and $Q_{V,\Omega}$ the associated non cylindrical evolution set, we consider

$$(18)\qquad \begin{aligned} \mathcal{V}(V,\Omega) \ = \{ \quad & W \in L^2(0,\tau,L^2(D,R^3)) \\ & s.t.\ a.e.t divW(t,.) = 0, \\ & < W(t,.), n_{\partial D} >= 0 \text{ in } H^{-\frac{1}{2}}(\partial D), \quad Q_{W,\Omega} = Q_{V,\Omega}\} \end{aligned}$$

Obviously $V \in \mathcal{V}(V,\Omega)$, and

Proposition 2.4. *$\mathcal{V}(V,\Omega)$ is a non empty closed convex set in $L^2(0,\tau,L^2(D,R^3))$*

We consider the unique field defined by

$$(19)\qquad V_\Omega = argMin\{\ ||W||_{L^2(0,\tau,L^2(D,R^3))} \ \mid W \in \mathcal{V}(V,\Omega) \ \ \}$$

We shall say that a divergence free field $V \in L^2(0,\tau,L^2(D,R^3))$ is $\Omega-$minimal if, given the set Ω in D, the field verifies $V = V_\Omega$. We consider the positive number

$$e_D(Q_V) = Min\{\ ||W||^2_{L^2(0,\tau,L^2(D,R^3))} \ \mid W \in \mathcal{V}(V,\Omega) \ \ \} = \ ||V_\Omega||^2_{L^2(0,\tau,L^2(D,R^3))}$$

as an energy term associated with the non cylindrical set Q_V. That term effectively depends on the "Universe D" through the boundary condition $< V, n >= 0$ on ∂D. For any smooth tube Q given in $[0,\tau] \times D$ with lateral boundary Σ we denote by $v(t,x)$ the time component of the normal field ν and we can exhibit an element in $\mathcal{V}_Q$ in the following way: Let, for all $t \in [0,\tau]$, $\phi \in L^2(,R^3)$ be the piecewise smooth function defined by

$$\Delta\phi = 0 \text{ in } \ \Omega_t \cup \Omega_t^c, \ \ \frac{\partial}{\partial n_t}(\phi|_{\Omega_t}) = \frac{\partial}{\partial n_t}(\phi|_{\Omega_t^c}) = v(t,.), \ \frac{\partial}{\partial n_t}(\phi|_{\Omega_t^c}) = 0 \text{ on } \partial D$$

As $\int_{\Gamma_t} v(t,.)\, d\Gamma_t = 0$, the previous problem is well posed in $H^1(\Omega_t)/R \times H^1(\Omega_t^c)/R$, the jump $[\frac{\partial}{\partial n_t}(\phi]$ is zero on Γ_t while the jump of ϕ itself, $[\phi]$, is defined as an element of $H^{\frac{1}{2}}(\Gamma_t)$ up to an additive constant. We set $V = \nabla\phi \in L^2([0,\tau] \times D, R^3)$, we get

$divV = 0$ in D and $Q = Q_V$, as $\Omega_t = T_t(V)(\Omega_0)$. As any element in $L^2(D, R^3)$, the field $V(t)$ can be decomposed in the following form: $V = \nabla\Phi + curl(u)$ with

$$\Delta\Phi = divV(t) \text{ in } D, \quad curlcurl\,(u) = curl\,(V(t)) \text{ in } D$$

and the boundary conditions $\frac{\partial}{\partial n}(\Phi) = curl(u).n =$ on ∂D. Here from the divergence free of $V(t)$ we derive that the potential field Φ vanishes while $curl(V(t)) = \gamma^*_{\Gamma_t}(\ \vec{n}_t \Lambda \nabla_{\Gamma_t}[\,\Phi\,]\ )$, where again [.] designates the jump accross the smooth manifold Γ_t. Obviously the L^2 norm of the field $V(t)$ is given by

$$||V(t)||^2_{L^2(D,R^3)} = \int (|\nabla\Phi|^2 + |curl(u)|^2)dx = \int_D |curl(u)|^2 dx$$

where u solves the variational problem:

$$\forall\zeta \in H^1(D, R^3)/R \quad \int_D curlu.curl\zeta\ dx = \int_{\Gamma_t} < \vec{n}_t \Lambda \nabla_{\Gamma_t}[\phi],\ \zeta > dx$$

Which gives an upper bound: $e_D(Q_V) \leq ||curlu||^2_{L^2(]0,\tau[\times D,R^3)}$.

2.2. Stability Result and Strong Convergence

When V is a smooth free divergence vector field $\chi(t) = \chi_\Omega o T_t(V)^{-1}$ is the solution to the convection problem. We consider a converging sequence V_n converging to an element V in $W^{1,\infty}([0,\tau], L^3(D, R^N))$, we deduce that the corresponding elements χ^n converge to the unique characteristic function χ defined in the previous section weakly in $L^2(Q)$, with $Q =]0,\tau[\times D$. As $(\chi^n(t))^2 = \chi^n(t)$ and $\chi(t)^2 = \chi(t)$ we also deduce the convergence of the $L^2(Q)$ norms so that the convergences stand in strong $L^2(Q)$ topology.

Proposition 2.5. *Let V^n and V in $W^{1,\infty}(0,\tau, L^\infty(D, R^N))$ be divergence free elements such that $V^n \to V$ in $L^2(0,\tau, L^2(D, R^N))$. Let Ω be a measurable subset in D and χ^n (resp. χ) be the solution of the convection problem (5) with characteristic initial condition χ_Ω and vector field V^n (resp. V). Then $\chi^n \longrightarrow \chi,\quad n \to \infty$, strongly in $L^2(0,\tau, L^2(D)) \cap H^1(0,\tau, H^{-1}(D))$ and strongly in $C^0([0,\tau], H^{-\frac{1}{2}}(D))$.*

Proof. We already know that the convergence stands strongly in $L^2(0,\tau, L^2(D))$. But we have

$$\forall\,\phi \in L^2(0,\tau, H^1_0(D)),\ < \frac{\partial}{\partial t}\chi^n, \phi >_{L^2(0,\tau,H^{-1}(D))\times L^2(0,\tau,H^1_0(D))}$$
$$= \int_0^\tau \int_D \chi^n\ \nabla\phi.V^n\ dxdt$$

and then we have:

$$< \frac{\partial}{\partial t}(\chi^n - \chi)\,, \phi >_{L^2(0,\tau,H^{-1}(D))\times L^2(0,\tau,H^1_0(D))}$$
$$\leq (\int_0^\tau \int_D ||\chi^n V^n\ -\ \chi V||^2\ dxdt)^{\frac{1}{2}}\ \ (\int_0^\tau \int_D ||\nabla\phi||^2\ dxdt)^{\frac{1}{2}}$$

We deduce the strong convergence to zero of $\frac{\partial}{\partial t}(\chi^n - \chi)$ in $L^2(0,\tau, H^{-1}(D))$.

3. First Variation of the Flow with Respect to the Field

3.1. The Tranverse Field Z

We determine the derivative, when it exists, of the mapping $V \to T_t(V)$. Given a second vector field W we consider for any real number s the flow mapping

$$\mathbf{T}_s^t = T_t(V + sW)oT_t(V)^{-1} \tag{20}$$

For any fixed t, the transformation $\mathbf{T}_s^t$ maps D onto D and the domain $\Omega_t(V)$ onto $\Omega_t(V + sW)$. It derives that it is the flow mapping of a speed vector field $\mathbf{Z}^t$. Of course $\mathbf{Z}^t$ is a non autonomous vector field and is defined as being the speed field associated with the transformation $x \to \mathbf{T}_s^t(x)$ that is to say that we have

$$\mathbf{Z}^t(s, x) = [\frac{d}{ds}(\mathbf{T}_s^t(x))]_{s=0} \, o \, \mathbf{T}_s^t(x)^{-1} \tag{21}$$

It turns out that when the fields V and W are smooth enough the field $\mathbf{Z}^t(s, x)$ exists and is characterized as the solution of a linear dynamical system associated with V and W.

Theorem 3.1. *([3]) Consider two non autonomous vector fields V and W given in E and set*

$$Z(t, x) = \mathbf{Z}^t(0, x) \tag{22}$$

The vector field Z verifies $Z(0, x) = 0$, and solves the following first order dynamical system in the cylindrical evolution domain $Q = [0, \tau] \times D$

(23)
$$\frac{d}{dt}Z(t, x) + DZ(t, x).V(t, x) - DV(t, x).Z(t, x) = W(t, x), \; for(t, x) \in [0, \tau] \times D$$

Notice that as $V(t)$ is a free divergence field then we also have $divZ(t, .) = 0$ in D. Effectively the "transverse" variation of the domain $s \to T_s(\mathcal{Z}^t)(\Omega_t(V)) = T_t(V + sW)(\Omega)$ preserves the measure of any domain $T_t(V)(\Omega_0)$ as V and W are divergence free fields. Let E be given by $E(V)(t) = DT_t(V)^{-1}oT_t(V)^{-1}$. We have the following derivatives:

Theorem 3.2.

$$\frac{\partial}{\partial s}(T_t(V + sW))|_{s=0} = Z(t)oT_t(V)^{-1}, \quad \frac{\partial}{\partial s}(T_t(V + sW)^{-1})|_{s=0} = -E(V)(t).Z(t)$$

4. Variational Formulation of the Euler Equation

Consider the following functional:

$$J(V) = \int_0^T \int_{\Omega_t(V)} ||V(t, x)||^2 - f.V \, dtdx \tag{24}$$

In the following computation we drop f for simplicity.

$$J'(V,W) = (\frac{\partial}{\partial s}\int_0^T \int_{\Omega_t(V+sW)} ||V+sW||^2 dxdt)|_{s=0}$$

That is

$$J'(V,W) = (\frac{\partial}{\partial s}\int_0^T \int_{T_s(\mathcal{Z}^t)(\Omega_t(V))} ||V+sW||^2 \ dxdt \ )|_{s=0}$$

$$J'(V,W) = (\frac{\partial}{\partial s}\int_0^T \int_{\Omega_t(V)} (||V+sW||^2) o T_s(\mathcal{Z}^t) \ det(DT_s(\mathcal{Z}^t)) \ dxdt \ )|_{s=0}$$

And we get

(25)

$$J'(V,W) = \int_0^T \int_{\Omega_t(V)} (2 < V, W > + < \nabla(||V||^2), \mathcal{Z}^t(0) > + ||V||^2 div \mathcal{Z}^t(0) \ )dxdt$$

but we have:

$$div\mathcal{Z}^t(0) = 0 \quad , \quad \nabla(||V||^2) = 2D^*V.V \, , \ \ \mathcal{Z}^t(0) = Z(t) \tag{26}$$

Then we get

$$J'(V,W) = 2\int_0^T \int_{\Omega_t(V)} (< V, W > + < D^*V.V, Z(t) >)dxdt \tag{27}$$

As the field Z is solution of (28) consider the formal adjoint operator defined by

$$H_V^*.\psi \ = \ -\partial_t\psi \ - div V \ \psi \ - D\psi.V \ - D^*V.\psi \ \text{ in } [0,T]\times D \tag{28}$$

Introduce the adjoint variable λ solution of

$$H_V^*.\lambda = \chi_{\Omega_t(V)} \ \ D^*V.V \ \ \lambda(\tau) = 0. \tag{29}$$

We get:

$$J'(V,W) = 2\int_0^T \int_D (\chi_{\Omega_t(V)} \ < V, W > + < H_V^*.\lambda \ , \ \ Z(t) \ >)dxdt \tag{30}$$

That is

$$J'(V,W) = 2\int_0^T \int_D < \chi_{\Omega_t(V)} V + \lambda \ , \ \ W(t) \ >)dxdt \tag{31}$$

Then the necessary condition for stationarity of the field V would be:

$$\text{there exist } \phi \, , \ s.t. \quad \lambda = -\chi_{\Omega_t(V)} V \ - \nabla\phi, \quad \text{ in } [0,T]\times D \tag{32}$$

$$\text{there exist } \phi \, , \ s.t. \quad \lambda = \ -(V + \nabla\phi), \quad \text{in}\, Q = \bigcup_{0\leq t\leq T} \ \{t\}\times\Omega_t(V) \quad \subset [0,T]\times D$$

Through the adjoint equation we get, as $divV = 0$, and assuming the non cylindrical evolution domain Q to be open, the following equation in Q:

$$H_V^*.\lambda = \partial_t V + DV.V + D^*V.V + (\ \ \nabla(\partial_t\phi) + D^*\nabla\phi.V + D^*V.\nabla\phi \ ) \tag{33}$$

And

$$H_V^*.\lambda = D^*V.V \tag{34}$$

That is:

$$\partial_t V + DV.V + \quad \nabla(\ \partial_t \phi + \frac{1}{2} \ < \nabla\phi, V >) = f \tag{35}$$

We introduce the "pressure" term

$$p = \ \partial_t \phi + \frac{1}{2} \ < \nabla\phi, V > \tag{36}$$

Then the optimality for a "smooth" extremal field V of the functional J_σ is that in the open non cylindrical evolution domain built by V itself, the field V solves the Euler equation:

$$\partial_t V + DV.V + \quad \nabla p = f \quad \text{in} \quad Q_V \ = \ \bigcup_{0<t<T} \{t\} \times \Omega_t(V) \tag{37}$$

4.1. Time Space Perimeter

We introduce the perimeter of the non cylindrical evolution domain Q_V associated with the field V in the time space cylinder $]0, \tau[\times D$.

$$p_{I\times D}(Q_V) = \sup\{ \int_0^\tau \int_D \ (div g_x \ + \ \partial_t g_t) \, dtdx \, | \ \} \tag{38}$$

The supremum being taken over

$$\{ \ g = (g_x, g_t) \in C^\infty_{comp}(D, R^{N+1}), \forall (t,x), \ ||g(t,x)|| \leq 1\}.$$

It turns out that for smooth lateral boundary Σ we get

$$p_{I\times D}(Q_V) = \int_0^\tau \int_{\Gamma_t(V)} \ (1 + (< V(t,x), n_t(x) >)^2 \ )^{\frac{1}{2}} \ dt \, d\Gamma_t \tag{39}$$

The derivative is

$$\frac{\partial}{\partial s}(p_{I\times D}(Q_{V+sW}) \ )|_{s=0} \tag{40}$$

$$= \int_0^\tau \{ \frac{\partial}{\partial s} \int_{\Gamma_t(V+sW)} \ (1 + (< V(t,x) + sW(t,x), n_{t,s}(x) >)^2 \ )^{\frac{1}{2}} \ dt \, d\Gamma_{t,s} \ \}_{s=0}$$

Where $n_{t,s}$ is the normal field on the boundary $\Gamma_t(V + sW)$ while $\Gamma_{t,s}$ is its surface measure. We get the existence of three functions $\tilde{g}_1$, g_2, g_3 in $L^1(\Sigma)$ depending explicitely on the curvature matrix $D^2 b$ (for simplicity we escape the explicit expressions of these functions which are not important for the present purpose) such that:

Corollary 4.1.

$$\begin{aligned} \frac{\partial}{\partial s}(p_{I\times D}(Q_{V+sW}) \ )|_{s=0} = \int_0^\tau \int_{\Gamma_t(V)} \quad & [\tilde{g}_1 \ < W(t,x), n_t(x) > \\ & + < \nabla_{\Gamma_t} g_2, W_{\Gamma_t}(t,x) > \\ & + g_3 < Z(t,x), n_t(x) >] dt d\Gamma_t \end{aligned} \tag{41}$$

4.2. J_σ Derivative

is in the following form:

$$J'_\sigma(V,W) = 2\int_0^T \int_{\Omega_t(V)} (< V, W > + < D^*V.V, Z(t) >) dx dt$$

$$+\sigma \int_0^T \int_{\Gamma_t} \tilde{g}_1 < W, n_t > + < \nabla_{\Gamma_t} g_2, W(t,x)_{\Gamma_t} > + g_3 < Z(t), n_t > \quad dt d\Gamma_t$$

Let $\tilde{G}_1,\ G_3$ be a smooth extension of $\tilde{g}_1$ and of g_3, as $divZ(t,.) = 0$, we get

$$J'_\sigma(V,W) = 2\int_0^T \int_{\Omega_t(V)} (< V + \sigma\,\nabla \tilde{G}_1, W > + < D^*V.V + \sigma\,\nabla G_3,\ Z(t) > \quad dt dx$$

$$+\sigma \int_0^T \int_{\Gamma_t} < \nabla_{\Gamma_t} g_2, W(t,x)_{\Gamma_t} > \quad , dt d\Gamma_t$$

We introduce the adjoint variable $\lambda_\sigma = Q$ solution of

$$H_V^*.Q = \chi_{Q_V}\ (\quad D^*V.V + \sigma\,\nabla G_3 \quad), \qquad Q(T) = 0$$

Then

$$J'_\sigma(V,W) = 2\int_0^T \int_{\Omega_t(V)} (< V + \sigma\,\nabla G_1, W > dx dt + 2\int_0^T \int_D < Q,\ W > \quad dt dx$$

$$+\sigma \int_0^T \int_{\Gamma_t} < \nabla_{\Gamma_t} g_2, W(t,x)_{\Gamma_t} > \quad , dt d\Gamma_t$$

Denote by γ_{Γ_t} the trace operator on the manifold Γ_t, then

$$\int_{\Gamma_t} < \nabla_{\Gamma_t} g_2, W(t,x)_{\Gamma_t} > \quad d\Gamma_t = \int_{\Gamma_t} < \nabla_{\Gamma_t} g_2, \gamma_{\Gamma_t}.W(t,x) > \quad d\Gamma_t$$

$$\int_D < \gamma_{\Gamma_t}^*.\nabla_{\Gamma_t} g_2, W(t,x) > \quad dx$$

$$J'_\sigma(V,W) = 2\int_0^T \int_D < \quad \chi_{\Omega_t(V)}(V + \sigma\,\nabla G_1) + \ \gamma_{\Gamma_t}^*.\nabla_{\Gamma_t} g_2 \ + Q\ , W \ > dt dx$$

4.3. Necessary Condition for Extremality

Consider a divergence free field V such that, for any divergence free field W we have $J'_\sigma(V,W) = 0$, then there exists a distribution g such that

$$\chi_{\Omega_t(V)}(V + \sigma\,\nabla G_1) + \ \gamma_{\Gamma_t}^*.\nabla_{\Gamma_t} g_2 \ + Q \ = f - \nabla g \quad \text{in }]0,\tau[\times D$$

That is

$$-Q \ = \ \chi_{\Omega_t(V)}(V + \sigma\,\nabla G_1) \ + \ \gamma_{\Gamma_t}^*.\nabla_{\Gamma_t} g_2 + \nabla g \quad \text{in }]0,\tau[\times D$$

Using the adjoint problem whose Q is solution we obtain

$$\begin{aligned}
&\frac{\partial}{\partial t}[\ \chi_{\Omega_t(V)}(V+\sigma\,\nabla G_1)\ +\ \gamma^*_{\Gamma_t}.\nabla_{\Gamma_t}g_2\ +\nabla g\ ]\\
&+D[\ \chi_{\Omega_t(V)}(V+\sigma\,\nabla G_1)\ +\ \gamma^*_{\Gamma_t}.\nabla_{\Gamma_t}g_2\ +\nabla g\ ].V\\
&+D^*V.[\ \chi_{\Omega_t(V)}(V+\sigma\,\nabla G_1)\ +\ \gamma^*_{\Gamma_t}.\nabla_{\Gamma_t}g_2\ +\nabla g\ ]\\
&\quad=\ -\chi_{Q_V}\ (\quad D^*V.V+\sigma\,\nabla G_3\quad)
\end{aligned}$$

If $\Omega_t(V)$ is an open subset in D, then there exists a distribution p such that in the non cylindrical evolution domain Q_V, we have

$$\frac{\partial}{\partial t}V+DV.V+\nabla p=f$$

5. Existence result

Let V_n be a minimizing sequence for the problem

$$Inf\ \{\ J_{\sigma,\epsilon}(V)=J_{\sigma(V)}+\epsilon\int_0^T\int_D||V||^2dxdt\ \mid v\in E\ \} \tag{42}$$

As we assume $\sigma>0$, we get $p_\sigma(Q_n)\leq M$, then there exists a subsequence, still denoted V_n, and an element Q such that:

(43)

$$\chi_{Q_{V_n}}\ \longrightarrow\ \chi_Q\ \text{ in }\ L^2(I\times D),\ \ P_{[0,T]\times D}(Q)\leq liminf\ \ P_{[0,T]\times D}(Q_{V_n})\leq M$$

Moreover we have the following boundedness:

$$\int\int_{[0,T]\times D}\chi_{\Omega_t(V_n)}\,|V_n|^2\ dxdt\leq\ M,\ \ ||V_n||_{L^2(I\times D)}\leq M \tag{44}$$

So that V_n weakly converges to a divergence free field V and that

$$\begin{aligned}\mu_n=\chi_{\Omega_t(V_n)}\,V_n\ \ &\text{weakly converges in }\ L^2(I\times D,R^3)\\ &,\text{ to an element }\ \mu\in L^2(I\times D,R^3)\end{aligned} \tag{45}$$

As the L^2 norm is weakly lower semi continuous we have in the limit:

$$\int\int_{[0,T]\times D}|\mu|^2\,dxdt\ \leq\ liminf\ \int\int_{[0,T]\times D}|\mu_n|^2\,dxdt\ \leq M \tag{46}$$

We shall verify that the limiting element μ is in the needed form:

$$<\mu_m,\psi>=\int\int_{[0,T]\times D}\chi_{Q_{V_n}}\ <\mu_n,\psi>\ dxdt\ \rightarrow\ \int\int_{[0,T]\times D}\chi_Q\ <\mu,\psi>\ dxdt$$

$$<\mu,\psi>=\int\int_{[0,T]\times D}\chi_Q\ <\mu,\psi>\ dxdt \tag{47}$$

That is:

$$\mu=\chi_Q\ \mu \tag{48}$$

We must now verify that $Q = Q_\mu$. We have:

$$\partial_t \chi_{\Omega_t(V_n)} + \nabla \chi_{\Omega_t(V_n)}.V_n = 0 \tag{49}$$

That is in weak form:

$$\forall\, \psi \in C_0^\infty([0,T] \times \bar{D}),\ \int\int_{[0,T]\times D} \chi_{\Omega_t(V_n)}\ (\partial_t \psi\ +\ \chi_{\Omega_t(V_n)}\, V_n.\nabla\psi\ )dxdt = f$$

As $\chi_{\Omega_t(V_n)} V_n \nabla\psi$ weakly converges to $\mu\nabla\psi$, we get$\mu = \chi_{Q_V} V$ and

$$\forall\, \psi \in C_0^\infty([0,T] \times \bar{D}),\ \int\int_{[0,T]\times D} \chi_Q\ (\partial_t \psi\ +\ \mu.\nabla\psi\ )dxdt = 0 \tag{50}$$

References

[1] J.P. Zolésio. *Weak Shape Formulation of Free Boundary Problems.* Annali Della Scuola Normale Superiore Di Pisa, vol.XXI, 1 pp. 11–45, 1994.

[2] ———. *Identification de domaine par déformation*, Doctorat d'état dissertation, University of Nice, 1979.

[3] ———. Shape Differential Equation with Non Smooth Field in "Computational Methods for Optimal Design and Control". J. Borggaard, J. Burns, E. Cliff and S. Schreck eds., PSCT 24, Birkhäuser, pp. 427–460, 1998.

CNRS Institut Non Linéaire de Nice
1361 route des lucioles
06560 Valbonne, France
and
Centre de Mathématiques Appliqueées
Ecole des Mines de Paris – INRIA
2004 route des Lucioles, B.P. 93
06902 Sophia Antipolis Cedex, France
E-mail address: Jean-Paul.Zolesio@sophia.inria.fr

International Series of Numerical Mathematics

Edited by
K.–H. Hoffmann, Technische Universität München, Germany
H.D. Mittelmann, Arizona State University, Tempe, CA, USA

International Series of Numerical Mathematics is open to all aspects of numerical mathematics. Some of the topics of particular interest include free boundary value problems for differential equations, phase transitions, problems of optimal control and optimization, other nonlinear phenomena in analysis, nonlinear partial differential equations, efficient solution methods, bifurcation problems and approximation theory. When possible, the topic of each volume is discussed from three different angles, namely those of mathematical modeling, mathematical analysis, and numerical case studies.

ISNM 106 Antontsev, S.N. / Hoffmann, K.-H. / Khludnev, A.M. (Ed.), **Free Boundary Problems in Continuum Mechanics.**
1992. 358 pages. Hardcover

ISNM 107 Barbu, V. / Bonnans, F.J. / Tiba, D. (Ed.), **Optimization, Optimal Control and Partial Differential Equations.**
1992. 348 pages. Hardcover

ISNM 108 Antes, H. / Panagiotopoulos, P.D., **The Boundary Integral Approach to Static and Dynamic Contact Problems.**
Equality and Inequality Methods.
1992. 308 pages. Hardcover

ISNM 109 Kuz'min, A.G., **Non-Classical Equations of Mixed Type and their Applications in Gas Dynamics.**
1992. 288 pages. Hardcover

ISNM 110 Hörnlein, H.R. / Schittkowski ,K., **Software Systems for Structural Optimization.**
1993. 284 pages. Hardcover. ISBN 3-7643-2836-3

ISNM 111 Bulirsch, R. et al. (Ed.), **Optimal Control.** Calculus of Variations, Optimal Control Theory and Numerical Methods
1993. 368 pages. Hardcover. ISBN 3-7643-2887-8

ISNM 112 Brass, H. / Hämmerlin, G. (Ed.), **Numerical Integration IV.**
Proceedings of the Confrence at the Mathematical Research Institute, Oberwolfach, November 8-14, 1992.
1993. 396 pages. Hardcover. ISBN 3-7643-2922-X

ISNM 113 Quartapelle, L., **Numerical Solution of the Incompressible Navier-Stokes Equations.**
1993. 291 pages. Hardcover. ISBN 3-7643-2935-1

ISNM 114 Douglas, J.jr. / Hornung, U. (Ed.), **Flow in Porous Media.** Proceedings of the Oberwolfach Conference, June 21-27, 1992
1993. 184 pages. Hardcover. ISBN 3-7643-2949-1

ISNM 115 Bulirsch, R. / Kraft, D. (Ed.), **Computational Optimal Control.**
1994. 388 pages. Hardcover. ISBN 3-7643-5015-6

ISNM 116 Hemker, P.W. / Wesseling, P. (Ed.), **Multigrid Methods IV.**
Proceedings of the Fourth European Multigrid Conference, Amsterdam, July 6-9, 1993.
1994. 358 pages. Hardcover. ISBN 3-7643-5030-X

ISNM 117 Bank, R.E. et al. (Ed.), **Math. Modelling and Simulation of Electrical Circuits and Semiconductor Devices.**
1994. 330 pages. Hardcover. ISBN 3-7643-5053-9

ISNM 118 Desch, W. / Kappel, F. / Kunisch, K. (Ed.), **Control and Estimation of Distributed Parameter Systems: Nonlinear Phenomena.** International Conference in Vorau, July 18-24, 1993
1994. 416 pages. Hardcover. ISBN 3-7643-5098-9

ISNM 119 Zahar, R. (Ed.), **Approximation and Computation.** A Festschrift in Honor of Walter Gautschi
1994. 640 pages. Hardcover. ISBN 3-7643-3753-2

ISNM 120 Hackbusch, W., **Integral Equations.** Theory and Numerical Treatment
1995. 376 pages. Hardcover. ISBN 3-7643-2871-1

ISNM 121 Jeltsch, R. / Mansour, M. (Ed.), **Stability Theory.** Hurwitz Centenary Conference, Centro Stefano Franscini, Ascona, 1995
1996. 258 pages. Hardcover. ISBN 3-7643-5474-7

ISNM 122 Khludnev, A.M. / Sokolowski, J., **Modelling and Control in Solid Mechanics.**
1997. 374 pages. Hardcover. ISBN 3-7643-5238-8

ISNM 123 Bandle, C. et al. (Ed.), **General Inequalities 7.** 7th International Conference, Oberwolfach, November 13–18, 1995
1997. 416 pages. Hardcover. ISBN 3-7643-5722-3

ISNM 124 Schmidt, W. et al. (Ed.), **Variational Calculus, Optimal Control and Applications.**
International Conference in Honour of L. Bittner and R. Klötzler, Trassenheide, Germany, September 23-27, 1996
1998. 360 pages. Hardcover. ISBN 3-7643-5906-4

ISNM 125 Nürnberger, G. / Schmidt, J.W. / Walz, G. (Ed.), **Multivariate Approximation and Splines.**
1997. 336 pages. Hardcover. ISBN 3-7643-5654-5

ISNM 126 Desch, W. / Kappel, F. / Kunisch, K. (Ed.), **Control and Estimation of Distributed Parameter Systems.**
International Conference in Vorau (Austria), July 14–20, 1996
1998. 320 pages. Hardcover. ISBN 3-7643-5835-1

ISNM 127 Gerke, H.H. et al., **Optimal Control of Soil Venting: Mathematical Modeling and Applications.**
1999.168 pages. Hardcover. ISBN 3-7643-6041-0

ISNM 128 Meyer-Spasche, R., **Pattern Formation in Viscous Flows.** The Taylor-Couette Problem and Rayleigh-Bénard Convection
1999. 220 pages. Hardcover. ISBN 3-7643-6047-X

ISNM 129/130 Fey, M. / Jeltsch, R. (Ed.), **Hyperbolic Problems: Theory, Numerics, Applications.**
Seventh International Conference in Zürich, February 1998. 2 Vols. Set
1999. 1034 pages. Hardcover. ISBN 3-7643-6123-9

ISNM 131 Gautschi, W. / Golub, G.H. / Opfer, G. (Ed.), **Applications and Computation of Orthogonal Polynomials.**
Conference at the Mathematical Research Institute Oberwolfach, Germany, March 1998
1999. Approx. 284 pages. Hardcover. ISBN 3-7643-6137-9

ISNM 132 Müller, M.W. et al. (Ed.), **New Developments in Approximation Theory.** Second International Dortmund Meeting (IDoMAT) '98, February 23–27, 1998. 1999. Approx. 240 pages. Hardcover. ISBN 3-7643-6143-3